COMPLETE DIGITAL DESIGN

COMPLETE DIGITAL DESIGN

A Comprehensive Guide to Digital Electronics and Computer System Architecture

Mark Balch

McGRAW-HILL

New York Chicago San Francisco
Lisbon London Madrid Mexico City Milan
New Delhi San Juan Seoul Singapore
Sydney Toronto

Library of Congress Cataloging-in-Publication Data

Balch, Mark.
Complete digital design : a comprehensive guide to digital electronics and computer system architecture / by Mark Balch.
p. cm.
Includes bibliographical references and index.
ISBN 0-07-140927-0
1. Digital electronics. 2. Computer architecture. I. Title.

TK7868.D5B37 2003
621.81—dc21 2003046465

1 2 3 4 5 6 7 8 9 0 DOC/DOC 9 8 7 6 5 4 3

ISBN 0-07-140927-0

The sponsoring editor for this book was Steve Chapman and the production supervisor was Pamela A. Pelton. It was set in Century Schoolbook by J. K. Eckert & Company, Inc.

Printed and bound by RR Donnelley.

This book is printed on recycled, acid-free paper containing a minimum of 50% recycled, de-inked fiber.

for Neil

CONTENTS

PREFACE

Digital systems are created to perform data processing and control tasks. What distinguishes one system from another is an architecture tailored to efficiently execute the tasks for which it was designed. A desktop computer and an automobile's engine controller have markedly different attributes dictated by their unique requirements. Despite these differences, they share many fundamental building blocks and concepts. Fundamental to digital system design is the ability to choose from and apply a wide range of technologies and methods to develop a suitable system architecture. Digital electronics is a field of great breadth, with interdependent topics that can prove challenging for individuals who lack previous hands-on experience in the field.

This book's focus is explaining the real-world implementation of complete digital systems. In doing so, the reader is prepared to immediately begin design and implementation work without being left to wonder about the myriad ancillary topics that many texts leave to independent and sometimes painful discovery. A complete perspective is emphasized, because even the most elegant computer architecture will not function without adequate supporting circuits.

A wide variety of individuals are intended to benefit from this book. The target audiences include

- *Practicing electrical engineers seeking to sharpen their skills in modern digital system design.* Engineers who have spent years outside the design arena or in less-than-cutting-edge areas often find that their digital design skills are behind the times. These professionals can acquire directly relevant knowledge from this book's practical discussion of modern digital technologies and design practices.
- *College graduates and undergraduates seeking to begin engineering careers in digital electronics.* College curricula provide a rich foundation of theoretical understanding of electrical principles and computer science but often lack a practical presentation of how the many pieces fit together in real systems. Students may understand conceptually how a computer works while being incapable of actually building one on their own. This book serves as a bridge to take readers from the theoretical world to the everyday design world where solutions must be complete to be successful.
- *Technicians and hobbyists seeking a broad orientation to digital electronics design.* Some people have an interest in understanding and building digital systems without having a formal engineering degree. Their need for practical knowledge in the field is as strong as for degreed engineers, but their goals may involve laboratory support, manufacturing, or building a personal project.

There are four parts to this book, each of which addresses a critical set of topics necessary for successful digital systems design. The parts may be read sequentially or in arbitrary order, depending on the reader's level of knowledge and specific areas of interest.

A complete discussion of digital logic and microprocessor fundamentals is presented in the first part, including introductions to basic memory and communications architectures. More advanced computer architecture and logic design topics are covered in Part 2, including modern microprocessor architectures, logic design methodologies, high-performance memory and networking technologies, and programmable logic devices.

Part 3 steps back from the purely digital world to focus on the critical analog support circuitry that is important to any viable computing system. These topics include basic DC and AC circuit analysis, diodes, transistors, op-amps, and data conversion techniques. The fundamental topics from the first three parts are tied together in Part 4 by discussing practical digital design issues, including clock distribution, power regulation, signal integrity, design for test, and circuit fabrication techniques. These chapters deal with nuts-and-bolts design issues that are rarely covered in formal electronics courses.

More detailed descriptions of each part and chapter are provided below.

PART 1 DIGITAL FUNDAMENTALS

The first part of this book provides a firm foundation in the concepts of digital logic and computer architecture. Logic is the basis of computers, and computers are intrinsically at the heart of digital systems. We begin with the basics: logic gates, integrated circuits, microprocessors, and computer architecture. This framework is supplemented by exploring closely related concepts such as memory and communications that are fundamental to any complete system. By the time you have completed Part 1, you will be familiar with exactly how a computer works from multiple perspectives: individual logic gates, major architectural building blocks, and the hardware/software interface. You will also have a running start in design by being able to thoughtfully identify and select specific off-the-shelf chips that can be incorporated into a working system. A multilevel perspective is critical to successful systems design, because a system architect must simultaneously consider high-level feature trade-offs and low-level implementation possibilities. Focusing on one and not the other will usually lead to a system that is either impractical (too expensive or complex) or one that is not really useful.

Chapter 1, "Digital Logic," introduces the fundamentals of Boolean logic, binary arithmetic, and flip-flops. Basic terminology and numerical representations that are used throughout digital systems design are presented as well. On completing this chapter, the awareness gained of specific logical building blocks will help provide a familiarity with supporting logic when reading about higher-level concepts in later chapters.

Chapter 2, "Integrated Circuits and the 7400 Logic Families," provides a general orientation to integrated circuits and commonly used logic ICs. This chapter is where the rubber meets the road and the basics of logic design become issues of practical implementation. Small design examples provide an idea of how various logic chips can be connected to create functional subsystems. Attention is paid to readily available components and understanding IC specifications, without which chips cannot be understood and used. The focus is on design with real off-the-shelf components rather than abstract representations on paper.

Chapter 3, "Basic Computer Architecture," cracks open the heart of digital systems by explaining how computers and microprocessors function. Basic concepts, including instruction sets, memory, address decoding, bus interfacing, DMA, and assembly language, are discussed to create a complete picture of what a computer is and the basic components that go into all computers. Questions are not left as exercises for the reader. Rather, each mechanism and process in a basic computer is discussed. This knowledge enables you to move ahead and explore the individual concepts in more depth while maintaining an overall system-level view of how everything fits together.

Chapter 4, "Memory," discusses this cornerstone of digital systems. With the conceptual understanding from Chapter 3 of what memory is and the functions that it serves, the discussion progresses to explain specific types of memory devices, how they work, and how they are applicable to different computing applications. Trade-offs of various memory technologies, including SRAM, DRAM, flash, and EPROM, are explored to convey an understanding of why each technology has its place in various systems.

Chapter 5, "Serial Communications," presents one of the most basic aspects of systems design: moving data from one system to another. Without data links, computers would be isolated islands. Communication is key to many applications, whether accessing the Internet or gathering data from a remote sensor. Topics including RS-232 interfaces, modems, and basic multinode networking are discussed with a focus on implementing real data links.

Chapter 6, "Instructive Microprocessors and Microcomputer Elements," walks through five examples of real microprocessors and microcontrollers. The devices presented are significant because of their trail-blazing roles in defining modern computing architecture, as exhibited by the fact that, decades later, they continue to turn up in new designs in one form or another. These devices are used as vehicles to explain a wide range of computing issues from register, memory, and bus architectures to interrupt vectoring and operating system privilege levels.

PART 2 ADVANCED DIGITAL SYSTEMS

Digital systems operate by acquiring data, manipulating that data, and then transferring the results as dictated by the application. Part 2 builds on the foundations of Part 1 by exploring the state of the art in microprocessor, memory, communications, and logic implementation technologies. To effectively conceive and implement such systems requires an understanding of what is possible, what is practical, and what tools and building blocks exist with which to get started. On completing Parts 1 and 2, you will have acquired a broad understanding of digital systems ranging from small microcontrollers to 32-bit microcomputer architecture and high-speed networking, and the logic design methodologies that underlie them all. You will have the ability to look at a digital system, whether pre-existing or conceptual, and break it into its component parts—the first step in solving a problem.

Chapter 7, "Advanced Microprocessor Concepts," discusses the key architectural topics behind modern 32- and 64-bit computing systems. Basic concepts including RISC/CISC, floating-point arithmetic, caching, virtual memory, pipelining, and DSP are presented from the perspective of what a digital hardware engineer needs to know to understand system-wide implications and design useful circuits. This chapter does not instruct the reader on how to build the fastest microprocessors, but it does explain how these devices operate and, more importantly, what system-level design considerations and resources are necessary to achieve a functioning system.

Chapter 8, "High-Performance Memory Technologies," presents the latest SDR/DDR SDRAM and SDR/DDR/QDR SSRAM devices, explains how they work and why they are useful in high-performance digital systems, and discusses the design implications of each. Memory is used by more than just microprocessors. Memory is essential to communications and data processing systems. Understanding the capabilities and trade-offs of such a central set of technologies is crucial to designing a practical system. Familiarity with all mainstream memory technologies is provided to enable a firm grasp of the applications best suited to each.

Chapter 9, "Networking," covers the broad field of digital communications from a digital hardware perspective. Network protocol layering is introduced to explain the various levels at which hardware and software interact in modern communication systems. Much of the hardware responsibility for networking lies at lower levels in moving bits onto and off of the communications medium. This chapter focuses on the low-level details of twisted-pair and fiber-optic media, transceiver technologies, 8B10B channel coding, and error detection with CRC and checksum logic. A brief presentation of Ethernet concludes the chapter to show how a real networking standard functions.

Chapter 10, "Logic Design and Finite State Machines," explains how to implement custom logic to make a fully functional system. Most systems use a substantial quantity of off-the-shelf logic products to solve the problem at hand, but almost all require some custom support logic. This chapter begins by presenting hardware description languages, and Verilog in particular, as an efficient means of designing synchronous and combinatorial logic. Once the basic methodology of designing logic has been discussed, common support logic solutions, including address decoding, control/status registers, and interrupt control logic, are shown with detailed design examples. Designing logic to handle asynchronous inputs across multiple clock domains is presented with specific examples. More complex logic circuits capable of implementing arbitrary algorithms are built from finite state machines—a topic explored in detail with design examples to ensure that the concepts are properly translated into reality. Finally, state machine optimization techniques, including pipelining, are discussed to provide an understanding of how to design logic that can be reliably implemented.

Chapter 11, "Programmable Logic Devices," explains the various logic implementation technologies that are used in a digital system. GALs, PALs, CPLDs, and FPGAs are presented from the perspectives of how they work, how they are used to implement arbitrary logic designs, and the capabilities and features of each that make them suitable for various types of designs. These devices represent the glue that holds some systems together and the core operational elements of others. This chapter aids in deciding which technology is best suited to each logic application and how to select the right device to suit a specific need.

PART 3 ANALOG BASICS FOR DIGITAL SYSTEMS

All electrical systems are collections of analog circuits, but digital systems masquerade as discrete binary entities when they are properly designed. It is necessary to understand certain fundamental topics in circuit analysis so that digital circuits can be made to behave in the intended binary manner. Part 3 addresses many essential analog topics that have direct relevance to designing successful digital systems. Many digital engineers shrink away from basic DC and AC circuit analysis either for fear of higher mathematics or because it is not their area of expertise. This needn't be the case, because most day-to-day analysis required for digital systems can be performed with basic algebra. Furthermore, a digital systems slant on analog electronics enables many simplifications that are not possible in full-blown analog design. On completing this portion of the book, you will be able to apply passive components, discrete diodes and transistors, and op-amps in ways that support digital circuits.

Chapter 12, "Electrical Fundamentals," addresses basic DC and AC circuit analysis. Resistors, capacitors, inductors, and transformers are explained with straightforward means of determining voltages and currents in simple analog circuits. Nonideal characteristics of passive components are discussed, which is a critical aspect of modern, high-speed digital systems. Many a digital system has failed because its designers were unaware of increasingly nonideal behavior of components as operating frequencies get higher. Frequency-domain analysis and basic filtering are presented to explain common analog structures and how they can be applied to digital systems, especially in minimizing noise, a major contributor to transient and hard-to-detect problems.

Chapter 13, "Diodes and Transistors," explains the basic workings of discrete semiconductors and provides specific and fully analyzed examples of how they are easily applied to digital applications. LEDs are covered as well as bipolar and MOS transistors. An understanding of how diodes and transistors function opens up a great field of possible solutions to design problems. Diodes are essential in power-regulation circuits and serve as voltage references. Transistors enable electrical loads to be driven that are otherwise too heavy for a digital logic chip to handle.

Chapter 14, "Operational Amplifiers," discusses this versatile analog building block with many practical applications in digital systems. The design of basic amplifiers and voltage comparators is offered with many examples to illustrate all topics presented. All examples are thoroughly analyzed in a step-by-step process so that you can learn to use op-amps effectively on your own. Op-amps are useful in data acquisition and interface circuits, power supply and voltage monitoring circuits, and for implementing basic amplifiers and filters. This chapter applies the basic AC analysis skills explained previously in designing hybrid analog/digital circuits to support a larger digital system.

Chapter 15, "Analog Interfaces for Digital Systems," covers the basics of analog-to-digital and digital-to-analog conversion techniques. Many digital systems interact with real-world stimuli including audio, video, and radio frequencies. Data conversion is a key portion of these systems, enabling continuous analog signals to be represented and processed as binary numbers. Several common means of performing data conversion are discussed along with fundamental concepts such as the Nyquist frequency and anti-alias filtering.

PART 4 DIGITAL SYSTEM DESIGN IN PRACTICE

When starting to design a new digital system, high-profile features such as the microprocessor and memory architecture often get most of the attention. Yet there are essential support elements that may be overlooked by those unfamiliar with them and unaware of the consequences of not taking time to address necessary details. All too often, digital engineers end up with systems that almost work. A microprocessor may work properly for a few hours and then quit. A data link may work fine one day and then experience inexplicable bit errors the next day. Sometimes these problems are the result of logic bugs, but mysterious behavior may be related to a more fundamental electrical flaw. The final part of this book explains the supporting infrastructure and electrical phenomena that must be understood to design and build reliable systems.

Chapter 16, "Clock Distribution," explores an essential component of all digital systems: proper generation and distribution of clocks. Many common clock generation and distribution methods are presented with detailed circuit implementation examples including low-skew buffers, termination, and PLLs. Related subjects, including frequency synthesis, DLLs, and source-synchronous clocking, are presented to lend a broad perspective on system-level clocking strategies.

Chapter 17, "Voltage Regulation and Power Distribution" discusses the fundamental power infrastructure necessary for system operation. An introduction to general power handling is provided that covers issues such as circuit specifications and safety issues. Thermal analysis is emphasized for safety and reliability concerns. Basic regulator design with discrete components and integrated circuits is explained with numerous illustrative circuits for each topic. The remainder of the chapter addresses power distribution topics including wiring, circuit board power planes, and power supply decoupling capacitors.

Chapter 18, "Signal Integrity," delves into a set of topics that addresses the nonideal behavior of high-speed digital signals. The first half of this chapter covers phenomena that are common causes of corrupted digital signals. Transmission lines, signal reflections, crosstalk, and a wide variety of termination schemes are explained. These topics provide a basic understanding of what can go wrong and how circuits and systems can be designed to avoid signal integrity problems. Electromagnetic radiation, grounding, and static discharge are closely related subjects that are presented in the second half of the chapter. An overview is presented of the problems that can arise and their possible solutions. Examples illustrate concepts that apply to both circuit board design and overall system enclosure design—two equally important matters for consideration.

Chapter 19, "Designing for Success," explores a wide range of system-level considerations that should be taken into account during the product definition and design phases of a project. Component selection and circuit fabrication must complement the product requirements and available development and manufacturing resources. Often considered mundane, these topics are discussed because a successful outcome hinges on the availability and practicality of parts and technologies that are designed into a system. System testability is emphasized in this chapter from several perspectives, because testing is prominent in several phases of product development. Test mechanisms including boundary scan (JTAG), specific hardware features, and software diagnostic routines enable more efficient debugging and fault isolation in both laboratory and assembly line environments. Common computer-aided design software for digital systems is presented with an emphasis on Spice analog circuit simulation. Spice applications are covered and augmented by complete examples that start with circuits, proceed with Spice modeling, and end with Spice simulation result analysis. The chapter closes with a brief overview of common test equipment that is beneficial in debugging and characterizing digital systems.

Following the main text is Appendix A, a brief list of recommended resources for further reading and self-education. Modern resources range from books to trade journals and magazines to web sites.

Many specific vendors and products are mentioned throughout this book to serve as examples and as starting points for your exploration. However, there are many more companies and products than can be practically listed in a single text. Do not hesitate to search out and consider manufacturers not mentioned here, because the ideal component for your application might otherwise lie undiscovered. When specific components are described in this book, they are described in the context of the discussion at hand. References to component specifications cannot substitute for a vendor's data sheet, because there is not enough room to exhaustively list all of a component's attributes, and such specifications are always subject to revision by the manufacturer. Be sure to contact the manufacturer of a desired component to get the latest information on that product. Component manufacturers have a vested interest in providing you with the necessary information to use their products in a safe and effective manner. It is wise to take advantage of the resources that they offer. The widespread use of the Internet has greatly simplified this task.

True proficiency in a trade comes with time and practice. There is no substitute for experience or mentoring from more senior engineers. However, help in acquiring this experience by being pointed in the right direction can not only speed up the learning process, it can make it more enjoyable as well. With the right guide, a motivated beginner's efforts can be more effectively channeled through the early adoption of sound design practices and knowing where to look for necessary information. I sincerely hope that this book can be your guide, and I wish you the best of luck in your endeavors.

Mark Balch

ACKNOWLEDGMENTS

On completing a book of this nature, it becomes clear to the author that the work would not have been possible without the support of others. I am grateful for the many talented engineers that I have been fortunate to work with and learn from over the years. More directly, several friends and colleagues were kind enough to review the initial draft, provide feedback on the content, and bring to my attention details that required correction and clarification. I am indebted to Rich Chernock and Ken Wagner, who read through the entire text over a period of many months. Their thorough inspection provided welcome and valuable perspectives on everything ranging from fine technical points to overall flow and style. I am also thankful for the comments and suggestions provided by Todd Goldsmith and Jim O'Sullivan, who enabled me to improve the quality of this book through their input.

I am appreciative of the cooperation provided by several prominent technology companies. Mentor Graphics made available their ModelSim Verilog simulator, which was used to verify the correctness of certain logic design examples. Agilent Technologies, Fairchild Semiconductor, and National Semiconductor gave permission to reprint portions of their technical literature that serve as examples for some of the topics discussed.

Becoming an electrical engineer and, consequently, writing this book was spurred on by my early interest in electronics and computers that was fully supported and encouraged by my family. Whether it was the attic turned into a laboratory, a trip to the electronic supply store, or accompaniment to the science fair, my family has always been behind me.

Finally, I haven't sufficient words to express my gratitude to my wife Laurie for her constant emotional support and for her sound all-around advice. Over the course of my long project, she has helped me retain my sanity. She has served as editor, counselor, strategist, and friend. Laurie, thanks for being there for me every step of the way.

ABOUT THE AUTHOR

Mark Balch is an electrical engineer in the Silicon Valley who designs high-performance computer-networking hardware. His responsibilities have included PCB, FPGA, and ASIC design. Prior to working in telecommunications, Mark designed products in the fields of HDTV, consumer electronics, and industrial computers.

In addition to his work in product design, Mark has actively participated in industry standards committees and has presented work at technical conferences. He has also authored magazine articles on topics in hardware and system design. Mark holds a bachelor's degree in electrical engineering from The Cooper Union in New York City.

P · A · R · T · 1

DIGITAL FUNDAMENTALS

CHAPTER 1
Digital Logic

All digital systems are founded on logic design. Logic design transforms algorithms and processes conceived by people into computing machines. A grasp of digital logic is crucial to the understanding of other basic elements of digital systems, including microprocessors. This chapter addresses vital topics ranging from Boolean algebra to synchronous logic to timing analysis with the goal of providing a working set of knowledge that is the prerequisite for learning how to design and implement an unbounded range of digital systems.

Boolean algebra is the mathematical basis for logic design and establishes the means by which a task's defining rules are represented digitally. The topic is introduced in stages starting with basic logical operations and progressing through the design and manipulation of logic equations. Binary and hexadecimal numbering and arithmetic are discussed to explain how logic elements accomplish significant and practical tasks.

With an understanding of how basic logical relationships are established and implemented, the discussion moves on to explain flip-flops and synchronous logic design. Synchronous logic complements Boolean algebra, because it allows logic operations to store and manipulate data over time. Digital systems would be impossible without a deterministic means of advancing through an algorithm's sequential steps. Boolean algebra defines algorithmic steps, and the progression between steps is enabled by synchronous logic.

Synchronous logic brings time into play along with the associated issue of how fast a circuit can reliably operate. Logic elements are constructed using real electrical components, each of which has physical requirements that must be satisfied for proper operation. Timing analysis is discussed as a basic part of logic design, because it quantifies the requirements of real components and thereby establishes a digital circuit's practical operating conditions.

The chapter concludes with a presentation of higher-level logic constructs that are built up from the basic logic elements already discussed. These elements, including multiplexers, tri-state buffers, and shift registers, are considered to be fundamental building blocks in digital system design. The remainder of this book, and digital engineering as a discipline, builds on and makes frequent reference to the fundamental items included in this discussion.

1.1 BOOLEAN LOGIC

Machines of all types, including computers, are designed to perform specific tasks in exact well defined manners. Some machine components are purely physical in nature, because their composition and behavior are strictly regulated by chemical, thermodynamic, and physical properties. For example, an engine is designed to transform the energy released by the combustion of gasoline and oxygen into rotating a crankshaft. Other machine components are algorithmic in nature, because their designs primarily follow constraints necessary to implement a set of logical functions as defined by human beings rather than the laws of physics. A traffic light's behavior is predominantly defined by human beings rather than by natural physical laws. This book is concerned with the design of digital systems that are suited to the algorithmic requirements of their particular range of applications. Digital logic and arithmetic are critical building blocks in constructing such systems.

An algorithm is a procedure for solving a problem through a series of finite and specific steps. It can be represented as a set of mathematical formulas, lists of sequential operations, or any combination thereof. Each of these finite steps can be represented by a *Boolean logic* equation. Boolean logic is a branch of mathematics that was discovered in the nineteenth century by an English mathematician named George Boole. The basic theory is that logical relationships can be modeled by algebraic equations. Rather than using arithmetic operations such as addition and subtraction, Boolean algebra employs logical operations including AND, OR, and NOT. Boolean variables have two enumerated values: true and false, represented numerically as 1 and 0, respectively.

The AND operation is mathematically defined as the product of two Boolean values, denoted A and B for reference. *Truth tables* are often used to illustrate logical relationships as shown for the AND operation in Table 1.1. A truth table provides a direct mapping between the possible inputs and outputs. A basic AND operation has two inputs with four possible combinations, because each input can be either 1 or 0 — true or false. Mathematical rules apply to Boolean algebra, resulting in a nonzero product only when both inputs are 1.

TABLE 1.1 AND Operation Truth Table

A	B	A AND B
0	0	0
0	1	0
1	0	0
1	1	1

Summation is represented by the OR operation in Boolean algebra as shown in Table 1.2. Only one combination of inputs to the OR operation result in a zero sum: $0 + 0 = 0$.

TABLE 1.2 OR Operation Truth Table

A	B	A OR B
0	0	0
0	1	1
1	0	1
1	1	1

AND and OR are referred to as *binary operators,* because they require two operands. NOT is a *unary operator*, meaning that it requires only one operand. The NOT operator returns the complement of the input: 1 becomes 0, and 0 becomes 1. When a variable is passed through a NOT operator, it is said to be *inverted*.

Boolean variables may not seem too interesting on their own. It is what they can be made to represent that leads to useful constructs. A rather contrived example can be made from the following logical statement:

"If today is Saturday or Sunday and it is warm, then put on shorts."

Three Boolean inputs can be inferred from this statement: Saturday, Sunday, and warm. One Boolean output can be inferred: shorts. These four variables can be assembled into a single logic equation that computes the desired result,

shorts = (Saturday OR Sunday) AND warm

While this is a simple example, it is representative of the fact that any logical relationship can be expressed algebraically with products and sums by combining the basic logic functions AND, OR, and NOT.

Several other logic functions are regarded as elemental, even though they can be broken down into AND, OR, and NOT functions. These are not–AND (NAND), not–OR (NOR), exclusive–OR (XOR), and exclusive–NOR (XNOR). Table 1.3 presents the logical definitions of these other basic functions. XOR is an interesting function, because it implements a sum that is distinct from OR by taking into account that 1 + 1 does not equal 1. As will be seen later, XOR plays a key role in arithmetic for this reason.

TABLE 1.3 NAND, NOR, XOR, XNOR Truth Table

A	B	A NAND B	A NOR B	A XOR B	A XNOR B
0	0	1	1	0	1
0	1	1	0	1	0
1	0	1	0	1	0
1	1	0	0	0	1

All binary operators can be chained together to implement a wide function of any number of inputs. For example, the truth table for a ten-input AND function would result in a 1 output only when all inputs are 1. Similarly, the truth table for a seven-input OR function would result in a 1 output if any of the seven inputs are 1. A four-input XOR, however, will only result in a 1 output if there are an odd number of ones at the inputs. This is because of the logical daisy chaining of multiple binary XOR operations. As shown in Table 1.3, an even number of 1s presented to an XOR function cancel each other out.

It quickly grows unwieldy to write out the names of logical operators. Concise algebraic expressions are written by using the graphical representations shown in Table 1.4. Note that each operation has multiple symbolic representations. The choice of representation is a matter of style when handwritten and is predetermined when programming a computer by the syntactical requirements of each computer programming language.

A common means of representing the output of a generic logical function is with the variable Y. Therefore, the AND function of two variables, A and B, can be written as Y = A & B or Y = A*B. As with normal mathematical notation, products can also be written by placing terms right next to each other, such as Y = AB. Notation for the inverted functions, NAND, NOR, and XNOR, is achieved by

TABLE 1.4 Symbolic Representations of Standard Boolean Operators

Boolean Operation	Operators
AND	*, &
OR	+, |, #
XOR	$\oplus$, ^
NOT	!, ~, $\overline{A}$

inverting the base function. Two equally valid ways of representing NAND are $Y = \overline{A \,\&\, B}$ and Y = !(AB). Similarly, an XNOR might be written as $Y = \overline{A \oplus B}$.

When logical functions are converted into circuits, graphical representations of the seven basic operators are commonly used. In circuit terminology, the logical operators are called *gates*. Figure 1.1 shows how the basic logic gates are drawn on a circuit diagram. Naming the inputs of each gate A and B and the output Y is for reference only; any name can be chosen for convenience. A small bubble is drawn at a gate's output to indicate a logical inversion.

More complex Boolean functions are created by combining Boolean operators in the same way that arithmetic operators are combined in normal mathematics. Parentheses are useful to explicitly convey precedence information so that there is no ambiguity over how two variables should be treated. A Boolean function might be written as

$$Y = (AB + \overline{C} + D)\&\overline{E \oplus F}$$

This same equation could be represented graphically in a circuit diagram, also called a *schematic diagram*, as shown in Fig. 1.2. This representation uses only two-input logic gates. As already mentioned, binary operators can be chained together to implement functions of more than two variables.

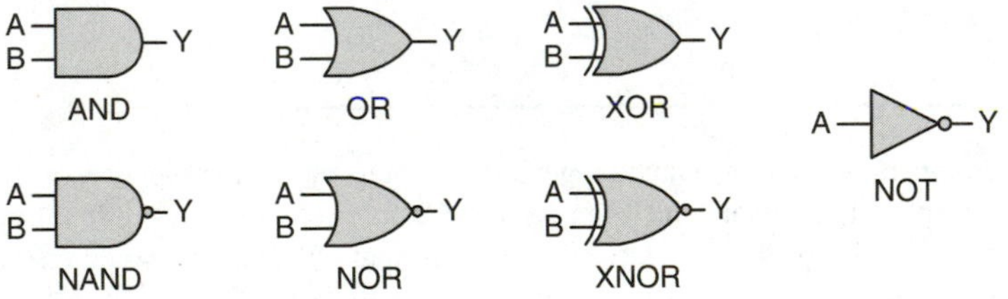

FIGURE 1.1 Graphical representation of basic logic gates.

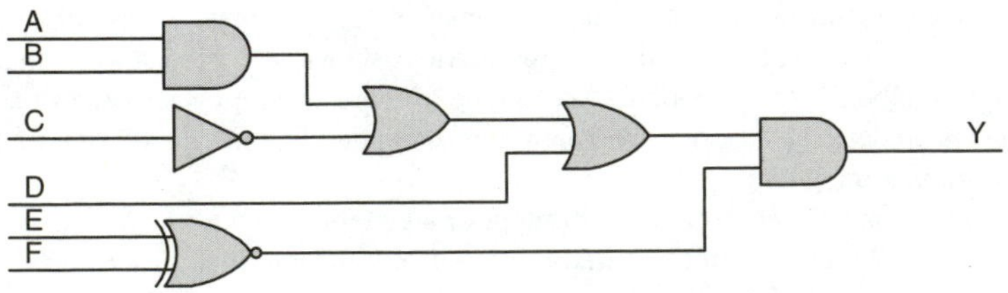

FIGURE 1.2 Schematic diagram of logic function.

An alternative graphical representation would use a three-input OR gate by collapsing the two-input OR gates into a single entity.

1.2 BOOLEAN MANIPULATION

Boolean equations are invaluable when designing digital logic. To properly use and devise such equations, it is helpful to understand certain basic rules that enable simplification and re-expression of Boolean logic. Simplification is perhaps the most practical final result of Boolean manipulation, because it is easier and less expensive to build a circuit that does not contain unnecessary components. When a logical relationship is first set down on paper, it often is not in its most simplified form. Such a circuit will function but may be unnecessarily complex. Re-expression of a Boolean equation is a useful skill, because it can enable you to take better advantage of the logic resources at your disposal instead of always having to use new components each time the logic is expanded or otherwise modified to function in a different manner. As will soon be shown, an OR gate can be made to behave as an AND gate, and vice versa. Such knowledge can enable you to build a less-complex implementation of a Boolean equation.

First, it is useful to mention two basic identities:

$$A \,\&\, \overline{A} = 0 \text{ and } A + \overline{A} = 1$$

The first identity states that the product of any variable and its logical negation must always be false. It has already been shown that both operands of an AND function must be true for the result to be true. Therefore, the first identity holds true, because it is impossible for both operands to be true when one is the negation of the other. The second identity states that the sum of any variable and its logical negation must always be true. At least one operand of an OR function must be true for the result to be true. As with the first identity, it is guaranteed that one operand will be true, and the other will be false.

Boolean algebra also has commutative, associative, and distributive properties as listed below:

- Commutative: A & B = B & A and A + B = B + A
- Associative: (A & B) & C = A & (B & C) and (A + B) + C = A + (B + C)
- Distributive: A & (B + C) = A & B + A & C

The aforementioned identities, combined with these basic properties, can be used to simplify logic. For example,

$$A \,\&\, B \,\&\, C + A \,\&\, B \,\&\, \overline{C}$$

can be re-expressed using the distributive property as

$$A \,\&\, B \,\&\, (C + \overline{C})$$

which we know by identity equals

$$A \,\&\, B \,\&\, (1) = A \,\&\, B$$

Another useful identity, $A + \overline{A}B = A + B$, can be illustrated using the truth table shown in Table 1.5.

TABLE 1.5 $A + \overline{A}B = A + B$ Truth Table

A	B	$\overline{A}B$	$A + \overline{A}B$	$A + B$
0	0	0	0	0
0	1	1	1	1
1	0	0	1	1
1	1	0	1	1

Augustus DeMorgan, another nineteenth century English mathematician, worked out a logical transformation that is known as DeMorgan's law, which has great utility in simplifying and re-expressing Boolean equations. Simply put, DeMorgan's law states

$$A + B = \overline{\overline{A}\&\overline{B}} \quad \text{and} \quad A\&B = \overline{\overline{A} + \overline{B}}$$

These transformations are very useful, because they show the direct equivalence of AND and OR functions and how one can be readily converted to the other. XOR and XNOR functions can be represented by combining AND and OR gates. It can be observed from Table 1.3 that $A \oplus B = A\overline{B} + \overline{A}B$ and that $\overline{A \oplus B} = AB + \overline{A}\,\overline{B}$. Conversions between XOR/XNOR and AND/OR functions are helpful when manipulating and simplifying larger Boolean expressions, because simpler AND and OR functions are directly handled with DeMorgan's law, whereas XOR/XNOR functions are not.

1.3 THE KARNAUGH MAP

Generating Boolean equations to implement a desired logic function is a necessary step before a circuit can be implemented. Truth tables are a common means of describing logical relationships between Boolean inputs and outputs. Once a truth table has been created, it is not always easy to convert that truth table directly into a Boolean equation. This translation becomes more difficult as the number of variables in a function increases. A graphical means of translating a truth table into a logic equation was invented by Maurice Karnaugh in the early 1950s and today is called the *Karnaugh map*, or *K-map*. A K-map is a type of truth table drawn such that individual product terms can be picked out and summed with other product terms extracted from the map to yield an overall Boolean equation. The best way to explain how this process works is through an example. Consider the hypothetical logical relationship in Table 1.6.

TABLE 1.6 Function of Three Variables

A	B	C	Y
0	0	0	1
0	0	1	1
0	1	0	0
0	1	1	1
1	0	0	1
1	0	1	1
1	1	0	0
1	1	1	0

If the corresponding Boolean equation does not immediately become clear, the truth table can be converted into a K-map as shown in Fig. 1.3. The K-map has one box for every combination of inputs, and the desired output for a given combination is written into the corresponding box. Each axis of a K-map represents up to two variables, enabling a K-map to solve a function of up to four variables. Individual grid locations on each axis are labeled with a unique combination of the variables represented on that axis. The labeling pattern is important, because only one variable per axis is permitted to differ between adjacent boxes. Therefore, the pattern "00, 01, 10, 11" is not proper, but the pattern "11, 01, 00, 10" would work as well as the pattern shown.

K-maps are solved using the *sum of products* principle, which states that any relationship can be expressed by the logical OR of one or more AND terms. Product terms in a K-map are recognized by picking out groups of adjacent boxes that all have a state of 1. The simplest product term is a single box with a 1 in it, and that term is the product of all variables in the K-map with each variable either inverted or not inverted such that the result is 1. For example, a 1 is observed in the box that corresponds to A = 0, B = 1, and C = 1. The product term representation of that box would be $\overline{A}BC$. A brute force solution is to sum together as many product terms as there are boxes with a state of 1 (there are five in this example) and then simplify the resulting equation to obtain the final result. This approach can be taken without going to the trouble of drawing a K-map. The purpose of a K-map is to help in identifying minimized product terms so that lengthy simplification steps are unnecessary.

Minimized product terms are identified by grouping together as many adjacent boxes with a state of 1 as possible, subject to the rules of Boolean algebra. Keep in mind that, to generate a valid product term, all boxes in a group must have an identical relationship to all of the equation's input variables. This requirement translates into a rule that product term groups must be found in power-of-two quantities. For a three-variable K-map, product term groups can have only 1, 2, 4, or 8 boxes in them.

Going back to our example, a four-box product term is formed by grouping together the vertically stacked 1s on the left and right edges of the K-map. An interesting aspect of a K-map is that an edge wraps around to the other side, because the axis labeling pattern remains continuous. The validity of this wrapping concept is shown by the fact that all four boxes share a common relationship with the input variables: their product term is $\overline{B}$. The other variables, A and C, can be ruled out, because the boxes are 1 regardless of the state of A and C. Only variable B is a determining factor, and it must be 0 for the boxes to have a state of 1. Once a product term has been identified, it is marked by drawing a ring around it as shown in Fig. 1.4. Because the product term crosses the edges of the table, half-rings are shown in the appropriate locations.

There is still a box with a 1 in it that has not yet been accounted for. One approach could be to generate a product term for that single box, but this would not result in a fully simplified equation, because a larger group can be formed by associating the lone box with the adjacent box corresponding to A = 0, B = 0, and C = 1. K-map boxes can be part of multiple groups, and forming the largest groups possible results in a fully simplified equation. This second group of boxes is circled in Fig. 1.5 to complete the map. This product term shares a common relationship where A = 0, C = 1, and B

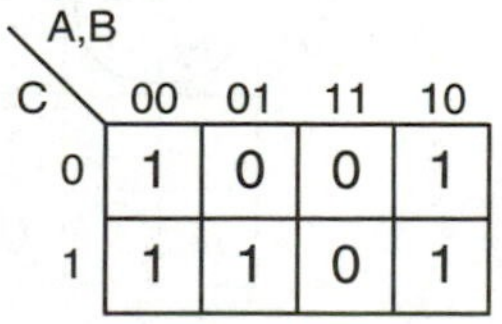

FIGURE 1.3 Karnaugh map for function of three variables.

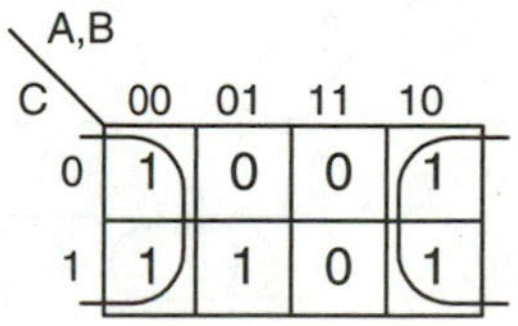

FIGURE 1.4 Partially completed Karnaugh map for a function of three variables.

is irrelevant: $\overline{A}C$. It may appear tempting to create a product term consisting of the three boxes on the bottom edge of the K-map. This is not valid because it does not result in all boxes sharing a common product relationship, and therefore violates the power-of-two rule mentioned previously. Upon completing the K-map, all product terms are summed to yield a final and simplified Boolean equation that relates the input variables and the output: $Y = \overline{B} + \overline{A}C$.

Functions of four variables are just as easy to solve using a K-map. Beyond four variables, it is preferable to break complex functions into smaller subfunctions and then combine the Boolean equations once they have been determined. Figure 1.6 shows an example of a completed Karnaugh map for a hypothetical function of four variables. Note the overlap between several groups to achieve a simplified set of product terms. The lager a group is, the fewer unique terms will be required to represent its logic. There is nothing to lose and something to gain by forming a larger group whenever possible. This K-map has four product terms that are summed for a final result: $Y = \overline{A}\,\overline{C} + \overline{B}\,\overline{C} + \overline{A}B\overline{D} + ABCD$.

In both preceding examples, each result box in the truth table and Karnaugh map had a clearly defined state. Some logical relationships, however, do not require that every possible result necessarily be a one or a zero. For example, out of 16 possible results from the combination of four variables, only 14 results may be mandated by the application. This may sound odd, but one explanation could be that the particular application simply cannot provide the full 16 combinations of inputs. The specific reasons for this are as numerous as the many different applications that exist. In such circumstances these so-called *don't care* results can be used to reduce the complexity of your logic. Because the application does not care what result is generated for these few combinations, you can arbitrarily set the results to 0s or 1s so that the logic is minimized. Figure 1.7 is an example that modifies the Karnaugh map in Fig. 1.6 such that two don't care boxes are present. Don't care values are most commonly represented with "x" characters. The presence of one x enables simplification of the resulting logic by converting it to a 1 and grouping it with an adjacent 1. The other x is set to 0 so that it does not waste additional logic terms. The new Boolean equation is simplified by removing B from the last term, yielding $Y = \overline{A}\,\overline{C} + \overline{B}\,\overline{C} + \overline{A}B\overline{D} + ACD$. It is helpful to remember that x values can generally work to your benefit, because their presence imposes fewer requirements on the logic that you must create to get the job done.

1.4 BINARY AND HEXADECIMAL NUMBERING

The fact that there are only two valid Boolean values, 1 and 0, makes the *binary* numbering system appropriate for logical expression and, therefore, for digital systems. Binary is a base-2 system in

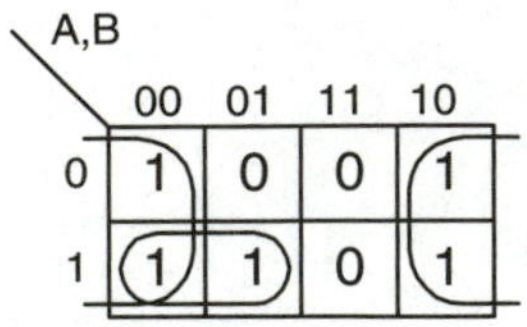

FIGURE 1.5 Completed Karnaugh map for a function of three variables.

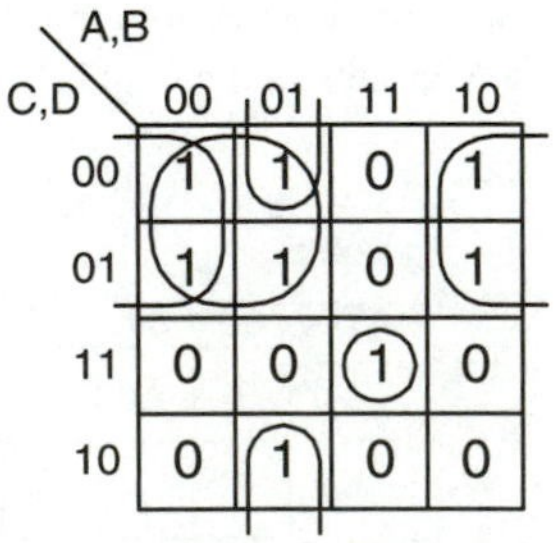

FIGURE 1.6 Completed Karnaugh map for function of four variables.

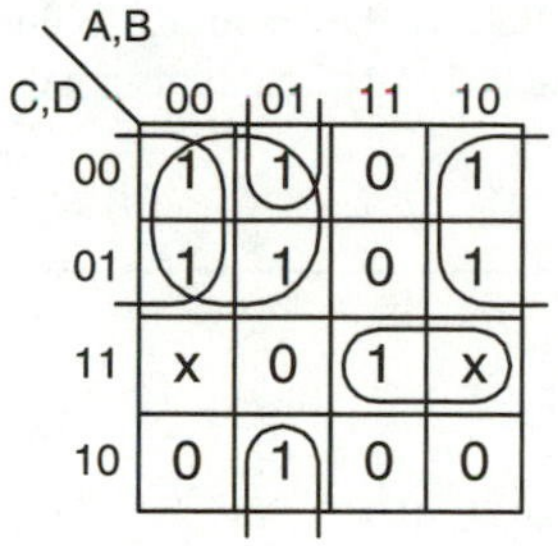

FIGURE 1.7 Karnaugh map for function of four variables with two "don't care" values.

which only the digits 1 and 0 exist. Binary follows the same laws of mathematics as decimal, or base-10, numbering. In decimal, the number 191 is understood to mean one hundreds plus nine tens plus one ones. It has this meaning, because each digit represents a successively higher power of ten as it moves farther left of the decimal point. Representing 191 in mathematical terms to illustrate these increasing powers of ten can be done as follows:

$$191 = 1 \times 10^2 + 9 \times 10^1 + 1 \times 10^0$$

Binary follows the same rule, but instead of powers of ten, it works on powers of two. The number 110 in binary (written as 110_2 to explicitly denote base 2) does not equal 110_{10} (decimal). Rather, $110_2 = 1 \times 2^2 + 1 \times 2^1 + 0 \times 2^0 = 6_{10}$. The number 191_{10} can be converted to binary by performing successive division by decreasing powers of 2 as shown below:

$$191 \div 2^7 = 191 \div 128 = 1 \text{ remainder } 63$$

$$63 \div 2^6 = 63 \div 64 = 0 \text{ remainder } 63$$

$$63 \div 2^5 = 63 \div 32 = 1 \text{ remainder } 31$$

$$31 \div 2^4 = 31 \div 16 = 1 \text{ remainder } 15$$

$$15 \div 2^3 = 15 \div 8 = 1 \text{ remainder } 7$$

$$7 \div 2^2 = 7 \div 4 = 1 \text{ remainder } 3$$

$$3 \div 2^1 = 3 \div 2 = 1 \text{ remainder } 1$$

$$1 \div 2^0 = 1 \div 1 = 1 \text{ remainder } 0$$

The final result is that $191_{10} = 10111111_2$. Each binary digit is referred to as a *bit*. A group of N bits can represent decimal numbers from 0 to $2^N - 1$. There are eight bits in a *byte*, more formally called an *octet* in certain circles, enabling a byte to represent numbers up to $2^8 - 1 = 255$. The preceding example shows the eight power-of-two terms in a byte. If each term, or bit, has its maximum value of 1, the result is $128 + 64 + 32 + 16 + 8 + 4 + 2 + 1 = 255$.

While binary notation directly represents digital logic states, it is rather cumbersome to work with, because one quickly ends up with long strings of ones and zeroes. *Hexadecimal*, or base 16 (*hex* for short), is a convenient means of representing binary numbers in a more succinct notation. Hex matches up very well with binary, because one hex digit represents four binary digits, given that

$2^4 = 16$. A four-bit group is called a *nibble*. Because hex requires 16 digits, the letters "A" through "F" are borrowed for use as hex digits beyond 9. The 16 hex digits are defined in Table 1.7.

TABLE 1.7 Hexadecimal Digits

Decimal value	0	1	2	3	4	5	6	7	8	9	10	11	12	13	14	15
Hex digit	0	1	2	3	4	5	6	7	8	9	A	B	C	D	E	F
Binary nibble	0000	0001	0010	0011	0100	0101	0110	0111	1000	1001	1010	1011	1100	1101	1110	1111

The preceding example, $191_{10} = 10111111_2$, can be converted to hex easily by grouping the eight bits into two nibbles and representing each nibble with a single hex digit:

$$1011_2 = (8 + 2 + 1)_{10} = 11_{10} = B_{16}$$

$$1111_2 = (8 + 4 + 2 + 1)_{10} = 15_{10} = F_{16}$$

Therefore, $191_{10} = 10111111_2 = BF_{16}$. There are two common prefixes, 0x and $, and a common suffix, h, that indicate hex numbers. These styles are used as follows: BF_{16} = 0xBF = $BF = BFh. All three are used by engineers, because they are more convenient than appending a subscript "16" to each number in a document or computer program. Choosing one style over another is a matter of preference.

Whether a number is written using binary or hex notation, it remains a string of bits, each of which is 1 or 0. Binary numbering allows arbitrary data processing algorithms to be reduced to Boolean equations and implemented with logic gates. Consider the equality comparison of two four-bit numbers, M and N.

"If M = N, then the equality test is true."

Implementing this function in gates first requires a means of representing the individual bits that compose M and N. When a group of bits are used to represent a common entity, the bits are numbered in ascending or descending order with zero usually being the smallest index. The bit that represents 2^0 is termed the *least-significant bit*, or LSB, and the bit that represents the highest power of two in the group is called the *most-significant bit*, or MSB. A four-bit quantity would have the MSB represent 2^3. M and N can be ordered such that the MSB is bit number 3, and the LSB is bit number 0. Collectively, M and N may be represented as M[3:0] and N[3:0] to denote that each contains four bits with indices from 0 to 3. This presentation style allows any arbitrary bit of M or N to be uniquely identified with its index.

Turning back to the equality test, one could derive the Boolean equation using a variety of techniques. Equality testing is straightforward, because M and N are equal only if each digit in M matches its corresponding bit position in N. Looking back to Table 1.3, it can be seen that the XNOR gate implements a single-bit equality check. Each pair of bits, one from M and one from N, can be passed through an XNOR gate, and then the four individual equality tests can be combined with an AND gate to determine overall equality,

$$Y = \overline{M[3] \oplus N[3]} \& \overline{M[2] \oplus N[2]} \& \overline{M[1] \oplus N[1]} \& \overline{M[0] \oplus N[0]}$$

The four-bit equality test can be drawn schematically as shown in Fig. 1.8.

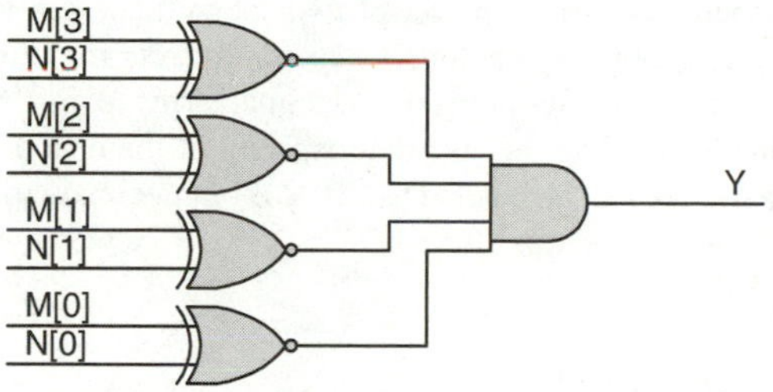

FIGURE 1.8 Four-bit equality logic.

Logic to compare one number against a constant is simpler than comparing two numbers, because the number of inputs to the Boolean equation is cut in half. If, for example, one wanted to compare M[3:0] to a constant 1001_2 (9_{10}), the logic would reduce to just a four-input AND gate with two inverted inputs:

$$y = M[3]\&\overline{M[2]}\&\overline{M[1]}\&M[0]$$

When working with computers and other digital systems, numbers are almost always written in hex notation simply because it is far easier to work with fewer digits. In a 32-bit computer, a value can be written as either 8 hex digits or 32 bits. The computer's logic always operates on raw binary quantities, but people generally find it easier to work in hex. An interesting historical note is that hex was not always the common method of choice for representing bits. In the early days of computing, through the 1960s and 1970s, *octal* (base-8) was used predominantly. Instead of a single hex digit representing four bits, a single octal digit represents three bits, because $2^3 = 8$. In octal, $191_{10} = 277_8$. Whereas bytes are the *lingua franca* of modern computing, groups of two or three octal digits were common in earlier times.

Because of the inherent binary nature of digital systems, quantities are most often expressed in orders of magnitude that are tied to binary rather than decimal numbering. For example, a "round number" of bytes would be 1,024 (2^{10}) rather than 1000 (10^3). Succinct terminology in reference to quantities of data is enabled by a set of standard prefixes used to denote order of magnitude. Furthermore, there is a convention of using a capital B to represent a quantity of bytes and using a lowercase b to represent a quantity of bits. Commonly observed prefixes used to quantify sets of data are listed in Table 1.8. Many memory chips and communications interfaces are expressed in units of bits. One must be careful not to misunderstand a specification. If you need to store 32 MB of data, be sure to use a 256 Mb memory chip rather than a 32 Mb device!

TABLE 1.8 Common Binary Magnitude Prefixes

Prefix	Definition	Order of Magnitude	Abbreviation	Usage
Kilo	$(1,024)^1 = 1,024$	2^{10}	k	kB
Mega	$(1,024)^2 = 1,048,576$	2^{20}	M	MB
Giga	$(1,024)^3 = 1,073,741,824$	2^{30}	G	GB
Tera	$(1,024)^4 = 1,099,511,627,776$	2^{40}	T	TB
Peta	$(1,024)^5 = 1,125,899,906,842,624$	2^{50}	P	PB
Exa	$(1,024)^6 = 1,152,921,504,606,846,976$	2^{60}	E	EB

The majority of digital components adhere to power-of-two magnitude definitions. However, some industries break from these conventions largely for reasons of product promotion. A key example is the hard disk drive industry, which specifies prefixes in decimal terms (e.g., 1 MB = 1,000,000 bytes). The advantage of doing this is to inflate the apparent capacity of the disk drive: a drive that provides 10,000,000,000 bytes of storage can be labeled as "10 GB" in decimal terms, but it would have to be labeled as only 9.31 GB in binary terms ($10^{10} \div 2^{30} = 9.31$).

1.5 BINARY ADDITION

Despite the fact that most engineers use hex data representation, it has already been shown that logic gates operate on strings of bits that compose each unit of data. Binary arithmetic is performed according to the same rules as decimal arithmetic. When adding two numbers, each column of digits is added in sequence from right to left and, if the sum of any column is greater than the value of the highest digit, a carry is added to the next column. In binary, the largest digit is 1, so any sum greater than 1 will result in a carry. The addition of 111_2 and 011_2 (7 + 3 = 10) is illustrated below.

	1	*1*	*1*	*0*	*carry bits*
		1	1	1	
+		0	1	1	
	1	0	1	0	

In the first column, the sum of two ones is 2_{10}, or 10_2, resulting in a carry to the second column. The sum of the second column is 3_{10}, or 11_2, resulting in both a carry to the next column and a one in the sum. When all three columns are completed, a carry remains, having been pushed into a new fourth column. The carry is, in effect, added to leading 0s and descends to the sum line as a 1.

The logic to perform binary addition is actually not very complicated. At the heart of a 1-bit adder is the XOR gate, whose result is the sum of two bits without the associated carry bit. An XOR gate generates a 1 when either input is 1, but not both. On its own, the XOR gate properly adds 0 + 0, 0 + 1, and 1 + 0. The fourth possibility, 1 + 1 = 2, requires a carry bit, because $2_{10} = 10_2$. Given that a carry is generated only when both inputs are 1, an AND gate can be used to produce the carry. A so-called *half-adder* is represented as follows:

$$\text{sum} = A \oplus B$$

$$\text{carry} = AB$$

This logic is called a *half-adder* because it does only part of the job when multiple bits must be added together. Summing multibit data values requires a carry to ripple across the bit positions starting from the LSB. The half-adder has no provision for a carry input from the preceding bit position. A *full-adder* incorporates a carry input and can therefore be used to implement a complete summation circuit for an arbitrarily large pair of numbers. Table 1.9 lists the complete full-adder input/output relationship with a carry input (C_{IN}) from the previous bit position and a carry output (C_{OUT}) to the next bit position. Note that all possible sums from zero to three are properly accounted for by combining C_{OUT} and sum. When $C_{IN} = 0$, the circuit behaves exactly like the half-adder.

TABLE 1.9 1-Bit Full-Adder Truth Table

C_{IN}	A	B	C_{OUT}	Sum
0	0	0	0	0
0	0	1	0	1
0	1	0	0	1
0	1	1	1	0
1	0	0	0	1
1	0	1	1	0
1	1	0	1	0
1	1	1	1	1

Full-adder logic can be expressed in a variety of ways. It may be recognized that full-adder logic can be implemented by connecting two half-adders in sequence as shown in Fig. 1.9. This full-adder directly generates a sum by computing the XOR of all three inputs. The carry is obtained by combining the carry from each addition stage. A logical OR is sufficient for C_{OUT}, because there can never be a case in which both half-adders generate a carry at the same time. If the A + B half-adder generates a carry, the partial sum will be 0, making a carry from the second half-adder impossible. The associated logic equations are as follows:

$$\text{sum} = A \oplus B \oplus C_{IN}$$

$$C_{OUT} = AB + [(A \oplus B)C_{IN}]$$

Equivalent logic, although in different form, would be obtained using a K-map, because XOR/XNOR functions are not direct results of K-map AND/OR solutions.

1.6 SUBTRACTION AND NEGATIVE NUMBERS

Binary subtraction is closely related to addition. As with many operations, subtraction can be implemented in a variety of ways. It is possible to derive a Boolean equation that directly subtracts two numbers. However, an efficient solution is to add the negative of the subtrahend to the minuend

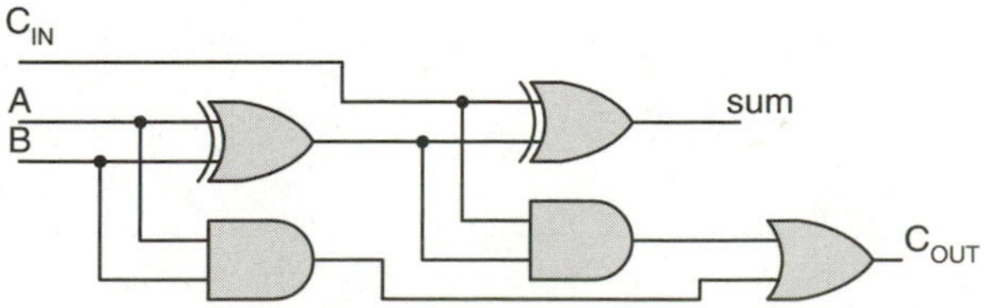

FIGURE 1.9 Full-adder logic diagram.

rather than directly subtracting the subtrahend from the minuend. These are, of course, identical operations: A – B = A + (–B). This type of arithmetic is referred to as subtraction by addition of the *two's complement*. The two's complement is the negative representation of a number that allows the identity A – B = A + (–B) to hold true.

Subtraction requires a means of expressing negative numbers. To this end, the most-significant bit, or left-most bit, of a binary number is used as the sign-bit when dealing with signed numbers. A negative number is indicated when the sign-bit equals 1. Unsigned arithmetic does not involve a sign-bit, and therefore can express larger absolute numbers, because the MSB is merely an extra digit rather than a sign indicator.

The first step in performing two's complement subtraction is to convert the subtrahend into a negative equivalent. This conversion is a two-step process. First, the binary number is inverted to yield a *one's complement*. Then, 1 is added to the one's complement version to yield the desired two's complement number. This is illustrated below:

	0	1	0	1	Original number (5)
	1	0	1	0	One's complement
+	0	0	0	1	Add one
	1	0	1	1	Two's complement (–5)

Observe that the unsigned four-bit number that can represent values from 0 to 15_{10} now represents signed values from –8 to 7. The range about zero is asymmetrical because of the sign-bit and the fact that there is no negative 0. Once the two's complement has been obtained, subtraction is performed by adding the two's complement subtrahend to the minuend. For example, 7 – 5 = 2 would be performed as follows, given the –5 representation obtained above:

1	*1*	*1*	*1*	*0*	*Carry bits*
	0	1	1	1	Minuend (7)
+	1	0	1	1	"Subtrahend" (–5)
	0	0	1	0	Result (2)

Note that the final carry-bit past the sign-bit is ignored. An example of subtraction with a negative result is 3 – 5 = –2.

		1	*1*	*0*	*Carry bits*
	0	0	1	1	Minuend (3)
+	1	0	1	1	"Subtrahend" (–5)
	1	1	1	0	Result (–2)

Here, the result has its sign-bit set, indicating a negative quantity. We can check the answer by calculating the two's complement of the negative quantity.

	1	1	1	0	Original number (–2)
	0	0	0	1	One's complement
+	0	0	0	1	Add one
	0	0	1	0	Two's complement (2)

This check succeeds and shows that two's complement conversions work "both ways," going back and forth between negative and positive numbers. The exception to this rule is the asymmetrical case in which the largest negative number is one more than the largest positive number as a result of the presence of the sign-bit. A four-bit number, therefore, has no positive counterpart of –8. Similarly, an 8-bit number has no positive counterpart of –128.

1.7 MULTIPLICATION AND DIVISION

Multiplication and division follow the same mathematical rules used in decimal numbering. However, their implementation is substantially more complex as compared to addition and subtraction. Multiplication can be performed inside a computer in the same way that a person does so on paper. Consider 12 × 12 = 144.

		1	2	
	X	1	2	
		2	4	Partial product × 10^0
+	1	2		Partial product × 10^1
	1	4	4	Final product

The multiplication process grows in steps as the number of digits in each multiplicand increases, because the number of partial products increases. Binary numbers function the same way, but there easily can be many partial products, because numbers require more digits to represent them in binary versus decimal. Here is the same multiplication expressed in binary (1100 × 1100 = 10010000):

				1	1	0	0	
			X	1	1	0	0	
				0	0	0	0	Partial product × 2^0
			0	0	0	0		Partial product × 2^1
		1	1	0	0			Partial product × 2^2
+	1	1	0	0				Partial product × 2^3
1	0	0	1	0	0	0	0	Final product

Walking through these partial products takes extra logic and time, which is why multiplication and, by extension, division are considered advanced operations that are not nearly as common as addition and subtraction. Methods of implementing these functions require trade-offs between logic complexity and the time required to calculate a final result.

1.8 FLIP-FLOPS AND LATCHES

Logic alone does not a system make. Boolean equations provide the means to transform a set of inputs into deterministic results. However, these equations have no ability to store the results of previous calculations upon which new calculations can be made. The preceding adder logic continually recalculates the sum of two inputs. If either input is removed from the circuit, the sum disappears as well. A series of numbers that arrive one at a time cannot be summed, because the adder has no means of storing a running total. Digital systems operate by maintaining *state* to advance through sequential steps in an algorithm. State is the system's ability to keep a record of its progress in a particular sequence of operations. A system's state can be as simple as a counter or an accumulated sum.

State-full logic elements called *flip-flops* are able to indefinitely hold a specific state (0 or 1) until a new state is explicitly loaded into them. Flip-flops load a new state when triggered by the transition of an input *clock*. A clock is a repetitive binary signal with a defined period that is composed of 0 and 1 phases as shown in Fig. 1.10. In addition to a defined period, a clock also has a certain *duty cycle*, the ratio of the duration of its 0 and 1 phases to the overall period. An ideal clock has a 50/50 duty cycle, indicating that its period is divided evenly between the two states. Clocks regulate the operation of a digital system by allowing time for new results to be calculated by logic gates and then capturing the results in flip-flops.

There are several types of flip-flops, but the most common type in use today is the *D flip-flop*. Other types of flip-flops include RS and JK, but this discussion is restricted to D flip-flops because of their standardized usage. A D flip-flop is often called a *flop* for short, and this terminology is used throughout the book. A basic rising-edge triggered flop has two inputs and one output as shown in Fig. 1.11a. By convention, the input to a flop is labeled D, the output is labeled Q, and the clock is represented graphically by a triangle. When the clock transitions from 0 to 1, the state at the D input is propagated to the Q output and stored until the next rising edge. State-full logic is often described through the use of a timing diagram, a drawing of logic state versus time. Figure 1.11b shows a basic flop timing diagram in which the clock's rising edge triggers a change in the flop's state. Prior to the rising edge, the flop has its initial state, Q_0, and an arbitrary 0 or 1 input is applied as D_0. The rising edge loads D_0 into the flop, which is reflected at the output. Once triggered, the flop's input can change without affecting the output until the next rising edge. Therefore, the input is labeled as "don't care," or "xxx" following the clock's rising edge.

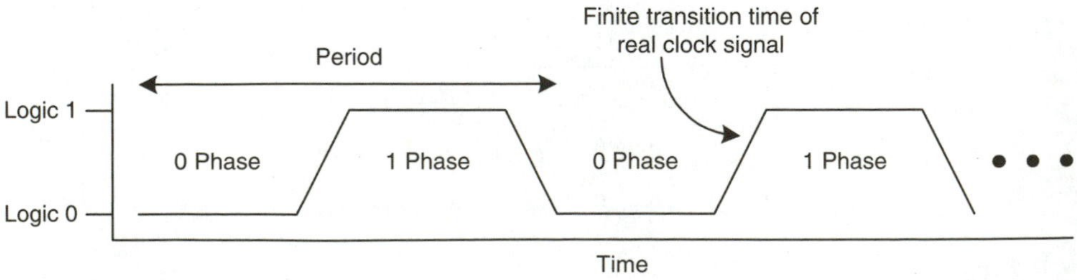

FIGURE 1.10 Digital clock signal.

Rising-edge flops are the norm, although some flops are falling-edge triggered. A falling-edge triggered flop is indicated by placing an inversion bubble at the clock input as shown in Fig. 1.12. Operation is the same, with the exception that the polarity of the clock is inverted. The remainder of this discussion assumes rising-edge triggered flops unless explicitly stated otherwise.

There are several common feature enhancements to the basic flop, including clock-enable, set, and clear inputs and a complementary output. Clock enable is used as a triggering qualifier each time a rising clock edge is detected. The D input is loaded only if clock enable is set to its active state. Inputs in general are defined by device manufacturers to be either active-low or active-high. An active-low signal is effective when set to 0, and an active-high signal is effective when set to 1. Signals are assumed to be active-high unless otherwise indicated. Active-low inputs are commonly indicated by the same inversion bubble used to indicate a falling-edge clock. When a signal is driven to its active state, it is said to be *asserted*. A signal is *de-asserted* when driven to its inactive state. Set and clear inputs explicitly force a flop to a 1 or 0 state, respectively. Such inputs are often used to initialize a digital system to a known state when it is first turned on. Otherwise, the flop powers up in a random state, which can cause problems for certain logic. Set and clear inputs can be either *synchronous* or *asynchronous*. Synchronous inputs take effect only on the rising clock edge, while asynchronous inputs take effect immediately upon being asserted. A complementary output is simply an inverted copy of the main output.

A truth table for a flop enhanced with the features just discussed is shown in Table 1.10. The truth table assumes a synchronous, active-high clock enable (EN) and synchronous, active-low set and clear inputs. The rising edge of the clock is indicated by the ↑ symbol. When the clock is at either static value, the outputs of the flop remain in their existing states. When the clock rises, the D, EN, $\overline{\text{CLR}}$, and $\overline{\text{SET}}$ inputs are sampled and acted on accordingly. As a general rule, conflicting information such as asserting $\overline{\text{CLR}}$ and $\overline{\text{SET}}$ at the same time should be avoided, because unknown results may arise. The exact behavior in this case depends on the specific flop implementation and may vary by manufacturer.

A basic application of flops is a binary ripple counter. Multiple flops can be cascaded as shown in Fig. 1.13 such that each complementary output is fed back to that flop's input and also used to clock the next flop. The current count value is represented by the noninverted flop outputs with the first flop representing the LSB. A three-bit counter is shown with an active-low reset input so that the counter can be cleared to begin at zero. The counter circuit diagram uses the standard convention of

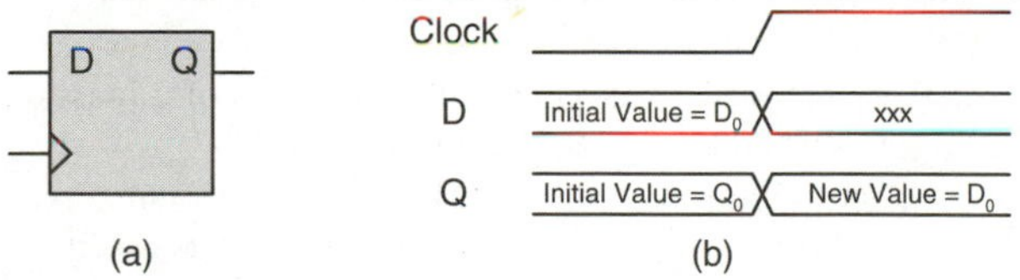

FIGURE 1.11 Rising-edge triggered flop.

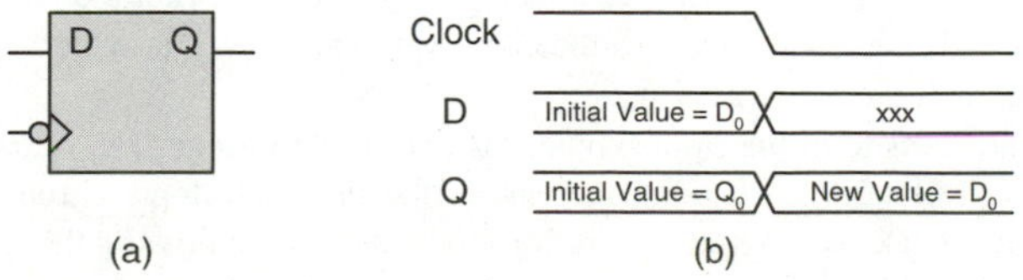

FIGURE 1.12 Falling-edge triggered flop.

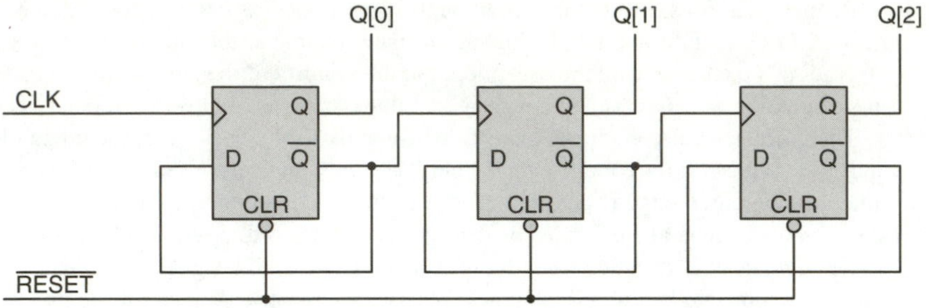

FIGURE 1.13 Three-bit ripple counter.

TABLE 1.10 Enhanced Flop Truth Table

Clock	D	EN	$\overline{CLR}$	$\overline{SET}$	Q	$\overline{Q}$
0	X	X	X	X	Q_{static}	$\overline{Q_{static}}$
↑	0	0	1	1	Q_{static}	$\overline{Q_{static}}$
↑	0	1	1	1	0	1
↑	1	1	1	1	1	0
↑	X	X	0	1	0	1
↑	X	X	1	0	1	0
↑	X	X	0	0	?	?
1	X	X	X	X	Q_{static}	$\overline{Q_{static}}$

showing electrical connectivity between intersecting wires by means of a junction dot. Wires that cross without a dot at their intersection are not electrically connected.

The ripple counter's operation is illustrated in Fig. 1.14. Each bit starts out at zero if $\overline{RESET}$ is asserted. Counting begins on the first rising edge of CLK following the de-assertion of $\overline{RESET}$. The LSB, Q[0], increments from 0 to 1, because its D input is driven by the complementary output, which is 1. The complementary output transitions to 0, which does not trigger the Q[1] rising-edge flop, but IT does set up the conditions for a trigger after the next CLK rising edge. When CLK rises again, Q[0] transitions back to 0, and $\overline{Q[0]}$ transitions to 1, forming a rising edge to trigger Q[1], which loads a 1. This sequence continues until the count value reaches 7, at which point the counter rolls over to zero, and the sequence begins again.

An undesirable characteristic of the ripple counter is that it takes longer for a new count value to stabilize as the number of bits in the counter increases. Because each flop's output clocks the next flop in the sequence, it can take some time for all flops to be updated following the CLK rising edge. Slow systems may not find this burdensome, but the added ripple delay is unacceptable in most high-speed applications. Ways around this problem will be discussed shortly.

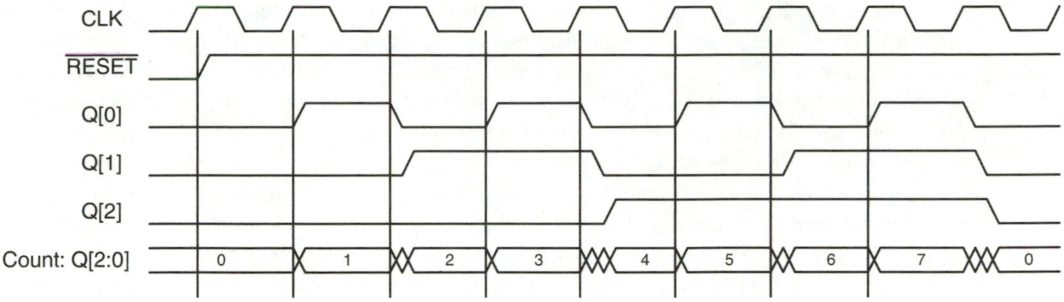

FIGURE 1.14 Ripple counter timing diagram.

A relative of the flop is the D-type *latch*, which is also capable of retaining its state indefinitely. A latch has a D input, a Q output, and an enable (EN) signal. Whereas a flop transfers its input to its output only on the active clock edge, a latch continuously transfers D to Q while EN is active. Latches are level sensitive, whereas flops are edge sensitive. A latch retains its state while EN is inactive. Table 1.11 shows the latch's truth table. Latches are simpler than flops and are unsuited to many applications in which flops are used. Latches would not substitute for flops in the preceding ripple counter example because, while the enable input is high, a continuous loop would be formed between the complementary output and input. This would result in rapid, uncontrolled oscillation at each latch during the time that the enable is held high.

TABLE 1.11 D-Latch Truth Table

EN	D	Q
0	X	Q_0
1	0	0
1	1	1

Latches are available as discrete logic elements and can also be assembled from simpler logic gates. The Boolean equation for a latch requires feeding back the output as follows:

$$Q = (EN\&D) + (\overline{EN}\&Q)$$

When EN is high, D is passed to Q. Q then feeds back to the second AND function, which maintains the state when EN is low. Latches are used in designs based on older technology that was conceived when the latch's simplicity yielded a cost savings or performance advantage. Most state-full elements today are flops unless there is a specific benefit to using a latch.

1.9 SYNCHRONOUS LOGIC

It has been shown that clock signals regulate the operation of a state-full digital system by causing new values to be loaded into flops on each active clock edge. *Synchronous logic* is the general term for a collection of logic gates and flops that are controlled by a common clock. The ripple counter is not synchronous, even though it is controlled by a clock, because each flop has its own clock, which leads to the undesirable ripple output characteristic previously mentioned. A synchronous circuit has all of its flops transition at the same time so that they settle at the same time, with a resultant improvement in performance. Another benefit of synchronous logic is easier circuit analysis, because all flops change at the same time.

Designing a synchronous counter requires the addition of logic to calculate the next count value based on the current count value. Figure 1.15 shows a high-level block diagram of a synchronous counter and is also representative of synchronous logic in general. Synchronous circuits consist of state-full elements (flops), with *combinatorial logic* providing feedback to generate the next state based on the current state. *Combinatorial logic* is the term used to describe logic gates that have no state on their own. Inputs flow directly through combinatorial logic to outputs and must be captured by flops to preserve their state.

An example of synchronous logic design can be made of converting the three-bit ripple counter into a synchronous equivalent. Counters are a common logic structure, and they can be designed in a variety of ways. The Boolean equations for small counters may be directly solved using a truth table and K-map. Larger counters may be assembled in regular structures using binary adders that generate the next count value by adding 1 to the current value. A three-bit counter is easily handled with a truth-table methodology. The basic task is to create a truth table relating each possible current state to a next state as shown in Table 1.12.

TABLE 1.12 Three-Bit Counter Truth Table

Reset	Current State	Next State
1	XXX	000
0	000	001
0	001	010
0	010	011
0	011	100
0	100	101
0	101	110
0	110	111
0	111	000

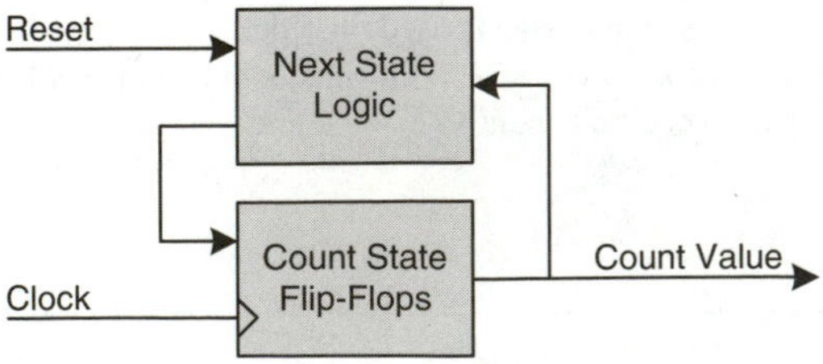

FIGURE 1.15 Synchronous counter block diagram.

Three Boolean equations are necessary, one for each bit that feeds back to the count state flops. If the flop inputs are labeled D[2:0], the outputs are labeled Q[2:0], and an active-high synchronous reset is defined, the following equations can be developed:

$$D[0] = \overline{Q[0]} \& \overline{RESET}$$

$$D[1] = \{(\overline{Q[0]} \& Q[1]) + (Q[0] \& \overline{Q[1]})\} \& \overline{RESET} = (Q[0] \oplus Q[1]) \& \overline{RESET}$$

$$D[2] = \{(\overline{Q[2]} \& Q[1] \& Q[0]) + (Q[2] \& \overline{Q[1]}) + (Q[2] \& \overline{Q[0]})\} \& \overline{RESET}$$

Each equation's output is forced to 0 when RESET is asserted. Otherwise, the counter increments on each rising clock edge. Synchronous logic design allows any function to be implemented by changing the feedback logic. It would not be difficult to change the counter logic to count only odd or even numbers, or to count only as high as 5 before rolling over to 0. Unlike the ripple counter, whose structure supports a fixed counting sequence, next state logic can be defined arbitrarily according to an application's needs.

1.10 SYNCHRONOUS TIMING ANALYSIS

Logic elements, including flip-flops and gates, are physical devices that have finite response times to stimuli. Each of these elements exhibits a certain propagation delay between the time that an input is presented and the time that an output is generated. As more gates are chained together to create more complex logic functions, the overall propagation delay of signals between the end points increases. Flip-flops are triggered by the rising edge of a clock to load their new state, requiring that the input to the flip-flop is stable prior to the rising edge. Similarly, a flip-flop's output stabilizes at a new state some time after the rising edge. In between the output of a flip-flop and the input of another flip-flop is an arbitrary collection of logic gates, as seen in the preceding synchronous counter circuit. Synchronous timing analysis is the study of how the various delays in a synchronous circuit combine to limit the speed at which that circuit can operate. As might be expected, circuits with lesser delays are able to run faster.

A clock breaks time into discrete intervals that are each the duration of a single clock period. From a timing analysis perspective, each clock period is identical to the last, because each rising clock edge is a new flop triggering event. Therefore, timing analysis considers a circuit's delays over one clock period, between successive rising (or falling) clock edges. Knowing that a wide range of clock frequencies can be applied to a circuit, the question of time arises of how fast the clock can go before the circuit stops working reliably. The answer is that the clock must be slow enough to allow sufficient time for the output of a flop to stabilize, for the signal to propagate through the combinatorial logic gates, and for the input of the destination flop to stabilize. The clock must also be slow enough for the flop to reliably detect each edge. Each flop circuit is characterized by a minimum clock pulse width that must be met. Failing to meet this minimum time can result in the flop missing clock events.

Timing analysis revolves around the basic timing parameters of a flop: *input setup time* (t_{SU}), *input hold time* (t_H), and *clock-to-out time* (t_{CO}). Setup time specifies the time immediately preceding the rising edge of the clock by which the input must be stable. If the input changes too soon before the clock edge, the electrical circuitry within the flop will not have enough time to properly recognize the state of the input. Hold time places a restriction on how soon after the clock edge the input may begin to change. Again, if the input changes too soon after the clock edge, it may not be properly detected by the circuitry. Clock-to-out time specifies how soon after the clock edge the output will be updated to the state presented at the input. These parameters are very brief in duration and are usually measured in nanoseconds. One nanosecond, abbreviated "ns," is one billionth of a second. In very fast microchips, they may be measured in picoseconds, or one trillionth or a second.

Consistent terminology is necessary when conducting timing analysis. Timing is expressed in units of both clock frequency and time. Clock frequency, or speed, is quantified in units of *hertz*, named after the twentieth century German physicist, Gustav Hertz. One hertz is equivalent to one clock cycle per second—one transition from low to high and a second transition from high to low. Units of hertz are abbreviated as Hz and are commonly accompanied by prefixes that denote an order of magnitude. Commonly observed prefixes used to quantify clock frequency and their definitions are listed in Table 1.13. Unlike quantities of bytes that use binary-based units, clock frequency uses decimal-based units.

TABLE 1.13 Common Clock Frequency Magnitude Prefixes

Prefix	Definition	Order of Magnitude	Abbreviation	Usage
Kilo	Thousand	10^3	K	kHz
Mega	Million	10^6	M	MHz
Giga	Billion	10^9	G	GHz
Tera	Trillion	10^{12}	T	THz

Units of time are used to express a clock's period as well as basic logic element delays such as the aforementioned t_{SU}, t_H, and t_{CO}. As with frequency, standard prefixes are used to indicate the order of magnitude of a time specification. However, rather than expressing positive powers of ten, the exponents are negative. Table 1.14 lists the common time magnitude prefixes employed in timing analysis.

TABLE 1.14 Common Time Magnitude Prefixes

Prefix	Definition	Order of Magnitude	Abbreviation	Usage
Milli	One-thousandth	10^{-3}	m	ms
Micro	One-millionth	10^{-6}	μ	μs
Nano	One-billionth	10^{-9}	n	ns
Pico	One-trillionth	10^{-12}	p	ps

Aside from basic flop timing characteristics, timing analysis must take into consideration the finite propagation delays of logic gates and wires that connect flop outputs to flop inputs. All real components have nonzero propagation delays (the time required for an electrical signal to move from an input to an output on the same component). Wires have an approximate propagation delay of 1 ns for every 6 in of length. Logic gates can have propagation delays ranging from more than

10 ns down to the picosecond range, depending on the technology being used. Newly designed logic circuits should be analyzed for timing to ensure that the inherent propagation delays of the logic gates and interconnect wiring do not cause a flop's t_{SU} and t_H specifications to be violated at a given clock frequency.

Basic timing analysis can be illustrated with the example logic circuit shown Fig. 1.16. There are two flops connected by two gates. The logic inputs shown unconnected are ignored in this instance, because timing analysis operates on a single path at a time. In reality, other paths exist through these unconnected inputs, and each path must be individually analyzed. Each gate has a finite propagation delay, t_{PROP}, which is assumed to be 5 ns for the sake of discussion. Each flop has t_{CO} = 7 ns, t_{SU} = 3 ns, and t_H = 1 ns. For simplicity, it is assumed that there is zero delay through the wires that connect the gates and flops.

The timing analysis must cover one clock period by starting with one rising clock edge and ending with the next rising edge. How fast can the clock run? The first delay encountered is t_{CO} of the source flop. This is followed by t_{PROP} of the two logic gates. Finally, t_{SU} of the destination flop must be met. These parameters may be summed as follows:

$$t_{CLOCK} = t_{CO} + 2 \times t_{PROP} + t_{SU} = 20 \text{ ns}$$

The frequency and period of a clock are inversely related such that $F = 1/t$. A 20-ns clock period corresponds to a 50-MHz clock frequency: $1/(20 \times 10^{-9}) = 50 \times 10^6$. Running at exactly the calculated clock period leaves no room for design margin. Increasing the period by 5 ns reduces the clock to 40 MHz and provides headroom to account for propagation delay through the wires.

Hold time compliance can be verified following setup time analysis. Meeting a flop's hold time is often not a concern, especially in slower circuits as shown above. The 1 ns t_H specification is easily met, because the destination flop's D-input will not change until $t_{CO} + 2 \times t_{PROP}$ = 17 ns after the rising clock edge. Actual timing parameters have variance associated with them, and the best-case t_{CO} and t_{PROP} would be somewhat smaller numbers. However, there is so much margin in this case that t_H compliance is not a concern.

Hold-time problems sometimes arise in fast circuits where t_{CO} and t_{PROP} are very small. When there are no logic gates between two flops, t_{PROP} can be nearly zero. If the minimum t_{CO} is nearly equal to the maximum t_H, the situation should be carefully investigated to ensure that the destination flop's input remains stable for a sufficient time period after the active clock edge.

1.11 CLOCK SKEW

The preceding timing analysis example is simplified for ease of presentation by assuming that the source and destination flops in a logic path are driven by the same clock signal. Although a synchronous circuit uses a common clock for all flops, there are small, nonzero variances in clock timing at individual flops. Wiring delay variances are one source of this nonideal behavior. When a clock source drives two flops, the two wires that connect to each flop's clock input are usually not identical

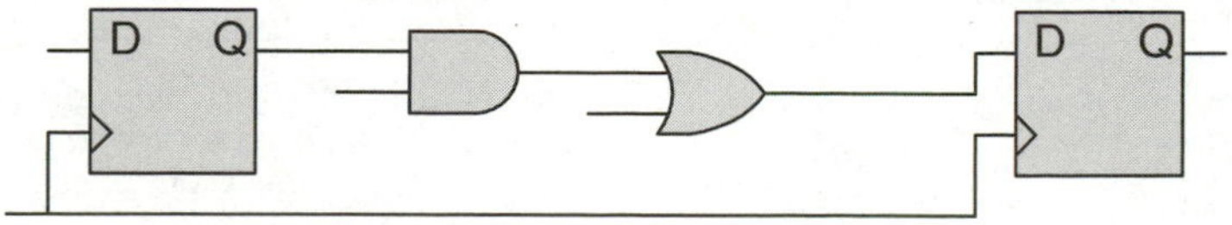

FIGURE 1.16 Hypothetical logic circuit.

in length. This length inequality causes one flop's clock to arrive slightly before or after the other flop's clock.

Clock skew is the term used to characterize differences in edge timing between multiple clock inputs. Skew caused by wiring delay variance can be effectively minimized by designing a circuit so that clock distribution wires are matched in length. A more troublesome source of clock skew arises when there are too many clock loads to be driven by a single source. Multiple clock drivers are necessary in these situations, with small variations in electrical characteristics between each driver. These driver variances result in clock skew across all the flops in a synchronous design. As might be expected, clock skew usually reduces the frequency at which a synchronous circuit can operate.

Clock skew is subtracted from the nominal clock period for setup time analysis purposes, because the worst-case scenario shown in Fig. 1.17 must be considered. This scenario uses the same logic circuit in Fig. 1.16 but shows two separate clocks with 1 ns of skew between them. The worst timing occurs when the destination flop's clock arrives before that of the source flop, thereby reducing the amount of time available for the D-input to stabilize. Instead of the circuit having zero margin with a 20-ns period, clock skew increases the minimum period to 21 ns. The extra 1 ns compensates for the clock skew to restore a minimum source to destination period time of 20 ns. A slower circuit such as this one is not very sensitive to clock skew, especially after backing off to 40 MHz for timing margin as shown previously. Digital systems that run at relatively low frequencies may not be affected by clock skew, because they often have substantial margins built into their timing analyses. As clock speeds increase, the margin decreases to the point at which clock skew and interconnect delay become important limiting factors in system design.

Hold time compliance can become more difficult in the presence of clock skew. The basic problem occurs when clock skew reduces the source flop's apparent t_{CO} from the destination flop's perspective, causing the destination's input to change before t_H is satisfied. Such problems are more prone in high-speed systems, but slower systems are not immune. Figure 1.18 shows a timing diagram for a circuit with 1 ns of clock skew where two flops are connected by a short wire with nearly zero propagation delay. The flops have $t_{CO} = 2$ ns and $t_H = 1.5$ ns. A scenario like this may be experienced when connecting two chips that are next to each other on a circuit board. In the absence of clock skew, the destination flop's input would change t_{CO} after the rising clock edge, exceeding t_H by 0.5 ns. The worst-case clock skew causes the source flop clock to arrive before that of the destination flop, resulting in an input change just 1 ns after the rising clock edge and violating t_H.

Solutions to skew-induced t_H violations include reducing the skew or increasing the delay between source and destination. Unfortunately, increasing a signal's propagation delay may cause t_{SU} violations in high-speed systems.

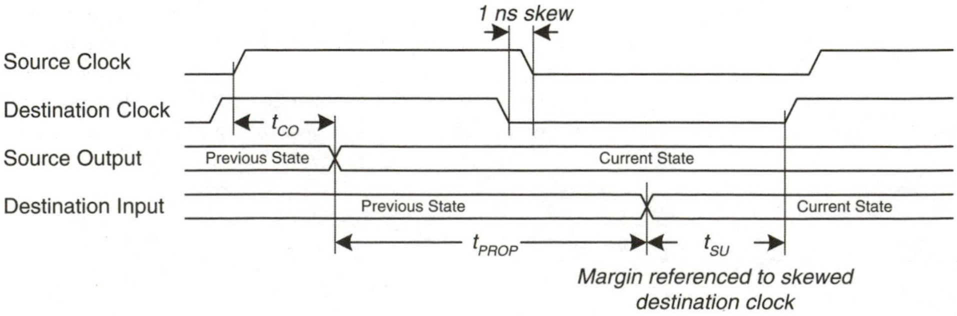

FIGURE 1.17 Clock skew influence on setup time analysis.

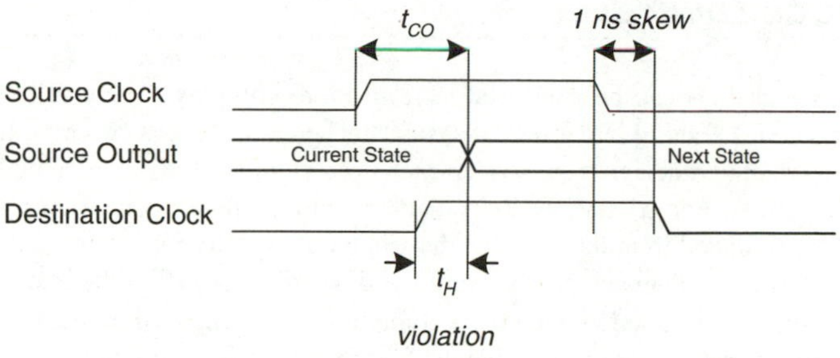

FIGURE 1.18 Hold-time violation caused by clock skew.

Hold time may not be a problem in slower circuits, because slower circuits often have paths between flops with sufficiently long propagation delays to offset clock skew problems. However, even slow circuits can experience hold-time problems if flops are connected with wires or components that have small propagation delays. It is also important to remember that hold-time compliance is not a function of clock period but of clock skew, t_{CO}, and t_H. Therefore, a slow system that uses fast components may have problems if the clock skew exceeds the difference between t_{CO} and t_H.

1.12 CLOCK JITTER

An ideal clock signal has a fixed frequency and duty cycle, resulting in its edges occurring at the exact time each cycle. Real clock signals exhibit slight variations in the timing of successive edges. This variation is known as *jitter* and is illustrated in Fig. 1.19. Jitter is caused by nonideal behavior of clock generator circuitry and results in some cycles being longer than nominal and some being shorter. The average clock frequency remains constant, but the cycle-to-cycle variance may cause timing problems.

Just as clock skew worsens the analysis for both t_{SU} and t_H, so does jitter. Jitter must be subtracted from calculated timing margins to determine a circuit's actual operating margin. Some systems are more sensitive to jitter than others. As operating frequencies increase, jitter becomes more of a problem, because it becomes a greater percentage of the clock period and flop timing specifications. Jitter specifications vary substantially. Many systems can tolerate 0.5 ns of jitter and more. Very sensitive systems may require high-quality clock circuitry that can reduce jitter to below 100 ps.

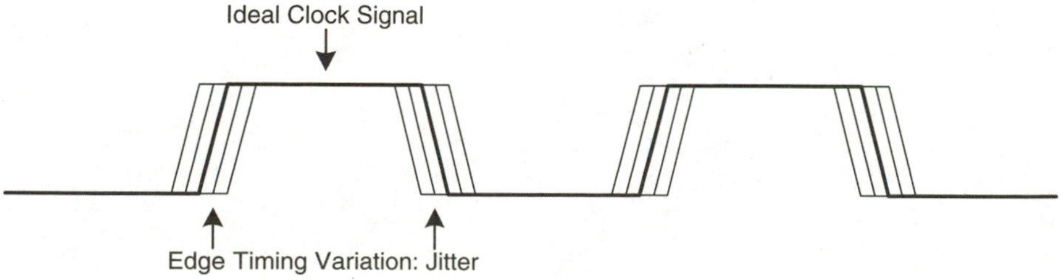

FIGURE 1.19 Clock jitter.

1.13 DERIVED LOGICAL BUILDING BLOCKS

Basic logic gates and flops can be combined to form more complex structures that are treated as building blocks when designing larger digital systems. There are various common functions that an engineer does not want to redesign from scratch each time. Some of the common building blocks are multiplexers, demultiplexers, tri-state buffers, registers, and shift registers. Counters represent another building block alluded to in the previous discussion of synchronous logic. A counter is a combination of flops and gates that can count either up or down, depending on the implementation.

Multiplexers, sometimes called *selectors*, are combinatorial elements that function as a multiposition logical switches to select one of many inputs. Figure 1.20 shows a common schematic representation of a multiplexer, often shortened to *mux*. A mux has an arbitrary number of data inputs, often an even power of two, and a smaller number of selector inputs. According to the binary state of the selector inputs, a specific data input is transferred to the output.

Muxes are useful, because logic circuits often need to choose between multiple data values. A counter, for example, may choose between loading a next count value or loading an arbitrary value from external logic. A possible truth table for a 4-to-1 mux is shown in Table 1.15. Each selector input value maps to one, and only one, data input.

TABLE 1.15 Four-to-One Multiplexer Truth Table

S1	S0	Y
0	0	A
0	1	B
1	0	C
1	1	D

A demultiplexer, also called a *demux*, performs the inverse operation of a mux by transferring a single input to the output that is selected by select inputs. A demux is drawn similarly to a mux, as shown in Fig. 1.21.

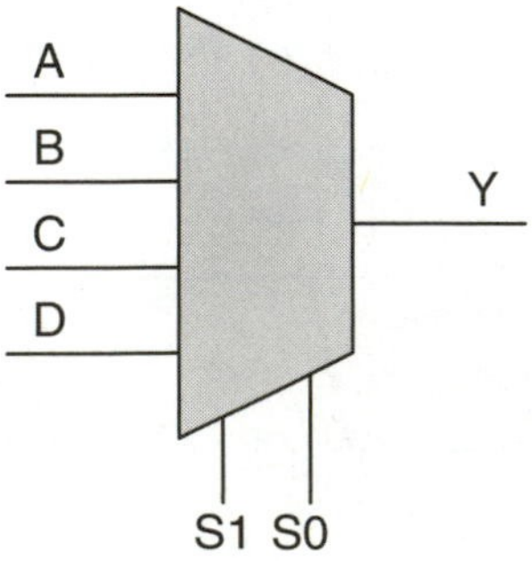

FIGURE 1.20 Four-to-one multiplexer.

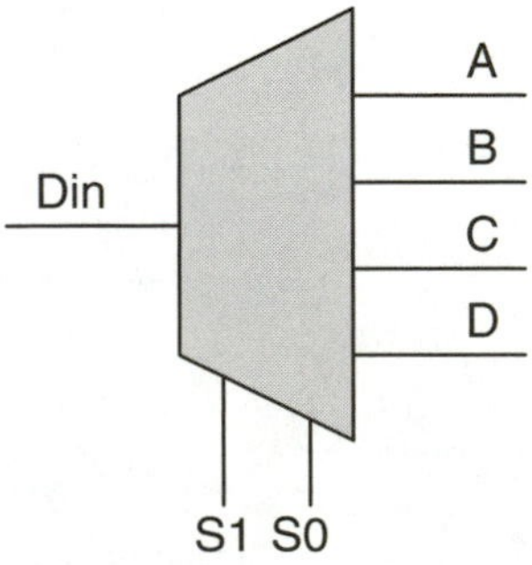

FIGURE 1.21 One-to-four demultiplexer.

A possible truth table for a 1-to-4 demux is shown in Table 1.16. Those outputs that are not selected are held low. The output that is selected assumes the state of the data input.

TABLE 1.16 One-to-Four Demultiplexer Truth Table

S1	S0	A	B	C	D
0	0	Din	0	0	0
0	1	0	Din	0	0
1	0	0	0	Din	0
1	1	0	0	0	Din

A popular use for a demux is as a decoder. The main purpose of a decoder is not so much to transfer an input to one of several outputs but simply to assert one output while not asserting those that are not selected. This function has great utility in microprocessor address decoding, which involves selecting one of multiple devices (e.g., a memory chip) at a time for access. The truth table for a 2-to-4 decoder is shown in Table 1.17. The decoder's outputs are active-low, because most memory and microprocessor peripheral chips use active-low enable signals.

TABLE 1.17 Two-to-Four Decoder Truth Table

S1	S0	A	B	C	D
0	0	0	1	1	1
0	1	1	0	1	1
1	0	1	1	0	1
1	1	1	1	1	0

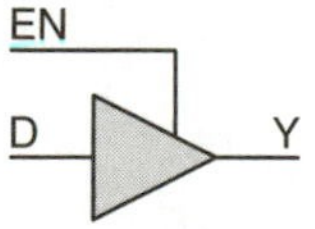

FIGURE 1.22 Tri-state buffer.

Tri-state buffers are combinatorial elements that can drive three output states rather than the standard 0 and 1 states. The third state is off, often referred to as high-impedance, *hi-Z*, or just Z. Tri-state buffers enable multiple devices to share a common output wire by cooperatively agreeing to have only one device drive the wire at any one time, during which all other devices remain in hi-Z. A tri-state buffer is drawn as shown in Fig. 1.22.

A tri-state buffer passes its D-input to Y-output when enabled. Otherwise, the output will be turned off as shown in Table 1.18.

Electrically, tri-state behavior allows multiple tri-state buffers to be connected to the same wire without *contention*. Contention normally results when multiple outputs are connected together because some want to drive high and some low. This creates potentially damaging electrical contention (a short circuit). However, if multiple tri-state buffers are connected, and only one at a time is enabled, there is no possibility of contention. The main advantage here is that digital *buses* in comput-

TABLE 1.18 Tri-state Buffer Truth Table

EN	D	Y
0	x	Z
1	0	0
1	1	1

ers can be arbitrarily expanded by adding more devices without the need to add a full set of input or output signals each time a new device is added. In a logical context, a *bus* is a collection of wires that serve a common purpose. For example, a computer's data bus might be eight wires that travel together and collectively represent a byte of data. Electrical contention on a bus is often called a *bus-fight*. Schematically, multiple tri-state buffers might be drawn as shown in Fig. 1.23.

Each tri-state buffer contains its own enable signal, which is usually driven by some type of decoder. The decoder guarantees that only one tri-state buffer is active at any one time, preventing contention on the common wire.

Registers are collections of multiple flops arranged in a group with a common function. They are a common synchronous-logic building block and are commonly found in multiples of 8-bit widths, thereby representing a byte, which is the most common unit of information exchange in digital systems. An 8-bit register provides a common clock and clock enable for all eight internal flops. The clock enable allows external control of when the flops get reloaded with new D-input values and when they retain their current values. It is common to find registers that have a built-in tri-state buffer, allowing them to be placed directly onto a shared bus without the need for an additional tri-state buffer component.

Whereas normal registers simply store values, synchronous elements called *shift registers* manipulate groups of bits. Shift registers exist in all permutations of serial and parallel inputs and outputs. The role of a shift register is to somehow change the sequence of bits in an array of bits. This includes creating arrays of bits from a single bit at a time (serial input) or distributing an array of bits one bit at a time (serial output). A serial-in, parallel-out shift register can be implemented by chaining several flops together as shown in Fig. 1.24.

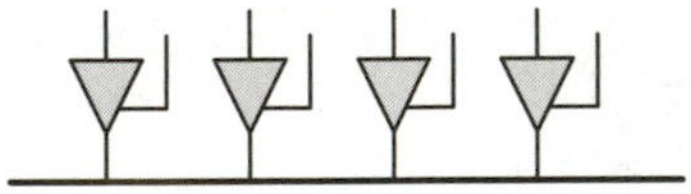

FIGURE 1.23 Multiple tri-state buffers on a single wire.

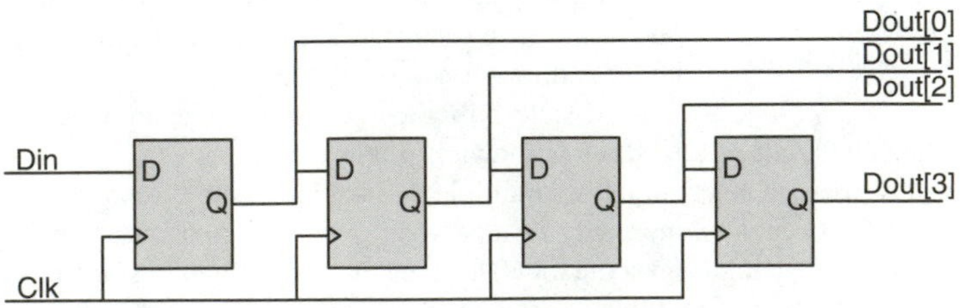

FIGURE 1.24 Serial-in, parallel-out shift register.

On each rising clock edge, a new serial input bit is clocked into the first flop, and each flop in succession loads its new value based on its predecessor's value. At any given time, the parallel output of an N-bit shift register reflects the state of the last N bits shifted in up to that time. In this example (N = 4), a serial stream of bits collected in four clock cycles can be operated upon as a unit of four bits once every fourth cycle. As shown, data is shifted in MSB first, because Dout[3] is shown in the last bit position. Such a simple transformation is useful, because it is often more practical to communicate digital data in serial form where only one bit of information is sent per clock cycle, but impractical to operate on that data serially. An advantage of serial communication is that fewer wires are required as compared to parallel. Yet, parallel representation is important because arithmetic logic can get overly cumbersome if it has to keep track of one bit at a time. A parallel-in, serial-out shift register is very similar, as shown in Fig. 1.25, with the signals connected for MSB first operation to match the previous example.

Four flops are used here as well. However, instead of taking in one bit at a time, all flops are loaded when the load signal is asserted. The 2-to-1 muxes are controlled by the load signal and determine if the flops are loaded with new parallel data or shifted serial data. Over each of the next four clock cycles, the individual bits are shifted out one at a time. If these two shift register circuits were connected together, a crude serial data communications link could be created whereby parallel data is converted to serial and then back to parallel at each end.

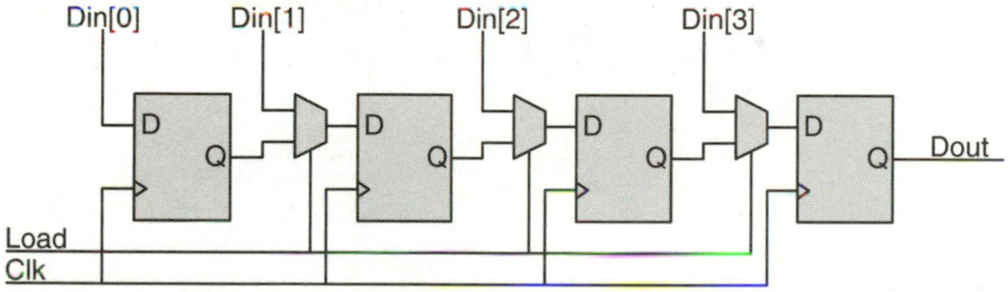

FIGURE 1.25 Parallel-in, serial-out shift register.

CHAPTER 2

Integrated Circuits and the 7400 Logic Families

Once basic logic design theory is understood, the next step is transferring that knowledge to a practical context that includes real components. This chapter explains what an integrated circuit is and how off-the-shelf components can be used to implement arbitrary logic functions.

Integrated circuits, called *chips* by engineers and laymen alike, are what enable digital systems as we know them. The chapter begins with an introduction to how chips are constructed. Familiarity with basic chip fabrication techniques and terminology enables an engineer to comprehend the distinctions between various products so that their capabilities can be more readily evaluated.

A survey of packaging technology follows to provide familiarity with the common physical characteristics of commercially available chips. Selecting a package that is appropriate for a particular design can be as critical as selecting the functional parameters of the chip itself. It is important to understand the variety of available chip packages and why different types of packages are used for different applications.

The chapter's major topic follows next: the 7400 logic families. These off-the-shelf logic chips have formed the basis of digital systems for decades and continue to do so, although in fewer numbers as a result of the advent of denser components. 7400 family features are presented along with complete examples of how the chips are applied in real designs. The purpose of this discussion is to impart a practical and immediately applicable understanding of how digital system design can be executed with readily available components. Although these devices are not appropriate for every application, many basic problems can be solved with 7400 chips once it is understood how to employ them.

Having seen how real chips can be used to solve actual design problems, a closely related topic is presented at the end of this chapter: the interpretation of data sheets. Manufacturers' data sheets contain critical information that must be understood to ensure a working design. An understanding of how data sheets are organized and the types of information that they contain is a necessary knowledge base for every engineer.

2.1 THE INTEGRATED CIRCUIT

Digital logic and electronic circuits derive their functionality from electronic switches called *transistors*. Roughly speaking, the transistor can be likened to an electronically controlled valve whereby energy applied to one connection of the valve enables energy to flow between two other connections. By combining multiple transistors, digital logic building blocks such as AND gates and flip-flops are formed. Transistors, in turn, are made from *semiconductors*. Consult a periodic table of elements in a college chemistry textbook, and you will locate semiconductors as a group of elements separating the metals and nonmetals. They are called semiconductors because of their ability to behave as both metals and nonmetals. A semiconductor can be made to conduct electricity like a metal or to insulate as a nonmetal does. These differing electrical properties can be accurately controlled by mixing the semiconductor with small amounts of other elements. This mixing is called *doping*. A semiconductor can be doped to contain more electrons (N-type) or fewer electrons (P-type). Examples of commonly used semiconductors are silicon and germanium. Phosphorous and boron are two elements that are used to dope N-type and P-type silicon, respectively.

A transistor is constructed by creating a sandwich of differently doped semiconductor layers. The two most common types of transistors, the *bipolar-junction transistor* (BJT) and the *field-effect transistor* (FET) are schematically illustrated in Fig. 2.1. This figure shows both the silicon structures of these elements and their graphical symbolic representation as would be seen in a circuit diagram. The BJT shown is an *NPN* transistor, because it is composed of a sandwich of N-P-N doped silicon. When a small current is injected into the *base* terminal, a larger current is enabled to flow from the *collector* to the *emitter*. The FET shown is an N-channel FET; it is composed of two N-type regions separated by a P-type substrate. When a voltage is applied to the insulated *gate* terminal, a current is enabled to flow from the *drain* to the *source*. It is called N-channel, because the gate voltage induces an N-channel within the substrate, enabling current to flow between the N-regions.

Another basic semiconductor structure shown in Fig. 2.1 is a *diode,* which is formed simply by a junction of N-type and P-type silicon. Diodes act like one-way valves by conducting current only from P to N. Special diodes can be created that emit light when a voltage is applied. Appropriately enough, these components are called *light emitting diodes*, or LEDs. These small lights are manufactured by the millions and are found in diverse applications from telephones to traffic lights.

The resulting small chip of semiconductor material on which a transistor or diode is fabricated can be encased in a small plastic package for protection against damage and contamination from the outside world. Small wires are connected within this package between the semiconductor sandwich and pins that protrude from the package to make electrical contact with other parts of the intended circuit. Once you have several discrete transistors, digital logic can be built by directly wiring these components together. The circuit will function, but any substantial amount of digital logic will be very bulky, because several transistors are required to implement each of the various types of logic gates.

At the time of the invention of the transistor in 1947 by John Bardeen, Walter Brattain, and William Shockley, the only way to assemble multiple transistors into a single circuit was to buy separate discrete transistors and wire them together. In 1959, Jack Kilby and Robert Noyce independently in-

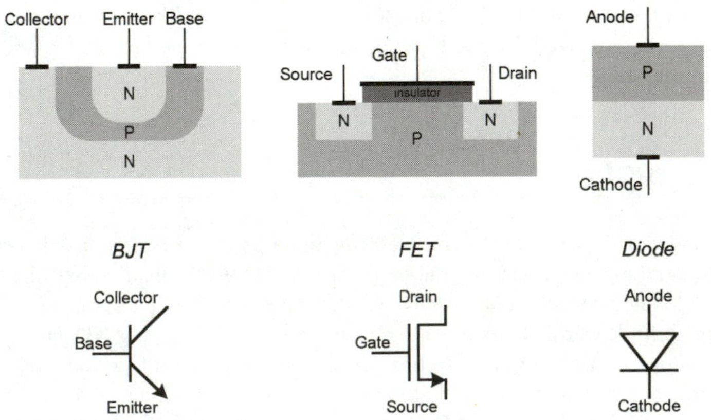

FIGURE 2.1 BJT, FET, and diode structural and symbolic representations.

vented a means of fabricating multiple transistors on a single slab of semiconductor material. Their invention would come to be known as the *integrated circuit*, or IC, which is the foundation of our modern computerized world. An IC is so called because it integrates multiple transistors and diodes onto the same small semiconductor chip. Instead of having to solder individual wires between discrete components, an IC contains many small components that are already wired together in the desired topology to form a circuit.

A typical IC, without its plastic or ceramic package, is a square or rectangular silicon die measuring from 2 to 15 mm on an edge. Depending on the level of technology used to manufacture the IC, there may be anywhere from a dozen to tens of millions of individual transistors on this small chip. This amazing density of electronic components indicates that the transistors and the wires that connect them are extremely small in size. Dimensions on an IC are measured in units of micrometers, with one micrometer (1 μm) being one millionth of a meter. To serve as a reference point, a human hair is roughly 100 μm in diameter. Some modern ICs contain components and wires that are measured in increments as small as 0.1 μm! Each year, researchers and engineers have been finding new ways to steadily reduce these feature sizes to pack more transistors into the same silicon area, as indicated in Fig. 2.2.

Many individual chemical process steps are involved in fabricating an IC. The process begins with a thin, clean, polished semiconductor wafer — most often silicon — that is usually one of three standard diameters: 100, 200, or 300 mm. The circular wafer is cut from a cylindrical ingot of solid silicon that has a perfect crystal structure. This perfect crystal base structure is necessary to promote the formation of other crystals that will be deposited by subsequent processing steps. Many dice are arranged on the wafer in a grid as shown in Fig. 2.3. Each die is an identical copy of a master pattern and will eventually be sliced from the wafer and packaged as an IC. An IC designer determines how different portions of the silicon wafer should be modified to create transistors, diodes, resistors, capacitors, and wires. This IC design layout can then be used to, in effect, draw tiny components onto the surface of the silicon. Sequential drawing steps are able to build sandwiches of differently doped silicon and metal layers.

Engineers realized that light provided the best way to faithfully replicate patterns from a template onto a silicon substrate, similar to what photographers have been doing for years. A photographer takes a picture by briefly exposing film with the desired image and then developing this film into a negative. Once this negative has been created, many identical photographs can be reproduced by briefly exposing the light-sensitive photographic paper to light that is focused through the negative. Portions of the negative that are dark do not allow light to pass, and these corresponding regions of the paper are not exposed. Those areas of the negative that are light allow the paper to be exposed.

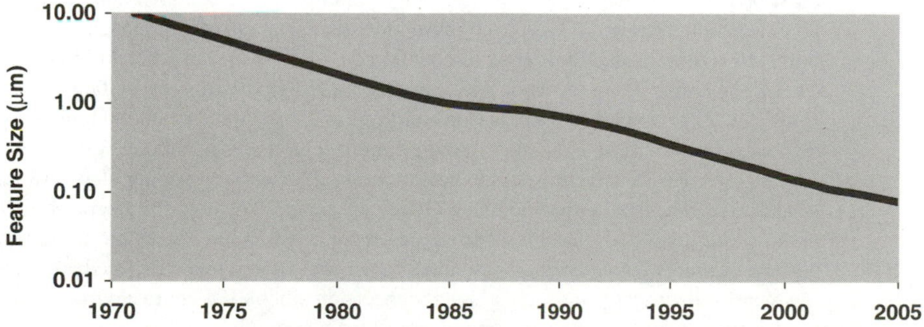

FIGURE 2.2 Decreasing IC feature size over time. *(Future data for years 2003 through 2005 compiled from* The International Technology Roadmap for Semiconductors, *Semiconductor Industry Association, 2001.)*

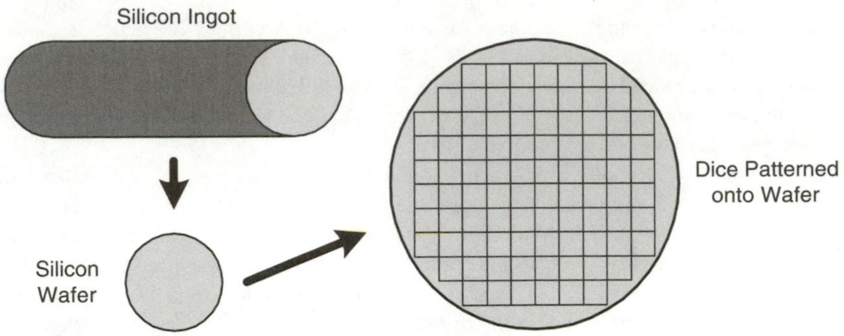

FIGURE 2.3 Silicon wafer.

When the paper is developed in a chemical bath, portions of the paper that were exposed change color and yield a visible image.

Photographic processes provide excellent resolution of detail. Engineers apply this same principle in fabricating ICs to create details that are fractions of a micron in size. Similar to a photographic negative, a mask is created for each IC processing step. Like a photographic negative, the mask does not have to be the same size as the silicon area it is to expose because, with lenses, light can be focused through the mask to an arbitrary area. Using a technique called *photolithography,* the silicon surface is first prepared with a light-sensitive chemical called *photoresist.* The prepared surface is then exposed to light through the mask. Depending on whether a positive or negative photoresist process is employed, the areas of photoresist that have been either exposed or not exposed to light are washed away in a chemical bath, resulting in a pattern of bare and covered areas of silicon. The wafer can then be exposed to chemical baths, high temperature metal vapors, and ion beams. Only the bare areas that have had photoresist washed away are affected in this step. In this way, specific areas of the silicon wafer can be doped according to the IC designers' specifications. Successive mask layers and process steps can continue to wash away and expose new layers of photoresist and then build sandwiches of semiconductor and metal material. A very simplified view of these process steps is shown in Fig. 2.4. The semiconductor fabrication process must be performed in a clean-room environment to prevent minute dust particles and other contaminants from disturbing the lithography and chemical processing steps.

In reality, dozens of such steps are necessary to fabricate an IC. The semiconductor structures that must be formed by layering different metals and dopants are complex and must be formed one thin layer at a time. Modern ICs typically have more than four layers of metal, each layer separated from others by a thin insulating layer of silicon dioxide. The use of more metal layers increases the cost of an IC, but it also increases its density, because more metal wires can be fabricated to connect more transistors. This complete process from start to finish usually takes one to four weeks. The chemical diffusion step (5) is an example of how different regions of the silicon wafer are doped to achieve varying electrical characteristics. In reality, several successive doping steps are required to create transistors. The metal deposition step (10) is an example of how the microscopic metal wires that connect the many individual transistors are created. Hot metal vapors are passed over the prepared surface of the wafer. Over time, individual molecules adhere to the exposed areas and form continuous wires. Historically, most metal interconnects on silicon ICs are made from aluminum. However, copper has become a common component of leading-edge ICs.

As IC feature sizes continue to shrink, the physical properties of light can become limiting factors in the resolution with which a wafer can be processed. Shorter light wavelengths are necessary to

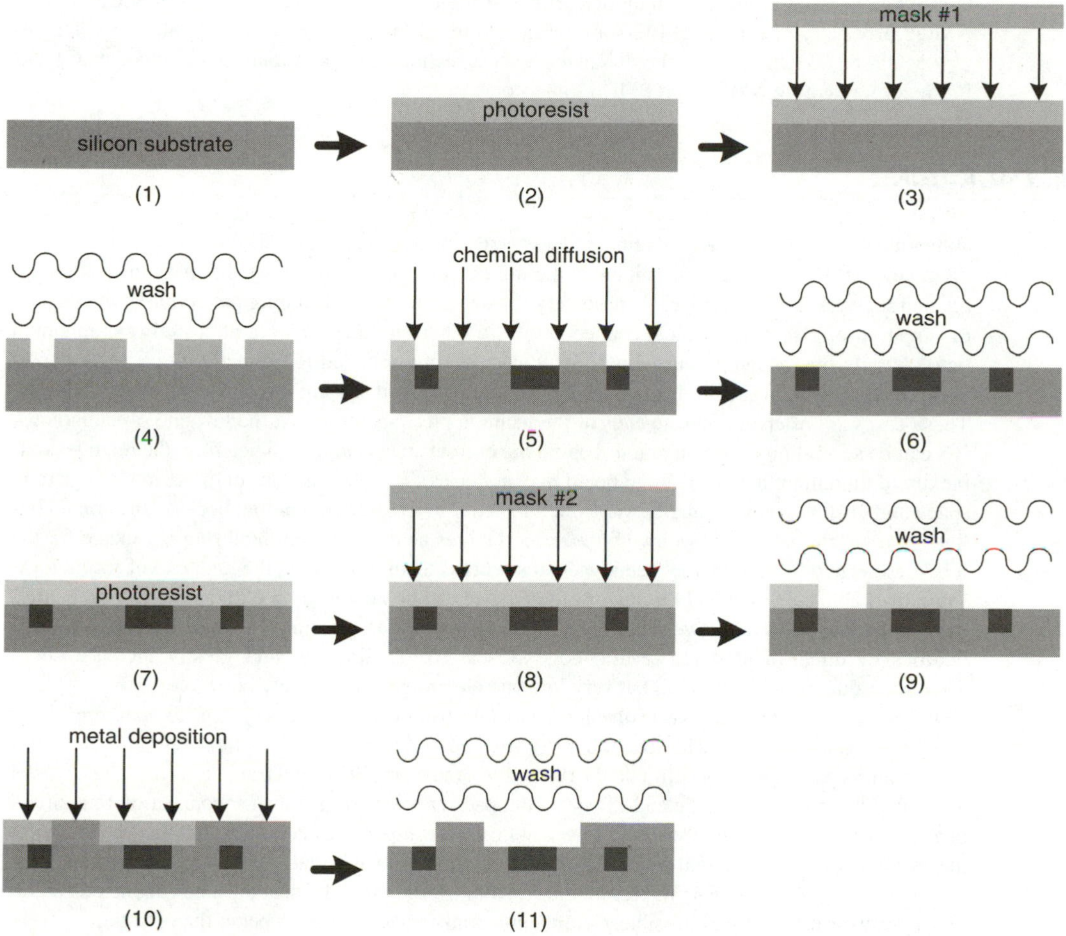

FIGURE 2.4 The IC fabrication process.

meet the demands of leading-edge IC process technology. The human eye can detect electromagnetic energy from about 700 nm (red) to 400 nm (violet). Whereas ultraviolet light (< 400 nm) was once adequate for IC fabrication, deep UV wavelengths are now in use, and shorter wavelengths below 200 nm are being explored.

Each of the process steps is applied to the entire wafer. The many dice on a single wafer are usually exposed to light through the same mask. The mask is either large enough to cover the entire wafer and therefore expose all dice at once, or the mask is stepped through the dice grid (using a machine appropriately called a *stepper*) such that each die location is exposed separately before the next processing step. In certain cases, such as small-volume or experimental runs, different die locations on the same wafer will be exposed with different masks. This is entirely feasible but may not be as efficient as creating a wafer on which all dice are identical.

When an IC is designed and fabricated, it generally follows one of two main transistor technologies: bipolar or metal-oxide semiconductor (MOS). Bipolar processes create BJTs, whereas MOS processes create FETs. Bipolar logic was more common before the 1980s, but MOS technologies have since accounted the great majority of digital logic ICs. N-channel FETs are fabricated in an NMOS process, and P-channel FETs are fabricated in a PMOS process. In the 1980s, complementary-MOS, or CMOS, became the dominant process technology and remains so to this day. CMOS ICs incorporate both NMOS and PMOS transistors.

2.2 IC PACKAGING

When the wafer has completed its final process step, it is tested and then sliced up to separate the individual dice. Dice that fail the initial testing are quickly discarded. Those that pass inspection are readied for packaging. A package is necessary for several reasons, including protection of the die and the creation of electromechanical connections with other circuitry. ICs are almost always mounted onto a circuit board, and it is usually difficult to mount unpackaged ICs directly to the board. However, there are special situations in which ICs are not packaged and are directly attached to the board. These cases are often at opposite ends of the technological spectrum. At the low end of technology, ICs can be several process generations behind the current state of the art. Therefore, the relative complexity of mounting them to a circuit board may not be as great. The savings of direct mounting are in space and cost. A common quartz wristwatch benefits from direct mounting, because the small confines of a watch match very well with the space savings achieved by not requiring a package for the IC. These watch ICs use mature semiconductor process technologies. At the high end of technology, some favorable electrical and thermal characteristics can be achieved by eliminating as much intermediate bulk as possible between individual ICs and supporting circuitry. However, the technical difficulties of direct-mounting a leading-edge IC can be challenging and greatly increase costs. Therefore, direct-mounting of all but very low-end electronics is relatively rare.

IC packaging technology has evolved dramatically from the early days, yet many mature package types still exist and are in widespread use. Plastic and ceramic are the two most common materials used in an IC package. They surround the die and its lead frame. The lead frame is a structure of metal wires that fan out from the die and extend to the package exterior as pins for connection to a circuit board. Plastic packages are generally lower in cost as compared to ceramics, but they have poorer thermal performance. Thermal characteristics are important for ICs that handle large currents and dissipate large quantities of heat. To prevent the IC from overheating, the heat must be conducted and radiated away as efficiently as possible. Ceramic material conducts heat far better than plastic.

A very common package is the *dual in-line package*, or DIP, shown in Fig. 2.5. A DIP has two parallel rows of pins that are spaced on 0.1-in centers. Each pin extends roughly 0.2 in below the bottom of the plastic or ceramic body. Pins are numbered sequentially from 1 going left to right along one side and resuming on the opposite side from right to left. There is usually at least one pin 1 marker at one end of the package. It is either a dot near pin 1 or a semicircular indentation on one edge of the package.

DIPs are commonly manufactured in standard sizes ranging from 6 to 48 pins, and some manufacturers go beyond 48 pins. Smaller pin-count devices have 0.3-in wide packages, and larger devices are 0.6 in wide. Because of the ubiquity of the DIP, there are many variations of pin counts and package widths. For many years, the DIP accounted for the vast majority of digital logic packages. Common logic ICs were manufactured in 14- and 16-pin DIPs. Memory ICs were manufactured in 16-, 18-, 24-, and 28-pin DIPs. Microprocessors were available in 40-, 44-, and 48-pin DIPs. DIPs are still widely available today, but their use as a percentage of the total IC market has declined markedly. However, the benefits of the DIP remain: they are inexpensive and easy to work with by hand, eliminating the need for costly assembly tools.

If you were to carefully crack open a DIP, you would be able to see the mechanical assembly of the die and lead frame. This is illustrated in Fig. 2.6. The die is cemented in the center of a stamped

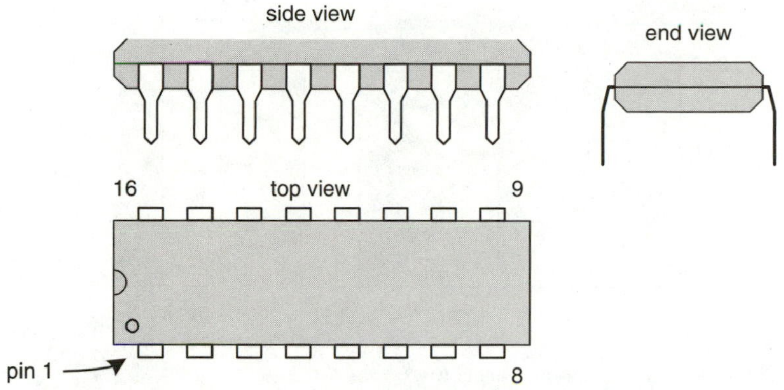

FIGURE 2.5 A 16-pin dual in-line package.

metal frame and is connected to the individual pins with extremely thin wires. Once the electrical connections are made, the fragile assembly is encased in a plastic or ceramic body for protection and the exterior portions of the pins are folded vertically.

All other IC packages are variations on this theme. Some packages use a similar lead-frame structure, whereas more advanced packages utilize very high-quality miniature circuit boards made from either ceramic or fiberglass.

An oft-quoted attribute of ICs is that their density doubles every 18 months as a result of improvements in process technology. This prediction was made in 1965 by Dr. Gordon Moore, a co-founder of Intel. It has since come to be known as *Moore's law,* because the semiconductor industry has matched this prediction over time. Before to the explosion of IC density, the semiconductor industry classified ICs into several categories depending on the number of logic gates on the device: *small-scale integration* (SSI), *medium-scale integration* (MSI), *large-scale integration* (LSI), and, finally, *very large-scale integration* (VLSI). Figure 2.7 provides a rough definition of these terms. As the density of ICs continued to grow at a rapid pace, it became rather ridiculous to keep adding words like "very" and "extra" to these categories, and the terms' widespread use declined. ICs are now often categorized based on their minimum feature size and metal process. For example, one might refer to an IC as "0.25 μm, three-layer metal (aluminum)" or "0.13 μm, six-layer copper."

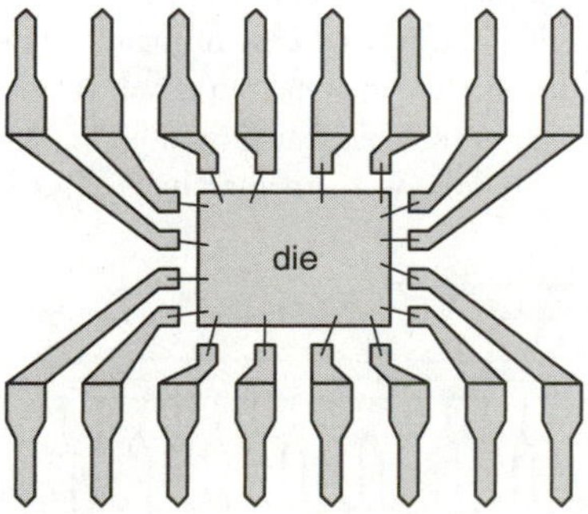

FIGURE 2.6 DIP lead frame.

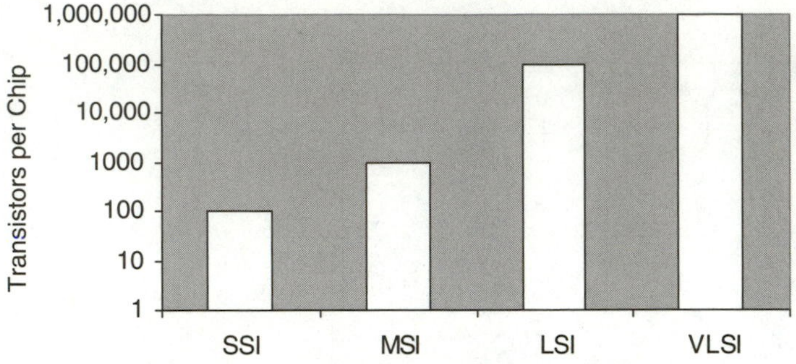

FIGURE 2.7 Relative component count of ICs.

As IC densities grew at this tremendous pace, the number of pins on each IC and the speed at which they operated began to increase as well. DIPs soon become a limiting factor in the performance of ICs. First, the addition of more pins made the package longer, because there are only two rows of pins. However, most chips are relatively square in shape to minimize on-chip interconnection distances. This creates a conflict: a long, narrow package that is unsuitable for increasing square die sizes. Second, the lengths of some pins in the DIP lead frame, especially those near the corners, are relatively long. This has an adverse impact on the quality of high-speed signals. Third, the 0.1-in pin spacing on DIPs keeps them artificially large as circuit board technologies continue improving to handle smaller contacts.

One solution to the pin density problem was the development of the *pin grid array*, or PGA, package. Shown in Fig. 2.8, the PGA is akin to a two-dimensional DIP with pins spaced on 0.1-in centers. Very high pin counts are achievable with a PGA, because all of its area is usable rather than just the perimeter. Being a square, the PGA is compatible with large ICs, because it more closely matches the proportions of a silicon chip.

The PGA provides high pin density, but its drawback is relatively high cost. Two lower-cost packages were developed for ICs that require more pins than DIPs but fewer pins than found on a PGA: the *small outline integrated circuit* (SOIC) and the *plastic leaded chip carrier* (PLCC). Examples of SOIC and PLCC packages are shown in Fig. 2.9. Both SOICs and PLCCs feature pins on a 0.05-in pitch — half that of a DIP or PGA. The SOIC is basically a shrunken DIP with shorter pins that are folded parallel to the plane of the package instead of protruding down vertically. This enables the SOIC to be surface mounted onto the circuit board by soldering the pins directly to metal pads on the board. By contrast, a DIP requires that holes be drilled in the board for the pins to be soldered into. The SOIC represents an improvement in packaging density and ease of manufacture over DIPs, but it is still limited to relatively simple ICs due to its one-dimensional pin arrangement.

PLCCs increase pin density and ease the design of the lead frame by utilizing a two-dimensional pin arrangement. Higher pin counts (68, 84, and 96 pins) were enabled by the PLCC, and its square

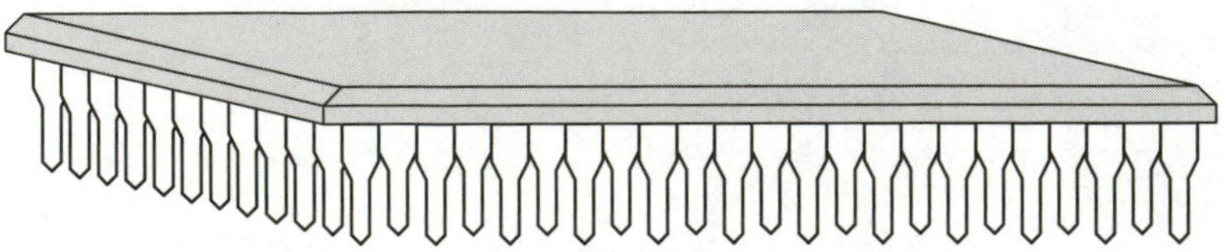

FIGURE 2.8 Pin grid array package.

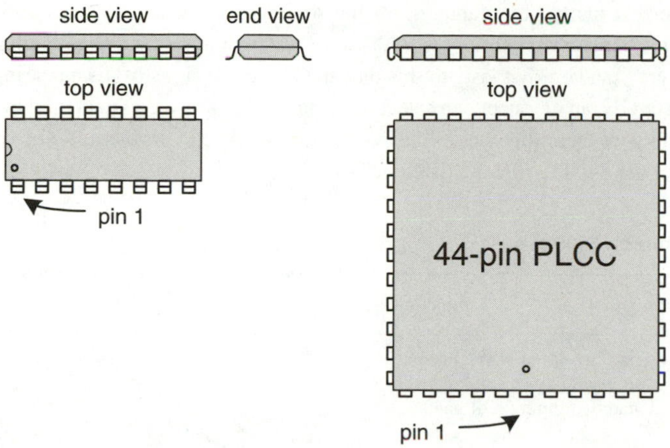

FIGURE 2.9 SOIC and PLCC.

design is more capable of accepting larger silicon dice than either the DIP or SOIC. PLCC leads are not bent outward, as in the case of a SOIC, but are curved inward in a "J" pattern. The more similar aspect ratio of the PLCC package and the dice that are placed into them enabled lead frames with shorter and more consistent pin lengths, reducing the degrading effects on high-speed signals.

A higher-density relative of the PLCC and SOIC is the *quad flat pack*, or QFP. A QFP resembles a PLCC in terms of its square or rectangular shape but has leads that are bent outward like an SOIC. Additionally, QFP leads are thinner and spaced at a smaller pitch to achieve more than twice the lead density of a comparably sized PLCC.

Perhaps the most widely used package for high-density ICs is the *ball grid array*, or BGA. The BGA is a surface mount analog to the PGA with significantly higher ball density. Contact is made between a BGA and a circuit board by means of many small preformed solder balls that adhere to contacts on the bottom surface of the BGA package. Figure 2.10 illustrates the general BGA form factor, but numerous variants on aspect ratio and ball pitch exist. Typical ball pitch ranges from 1.27 mm down to 0.8 mm, and higher densities are on the way.

There are many variations of the packaging technologies already mentioned. Most packages comply with industry standard dimensions, but others are proprietary. Semiconductor manufacturers provide detailed drawings of their packages to enable the proper design of circuit boards for their products.

2.3 THE 7400-SERIES DISCRETE LOGIC FAMILY

With the advent of ICs in the early 1960s, engineers needed ready access to a library of basic logic gates so that these gates could be wired together on circuit boards and turned into useful products. Rather than having to design a custom microchip for each new project, semiconductor companies

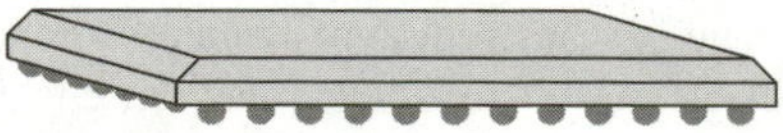

FIGURE 2.10 Ball grid array.

began to recognize a market for standard, off-the-shelf logic ICs. In 1963 and 1964, Sylvania and Texas Instruments began shipment of the 7400-series discrete logic family and unknowingly started a *de facto* industry standard that lasts to this day and shows no signs of disappearing anytime soon. Using the 7400 family, an engineer can select logic gates, flip-flops, counters, and buffers in individual packages and wire them together as desired to solve a specific problem. Some of the most common members of the 7400 family are listed in Table 2.1.

TABLE 2.1 Common 7400 ICs

Part Number	Function	Number of Pins
7400	Quad two-input NAND gates	14
7402	Quad two-input NOR gates	14
7404	Hex inverters	14
7408	Quad two-input AND gates	14
7432	Quad two-input OR gates	14
7447	BCD to seven-segment display decoder/driver	16
7474	Dual D-type positive edge triggered flip-flops	14
7490	Four-bit decade counter	14
74138	Three-to-eight decoder	16
74153	Dual 4-to-1 multiplexer	16
74157	Quad 2-to-1 multiplexers	16
74160	Four-bit binary synchronous counter	16
74164	Eight-bit parallel out serial shift registers	16
74174	Quad D-type flip-flops with complementary outputs	16
74193	Four-bit synchronous up/down binary counter	16
74245	Octal bus transceivers with tri-state outputs	20
74373	Octal D-type transparent latch	20
74374	Octal D-type flip-flops	20

These are just a few of the full set of 7400 family members. Many 7400 parts are no longer used, because their specific function is rarely required as a separate chip in modern digital electronics designs. However, the parts listed above, and many others that are not listed, are still readily available today and are commonly found in a broad range of digital designs ranging from low-end to high-tech devices. 7400-series logic has been available in DIPs for a long time, as well as (more recently) SOICs and other high-density surface mount packages. All flavors of basic logic gates are available with varying numbers of inputs. For example, there are 2-, 3-, and 4-input AND gates and 2-, 3-, 4-,

8-, 12-, and 13-input NAND gates. There are numerous varieties of flip-flops, counters, multiplexers, shift registers, and bus transceivers. Flip-flops exist with and without complementary outputs, preset/clear inputs, and independent clocks. Counters are available in 4-bit blocks that can both increment and decrement and count to either 15 (binary counter) or 9 (decade counter) before restarting the count at 0. Shift registers exist in all permutations of serial and parallel inputs and outputs. Bus transceivers in 4- and 8-bit increments exist with different types of output enables and capabilities to function in unidirectional or bidirectional modes. Bus transceivers enable the creation and expansion of tri-state buses on which multiple devices can communicate.

One interesting IC is the 7447 seven-segment display driver. This component allows the creation of graphical numeric displays in applications such as counters and timers. Seven-segment displays are commonly seen in automobiles, microwave ovens, watches, and consumer electronics. Seven independent on/off elements can represent all ten digits as shown in Fig. 2.11. The 7447 is able to drive an LED-based seven-segment display when given a *binary coded decimal* (BCD) input. BCD is a four-bit binary number that has valid values from 0 through 9. Hexadecimal values from 0xA through 0xF are not considered legal BCD values.

Familiarity with the 7400 series proves very useful no matter what type of digital system you are designing. For low-end systems, 7400-series logic may be the only type of IC at your disposal to solve a wide range of problems. At the high end, many people are often surprised to see a small 14-pin 7400-series IC soldered to a circuit board alongside a fancy 32-bit microprocessor running at 100 MHz. The fact is that the basic logic functions that the 7400 series offers are staples that have direct applications at all levels of digital systems design. It is time well spent to become familiar with the extensive capabilities of the simple yet powerful 7400 family. Manufacturers' logic data books, either in print or on line, are invaluable references. It can be difficult to know ahead of time if a design may call for one more gate to function properly; that is when a 40-year old logic family can save the day.

2.4 APPLYING THE 7400 FAMILY TO LOGIC DESIGN

Applications of the 7400 family are truly infinite, because the various ICs represent basic building blocks rather than complete solutions. Up through the early 1980s, it was common to see computer systems constructed mainly from interconnected 7400-series ICs along with a few LSI components such as a microprocessor and a few memory chips. These days, most commercial digital systems are designed using some form of higher-density logic IC, either fully custom or user programmable. However, the engineer or hobbyist who has a relatively small-scale logic problem to solve, and who may not have access to more expensive custom or programmable logic ICs, may be able to utilize only 7400 logic in an efficient and cost-effective solution. Two examples follow to provide insight into how 7400 building blocks can be assembled to solve logic design problems.

A hypothetical example is a logic circuit to examine three switches and turn on an LED if two and only two of the three switches are turned on. The truth table for such a circuit is as follows in

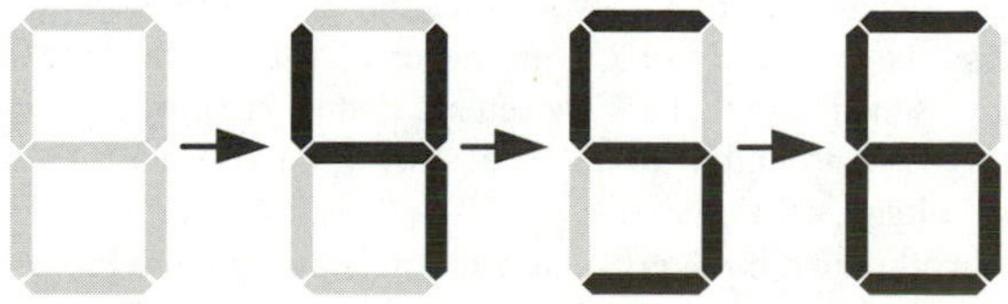

FIGURE 2.11 Seven-segment display.

Table 2.2, given that A, B, and C are the inputs, and an LED is the active-low output (assume that the LED is turned on by driving a logic 0 rather than a logic 1).

TABLE 2.2 LED Driver Logic Truth Table

A	B	C	$\overline{\text{LED}}$
0	0	0	1
0	0	1	1
0	1	0	1
0	1	1	0
1	0	0	1
1	0	1	0
1	1	0	0
1	1	1	1

This LED driver truth table can be converted into the following Boolean logic equation with a Karnaugh map or simply by inspection:

$$\overline{\text{LED}} = \overline{\overline{A}BC + A\overline{B}C + AB\overline{C}}$$

After consulting a list of available 7400 logic ICs, three become attractive for our application: the 7404 inverter, 7408 AND, and 7432 OR. The LED driver logic equation requires four inverters, six two-input AND gates, and two 2-input OR gates. Four ICs are required, because a 7404 provides six inverters, a 7408 provides four AND gates, and a 7432 contains four OR gates. These four ICs can be connected according to a *schematic diagram* as shown in Fig. 2.12. A schematic diagram illustrates the electrical connectivity scheme of various components. Each component is identified by a *reference designator* consisting of a letter followed by a number. ICs are commonly identified by reference designators beginning with the letter "U". Additionally, each component has numerous pins that are numbered on the diagram. These pin numbers conform to the IC manufacturer's numbering scheme. Each of these 7400-series ICs has 14 pins. Another convention that remains from bipolar logic days is the use of the label VCC to indicate the positive voltage supply node. GND represents ground—the common, or return, voltage supply node.

All ICs require connections to a power source. In this circuit, +5 V serves as the power supply, because the 7400 family is commonly manufactured in a bipolar semiconductor process requiring a +5-V supply. The four rectangular blocks at the top of the diagram represent this power connection information. Because this schematic diagram shows individual gates, the gates' reference designators contain an alphabetic suffix to identify unique instances of gates within the same IC. Not all gates in each IC are actually used. Those that are unused are tied inactive by connecting their inputs to a valid logic level—in this case, ground. It would be equally valid to connect the inputs of unused gates to the positive supply voltage, +5 V.

This logic circuit would work, but a more efficient solution is available to those who are familiar with the capabilities of the 7400 family. The 7411 provides three 3-input AND gates, which is perfect for this application, allowing a reduction in the part count to three ICs instead of four. This cir-

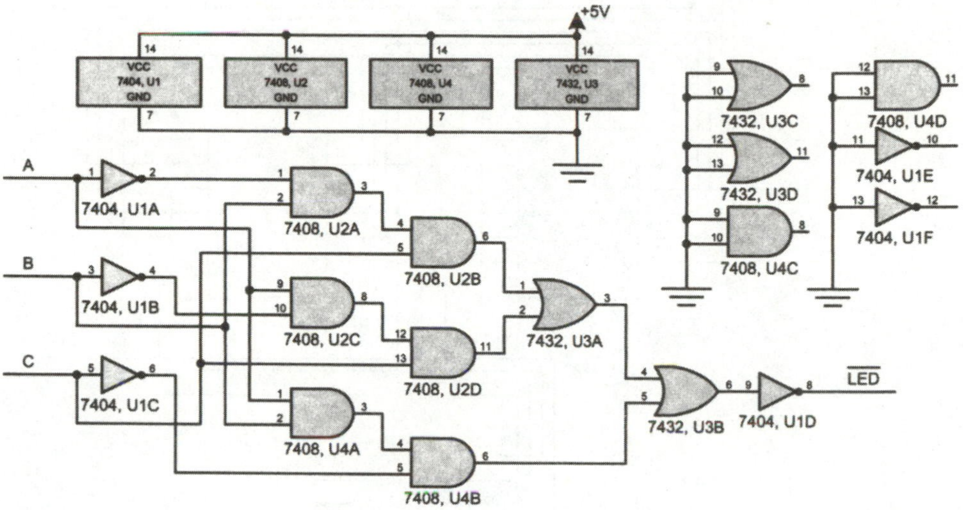

FIGURE 2.12 LED driver logic implementation.

cuit is shown in Fig. 2.13 with alternative notation to illustrate varying circuit presentation styles. Rather than drawing gates as separate elements, the complete 7400-series ICs are shown as monolithic blocks. Either notation is commonly accepted and depends on the engineer's preference.

2.5 SYNCHRONOUS LOGIC DESIGN WITH THE 7400 FAMILY

The preceding LED driver example shows how state-less logic (logic without flops and a clock) can be designed to implement an arbitrary logic equation. State-full logic is almost always required in a digital system, because it is necessary to advance one step at a time (one step each cycle) through an algorithm. Some 7400 ICs, such as counters, implement synchronous logic within the IC itself by combining Boolean logic gates and flops on the same die. Other 7400 ICs implement only flops that may be combined externally with logic to create the desired function.

An example of a synchronous logic application is a basic serial communications controller. Serial communications is the process of taking parallel data, perhaps a byte of information, and transmitting or receiving that byte at a rate of one bit per clock cycle. The obvious downside of doing this is that it will take longer to transfer the byte, because it would be faster to just send the entire byte during the same cycle. The advantage of serial communications is a reduction in the number of wires required to transfer information. Being able to string only a few wires between buildings instead of dozens usually compensates for the added serial transfer time. If the time required to serially transfer bits is too slow, the rate at which the bits are sent can be increased with some engineering work to achieve the desired throughput. Such speed improvements are beyond the scope of this presentation.

Real serial communications devices can get fairly complicated. For purposes of discussion, a fairly simplistic approach is taken. Once the decision is made to serialize a data byte, the problem arises of knowing when that byte begins and ends. *Framing* is the process of placing special patterns into the data stream to indicate the start and end of data units. Without some means to frame the individual bits as they are transmitted, the receiver would have no means of finding the first and last bits of each byte. In this example, a single *start bit* is used to mark the first bit. Once the first bit is

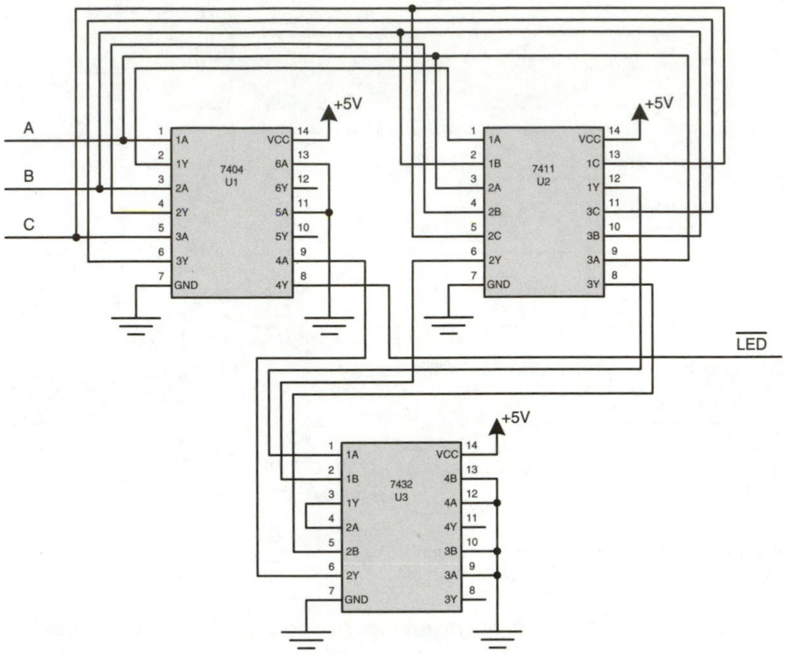

FIGURE 2.13 LED driver logic using 74111 with fewer ICs.

detected, the last bit is found by knowing that there are eight bits in a byte. During periods of inactivity, an idle communications interface is indicated by a persistent logic 0. When the transmitter is given a byte to send, it first drives a logic-1 start bit and then sends eight data bits. Each bit is sent in its own clock cycle. Therefore, nine clock cycles are required to transfer each byte. The serial interface is composed of two signals, *clock* and *serial data*, and functions as shown in Fig. 2.14.

The eight data bits are sent from least-significant bit, bit 0, to most-significant bit, bit 7, following the start bit. Following the transmission of bit 7, it is possible to immediately begin a new byte by inserting a new start bit. This timing diagram does not show a new start bit directly following bit 7. The corresponding output of the receiver is shown in Fig. 2.15. Here, *data out* is the eight-bit quan-

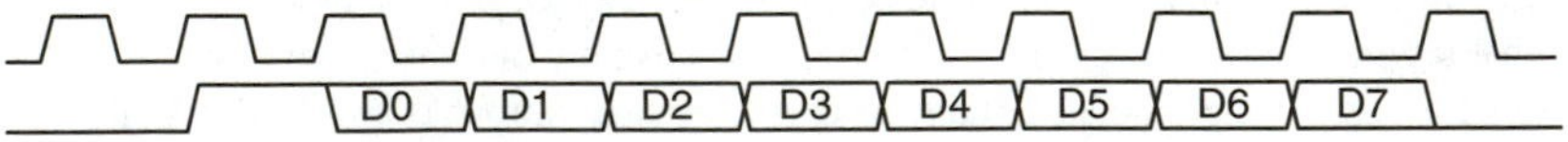

FIGURE 2.14 Serial interface bit timing.

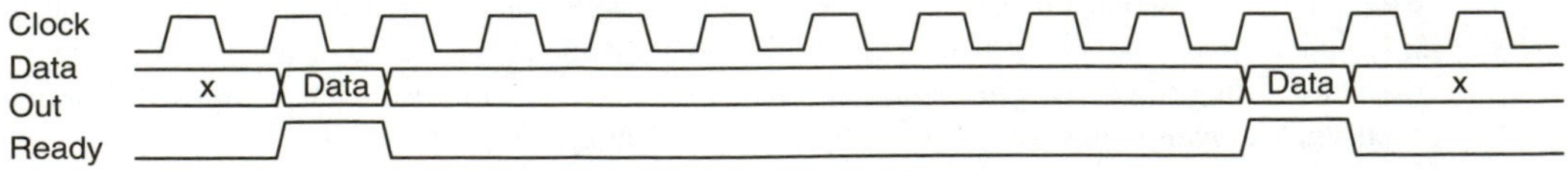

FIGURE 2.15 Serial receive output timing.

tity that has been reconstructed from the serialized bit stream of Fig. 2.14. *Ready* indicates when data out is valid and is active-high.

All that is required of this receiver is to assemble the eight data bits in their proper order and then generate a *ready* signal. This ready signal lasts only one cycle, and any downstream logic waiting for the newly arrived byte must process it immediately. In a real system, a register might exist to capture the received byte when ready goes active. This register would then pass the byte to the appropriate destination. This output timing shows two bytes transmitted back to back. They are separated by nine cycles, because each byte requires an additional start bit for framing.

In contemplating the design of the receive portion of the serial controller, the need for a serial-in/parallel-out shift register becomes apparent to assemble the individual bits into a whole byte. Additionally, some control logic is necessary to recognize the start bit, wait eight clocks to assemble the incoming byte, and then generate a ready signal. This receiver has two basic states, or modes, of operation: idle and receiving. When idling, no start bit has yet been detected, so there is no useful work to be done. When receiving, a start bit has been observed, incoming bits are shifted into the shift register, and then a ready signal is generated. As soon as the ready signal is generated, the receiver state may return to idle or remain in receiving if a new start bit is detected. Because there are two basic control logic states, the state can be stored in a single flip-flop, forming a two-state *finite state machine* (FSM). An FSM is formed by one or more *state flops* with accompanying logic to generate a new state for the next clock cycle based on the current cycle's state. The state is represented by the combined value of the state flops. An FSM with two state flops can represent four unique states. Each state can represent a particular step in an algorithm. The accompanying *state logic* controls the FSM by determining when it is time to transition to a new piece of the algorithm—a new state.

In the serial receive state machine, transitioning from idle to receiving can be done according to the serial data input, which is 0 when inactive and 1 when indicating a start bit. Transitioning back to idle must somehow be done nine cycles later. A counter could be used but would require some logic to sense a particular count value. Instead, a second shift register can be used to delay the start bit by nine cycles. When the start bit emerges from the last output bit in the shift register, the state machine can return to the idle state. Consider the logic in Fig. 2.16. The arrow-shaped boxes indicate connection points, or ports, of the circuit.

Under an idle condition, the input to the shift register is zero until the start bit appears at the data input, *din*. Nine cycles later, the ready bit emerges from the shift register. As soon as the start bit is observed, the state machine transitions to the receiving state, changing the *idle* input to 0, effectively masking further input to the shift register. This masking prevents nonzero data bits from entering the *ready* delay logic and causing false results.

Delaying the start bit by nine cycles solves one problem but creates another. The transition of the state machine back to idle is triggered by the emergence of *ready* from the shift register. Therefore, this transition will actually occur *ten* cycles after the start bit, because the state flop, like all D flip-flops, requires a single cycle of latency to propagate its input to its output. This additional cycle will prevent the control logic from detecting a new start bit immediately following the last data bit of the byte currently in progress. A solution is to design *ready* with its nine-cycle delay and *ready_next* with an eight-cycle delay by tapping off one stage earlier in the shift register. In doing so, the state

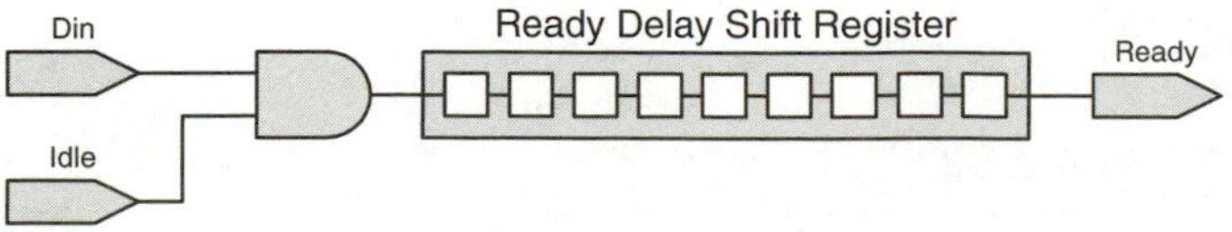

FIGURE 2.16 Serial receive ready delay.

machine can look ahead one cycle into the future and return to idle in time for a new start bit that may be arriving. With the logical details of the state machine now complete, the state machine can be represented with the *state transition diagram* in Fig. 2.17.

A state transition diagram, often called a *bubble diagram*, shows all the states of an FSM and the logical *arcs* that dictate how one state leads to another. When implemented, the arcs are translated into the state logic to make the FSM function. With a clearly defined state transition diagram, the logic to drive the state machine can be organized as shown in Table 2.3.

TABLE 2.3 Serial Receive State Machine Logic Truth Table

Current State	din	ready_next	Next State
1	0	X	1
1	1	X	0
0	X	0	0
0	X	1	1

When in the idle state (1), a high on *din* (the start bit) must be observed to transition to the receiving state (0). Once in the receiving state, *ready_next* must be high to return to idle. This logic is represented by the Boolean equation,

$$\text{Next} = (\text{State}\&\overline{\text{Din}}) + (\overline{\text{State}}\&\text{ready\_next})$$

As with most problems, there exists more than one solution. Depending on the components available, one may choose to design the logic differently to make more efficient use of those components. As a general rule, it is desirable to limit the number of ICs used. The 7451 provides two "AND-OR-INVERT" gates, each of which implements the Boolean function,

$$Y = \overline{AB + CD}$$

This function is tantalizingly close to what is required for the state machine. It differs in that the inversion of two inputs (*state* and *din*) and a NOR function rather than an OR are necessary. Both differences can be resolved using a 7404 inverter IC, but there is a more efficient solution using the 74175 quad flop. The 74175's four flops each provide both true and inverted outputs. Therefore, a separate 7404 is not necessary. An inverted version of *din* can be obtained by passing *din* through a flip-flop before feeding the remainder of the circuit's logic. For purposes of notation, we will refer to this "flopped" *din* as *din´*. Another flop will be used for the state machine. The inverted output of the state flop will compensate for the NOR vs. OR function of the 7451. A third flop will form the ninth bit of the *ready* delay shift register when combined with a 74164 eight-bit parallel-out shift register.

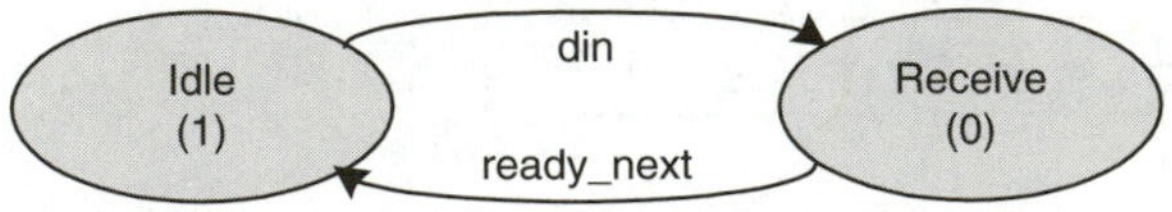

FIGURE 2.17 Serial receive state machine.

Conveniently, the 74164 contains an internal AND gate at its input to implement the idle-enable of the start bit into the shift register.

The total parts count for this serial receiver is four 7400-family ICs: two 74164 shift registers, one 7451 AND-OR-INVERT, and one 74175 quad flop. One flop and one-half of the 7451 are unused in this application. Figure 2.18 shows how these ICs are connected to implement the serial receive logic. Note that a mixed-style of IC representation is used: most ICs are shown in a single block, but the 74175 is broken into separate flops for clarity. Even if an IC is represented as a single block, it is not necessary to draw the individual pins in the order in which they physically appear. As with the previous example, the graphical representation of logic depends on individual discretion. In addition to being functionally and electrically correct, a schematic diagram should be easy to understand.

All synchronous elements, the shift registers and flops, are driven by an input clock signal, *clk*. The synchronous elements involved in the control path of the logic are also reset at the beginning of operation with the active-low *reset_* signal. *Reset_* is necessary to ensure that the state flop and the *ready_next* delay logic begin in an idle state when power is first applied. This is necessary, because flip-flops power up in a random, hence unknown, state. Once they are explicitly reset, they hold their state until the logic specifically changes their state. The shift register in the data path, U3, does not require a reset, because its contents are not used until eight valid data bits are shifted in, thereby flushing the eight bits with random power-up states. It would not hurt to connect U3's *clr_* pin to *reset_*, but this is not done to illustrate the option that is available. In certain logic implementations, adding reset capability to a flop may incur a penalty in terms of additional cost or circuit size. When a reset function is not free, it may be decided not to reset certain flops if their contents do not need to be guaranteed at power up, as is the case here.

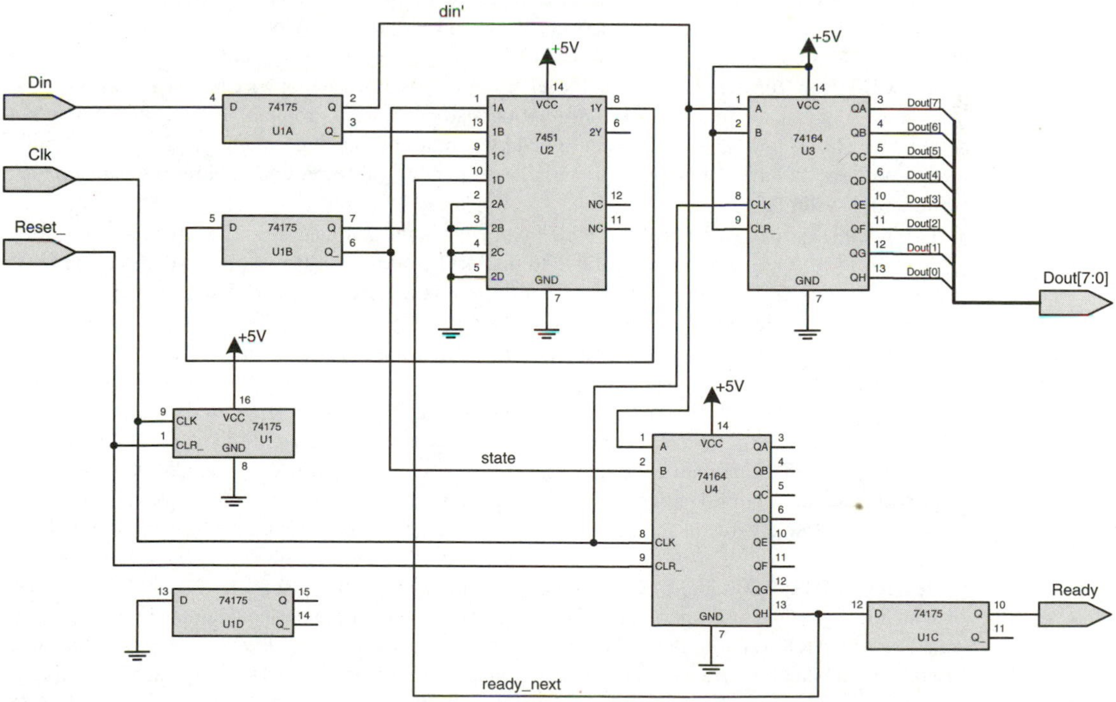

FIGURE 2.18 Serial receive logic schematic diagram.

In this logic circuit, the inverted output of the state flop, U1B, is used as the *state* bit to compensate for the 7451's NOR function. The unused *clr_* and *b* pins of U3 are connected to +5 V to render them neutral on the shift register's behavior. The shift register will not clear itself, because *clr_* is active-low and, similarly, the internal input AND-gate that combines *a* and *b*, will be logically bypassed by tying *b* to logic 1. The parallel byte output of this serial receiver is designated *Dout[7:0]* and is formed by grouping the eight outputs of the shift register into a single bus. One common notation for assigning members of a bus is to connect each individual member to a thicker line with some type of *bus-ripper* line. The bus ripper is often drawn in the schematic diagram as mitered or curved at the bus end to make its function more visually apparent.

Designing an accompanying serial transmitter follows a very similar design process to the preceding discussion. It is left as an exercise to the reader.

2.6 COMMON VARIANTS OF THE 7400 FAMILY

In the 1970s and 1980s, the 7400 family was commonly manufactured in a bipolar semiconductor process that operated using a +5-V power supply and was known as transistor-transistor logic (TTL). The discussion of the 7400 family thus far has included only the original +5-V bipolar type. The 7400's popularity and broad application to digital design has kept it relevant through many improvements in semiconductor process technology. As engineers learned to fabricate faster and more efficient ICs, the 7400 was redesigned in many different process generations beginning in the late 1960s. Some of the more common 7400 variants are briefly discussed here.

The original 7400 discrete TTL logic family featured typical propagation delays of 10 ns per gate and power consumption, also called *power dissipation*, of approximately 10 mW per gate. By modern standards, the 7400's speed is relatively slow, and its power dissipation is relatively high. Increasing system complexity dictates deeper logic: more gates chained together to implement more complex Boolean functions. Each added level of logic adds at least another gate's worth of propagation delay. At the same time, power consumption also becomes a problem. Ten milliwatts may not sound like a lot of power, but, when multiplied by several thousand gates, it represents a substantial design problem in terms of both supplying a large quantity of power and cooling the radiated heat from digital systems.

Two notable bipolar variants of the 7400 are the 74LS and 74F families. The 74LS, LS indicating *low-power Schottky*, has speed comparable to that of the original 7400, but it dissipates roughly 20 percent of its power. The 74F, F indicating fast, is approximately 80 percent faster than the 7400 and reduces power consumption by almost half. Whether the concern is reducing power or increasing speed, these two families are useful for applications requiring 5-V bipolar technology.

CMOS technology began to emerge in the 1980s as a popular process for fabricating digital ICs as a result of its lower power consumption as compared to bipolar. The low-power characteristics of CMOS logic stem from the fact that a FET requires essentially no current to keep it in an on or off state (unlike a BJT, which always draws some current when it is turned on). A CMOS gate, therefore, will draw current only when it switches. For this reason, the power consumption of a CMOS logic gate is extremely low in an idle, or quiescent, state and increases with the frequency at which it switches.

Several CMOS 7400 families were introduced, among them being the 74HCT and 74ACT, each of which has power consumption orders of magnitude less than bipolar equivalents at low frequencies. Earlier CMOS versions of the 7400 were not fully compatible with the bipolar devices, because of voltage threshold differences between the CMOS and bipolar processes. A typical TTL output is only guaranteed to rise above 2.5 V, depending on output loading. In contrast, a typical 5-V CMOS input requires a minimum level of around 3 V to guarantee detecting a logic 1. This inconsistency in voltage range causes a fundamental problem in which a TTL gate driving an ordinary CMOS gate cannot be guaranteed to operate in all situations. Both the 74HCT and 74ACT families possess the low-power benefits of CMOS technology and retain compatibility with bipolar ICs. A 74HCT device is somewhat slower than a 74LS equivalent, and the 74ACT is faster than a 74LS device.

There has been an explosion of 7400 variants. Most of the families introduced in the last decade are based on CMOS technology and are tailored to a broad set of applications ranging from simple speed to high-power bus drivers. Most types of 7400 devices share common pin-outs and functions, with the exception of some proprietary specialized parts that may be produced by only a single manufacturer. Most of the 7400 families still require +5-V supplies, but lower voltages such as 3.3 V, 2.5 V, 1.8 V, and 1.5 V are available as well. These lower-voltage families are important because of the general trend toward lower voltages for digital logic.

2.7 INTERPRETING A DIGITAL IC DATA SHEET

Semiconductor manufacturers publish data sheets for each of their products. Regardless of the specific family or device, all logic IC data sheets share common types of information. Once the basic data sheet terminology and organization is understood, it is relatively easy to figure out other data sheets even when their exact terminology changes. Data sheet structure is illustrated using the 74LS00 from Fairchild Semiconductor as an example. A page from its data sheet is shown in Fig. 2.19.

Digital IC data sheets should have at least two major sections: functional description and electrical specifications. The functional description usually contains the device pin assignment, or *pin-out*, as well as a detailed discussion of how the part logically operates. A simple IC such as the 74LS00 will have a very brief functional description, because there is not much to say about a NAND gate's operation. More complex ICs such as microprocessors can have functional descriptions that fill dozens or hundreds of pages and are broken into many chapters. Some data sheets add additional sections to present the mechanical dimensions of the package and its thermal properties. Digital IC electrical specifications are similar across most types of devices and often appear in the following four categories:

- *Absolute maximum ratings.* As the term implies, these parameters specify the absolute extremes that the IC may be subjected to without sustaining permanent damage. Manufacturers almost universally state that the IC should never be operated under these extreme conditions. These ratings are useful, because they indicate how the device may be stored and express the quality of design and manufacture of the physical chip. Manufacturers specify a storage temperature range within which the semiconductor structures will not break down. In the case of Fairchild's 74LS00, this range is –65 to 150°C. Maximum voltage levels are also specified, 7 V in the case of the 74LS00, indicating that the device may be subjected to a 7-V potential without destructing.
- *Recommended operating conditions.* These parameters specify the normal range of voltages and temperatures that the IC should be operated within such that its functionality is guaranteed to meet specifications set forth by the manufacturer. Two of the most important specifications in this section are the supply voltage (commonly labeled as either V_{CC} or V_{DD}, depending on whether a bipolar or MOS process) and the operating temperature. An IC may have multiple supply voltage specifications, because an IC can actually operate on several different voltages simultaneously. Each supply voltage may power a different portion of the chip. When the manufacturer specifies supply voltage, it does so with a certain tolerance, usually either ±5 or ±10 percent. Many 5-V logic ICs are guaranteed to operate only at a supply voltage from 4.75 to 5.25 V (±5 percent). Operating temperature is very important, because it affects the timing of the device. As a semiconduc-

Absolute Maximum Ratings (Note 1)

Supply Voltage	7V
Input Voltage	7V
Operating Free Air Temperature Range	0°C to +70°C
Storage Temperature Range	−65°C to +150°C

Note 1: The "Absolute Maximum Ratings" are those values beyond which the safety of the device cannot be guaranteed. The device should not be operated at these limits. The parametric values defined in the Electrical Characteristics tables are not guaranteed at the absolute maximum ratings. The "Recommended Operating Conditions" table will define the conditions for actual device operation.

Recommended Operating Conditions

Symbol	Parameter	Min	Nom	Max	Units
V_{CC}	Supply Voltage	4.75	5	5.25	V
V_{IH}	HIGH Level Input Voltage	2			V
V_{IL}	LOW Level Input Voltage			0.8	V
I_{OH}	HIGH Level Output Current			−0.4	mA
I_{OL}	LOW Level Output Current			8	mA
T_A	Free Air Operating Temperature	0		70	°C

Electrical Characteristics

over recommended operating free air temperature range (unless otherwise noted)

Symbol	Parameter	Conditions	Min	Typ (Note 2)	Max	Units
V_I	Input Clamp Voltage	V_{CC} = Min, I_I = −18 mA			−1.5	V
V_{OH}	HIGH Level Output Voltage	V_{CC} = Min, I_{OH} = Max, V_{IL} = Max	2.7	3.4		V
V_{OL}	LOW Level Output Voltage	V_{CC} = Min, I_{OL} = Max, V_{IH} = Min		0.35	0.5	V
		I_{OL} = 4 mA, V_{CC} = Min		0.25	0.4	
I_I	Input Current @ Max Input Voltage	V_{CC} = Max, V_I = 7V			0.1	mA
I_{IH}	HIGH Level Input Current	V_{CC} = Max, V_I = 2.7V			20	μA
I_{IL}	LOW Level Input Current	V_{CC} = Max, V_I = 0.4V			−0.36	mA
I_{OS}	Short Circuit Output Current	V_{CC} = Max (Note 3)	−20		−100	mA
I_{CCH}	Supply Current with Outputs HIGH	V_{CC} = Max		0.8	1.6	mA
I_{CCL}	Supply Current with Outputs LOW	V_{CC} = Max		2.4	4.4	mA

Note 2: All typicals are at V_{CC} = 5V, T_A = 25°C.

Note 3: Not more than one output should be shorted at a time, and the duration should not exceed one second.

Switching Characteristics

at V_{CC} = 5V and T_A = 25°C

Symbol	Parameter	R_L = 2 kΩ				Units
		C_L = 15 pF		C_L = 50 pF		
		Min	Max	Min	Max	
t_{PLH}	Propagation Delay Time LOW-to-HIGH Level Output	3	10	4	15	ns
t_{PHL}	Propagation Delay Time HIGH-to-LOW Level Output	3	10	4	15	ns

FIGURE 2.19 74LS00 manufacturer's specifications. *(Reprinted with permission from Fairchild Semiconductor and National Semiconductor.)*

tor heats up, it slows down. As it cools, its speed increases. Outside of the recommended operating temperature, the device is not guaranteed to function, because the effects of temperature become so severe that functionality is compromised. There are four common temperature ranges for ICs: commercial (0 to 70°C), industrial (–40 to 85°C), automotive (–40 to 125°C), and military (–55 to 125°C). It is more difficult to manufacture an IC that operates over wider temperature ranges. As such, more demanding temperature grades are often more expensive than the commercial grade.

Other parameters establish the safe operating limits for input signals as well as the applied voltage thresholds that represent logic 0 and 1 states. Minimum and maximum input levels are expressed as either absolute voltages or voltages relative to the supply voltage pins of the device. Exceeding these voltages may damage the device. Logic threshold specifications are provided to ensure that the logic input voltages are such that the device will function as intended and not confuse a 1 for a 0, or vice versa. There is also a limit to how must current a digital output can drive. Current output specifications should be known so that a chip is not overloaded, which could result in either permanent damage to the chip or the chip's failure to meet its published specifications.

- *DC electrical characteristics.* DC parameters specify the voltages and currents that the IC will present to other circuitry to which it is connected. Whereas recommended operating conditions specify the environment under which the chip will properly operate, DC electrical characteristics specify the environment that the chip itself will create. Output voltage specifications define the logic 0 and 1 thresholds that the chip is guaranteed to drive under all legal operating conditions. These specifications confirm that the chip is compatible with other chips in the same family and also allow an engineer to determine if the output levels are compatible with another chip that it may be driving.

  Input current specifications characterize the load that the chip presents to whatever circuit is driving it. When either logic state is applied to the chip, a small current flows between the driver and the chip in question. Quantifying these currents enables an engineer to ensure compatibility between multiple ICs. When one IC drives several other ICs, the sum of the input currents should not exceed the output current specification of the driver.

- *AC electrical characteristics or switching characteristics).* AC parameters often represent the greatest complexity and level of detail in a digital IC's specifications. They are the guaranteed timing parameters of inputs and outputs. If the IC is purely combinatorial (e.g., 74LS00), timing may just be matter of specifying propagation delays and rise and fall times. Logic ICs with synchronous elements (e.g., flops) have associated parameters such as setup, hold, clock frequency, and output valid times.

Keep in mind that each manufacturer has a somewhat different style of presenting these specifications. The necessary information should exist, but data sheet sections may be named differently; they may include certain information in different groupings, and terminology may be slightly different.

Specifications may be provided in mixed combinations of minimum, typical/nominal, and maximum. When a minimum or maximum limit is not specified, it is understood to be self-evident or subject to a physical limitation that is beyond the scope of the device. Using Fairchild's 74LS00 as an example, no minimum output current is specified, because the physical minimum is very near zero. The actual output current is determined by the load that is being driven, assuming that the load draws no more than the specified maximum. Other specifications are shown under certain operating conditions. A well written data sheet provides guaranteed specifications under worst-case conditions. Here, the logic 1 output voltage (V_{OH}) is specified as a minimum of 2.5 V under conditions of minimum supply voltage (V_{CC}), maximum output current (I_{OH}), and maximum logic-low input voltage (V_{IL}). These are worst-case conditions. When V_{CC} decreases, so will V_{OH}. When I_{OH} increases, it places a greater load on the output, dragging it down to its lowest level.

Timing specifications may also be incomplete. Manufacturers do not always guarantee minimum or maximum parameters, depending on the specific type of device and the particular specification. As with DC voltages, worst-case parameters should always be specified. When a minimum or maximum delay is not specified, it is generally because that parameter is of secondary importance, and the manufacturer was unable to control its process to a sufficient level of detail to guarantee that value. In many situations where incomplete specifications are given, there are acceptable reasons for doing so, and the lack of information does not hurt the quality of the design.

Typical timing numbers are not useful in many circumstances, because they do not represent a limit of the device's operation. A thorough design must take into account the best and worst performance of each IC in the circuit so that one can guarantee that the circuit will function under all conditions. Therefore, worst-case timing parameters are usually the most important to consider first, because they are the dominant limit of a digital system's performance in most cases. In more advanced digital systems, minimum parameters can become equally as important because of the need to meet hold time and thereby ensure that a signal does not disappear too quickly before the driven IC can properly sense the signal's logic level.

Output timing specifications are often specified with an assumed set of loading conditions, because the current drawn by the load has an impact on the output driver's ability to establish a valid logic level. A small load will enable the IC to switch its output faster, because less current is demanded of the output. A heavier load has the opposite effect, because it draws more current, which places a greater strain on the output driver.

CHAPTER 3
Basic Computer Architecture

Microprocessors are central components of almost all digital systems, because combinations of hardware and software are used to solve design problems. A computer is formed by combining a microprocessor with a mix of certain basic elements and customized logic. Software runs on a microprocessor and provides a flexible framework that orchestrates the behavior of hardware that has been customized to fit the application. When many people think about computers, images of desktop PCs and laptops come to their minds. Computers are much more diverse than the stereotypical image and permeate everyday life in increasing numbers. Small computers control microwave ovens, telephones, and CD players.

Computer architecture is fundamental to the design of digital systems. Understanding how a basic computer is designed enables a digital system to take shape by using a microprocessor as a central control element. The microprocessor becomes a programmable platform upon which the major components of an algorithm can be implemented. Digital logic can then be designed to surround the microprocessor and assist the software in carrying out a specific set of tasks.

The first portion of this chapter explains the basic elements of a computer, including the microprocessor, memory, and input/output devices. Basic microprocessor operation is presented from a hardware perspective to show how instructions are executed and how interaction with other system components is handled. Interrupts, registers, and stacks are introduced as well to provide an overall picture of how computers function. Following this basic introduction is a complete example of how an actual eight-bit computer might be designed, with detailed descriptions of bus operation and address decoding.

Once basic computer architecture has been discussed, common techniques for improving and augmenting microprocessor capabilities are covered, including direct memory access and bus expansion. These techniques are not relegated to high-end computing but are found in many smaller digital systems in which it is more economical to add a little extra hardware to achieve feature and performance goals instead of having to use a microprocessor that may be too complex and more expensive than desired.

The chapter closes with an introduction to assembly language and microprocessor addressing modes. Writing software is not a primary topic of this book, but basic software design is an inseparable part of digital systems design. Without software, a computer performs no useful function. Assembly language basics are presented in a general manner, because each microprocessor has its own instruction set and assembly language, requiring specific reading focused on that particular device. Basic concepts, however, are universal across different microprocessor implementations and serve to further explain how microprocessors actually function.

3.1 THE DIGITAL COMPUTER

A digital computer is a collection of logic elements that can execute arbitrary algorithms to perform data calculation and manipulation functions. A computer is composed of a microprocessor, memory, and some input/output (I/O) elements as shown in Fig. 3.1. The microprocessor, often called a microprocessor unit (MPU) or central processing unit (CPU), contains logic to step through an algorithm, called a *program*, that has been stored in the computer's program memory. The data used and manipulated by that program is held in the computer's data memory. Memory is a repository for data that is usually organized as a linear array of individually accessible locations. The microprocessor can access a particular location in memory by presenting a memory address (the index of the desired location) to the memory element. I/O elements enable the microprocessor to communicate with the outside world to acquire new data and present the results of its programmed computations. Such elements can include a keyboard or display controller.

Programs are composed of many very simple individual operations, called *instructions,* that specify in exact detail how the microprocessor should carry out an algorithm. A simple program may have dozens of instructions, whereas a complex program can have tens of millions of instructions. Collectively, the programs that run on microprocessors are called *software*, in contrast to the *hardware* on which they run. Each type of microprocessor has its own *instruction set* that defines the full set of unique, discrete operations that it is capable of executing. These instructions perform very narrow tasks that, on their own, may seem insignificant. However, when thousands or millions of these tiny instructions are strung together, they may create a video game or a word processor.

A microprocessor possesses no inherent intelligence or capability to spontaneously begin performing useful work. Each microprocessor is constructed with an instruction set that can be invoked in arbitrary sequences. Therefore, a microprocessor has the potential to perform useful work but will do nothing of the sort on its own. To make the microprocessor perform useful work, it requires explicit guidance in the form of software programming. A task of even moderate complexity must be broken down into many tiny steps to be implemented on a microprocessor. These steps include basic arithmetic, Boolean operations, loading data from memory or an input element such as a keyboard, and storing data back to memory or an output element such as a printer.

Memory structure is one of a computer's key characteristics, because the microprocessor is almost constantly accessing it to retrieve a new instruction, load new data to operate on, or store a calculated result. While program and data memory are logically distinct classifications, they may share the same physical memory resource. *Random access memory* (RAM) is the term used to describe a generic memory resource whose locations can be accessed, or *addressed*, in an arbitrary order and either read or written. A *read* is the process of retrieving data from a memory address and loading it into the microprocessor. A *write* is the process of storing data to a memory address from the microprocessor. Both programs and data can occupy RAM. Consider your desktop computer. When you

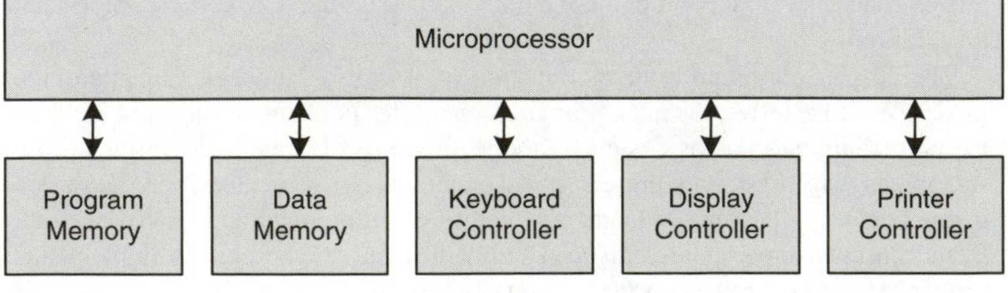

FIGURE 3.1 Generic computer block diagram.

execute a program that is located on the disk drive, that program is first loaded into the computer's RAM and then executed from a region set aside for program memory. As on a desktop computer, RAM is most often *volatile*—meaning that it loses its contents when the power is turned off.

Some software cannot be stored in volatile memory, because basic initialization instructions, or *boot code*, must be present when the computer is turned on. Remember that a microprocessor can do nothing useful without software being readily available. When power is first applied to a computer, the microprocessor must be able to quickly locate boot code so that it can get itself ready to accept input from a user or load a program from an input device. This startup sequence is called *booting*, hence the term *boot code*. When you turn your computer on, the first messages that it displays on the monitor are a product of its boot code. Eventually, the computer is able to access its disk drive and begins loading software into RAM as part of its normal operation. To ensure that boot code is ready at power-up, *nonvolatile* memory called *read only memory* (ROM) exists. ROM can be used to store both programs as well as any data that must be present at power-up and immediately accessible. Software contained in ROM is also known as *firmware*. As its name implies, ROM can only be read but not written. More complex computers contain a relatively small quantity of ROM to hold basic boot code that then loads main operating software from another device into RAM. Small computers may contain all of their software in ROM. Figure 3.2 shows how ROM and RAM complement each other in a typical computer architecture.

A microprocessor connects to devices such as memory and I/O via *data* and *address buses*. Collectively, these two buses can be referred to as the *microprocessor bus*. A bus is a collection of wires that serve a common purpose. The data bus is a bit array of sufficient size to communicate one complete data unit at a time. Most often, the data bus is one or more bytes in width. An eight-bit microprocessor, operating on one byte at time, almost always has an eight-bit data bus. A 32-bit microprocessor, capable of operating on up to 4 bytes at a time, can have a data bus that is 32, 16, or 8 bits wide. The exact data bus width is implementation specific and varies according to the intended application of the microprocessor. A narrower bus width means that it will take more time to communicate a quantity of data as compared to a wider bus. Common notation for a data bus is *D[7:0]* for an 8-bit bus and *D[31:0]* for a 32-bit bus, where 0 is the least-significant bit.

The address bus is a bit array of sufficient size to fully express the microprocessor's *address space*. Address space refers to the maximum amount of memory and I/O that a microprocessor can directly address. If a microprocessor has a 16-bit address bus, it can address up to $2^{16} = 65{,}536$ bytes. Therefore, it has a 64 kB address space. The entire address space does not have to be used; it simply establishes a maximum limit on memory size. Common notation for a 16-bit address bus is *A[15:0]*, where 0 is the least-significant bit. Figure 3.3 shows a typical microprocessor bus configuration in a computer. Note that the address bus is unidirectional (the microprocessor asserts requested addresses to the various devices), and the data bus is bidirectional (the microprocessor asserts data on a write and the devices assert data on reads).

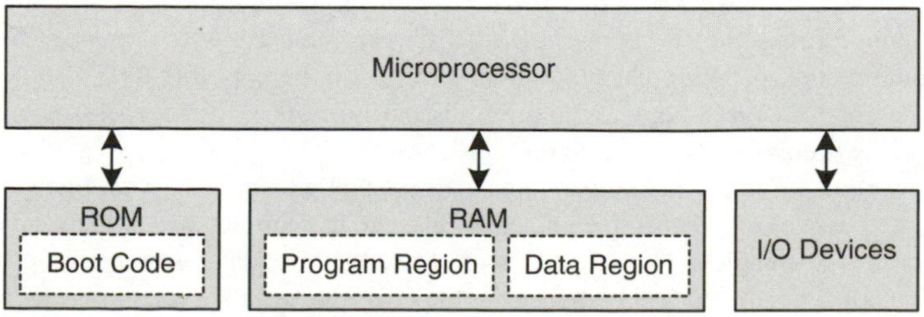

FIGURE 3.2 Basic ROM/RAM memory complement.

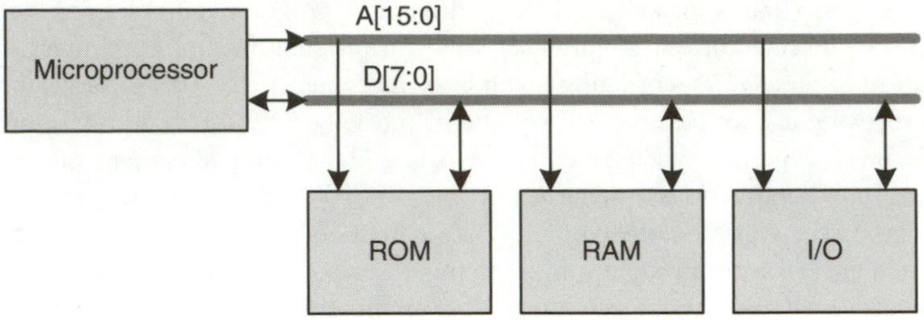

FIGURE 3.3 Microprocessor buses.

A microprocessor's entire address space is never occupied by a single function; rather, it is shared by ROM, RAM, and various I/Os. Each device is *mapped* into its own region of the address space and is enabled only when the microprocessor asserts an address within a device's mapped region. The process of recognizing that an address is within a desired region is called *decoding*. Address decoding logic is used to divide the overall address space into smaller sections in which memory and I/O devices can reside. This logic generates individual signals that enable the appropriate device based on the state of the address bus so that the devices themselves do not need any knowledge of the specific computer's unique address decoding.

3.2 MICROPROCESSOR INTERNALS

The multitude of complex tasks performed by computers can be broken down into sequences of simple operations that manipulate individual numbers and then make decisions based on those calculations. Certain types of basic instructions are common across nearly every microprocessor in existence and can be classified as follows for purposes of discussion:

- Arithmetic: add or subtract two values
- Logical: Boolean (e.g., AND, OR, XOR, NOT, etc.) manipulation of one or two values
- Transfer: retrieve a value from memory or store a value to memory
- Branch: jump ahead or back to a particular instruction if a specified condition is satisfied

Arithmetic and logical instructions enable the microprocessor to modify and manipulate specific pieces of data. Transfer instructions enable these data to be saved for later use and recalled when necessary from memory. Branch operations enable instructions to execute in different sequences, depending on the results of arithmetic and logical operations. For example, a microprocessor can compare two numbers and take one of two different actions if the numbers are equal or unequal.

Each unique instruction is represented as a binary value called an *opcode*. A microprocessor fetches and executes opcodes one at a time from program memory. Figure 3.4 shows a hypothetical microprocessor to serve as an example for discussing how a microprocessor actually advances through and executes the opcodes that form programs.

A microprocessor is a synchronous logic element that advances through opcodes on each clock cycle. Some opcodes may be simple enough to execute in a single clock cycle, and others may take multiple cycles to complete. Clock speed is often used as an indicator of a microprocessor's performance. It is a valid indicator but certainly not the only one, because each microprocessor requires a different number of cycles for each instruction, and each instruction represents a different quantity of useful work.

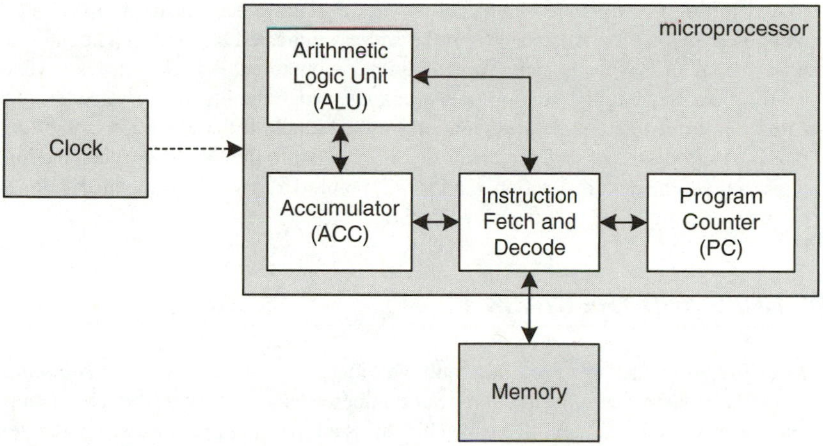

FIGURE 3.4 Simple microprocessor.

When an opcode is fetched from memory, it must be briefly examined to determine what needs to be done, after which the appropriate actions are carried out. This process is called *instruction decoding*. A central logic block coordinates the operation of the entire microprocessor by fetching instructions from memory, decoding them, and loading or storing any data as required. The *accumulator* is a register that temporarily holds data while it is being processed. Execution of an instruction to load the accumulator with a byte from memory would begin with a fetch of the opcode that represents this action. The instruction decoder would then recognize the opcode and initiate a memory read via the same microprocessor bus that was used to fetch the opcode. When the data returns from memory, it would be loaded into the accumulator. While there may be multiple distinct logical steps in decoding an instruction, the steps may occur simultaneously or sequentially, depending on the architecture of the microprocessor and its decoding logic.

The accumulator is sized to hold the largest data value that the microprocessor can handle in a single arithmetic or logical instruction. When engineers talk of an 8-bit or 32-bit microprocessor, they are usually referring to the internal *data-path* width—the size of the accumulator and the *arithmetic logic unit* (ALU). The ALU is sometimes the most complex single logic element in a microprocessor. It is responsible for performing arithmetic and logical operations as directed by the instruction decode logic. Not only does the ALU add or subtract data from the accumulator, it also keeps track of status flags that tell subsequent branch instructions whether the result was positive, negative, or zero, and whether an addition or subtraction operation created a carry or borrow bit. These status bits are also updated for logical operations such as AND or OR so that software can take different action if a logical comparison is true or false.

For ease of presentation, the microprocessor in Fig. 3.4 is shown having a single general-purpose accumulator register. Most real microprocessors contain more than one internal register that can be used for general manipulation operations. Some microprocessors have as few as one or two such registers, and some have dozens or more than a hundred. It is the concept of an accumulator that is discussed here, but there is no conceptual limitation on how many accumulators or registers a microprocessor can have.

A microprocessor needs a mechanism to keep track of its place in the instruction sequence. Like a bookmark that saves your place as you read through a book, the *program counter* (PC) maintains the address of the next instruction to be fetched from program memory. The PC is a counter that can be reloaded with a new value from the instruction decoder. Under normal operation, the microprocessor moves through instructions sequentially. After executing each instruction, the PC is incremented, and a new instruction is fetched from the address indicated by the PC. The major exception to this linear behavior is when branch instructions are encountered. Branch instructions exist specifically to override the sequential execution of instructions. When the instruction decoder fetches a branch instruction, it must determine the condition for the branch. If the condition is met (e.g., the ALU zero flag is asserted), the *branch target address* is loaded into the PC. Now, when the instruction decoder goes to fetch the next instruction, the PC will point to a new part of the instruction sequence instead of simply the next program memory location.

3.3 SUBROUTINES AND THE STACK

Most programs are organized into multiple blocks of instructions called *subroutines* rather than a single large sequence of instructions. Subroutines are located apart from the main program segment and are invoked by a subroutine call. This call is a type of branch instruction that temporarily jumps the microprocessor's PC to the subroutine, allowing it to be executed. When the subroutine has competed, control is returned to the program segment that called it via a return from subroutine instruction. Subroutines provide several benefits to a program, including modularity and ease of reuse. A modular subroutine is one that can be relocated in different parts of the same program while still performing the same basic function. An example of a modular subroutine is one that sorts a list of numbers in ascending order. This sorting subroutine can be called by multiple sections of a program and will perform the same operation on multiple lists. Reuse is related to modularity and takes the concept a step farther by enabling the subroutine to be transplanted from one program to another without modification. This concept greatly speeds the software development process.

Almost all microprocessors provide inherent support for subroutines in their architectures and instruction sets. Recall that the program counter keeps track of the next instruction to be executed and that branch instructions provide a mechanism for loading a new value into the PC. Most branch instructions simply cause a new value to be loaded into the PC when their specific branch condition is satisfied. Some branch instructions, however, not only reload the PC but also instruct the microprocessor to save the current value of the PC off to the side for later recall. This stored PC value, or *subroutine return address*, is what enables the subroutine to eventually return control to the program that called it. Subroutine call instructions are sometimes called *branch-to-subroutine* or *jump-to-subroutine,* and they may be unconditional.

When a branch-to-subroutine is executed, the PC is saved into a data structure called a *stack*. The stack is a region of data memory that is set aside by the programmer specifically for the main purpose of storing the microprocessor's state information when it branches to a subroutine. Other uses for the stack will be mentioned shortly. A stack is a *last-in, first-out* memory structure. When data is stored on the stack, it is *pushed* on. When data is removed from the stack, it is *popped* off. Popping the stack recalls the most recently pushed data. The first datum to be pushed onto the stack will be the last to be popped. A *stack pointer* (SP) holds a memory address that identifies the *top* of the stack at any given time. The SP decrements as entries are pushed on and increments at they are popped off, thereby growing the stack downward in memory as data is pushed on as shown in Fig. 3.5.

By pushing the PC onto the stack during a branch-to-subroutine, the microprocessor now has a means to return to the calling routine at any time by restoring the PC to its previous value by simply popping the stack. This operation is performed by a return-from-subroutine instruction. Many microprocessors push not only the PC onto the stack when calling a subroutine, but the accumulator and ALU status flags as well. While this increases the complexity of a subroutine call and return somewhat, it is useful to preserve the state of the calling routine so that it may resume control smoothly when the subroutine ends.

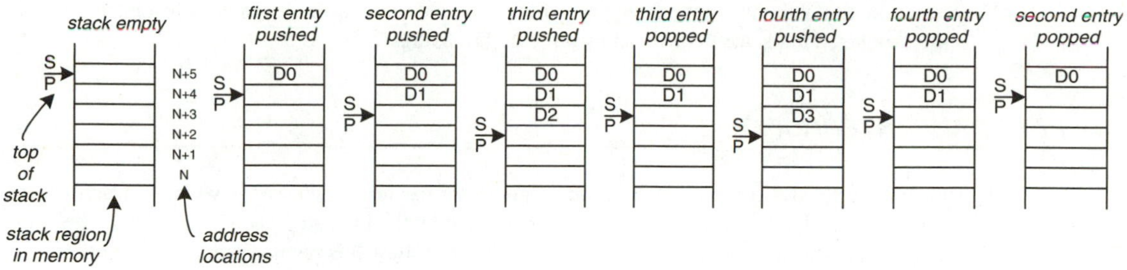

FIGURE 3.5 Generic stack operation.

The stack can store multiple entries, enabling multiple subroutines to be active at the same time. If one subroutine calls another, the microprocessor must keep track of both subroutines' return addresses in the order in which the subroutines have been called. This subroutine *nesting* process of one calling another subroutine, which calls another subroutine, naturally conforms to the last-in, first-out operation of a stack.

To implement a stack, a microprocessor contains a stack pointer register that is loaded by the programmer to establish the initial starting point, or top, of the stack. Figure 3.6 shows the hypothetical microprocessor in more complete form with a stack pointer register.

Like the PC, the SP is a counter that is automatically modified by certain instructions. Not only do subroutine branch and return instructions use the stack, there are also general-purpose push/pop instructions provided to enable the programmer to use the stack manually. The stack can make certain calculations easier by pushing the partial results of individual calculations and then popping them as they are combined into a final result.

The programmer must carefully manage the location and size of the stack. A microprocessor will freely execute subroutine call, subroutine return, push, and pop instructions whenever they are encountered in the software. If an empty stack is popped, the microprocessor will oblige by reading back whatever data value is present in memory at the time and then incrementing the SP. If a full stack is pushed, the microprocessor will write the specified data to the location pointed to by the SP and then decrement it. Depending on the exact circumstances, either of these operations can corrupt other parts of the program or data that happens to be in the memory location that gets overwritten. It is the programmer's responsibility to leave enough free memory for the desired stack depth and then to not nest too many subroutines simultaneously. The programmer must also ensure that there is symmetry between push/pop and subroutine call/return operations. Issuing a return-from-subroutine

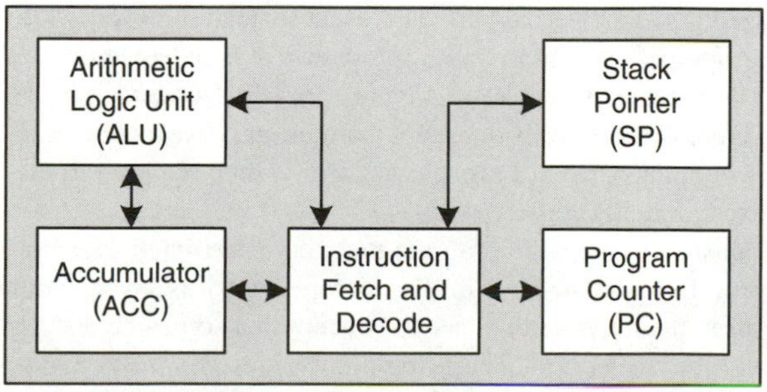

FIGURE 3.6 Microprocessor with stack pointer register.

instruction while already in the main program would lead to undesirable results when the microprocessor fetches reloads the PC with an incorrect return address.

3.4 RESET AND INTERRUPTS

Thus far, the steady-state operation of a microprocessor has been discussed in which instructions are fetched, decoded, and executed in an order determined by the PC and branch instructions. There are two special cases in which the microprocessor does not follow this regular pattern of operation. The first case is at power-up, when the microprocessor must transition from an idle state to executing instructions. This transition sequence is called *reset* and involves the microprocessor fetching its boot code from memory to begin the programmed software sequence. Reset is triggered by asserting a particular logic level onto a microprocessor pin and can occur either at power-up or at any arbitrary time when it is desired to restart, or reboot, the microprocessor from a known initial state. Some microprocessors have special instructions that can actually trigger a soft reset.

The question arises of how the microprocessor determines which instruction to execute first when it has just been reset. To solve this problem, each microprocessor has a *reset vector* that points it to a fixed, predetermined memory address where the programmer must locate the first instruction of the boot sequence. The reset vector is specified by the microprocessor's designer. Some microprocessors locate the reset vector at the beginning of memory and some place it toward the end of the address space. Sometimes the main body of the program will be located in another portion of memory, and the first instruction at the reset vector will contain a branch instruction to jump to the desired location.

The second case in which the microprocessor does not follow the normal instruction sequence is during normal operation when an event occurs and the programmer wishes the microprocessor to pause what it is currently doing and handle the event with a special software routine. Such an event is called an *interrupt*. A common application for an interrupt is the implementation of a periodic, timed operation such as monitoring the temperature of a room. Because the room temperature does not change often, the microprocessor can handle other tasks during normal operation. A timer can be set to expire every few seconds, causing an interrupt event. When the interrupt triggers, the microprocessor can read the room temperature, take any appropriate action (e.g., turn on a ventilation fan), and then resume its normal operation.

An interrupt can be triggered by asserting a special-purpose microprocessor interrupt signal. Interrupt events can also be triggered from within a microprocessor via special instructions. When an interrupt occurs, the microprocessor saves its state by pushing the PC and other registers onto the stack, and then the PC is loaded with an *interrupt vector* that points to an *interrupt service routine* (ISR) in memory. In this way, the interrupt process is similar to a branch-to-subroutine. However, the interrupt may be triggered by an external hardware event instead of by software. Like reset, each interrupt pin on the microprocessor has an interrupt vector associated with it. The programmer knows that an ISR is to be located at a specific memory location to service a particular interrupt. When the ISR has completed, a *return-from-interrupt* instruction is executed that restores the microprocessor's prior state by popping it from the stack. Control is then returned to the routine that was interrupted and normal execution proceeds.

As the interrupt mechanism executes, the program that gets interrupted does not necessarily have any knowledge of the event. Because the state of the microprocessor is saved and then restored during the return-from-interrupt, the main routine has no concept that somewhere along the way its execution was paused for an arbitrary period. The programmer may choose to make such knowledge available by sharing information between the ISR and other routines, but this is left to individual software implementations.

Multiple interrupt sources are common in microprocessors. Depending on the complexity of the microprocessor, there may be one, two, ten, or dozens of separate interrupt sources, each with its own vector. Conflicts in which multiple interrupt sources are activated at the same time are handled by assigning priorities to each interrupt. Interrupt priorities may be predetermined by the designer of the microprocessor or programmed by software. In a microprocessor with multiple interrupt priorities, once a higher-priority interrupt has taken control and its ISR is executing, lower-priority interrupts will remain pending until the current higher-priority ISR issues a return-from-interrupt.

Interrupts can usually be turned off, or *masked*, by writing to a control register within the microprocessor. Masking an interrupt is useful, because an interrupt should not be triggered before the program has had a chance to set up the ISR or otherwise get ready to handle the interrupt condition. If the program is not yet ready and the microprocessor takes an interrupt by jumping to the interrupt vector, the microprocessor will crash by executing invalid instructions.

Masking is also useful when performing certain time-critical operations. A task may be programmed into an ISR that must complete within 10 μs. Under normal circumstances, the task is easily accomplished in this period of time. However, if a competing interrupt is triggered during the time-critical ISR, there may be no guarantee of meeting the 10-μs requirements. One solution to this problem is to mask subsequent interrupts when the time-critical interrupt is triggered and then unmask interrupts when the ISR has completed. If an interrupt arrives while masked, the microprocessor will remember the interrupt request and trigger the interrupt when it is unmasked.

Certain microprocessors have one or more interrupts that are classified as nonmaskable. This means that the interrupt cannot be disabled. Therefore, the hardware design of the computer must ensure that such an interrupt is not activated unless the software is able to respond to it. Nonmaskable interrupts are generally used for low-level error recovery or debugging purposes where it must be guaranteed that the interrupt will be taken regardless of what the microprocessor is doing at the time. Nonmaskable ISRs are sometimes implemented in nonvolatile memory to ensure that they are always ready for execution.

3.5 IMPLEMENTATION OF AN EIGHT-BIT COMPUTER

Having discussed some of the basic principles of microprocessor architecture and operation, we can examine how a microprocessor fits into a system to form a computer. Microprocessors need external memory in which to store their programs and the data upon which they operate. In this context, external memory is viewed from a logical perspective. That is, the memory is always external to the core microprocessor element. Some processor chips on the market actually contain a certain quantity of memory within them, but, logically speaking, this memory is still external to the actual microprocessor core.

In the general sense, a computer requires a quantity of nonvolatile memory, or ROM, in which to store the boot code that will be executed on reset. The ROM may contain all or some of the microprocessor's full set of software. A small embedded computer, such as the one in a microwave oven, contains all its software in ROM. A desktop computer contains very little of its software in ROM. A computer also requires a quantity of volatile memory, or RAM, that can be used to store data associated with the various tasks running on the computer. RAM is where the microprocessor's stack is located. Additionally, RAM can be used to hold software that is loaded from an external source.

For purposes of discussion, consider the basic eight-bit computer shown in Fig. 3.7 with a small quantity of memory and a serial port with which to communicate with the outside world. Eight kilobytes of ROM is sufficient to store boot code and software, including a serial communications program. Eight kilobytes of RAM is sufficient to hold data associated with the ROM software, and it also enables loading additional software not already included in the ROM. The control signals in this

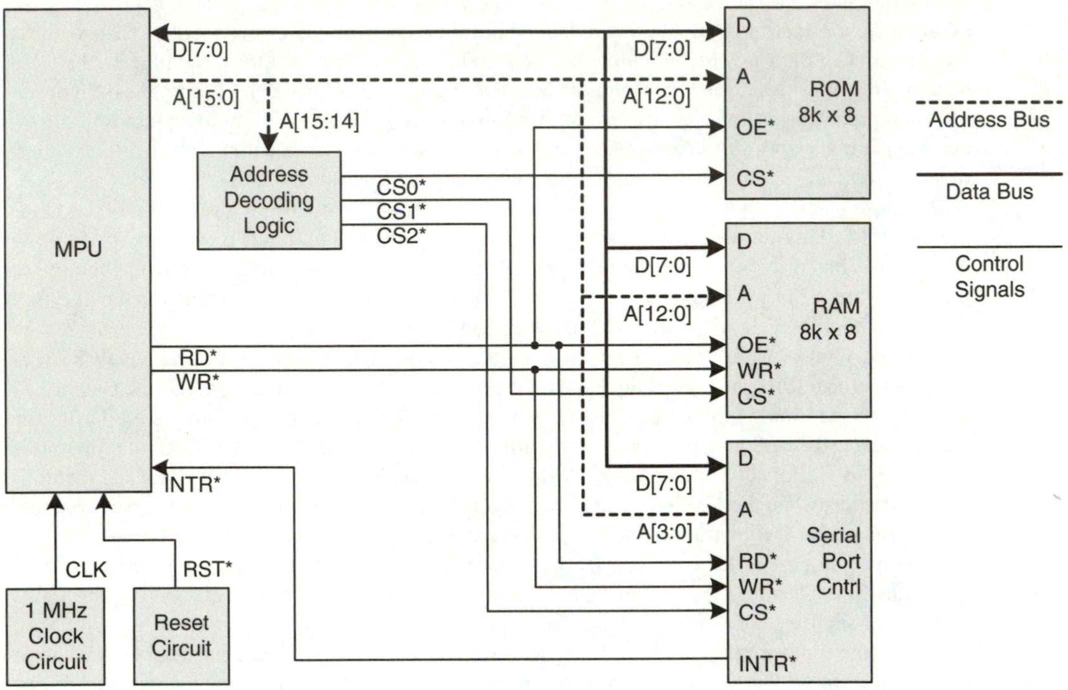

FIGURE 3.7 Eight-bit computer block diagram.

hypothetical computer are active-low, as are the control signals in most computer designs that, according to convention, have been in widespread use for the past few decades. Active-low signal names have some type of symbol as a prefix or suffix to the signal name that distinguishes them from active-high signals. Common symbols used for this purpose include #, *, –, and _. From a logical perspective, it is perfectly valid to use active-high signaling. However, because most memory and peripheral devices conform to the active-low convention, it is often easier to go along with the established convention.

While hypothetical, the microprocessor shown contains characteristics that are common in off-the-shelf eight-bit microprocessors. It contains an 8-bit data bus and a 16-bit address bus with a total address space of 64 kB. The combined MPU bus, consisting of address, data, and control signals, is asynchronous and is enabled by the assertion of read and write enable signals. When the microprocessor wants to read a location in memory, it asserts the appropriate address along with RD* and then takes the resulting value driven onto the data bus. As shown in the diagram, memory chips usually have *output enable* (OE*) signals that can be connected to a read enable. Such devices continuously decode the address bus and will emit data whenever OE* is active.

Not all 64 kB of address space is used in this computer. Address decoding logic breaks the single 64-kB space into four 16-kB regions. According to the state of A[15:14], one and only one of the chip select signals is activated. The address decoding follows the truth table shown in Table 3.1 and establishes four address ranges.

Once decoded into regions, A[13:0] provides unique address information to the memory and I/O devices connected to the MPU bus. One memory region, the upper 16 kB, is currently left unused. It may be used in the future if more memory or another I/O device is added. Each memory and I/O de-

TABLE 3.1 Address Decoding Truth Table

A[15]	A[14]	Chip Select	Address Range
0	0	CS0*	0x0000-0x3FFF
0	1	CS1*	0x4000-0x7FFF
1	0	CS2*	0x8000-0xBFFF
1	1	none	0xC000-0xFFFF

vice has a chip select input and will respond to a read or write command only when that select signal is active. Furthermore, each chip, including the microprocessor, contains internal tri-state buffers to prevent contention on the bus. The tri-state buffers are not enabled unless the chip's select signal is active and a read is being performed (a write, in the case of the microprocessor). Without external address decoding, none of these chips can share an address region with any other devices, because they do not have enough address bits to fully decode the entire 16-bit address bus.

Not all address bits are used by the memory and serial port chips. The ROM and RAM are each only 8k in size. Therefore, only 13 address bits, A[12:0], are required and, as a result, A[13] is left unconnected. The serial port has far fewer memory locations and therefore uses only A[3:0], for a maximum of 16 unique addresses.

When a device does not utilize all of the address bits that have been allocated for its particular address region, the potential for *aliasing* exists. The ROM occupies only 8k (13 bits) of the 16k (14 bits) address region. Therefore, the ROM has no knowledge of any additional addresses above 8k: the region from 0x2000 to 0x3FFFF. What happens if the MPU tries to read location 0x2000? 0x2000 differs from 0x0000 only in the state of A[13]. Because the ROM does not have any knowledge of A[13], it interprets 0x2000 to be 0x0000. In other words, 0x2000 *aliases* to 0x0000. Similarly, the entire upper 8k of the address region aliases to the lower 8k. In the case of the serial port controller, there is a greater degree of aliasing, because the serial port only uses A[3:0]. This means that there can be only 16 unique address locations in the entire 16k region. These 16 locations will therefore appear to be replicated 2^{10} = 1,024 times as indicated by the ten unused address bits, A[13:4].

As long as the software is properly written to understand the computer's memory map, it will properly access the memory locations that are available and will avoid aliased portions of the memory map. Aliasing is not a problem in itself but can lead to problems if software does not access memory and peripherals in the way in which the hardware engineer intended. If software is written for the hypothetical computer with the incorrect assumption that 16 kB of RAM is present, data may be unwittingly corrupted when addresses between 0x6000 and 0x7FFF are written, because they will alias to 0x4000-0x5FFF and overwrite any existing data.

When the MPU wants to read data from a particular memory location, it asserts that address onto A[15:0]. This causes the address decoder to update its chip select outputs, which enables the appropriate memory chip or the serial port. After allowing time for the chip select to propagate, the RD* signal is asserted, and the WR* signal is left unasserted. This informs the selected device that a read is requested. The device is then able to drive the data bus, *D[7:0]*, with the requested data. After allowing some time for the read data to be driven, the MPU captures the data and releases the RD* signal, ending the read request. The sequence of events, or timing, for the read transaction is shown in Fig. 3.8.

This type of MPU bus is asynchronous, because its sequence of events is not driven by a clock but rather by the assertion and removal of the various signals that are timed relative to one another by the

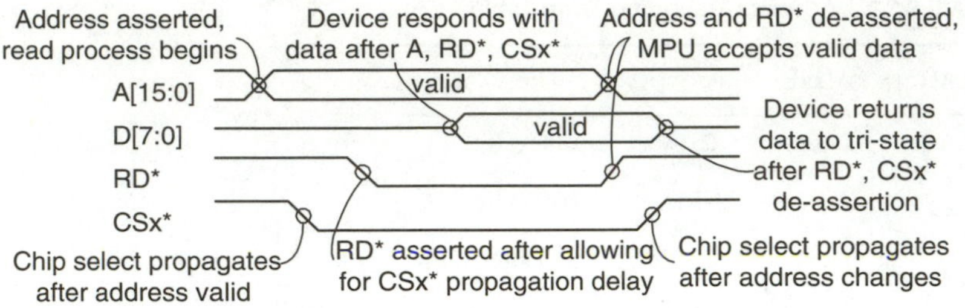

FIGURE 3.8 MPU read timing.

MPU and the devices with which it is communicating. For this interface to work properly, the MPU must allow enough time for the read to occur, regardless of the specific device with which it is communicating. In other words, it must operate according to the capabilities of the slowest device—the least common denominator.

Write timing is very similar, as seen in Fig. 3.9. Again, the MPU asserts the desired address onto A[15:0], and the appropriate chip select is decoded. At the same time, the write data is driven onto D[7:0]. Once the address and data have had time to stabilize, and after allowing time for the chip select to propagate, the WR* enable signal is asserted to actually trigger the write. The WR* signal is de-asserted while data, address, and chip select are still stable so that there is no possibility of writing to a different location and corrupting data. If the WR* signal is de-asserted at the same time as the others, a race condition could develop wherein a particular device may sense the address (or data or chip select) change just prior to WR* changing, resulting in a false write to another location or to the current location with wrong data. Being an asynchronous interface, the duration of all signal assertions must be sufficient for all devices to properly execute the write.

An MPU interrupt signal is asserted by the serial port controller to enable easier programming of the serial port communication routine. Rather than having software continually poll the serial port to see if data are waiting, the controller is configured to assert INTR* whenever a new byte arrives. The MPU is then able to invoke an ISR, which can transfer the data byte from the serial port to the RAM. The interrupt also helps when transmitting data, because the speed of the typical serial port (often 9,600 to 38,400 bps) is very slow as compared to the clock speed of even a slow MPU (1 to 10 MHz). When the software wants to send a set of bytes out the serial port, it must send one byte and then wait a relatively long time until the serial port is ready for the next byte. Instead of polling in a loop between bytes, the serial port controller asserts INTR* when it is time to send the next byte. The ISR can then respond with the next byte and return control to the main program that is run-

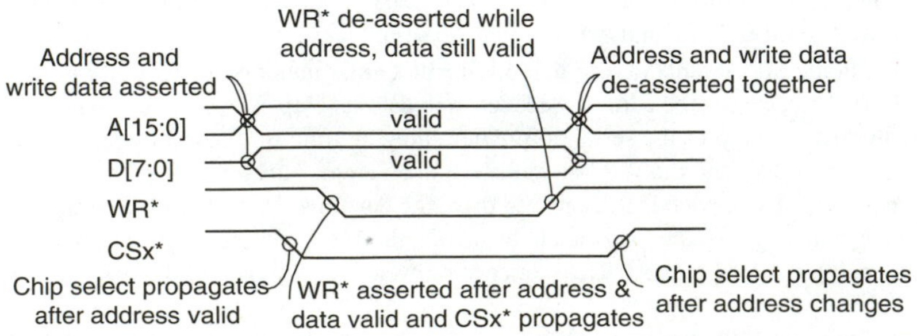

FIGURE 3.9 MPU write timing.

ning at the time. Each time INTR* is asserted and the ISR responds, the ISR must be sure to clear the interrupt condition in the serial port. Depending on the exact serial port device, a read or write to a specific register will clear the interrupt. If the interrupt is not cleared before the ISR issues a return-from-interrupt, the MPU may be falsely interrupted again for the same condition.

This computer contains two other functional elements: the clock and reset circuits. The 1-MHz clock must be supplied to the MPU continually for proper operation. In this example design, no other components in the computer require this clock. For fairly simple computers, this is a realistic scenario, because the buses and memory devices operate asynchronously. Many other computers, however, have synchronous buses, and the microprocessor clock must be distributed to other components in the system.

The reset circuit exists to start the MPU when the system is first turned on. Reset must be applied for a certain minimum duration after the power supply has stabilized. This is to ensure that the digital circuits properly settle to known states before they are released from reset and allowed to begin normal operation. As the computer is turned on, the reset circuit actively drives the RST* signal. Once power has stabilized, RST* is de-asserted and remains in this state indefinitely.

3.6 ADDRESS BANKING

A microprocessor's address space is normally limited by the width of its address bus, but supplemental logic can greatly expand address space, subject to certain limitations. Address banking is a technique that increases the amount of memory a microprocessor can address. If an application requires 1 MB of RAM for storing large data structures, and an 8-bit microprocessor is used with a 64-kB address space, address banking can enable the microprocessor to access the full 1 MB one small section at a time.

Address banking, also known as *paging*, takes a large quantity of memory, divides it into multiple smaller banks, and makes each bank available to the microprocessor one at a time. A *bank address register* is maintained by the microprocessor and determines which bank of memory is selected at any given time. The selected bank is accessed through a portion of the microprocessor's fixed address space, called a *window*, set aside for banked memory access. As shown in Fig. 3.10a, the upper 16 kB of address space provides direct access to one of many 16-kB pages in the larger banked memory structure. Figure 3.10b shows the logical implementation of this banked memory scheme. A

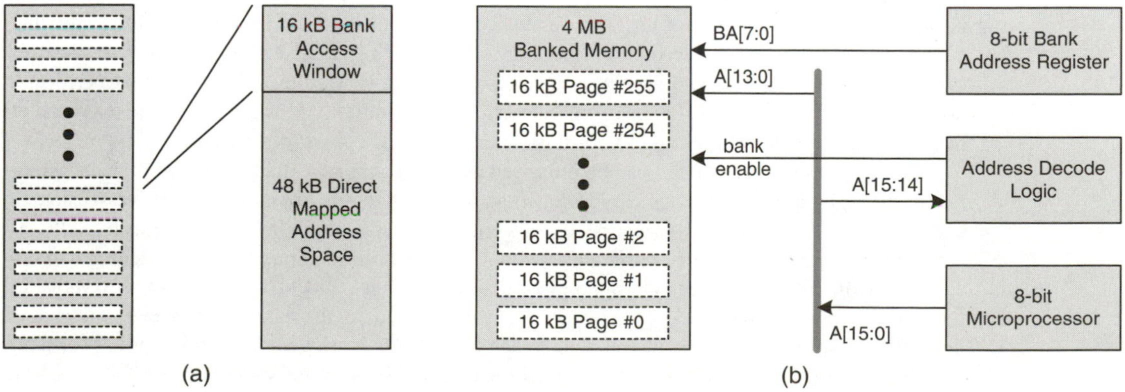

FIGURE 3.10 Address banking.

22-bit combined address is sent to the 4-MB banked memory structure: 256 pages × 16 kB per page = 4 MB. These 22 bits are formed through the concatenation of the 8-bit bank address register and 14 of the microprocessor's low-order address bits, A[13:0]. The eight bank-address bits are changed infrequently whenever the microprocessor is ready for a new page in memory. The 14 microprocessor-address bits can change each time the window is accessed.

The details of a banking scheme can be modified according to the application's requirements. The bank access window can be increased or decreased, and more or fewer pages can be defined. If an application operates on many small sets of data, a larger number of smaller pages may be suitable. If the data or software set is widely dispersed, it may be better to increase the window size as much as possible to minimize the bank address register update rate.

While address banking can greatly increase the memory available to a microprocessor, it does so with the penalties of increased access time on page switches and more complexity in managing the segmented address space. Each time the microprocessor wants to access a location in a different page, it must update the bank address register. This penalty is acceptable in some applications. However, if the application requires both consistently fast access time and large memory size, a faster, more expensive microprocessor may be required that suits these needs.

The complexity of managing the segmented address space dissuades some engineers from employing address banking. Software usually bears the brunt of recognizing when necessary data resides in a different page and then updating the bank address register to access that page. It is easier for software to deal with a large, continuous address space. With the easy availability and low cost of 32-bit microprocessors, address banking is not as common as it used to be. However, if an 8-bit microprocessor must be used for cost reduction or other limitations, address banking may be useful when memory demands increase beyond 64 kB.

3.7 DIRECT MEMORY ACCESS

Transferring data from one region of memory to another is a common task performed within a computer. Incoming data may be transferred from a serial communications controller into memory, and outgoing data may be transferred from memory to the controller. Memory-to-memory transfers are common, too, as data structures are moved between subprograms, each of which may have separate regions of memory set aside for its private use. The speed with which memory is transferred normally depends on the time that the microprocessor takes to perform successive read and write operations. Each byte transferred requires several microprocessor operations: load accumulator, store accumulator, update address for next byte, and check if there is more data. Instead of simply moving a stream of bytes without interruption, the microprocessor is occupied mostly by the overhead of calculating new addresses and checking to see if more data is waiting. Computers that perform a high volume of memory transfers may exhibit performance bottlenecks as a result of the overhead of having the microprocessor spend too much of its time reading and writing memory.

Memory transfer performance can be improved using a technique called *direct memory access*, or DMA. DMA logic intercedes at the microprocessor's request to directly move data between a source and destination. A *DMA controller* (DMAC) sits on the microprocessor bus and contains logic that is specifically designed to rapidly move data without the overhead of simultaneously fetching and decoding instructions. When the microprocessor determines that a block of data is ready to move, it programs the DMAC with the starting address of the source data, the number of bytes to move, and the starting address of the destination data. When the DMAC is triggered, the microprocessor temporarily relinquishes control of its bus so the DMAC can take over and quickly move the data. The DMAC serves as a surrogate processor by directly generating addresses and reading and writing data. From the microprocessor bus perspective, nothing has changed, and data transfers proceed normally despite being controlled by the DMAC rather than the microprocessor. Figure 3.11 shows the basic internal structure of a DMAC.

A DMA transfer can be initiated by either the microprocessor or an I/O device that contains logic to assert a request to the DMAC. DMA transfers are generally broken into two categories: peripheral/memory and memory/memory. Peripheral/memory transfers move data to a peripheral or retrieve data from a peripheral. A peripheral/memory transfer can be triggered by a DMA-aware I/O-device when it is ready to accept more outgoing data or incoming data has arrived. These are called *single-address transfers,* because the DMAC typically controls only a single address—that of the memory side of the transfer. The peripheral address is typically a fixed offset into its register set and is asserted by supporting control logic that assists in the connectivity between the peripheral and the DMAC.

DMA transfers do not have to be continuous, and they are often not in the case of a peripheral transfer. If the microprocessor sets up a DMA transfer from a serial communications controller to memory, it programs the DMAC to write a certain quantity of data into memory. However, the transfer does not begin until the serial controller asserts a DMA request indicating that data is ready. When this request occurs, the DMAC arbitrates for access to the microprocessor bus by asserting a bus request. Some time later, the microprocessor or its support logic will grant the bus to the DMAC and temporarily pause the microprocessor's bus activity. The DMAC can then transfer a single unit of data from the serial controller into memory. The unit of data transfer may be any number of bytes. When finished, the DMAC relinquishes control of the bus back to the microprocessor.

Memory/memory transfers move data from one region in memory to another. These are called *dual-address transfers,* because the DMAC controls two addresses into memory—source and destination. Memory/memory transfers are triggered by the microprocessor and can execute continuously, because the data block to be moved is ready and waiting in memory.

Even when DMA transfers execute one byte at a time, they are still more efficient than the microprocessor, because the DMAC is capable of transferring a byte or word (per the microprocessor's data bus width) in a single bus cycle rather than the microprocessor's load/store mechanism with additional overhead. There is some initial overhead in setting up the DMA transfer, so it is not efficient to use DMA for very short transfers. If the microprocessor needs to move only a few bytes, it should probably do so on its own. However, the DMAC initialization overhead is more than compensated for if dozens or hundreds of bytes are being moved.

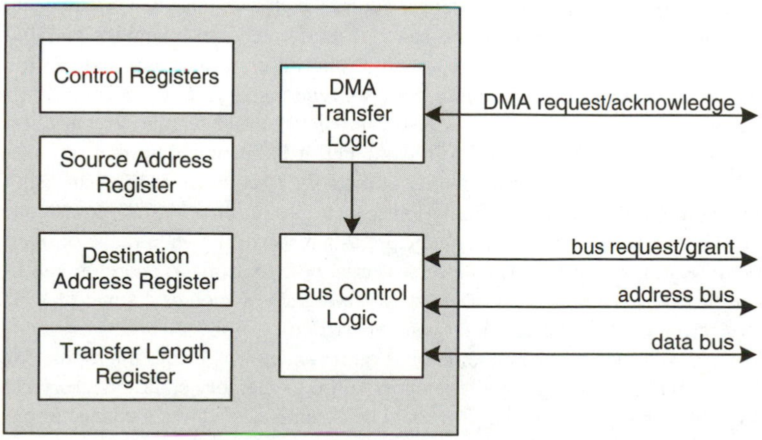

FIGURE 3.11 DMA controller block diagram.

A typical DMAC supports multiple channels, each of which controls a different DMA transfer. While only one transfer can execute at any given moment, multiple transfers can be interleaved to prevent one peripheral from being starved for data while another is being serviced. Because a typical peripheral transfer is not continuous, each DMA channel can be assigned to each active peripheral. A DMAC can have one channel configured to load incoming data from a serial controller, another to store data to a disk drive controller, and a third to move data from one region of memory to another. Once initialized by the microprocessor, the exact order and interleaving of multiple channels is resolved by the individual DMA request signals, and any priority information is stored in the DMAC.

When a DMAC channel has completed transferring the requested quantity of data, the DMAC asserts an interrupt to the microprocessor to signal that the data has been moved. At this point, the microprocessor can restart a new DMA transfer if desired and invoke any necessary routines to process data that has been moved.

External DMA support logic may be necessary, depending on the specific DMAC, microprocessor, and peripherals that are being used. Some microprocessors contain built-in DMAC arbitration logic. Some peripherals contain built-in DMA request logic, because they are specifically designed for these high-efficiency memory transfers. Custom arbitration logic typically functions by waiting for the DMAC to request the bus and then pausing the microprocessor's bus transfers until the DMAC relinquishes the bus. This pause operation is performed according to the specifications of the particular microprocessor. Custom peripheral control logic can include DMAC read/write interface logic to assert the correct peripheral address when a transfer begins and perform any other required mapping between the DMAC's transfer enable signaling and the peripheral's read/write interface.

3.8 EXTENDING THE MICROPROCESSOR BUS

A microprocessor bus is intended to directly connect to memory and I/O devices that are in close proximity to the microprocessor. As such, its electrical and functional properties are suited for relatively short interconnecting wires and relatively simple device interfaces that respond with data soon after the microprocessor issues a request. Many computers, however, require some mechanism to extend the microprocessor bus so that additional hardware, such as plug-in expansion cards or memory modules, can enhance the system with new capabilities. Supporting these modular extensions to the computer's architecture can be relatively simple or quite complex, depending on the required degree of expandability and the physical distances across which data must be communicated.

Expansion buses are generally broken into two categories, memory and I/O, because these groups' respective characteristics are usually quite different. General-purpose memory is a high-bandwidth resource to which the microprocessor requires immediate access so that it can maintain a high level of throughput. Memory is also a predictable and regular structure, both logically and physically. If more RAM is added to a computer, it is fairly certain that some known number of chips will be required for a given quantity of memory. In contrast, I/O by nature is very diverse, and its bandwidth requirements are usually lower than that of memory. I/O expansion usually involves cards of differing complexity and architecture as a result of the wide range of interfaces that can be supported (e.g., disk drive controller versus serial port controller). Therefore, an I/O expansion bus must be flexible enough to interface with a varying set of modules, some of which may not have been conceived of when the computer is first designed.

Memory expansion buses are sometimes direct extensions of the microprocessor bus. From the preceding 8-bit computer example, the upper 16 kB of memory could be reserved for future expansion. A provision for future expansion could be as simple as adding a connector or socket for an extra memory chip. In this case, no special augmentation of the microprocessor bus is required. However, in a larger system with more address space, provisions must be made for more than one additional memory chip. In these situations, a simple *buffered* extension of the microprocessor bus may suffice. A buffer, in this context, is an IC that passes data from one set of pins to another, thereby electrically separating two sections of a bus. As shown in Fig. 3.12, a buffer can extend a microprocessor bus so that its logical functionality remains unchanged, but its electrical characteristics are enhanced to provide connectivity across a greater distance (to a multichip memory expansion module). A unidirectional address buffer extends the address bus from the microprocessor to expansion memory devices. A bidirectional data buffer extends the bus away from the microprocessor on writes and toward the microprocessor on reads. The direction of the data buffer is controlled according to the state of read/write enable signals generated by the microprocessor.

More complex memory structures may contain dedicated memory control logic that sits between the microprocessor and the actual memory devices. Expanding such a memory architecture is generally accomplished by augmenting the "back-side" memory device bus as shown in Fig. 3.13 rather than by adding additional controllers onto an extended microprocessor bus. Such an expansion scheme may or may not require buffers, depending on the electrical characteristics of the bus in question.

I/O buses may also be direct extensions of the microprocessor bus. The original expansion bus in the IBM PC, developed in the early 1980s, is essentially an extended Intel 8088 microprocessor bus that came to be known as the *Industry Standard Architecture* (ISA) bus. Each I/O card on the ISA bus is mapped in a unique address range in the microprocessor's memory. Therefore, when software wants to read or write a register on an I/O card, it simply performs an access to the desired location. The ISA bus added a few features beyond the raw 8088 bus, including DMA and variable *wait states* for slow I/O devices. A wait state results when a device cannot immediately respond to the microprocessor's request and asserts a signal to stretch the access so that it can respond properly.

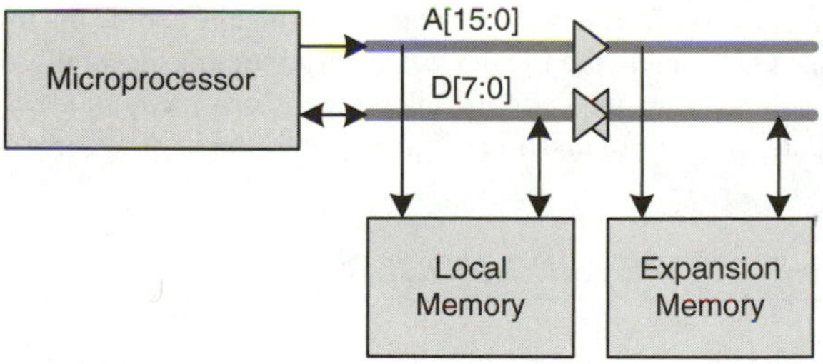

FIGURE 3.12 Buffered microprocessor bus for memory expansion.

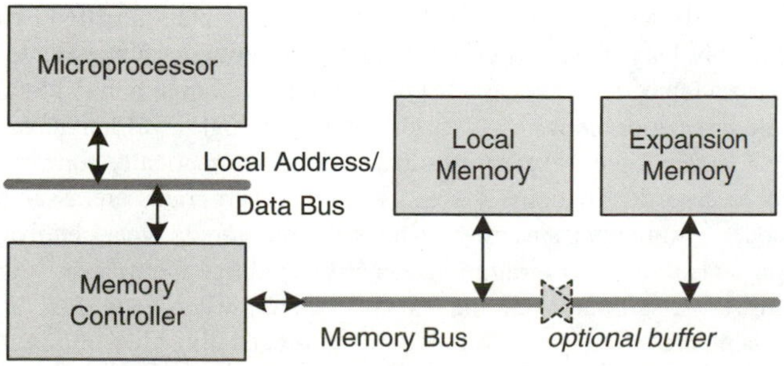

FIGURE 3.13 Extended memory controller bus.

Direct extensions such as the ISA bus are fairly easy to implement and serve well in applications where I/O response time does not unduly restrict microprocessor throughput. As computers have gotten faster, the throughput of microprocessors has rapidly outstripped the response times of all but the fastest I/O devices. In comparison to a modern microprocessor, a hard-disk controller is rather slow, with response times measured in microseconds rather than nanoseconds. Additionally, as bus signals become faster, the permissible length of interconnecting wires decreases, limiting their expandability. These and other characteristics motivate the decoupling of the microprocessor's local bus from the computer's I/O bus.

An I/O bus can be decoupled from the microprocessor bus by inserting an intermediate bus controller between them that serves as an interface, or translator, between the two buses. Once the buses are separated, activity on one bus does not necessarily obstruct activity on the other. If the microprocessor wants to write a block of data to a slow device, it can rapidly transfer that data to the bus controller and then continue with other operations at full speed while the controller slowly transfers the data to the I/O device. This mechanism is called a *posted-write,* because the bus controller allows the microprocessor to complete, or *post*, its write before the write actually completes. Separate buses also open up the possibility of multiple microprocessors or logic elements performing I/O operations without conflicting with the central microprocessor. In a multimaster system, a specialized DMA controller can transfer data between two peripherals such as disk controllers while the microprocessor goes about its normal business.

The *Peripheral Component Interconnect* (PCI) bus is the industry-standard follow-on to the ISA bus, and it implements such advanced features as posted-writes, multiple-masters, and multiple bus segments. Each PCI bus segment is separated from the others via a PCI bridge chip. Only traffic that must travel between buses crosses a bridge, thereby reducing congestion on individual PCI bus segments. One segment can be involved in a data transfer between two devices without affecting a simultaneous transfer between two other devices on a different segment. These performance-enhancing features do not come for free, however. Their cost is manifested by the need for dedicated PCI control logic in bridge chips and in the I/O devices themselves. It is generally simpler to implement an I/O device that is directly mapped into the microprocessor's memory space, but the overall performance of the computer may suffer under demanding applications.

3.9 ASSEMBLY LANGUAGE AND ADDRESSING MODES

With the hardware ready, a computer requires software to make it more than an inactive collection of components. Microprocessors fetch instructions from program memory, each consisting of an opcode and, optionally, additional operands following the opcode. These opcodes are binary data that are easy for the microprocessor to decode, but they are not very readable by a person. To enable a programmer to more easily write software, an instruction representation called *assembly language* was developed. Assembly language is a low-level language that directly represents each binary opcode with a human-readable text mnemonic. For example, the mnemonic for an unconditional branch-to-subroutine instruction could be `BSR`. In contrast, a high-level language such as C++ or Java contains more complex logical expressions that may be automatically converted by a compiler to dozens of microprocessor instructions. Assembly language programs are assembled, rather than compiled, into opcodes by directly translating each mnemonic into its binary equivalent.

Assembly language also makes programming easier by enabling the usage of text labels in place of hard-coded addresses. A subroutine can be named `FOO`, and when `BSR FOO` is encountered by the assembler, a suitable branch target address will be automatically calculated in place of the label `FOO`. Each type of assembler requires a slightly different format and syntax, but there are general assembly language conventions that enable a programmer to quickly adapt to specific implementations once the basics are understood. An assembly language program listing usually has three columns of text followed by an optional comment column as shown in Fig. 3.14. The first column is for labels that are placeholders for addresses to be resolved by the assembler. Instruction mnemonics are located in the second column. The third column is for instruction operands.

This listing uses the Motorola 6800 family's assembly language format. Though developed in the 1970s, 68xx microprocessors are still used today in embedded applications such as automobiles and industrial automation. The first line of this listing is not an instruction, but an assembler *directive* that tells the assembler to locate the program at memory location $100. When assembled, the listing is converted into a memory dump that lists a range of memory addresses and their corresponding contents—opcodes and operands. Assembler directives are often indicated with a period prefix.

The program in Fig. 3.14 is very simple: it counts to 30 ($1E) and then sends the "Z" character out the serial port. It continues in an infinite loop by returning to the start of the program when the serial port routine has completed its task. The subroutine to handle the serial port is not shown and is referenced with the `SEND_CHAR` label. The program begins by clearing accumulator A (the 6800 has two accumulators: ACCA and ACCB). It then enters an incrementing loop where the accumulator is incremented and then compared against the terminal count value, $1E. The # prefix tells the assembler to use the literal value $1E for the comparison. Other alternatives are possible and will soon be discussed. If ACCA is unequal to $1E, the microprocessor goes back to increment ACCA. If equal, the accumulator is loaded with the ASCII character to be transmitted, also a literal operand. The assumption here is that the `SEND_CHAR` subroutine transmits whatever is in ACCA. When the subroutine finishes, the program starts over with the branch-always instruction.

Each of the instructions in the preceding program contains at least one operand. `CLRA` and `INCA` have only one operand: ACCA. `CMPA` and `LDAA` each have two operands: ACCA and associated data. Complex microprocessors may reference three or more operands in a single instruction. Some instructions can reference different types of operands according to the requirements of the program being implemented. Both `CMPA` and `LDAA` reference literal operands in this example, but a programmer cannot always specify a predetermined literal data value directly in the instruction sequence.

Operands can be referenced in a variety of manners, called *addressing modes,* depending on the type of instruction and the type of operand. Some types of instructions inherently use only one addressing mode, and some types have multiple modes. The manners of referencing operands can be categorized into six basic addressing modes: *implied*, *immediate*, *direct*, *relative*, *indirect*, and *indexed*. To fully understand how a microprocessor works, and to efficiently utilize an instruction set, it is necessary to explore the various mechanisms used to reference data.

- *Implied addressing* specifies the operand of an instruction as an inherent property of that instruction. For example, `CLRA` implies the accumulator by definition. No additional addressing information following the opcode is needed.

```
                .ORIG       $100

BEGIN           CLRA
INC_LOOP        INCA
                CMPA        #$1E          ; compare ACCA = $1E
                BNE         INC_LOOP      ; if not equal, go back
                LDAA        #'Z'          ; else, load ASCII 'Z'
                BSR         SEND_CHAR     ; send ACCA to serial port
                BRA         BEGIN         ; start over again
```

FIGURE 3.14 Typical assembly language listing.

- *Immediate addressing* places an operand's value literally into the instruction sequence. `LDAA #'Z'` has its primary operand immediately available following the opcode. An immediate operand is indicated with the # prefix in some assembly languages. Eight-bit microprocessors with eight-bit instruction words cannot fit an immediate value into the instruction word itself and, therefore, require that an extra byte following the opcode be used to specify the immediate value. More powerful 32-bit microprocessors can often fit a 16-bit or 24-bit immediate value within the instruction word. This saves an additional memory fetch to obtain the operand.

- *Direct addressing* places the address of an operand directly into the instruction sequence. Instead of specifying `LDAA #'Z'`, the programmer could specify `LDAA $1234`. This version of the instruction would tell the microprocessor to read memory location $1234 and load the resulting value into the accumulator. The operand is directly available by looking into the memory address specified just following the instruction. Direct addressing is useful when there is a need to read a fixed memory location. Usage of the direct addressing mode has a slightly different impact on various microprocessors. A typical 8-bit microprocessor has a 16-bit address space, meaning that two bytes following the opcode are necessary to represent a direct address. The 8-bit microprocessor will have to perform two additional 8-bit fetch operations to load the direct address. A typical 32-bit microprocessor has a 32-bit address space, meaning that 4 bytes following the opcode are necessary. If the 32-bit microprocessor has a 32-bit data bus, only one additional 32-bit fetch operation is required to load the direct address.

- *Relative addressing* places an operand's relative address into the instruction sequence. A relative address is expressed as a signed offset relative to the current value of the PC. Relative addressing is often used by branch instructions, because the target of a branch is usually within a short distance of the PC, or current instruction. For example, `BNE INC_LOOP` results in a branch-if-not-equal backward by two instructions. The assembler automatically resolves the addresses and calculates a relative offset to be placed following the `BNE` opcode. This relative operation is performed by adding the offset to the PC. The new PC value is then used to resume the instruction fetch and execution process. Relative addressing can utilize both positive and negative deltas that are applied to the PC. A microprocessor's instruction format constrains the relative range that can be specified in this addressing mode. For example, most 8-bit microprocessors provide only an 8-bit signed field for relative branches, indicating a range of +127/–128 bytes. The relative delta value is stored into its own byte just after the opcode. Many 32-bit microprocessors allow a 16-bit delta field and are able to fit this value into the 32-bit instruction word, enabling the entire instruction to be fetched in a single memory read. Limiting the range of a relative operation is generally not an excessive constraint because of software's *locality* property. *Locality* in this context means that the set of instructions involved in performing a specific task are generally relatively close together in memory. The locality property covers the great majority of branch instructions. For those few branches that have their targets outside of the allowed relative range, it is necessary to perform a short relative branch to a long jump instruction that specifies a direct address. This reduces the efficiency of the microprocessor by having to perform two branches when only one is ideally desired, but the overall efficiency of saving extra memory accesses for the majority of short branches is worth the trade-off.

- *Indirect addressing* specifies an operand's direct address as a value contained in another register. The other register becomes a pointer to the desired data. For example, a microprocessor with two accumulators can load *ACCA* with the value that is at the address in ACCB. `LDAA (ACCB)` would tell the microprocessor to put the value of accumulator B onto the address bus, perform a read, and put the returned value into accumulator A. Indirect addressing allows writing software routines that operate on data at different addresses. If a programmer wants to read or write an arbitrary entry in a data table, the software can load the address of that entry into a microprocessor register and then perform an indirect access using that register as a pointer. Some microprocessors place constraints on which registers can be used as references for indirect addressing. In the case of a 6800 microprocessor, `LDAA (ACCB)` is not actually a supported operation but serves as a syntactical example for purposes of discussion.

- *Indexed addressing* is a close relative (no pun intended) of indirect addressing, because it also refers to an address contained in another register. However, indexed addressing also specifies an offset, or index, to be added to that register base value to generate the final operand address: base + offset = final address. Some microprocessors allow general accumulator registers to be used as base-address registers, but others, such as the 6800, provide special *index registers* for this purpose. In many 8-bit microprocessors, a full 16-bit address cannot be obtained from an 8-bit accumulator serving as the base address. Therefore, one or more separate index registers are present for the purpose of indexed addressing. In contrast, many 32-bit microprocessors are able to specify a full 32-bit address with any general-purpose register and place no limitations on which register serves as the index register. Indexed addressing builds upon the capabilities of indirect addressing by enabling multiple address offsets to be referenced from the same base address. `LDAA (X+$20)` would tell the microprocessor to add $20 to the index register, X, and use the resulting address to fetch data to be loaded into ACCA. One simple example of using indexed addressing is a subroutine to add a set of four numbers located at an arbitrary location in memory. Before calling the subroutine, the main program can set an index register to point to the table of numbers. Within the subroutine, four individual addition instructions use the indexed addressing mode to add the locations X+0, X+1, X+2, and X+3. When so written, the subroutine is flexible enough to be used for any such set of numbers. Because of the similarity of indexed and indirect addressing, some microprocessors merge them into a single mode and obtain indirect addressing by performing indexed addressing with an index value of zero.

The six conceptual addressing modes discussed above represent the various logical mechanisms that a microprocessor can employ to access data. It is important to realize that each individual microprocessor applies these addressing modes differently. Some combine multiple modes into a single mode (e.g., indexed and indirect), and some will create multiple submodes out of a single mode. The exact variation depends on the specifics of an individual microprocessor's architecture.

With the various addressing modes modifying the specific opcode and operands that are presented to the microprocessor, the benefits of using assembly language over direct binary values can be observed. The programmer does not have to worry about calculating branch target addresses or resolving different addressing modes. Each mnemonic can map to several unique opcodes, depending on the addressing mode used. For example, the `LDAA` instruction in Fig. 3.14 could easily have used *extended* addressing by specifying a full 16-bit address at which the ASCII transmit-value is located. Extended addressing is the 6800's mechanism for specifying a 16-bit direct address. (The 6800's direct addressing involves only an eight-bit address.) In either case, the assembler would determine the correct opcode to represent `LDAA` and insert the correct binary values into the memory dump. Additionally, because labels are resolved each time the program is assembled, small changes to the program can be made that add or remove instructions and labels, and the assembler will automatically adjust the resulting addresses accordingly.

Programming in assembly language is different from using a high-level language, because one must think in smaller steps and have direct knowledge about the microprocessor's operation and architecture. Assembly language is processor-specific instead of generic, as with a high-level language. Therefore, assembly language programming is usually restricted to special cases such as boot code or routines in which absolute efficiency and performance are demanded. A human programmer will usually be able to write more efficient assembly language than a high-level language compiler can generate. In large programs, the slight inefficiency of the compiler is well worth the trade-off for ease of programming in a high-level language. However, time-critical routines such as I/O drivers or ISRs may benefit from manual assembly language coding.

CHAPTER 4
Memory

Memory is as fundamental to computer architecture as any other element. The ability of a system's memory to transact the right quantity of data in the right span of time has a substantial impact on how that system fulfills its design goals. Digital engineers struggle with innovative ways to improve memory density and bandwidth in a way that is tailored to a specific application's performance and cost constraints.

Knowledge of prevailing memory technologies' strengths and weaknesses is a key requirement for designing digital systems. When memory architecture is chosen that complements the rest of the system, a successful design moves much closer to fruition. Conversely, inappropriate memory architecture can doom a good idea to the engineering doldrums of impracticality brought on by artificial complexity.

This chapter provides an introduction to various solid-state memory technologies and explains how they work from an internal structural perspective as well as an interface timing perspective. A memory's internal structure is important to an engineer, because it explains why that memory might be more suited for one application over another. Interface timing is where the rubber meets the road, because it defines how other elements in the system can access memory components' contents. The wrong interface on a memory chip can make it difficult for external logic such as a microprocessor to access that memory and still have time left over to perform the necessary processing on that data.

Basic memory organization and terminology are introduced first. This is followed by a discussion of the prevailing read-only memory technologies: EPROM, flash, and EEPROM. Asynchronous SRAM and DRAM technologies, the foundations for practically all random-access memories, are presented next. These asynchronous RAMs are no longer on the forefront of memory technology but still find use in many systems. Understanding their operation not only enables their application, it also contributes to an understanding of the most recent synchronous RAM technologies. (High-performance synchronous memories are discussed later in the book.) The chapter concludes with a discussion of two types of specialty memories: multiport RAMs and FIFOs. Multiport RAMs and FIFOs are found in many applications where memory serves less as a storage element and more as a communications channel between distinct logic blocks.

4.1 MEMORY CLASSIFICATIONS

Microprocessors require memory resources in which to store programs and data. Memory can be classified into two broad categories: volatile and nonvolatile. Volatile memory loses its contents when power is turned off. Nonvolatile memory retains its contents indefinitely, even when there is no power present. Nonvolatile memory can be used to hold the boot code for a computer so that the microprocessor can have a place to get started. Once the computer begins initializing itself from nonvolatile memory, volatile memory is used to store dynamic variables, including the stack and other programs that may be loaded from a disk drive. Figure 4.1 shows that a general memory device consists of a bit-storage array, address-decode logic, input/output logic, and control logic.

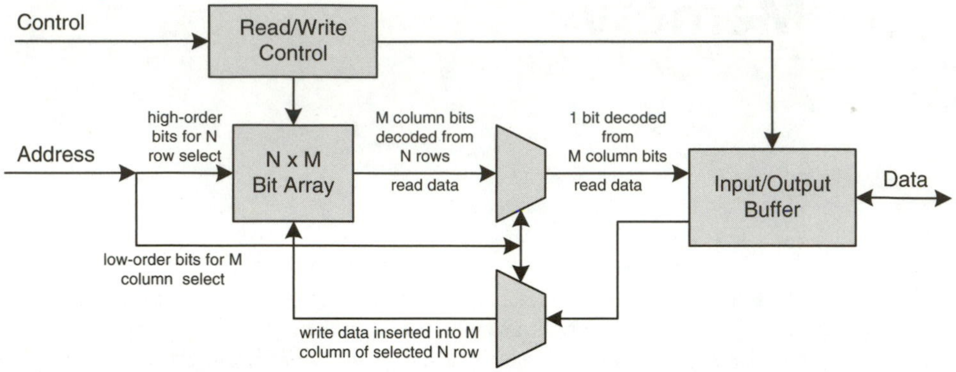

FIGURE 4.1 General memory device.

Despite the logical organization of the device, the internal bit array is usually less rectangular and more square in its aspect ratio. For example, a 131,072 × 8 memory (128 kB) may be implemented as 512 × 256 × 8. This aspect ratio minimizes the complexity of the address-decode logic and also has certain manufacturing process benefits. It takes more logic to generate 131,072 enable signals in one pass than to generate 512 and then 256 enables in two passes. The first decode is performed upfront in the memory array, and the second decode is performed by a multiplexer to pass the desired memory location.

Nonvolatile memory can be separated into two subcategories: devices whose contents are programmed at a factory without the expectation of the data changing over time, and devices whose contents are loaded during system manufacture with anticipation of in-circuit updates during the life of the product. The former devices are, for all practical purposes, write-once devices that cannot be erased easily, if at all. The latter devices are designed primarily to be nonvolatile, but special circuitry is designed into the devices to enable erasure and rewriting of the memory contents while the devices are functioning in a system. Most often, these circuits and their associated algorithms cause the erase/write cycle to be more lengthy and complex than simply reading the existing data out of the devices. This penalty on write performance reflects both the desire to secure the nonvolatile memory from accidental modification as well as the inherent difficulty in modifying a memory that is designed to retain its contents in the absence of power.

Volatile memory can also be separated into two subcategories: devices whose contents are nonvolatile for as long as power is applied (these devices are referred to as *static*) and devices whose contents require periodic refreshing to avoid loss of data even while power is present (these devices are referred to as *dynamic*). On first thought, the category of dynamic devices may seem absurd. What possible benefit is there to a memory chip that cannot retain its memory without assistance? The benefit is significantly higher density of memory per unit silicon area, and hence lower cost of dynamic versus static memory. One downside to dynamic memory is somewhat increased system complexity to manage its periodic update requirement. An engineer must weight the benefits and complexities of each memory type when designing a system. Some systems benefit from one memory type over the other, and some use both types in different proportions according to the needs of specific applications.

Memory chips are among the more complex integrated circuits that are standardized across multiple manufacturers through cooperation with an industry association called the Joint Electron Device Engineering Council (JEDEC). Standardization of memory chip pin assignments and functionality is important, because most memory chips are commodities that derive a large portion of their value by being interoperable across different vendors. Newer memory technologies introduced in the 1990s resulted in more proprietary memory architectures that did not retain the high degree of compatibility present in other mainstream memory components. However, memory devices still largely conform to JEDEC standards, making their use that much easier.

4.2 EPROM

Erasable-programmable read-only-memory, EPROM, is a basic type of nonvolatile memory that has been around since the late 1960s. During the 1970s and into the 1990s, EPROM accounted for the majority of nonvolatile memory chips manufactured. EPROM maintained its dominance for decades and still has a healthy market share because of its simplicity and low cost: a typical device is programmed once on an assembly line, after which it functions as a ROM for the rest of its life. An EPROM can be erased only by exposing its die to ultraviolet light for an extended period of time (typically, 30 minutes). Therefore, once an EPROM is assembled into a computer system, its contents are, for all practical purposes, fixed forever. Older ROM technologies included programmable-ROMs, or PROMs, that were fabricated with tiny fuses on the silicon die. These fuses could be burned only once, which prevented a manufacturer from testing each fuse before shipment. In contrast, EPROMs are fairly inexpensive to manufacture, and their erasure capability allows them to be completely tested by the semiconductor manufacturer before shipment to the customer. Only a full-custom mask-programmed chip, a true ROM, is cheaper to manufacture than an EPROM on a bit-for-bit basis. However, mask ROMs are rare, because they require a fixed data image that cannot be changed without modifying the chip design. Given that software changes are fairly common, mask ROMs are relatively uncommon.

An EPROM's silicon bit structure consists of a special MOSFET structure whose gate traps a charge that is applied to it during programming. Programming is performed with a higher than normal voltage, usually 12 V (older generation EPROMs required 21 V), that places a charge on the floating gate of a MOSFET as shown in Fig. 4.2.

When the programming voltage is applied to the control gate, a charge is induced on the floating gate, which is electrically isolated from both the silicon substrate as well as the control gate. This isolation enables the floating gate to function as a capacitor with almost zero current leakage across the dielectric. In other words, once a charge is applied to the floating gate, the charge remains almost indefinitely. A charged floating gate causes the silicon that separates the MOSFET's source and drain contacts to electrically conduct, creating a connection from logic ground to the bit output. This means that a programmed EPROM bit reads back as a 0. An unprogrammed bit reads back as a 1, because the lack of charge on the floating gate does not allow an electrical connection between the source and drain.

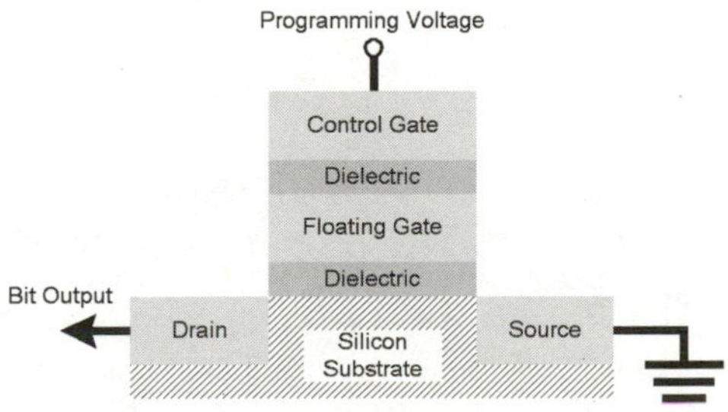

FIGURE 4.2 EPROM silicon bit structure.

Once programmed, the charge on the floating gate cannot be removed electrically. UV photons cause the dielectric to become slightly conductive, allowing the floating gate's charge to gradually drain away to its unprogrammed state. This UV erasure feature is the reason why many EPROMs are manufactured in ceramic packages with transparent quartz windows directly above the silicon die. These ceramic packages are generally either DIPs or PLCCs and are relatively expensive. In the late 1980s it became common for EPROMs to be manufactured in cheaper plastic packages without transparent windows. These EPROM devices are rendered one-time programmable, or OTP, because it is impossible to expose the die to UV light. OTP devices are attractive, because they are the least expensive nonmask ROM technology and provide a manufacturer with the flexibility to change software on the assembly line by using a new data image to program EPROMs.

The industry standard EPROM family is the 27xxx, where the "xxx" indicates the chip's memory capacity in kilobits. The 27256 and 27512 are very common and easily located devices. Older parts include the 2708, 2716, 2732, 2764, and 27128. There are also newer, higher-density EPROMs such as the 27010, 27020, and 27040 with 1 Mb, 2 Mb, and 4 Mb densities, respectively. 27xxx EPROM devices are most commonly eight bits wide (a 27256 is a 32,768 × 8 EPROM). Wider data words, such as 16 or 32 bits, are available but less common.

Older members of the 27xxx family, such as early NMOS 2716 and 2732 devices, required 21-V programming voltages, consumed more power, and featured access times of between 200 and 450 ns. Newer CMOS devices are designated 27Cxxx, require a 12-V programming voltage, consume less power, and have access times as fast as 45 ns, depending on the manufacturer and device density.

EPROMs are very easy to use because of their classic asynchronous interface. In most applications, the EPROM is treated like a ROM, so writes to the device are not an issue. Two programming control pins, V_{PP} and PGM*, serve as the high-voltage source and program enable, respectively. These two pins can be set to inactive levels and forgotten. What remains are a chip enable, CE*, an output enable, OE*, an address bus, and a data output bus as shown in Fig. 4.3, using a 27C64 (8K × 8) as an example.

When CE* is inactive, or high, the device is in a powered-down mode in which it consumes the least current—measured in microamps due to the quiescent nature of CMOS logic. When CE* and OE* are active simultaneously, D[7:0] follows A[12:0] subject to the device's access time, or propagation delay. This read timing is shown in Fig. 4.4.

When OE* is inactive, the data bus is held in a high-impedance state. A certain time after OE* goes active, t_{OE}, the data word corresponding to the given address is driven—assuming that A1 has been stable for at least t_{ACC}. If not, t_{ACC} will determine how soon D1 is available rather than t_{OE}. While OE* is active, the data bus transitions t_{ACC} ns after the address bus. As soon as OE* is removed, the data bus returns to a high-impedance state after t_{OEZ}.

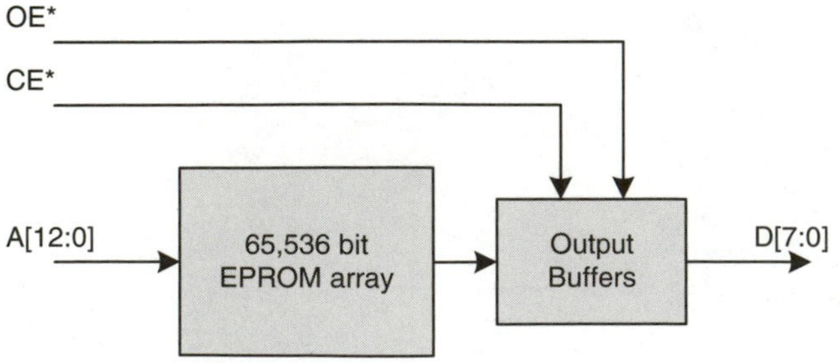

FIGURE 4.3 27C64 block diagram.

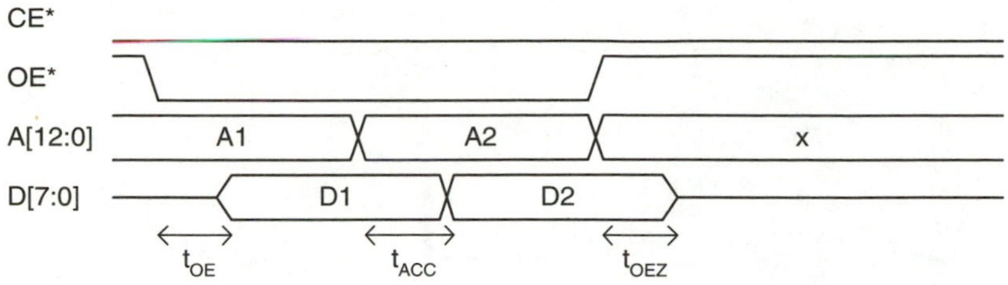

FIGURE 4.4 EPROM asynchronous read timing.

Many microprocessors are able to directly interface to an EPROM via this asynchronous bus because of its ubiquity. Most eight-bit microprocessors have buses that function solely in this asynchronous mode. In contrast, some high-performance 32-bit microprocessors may initially boot in a low-speed asynchronous mode and then configure themselves for higher performance operation after retrieving the necessary boot code and initialization data from the EPROM.

4.3 FLASH MEMORY

Flash memory captured the lion's share of the nonvolatile memory market from EPROMs in the 1990s and holds a dominant position as the industry leader to this day. Flash is an enhanced EPROM that can both program and erase electrically without time-consuming exposure to UV light, and it has no need for the associated expensive ceramic and quartz packaging. Flash does cost a small amount more to manufacture than EPROM, but its more flexible use in terms of electronic erasure more than makes up for a small cost differential in the majority of applications. Flash is found in everything from cellular phones to automobiles to desktop computers to solid-state disk drives. It has enabled a whole class of flexible computing platforms that are able to upgrade their software easily and "on the fly" during normal operation. Similar to EPROMs, early flash devices required separate programming voltages. Semiconductor vendors quickly developed single-supply flash devices that made their use easier.

A flash bit structure is very similar to that of an EPROM. Two key differences are an extremely thin dielectric between the floating gate and the silicon substrate and the ability to apply varying bias voltages to the source and control gate. A flash bit is programmed in the same way that an EPROM bit is programmed—by applying a high voltage to the control gate. Flash devices contain internal voltage generators to supply the higher programming voltage so that multiple external voltages are not required. The real difference appears when the bit is erased electrically. A rather complex quantum-mechanical behavior called *Fowler-Nordheim tunneling* is exploited by applying a negative voltage to the control gate and a positive voltage to the MOSFET's source as shown in Fig. 4.5.

The combination of the applied bias voltages and the thin dielectric causes the charge on the floating gate to drain away through the MOSFET's source. Flash devices cannot go through this program/erase cycle indefinitely. Early devices were rated for 100,000 erase cycles. Modern flash chips are often specified up to 1,000,000 erase cycles. One million cycles may sound like a lot, but remember that microprocessors run at tens or hundreds of millions of cycles per second. When a processor is capable of writing millions of memory locations each second, an engineer must be sure that the flash memory is used appropriately and not updated too often so as to maximize its operational life. Products that utilize flash memory generally contain some a management algorithm to ensure that the erasure limit is not reached during the product's expected lifetime. This algorithm can be as simple as performing software updates only several times per year. Alternatively, algorithms can be

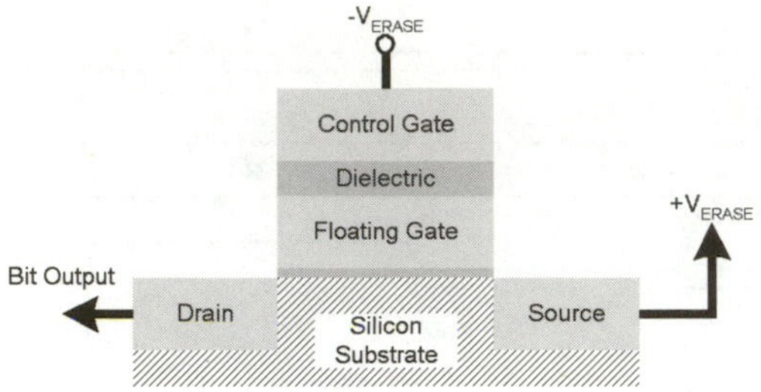

FIGURE 4.5 Flash bit erasure.

smart enough to track how many times each portion of a flash device has been erased and dynamically make decisions about where to place new data accordingly.

Flash chips are offered in two basic categories, NOR and NAND, named according to the circuits that make up each memory bit. NOR flash is a random access architecture that often functions like an EPROM when reading data. NOR memory arrays are directly accessed by a microprocessor and are therefore well suited for storing boot code and other programs. NAND flash is a sequential access architecture that segments the memory into many pages, typically 256 or 512 bytes. Each page is accessed as a discrete unit. As such, NAND flash does not provide the random access interface of a NOR flash. In return for added interface complexity and slower response time, NAND flash provides greater memory density than NOR flash. NAND's greater density makes it ideal for bulk data storage. If programs are stored in NAND flash, they must usually be loaded into RAM before they can be executed, because the NAND page architecture is not well suited to a microprocessor's read/write patterns. NAND flash is widely used in consumer electronic memory cards such as those used in digital cameras. NAND flash devices are also available in discrete form for dense, nonvolatile data storage in a digital system.

NOR flash is discussed here because of its direct microprocessor interface capability. When operating in read-only mode, many NOR flash devices function similarly to EPROMs with a simple asynchronous interface. More advanced flash devices implement high-performance synchronous burst transfer modes that increase their bandwidth for special applications. Most NOR flash chips, however, are used for general processor boot functions where high memory bandwidth is not a main concern. Therefore, an inexpensive asynchronous interface *a la* 27xxx is adequate.

Writing to flash memory is not as simple as presenting new data to the chip and then applying a write enable, as is done with a RAM. Like an EPROM, an already programmed bit must first be erased before it can be reprogrammed. This erasure process takes longer than a simple read access. As Fig. 4.5 shows, the programming and source contacts of each flash bit must be switched to special voltage levels for erasure. Instead of building switches for each individual bit, the complexity of the silicon implementation is reduced by grouping many bits together into blocks. Therefore, a flash device is not erased one bit or byte at a time, but rather a block at a time. Flash chips are segmented into multiple blocks, depending on the particular device and manufacturer. This block architecture is beneficial in that the whole device does not have to be erased, allowing sensitive information to be preserved. A good system design takes the flash block structure into account when deciding where to locate certain pieces of data or sections of software, thereby requiring the erasure of only a limited number of blocks when performing an update of system software or configuration. The block erasure process takes a relatively long time when measured in microprocessor clock cycles. Given that the erase procedure clears an entire range of memory, special algorithms are built into the chips to protect the blocks by requiring a special sequence of flash accesses before the actual erase process is initiated.

Flash chips are not as standard as EPROMs, because different manufacturers have created their own programming algorithms, memory organizations, and pin assignments. Many conventional parallel data-bus devices have part numbers with "28F" or "29F" prefixes. For example, Advanced Micro Devices' flash memory family is the 29Fxxx. Intel's family is the 28Fxxx. Aside from programming differences, the size and organization of blocks within a flash device is a key functional difference that may make one vendor's product better than another for a particular application. Two main attributes of flash chips are uniformity of block size and hardware protection of blocks.

Uniform-block devices divide the memory array into equally sized blocks. Boot-block devices divide the memory array into one or more small *boot blocks* and then divide the remainder of memory into equally sized blocks. Boot-block devices are popular, because the smaller boot blocks can be used to hold the rarely touched software that is used to initialize the system's microprocessor when it first turns on. Boot code is often a small fraction of the system's overall software. Due to its critical nature, boot code is often kept simple to reduce the likelihood of errors. Therefore, boot code is seldom updated. In contrast, other flash ROM contents, such as application code and any application data, may be updated more frequently. Using a boot-block device, a microprocessor's boot code can be stored away into its own block without wasting space and without requiring that it be disturbed during a more general software update. Applications that do not store boot code in flash may not want the complexity of dealing with nonuniform boot blocks and may therefore be better suited to uniform-block devices.

Hardware protection of blocks is important when some blocks hold very sensitive information whose loss could cause permanent damage to the system. A common example of this is boot code stored in a boot block; if the boot code is corrupted, the CPU will fail to initialize properly the next time it is reset. A flash device can implement a low-level protection scheme whereby write/erase operations to certain blocks can be disabled with special voltage levels and data patterns presented to the device.

Examples of real flash devices serve well to explain how this important class of nonvolatile memory functions. Advanced Micro Devices (AMD) manufactures two similar flash devices: the 29LV010B and the 29LV001B. Both devices are 3.3-V, 1-MB, 128k × 8 parts that offer hardware sector protection. The 29LV010B is a uniform-sector device, and the 29LV001B is a boot-sector device. AMD uses the term *sector* instead of block. Both chips have the same basic functional bock diagram shown in Fig. 4.6.

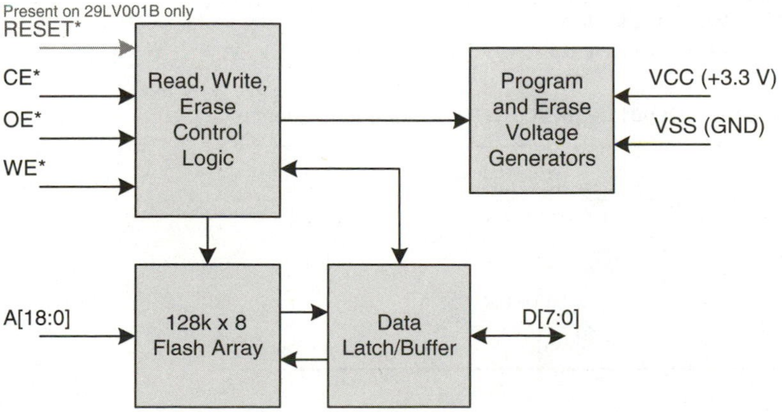

FIGURE 4.6 AMD 29LV010B/29LV001B block diagram.

Modern flash devices require only a single supply voltage and contain on-chip circuitry to create the nonstandard programming and erasure voltages required by the memory array. Control logic determines which block is placed into erase or program mode at any given time as requested by the microprocessor with a predefined flash control algorithm. AMD's algorithm consists of six special write transactions to the flash: two unlock cycles, a setup command, two more unlock cycles, and the specific erase command. This sequence is detailed in Table 4.1. If interrupted, the sequence must be restarted to ensure integrity of the command.

TABLE 4.1 29LV010B/29LV001B Erase Sequence*

Cycle	Write Address	Write Data
1	0x555	0xAA
2	0x2AA	0x55
3	0x555	0x80
4	0x555	0xAA
5	0x2AA	0x55
6	Erase address	Erase command

**Source:* Am29LV001B, Pub#21557, and Am29LV010B, Pub #22140, Advanced Micro Devices, 2000.

For a whole-chip erase, the address/data in cycle 6 is 0x555/0x10. For a single-sector erase, the address/data in cycle 6 is the sector address/0x30. Multiple erase commands may be queued together to reduce the total time spent by the internal control logic erasing its sectors. While executing commands, the data bus is converted into a status communication mechanism. The microprocessor is able to periodically poll the device by reading from any valid address. While the erase is in progress, a value other than 0xFF will be returned. As soon as the erase has completed, the microprocessor will read back 0xFF.

Writes to previously erased flash memory locations are accomplished with a similar technique. For each location to be programmed, a four-cycle program command sequence is performed as shown in Table 4.2. Again, the microprocessor polls for command completion by reading from the device. This time, however, the address polled must be the write address. When the microprocessor reads back the data that it has written, the command is known to have completed.

TABLE 4.2 29LV010B/29LV001B Programming Sequence

Cycle	Write Address	Write Data
1	0x555	0xAA
2	0x2AA	0x55
3	0x555	0xA0
4	Write address	Write data

Other ancillary commands are supported, including device reset and identification operations. The 29LV001B includes a hardware-reset signal in addition to the soft reset command. Identification enables the microprocessor to verify exactly which flash device it is connected to and which sectors have been hardware protected. Identification is useful for a removable flash module that can be built with different parts for specific capacities. Protection status is useful so that software running on the microprocessor can know if it is possible to program certain areas of the memory.

Hardware sector protection is accomplished during the time of system manufacture by applying a higher than normal voltage to designated pins on the flash device using special equipment. The designated pins on the 29LV010B/29LV001B are address bit 9, A9, and the output enable, OE*. These pins are driven to 12 V while the address of the sector to be protected is applied to other address pins. During normal operation, there is no way for 12 V to be driven onto these signals, preventing the protected sectors from being unprotected while in circuit. The exception to this is a feature on the 29LV001B that AMD calls *temporary sector unprotect*. Previously protected sectors can be temporarily unprotected by driving 12 V onto the RESET* pin with specific circuitry for this purpose. Taking advantage of this feature makes it possible to modify the most sensitive areas of the flash by locating a hardware unprotect enable signal in a logic circuit separate from the flash chip itself.

The major difference between the 29LV010B and 29LV001B is their sector organization. The 29LV010B contains 8 uniform sectors of 16 kB each. The 29LV001B contains 10 sectors of nonuniform size. Two variants of the 29LV010B are manufactured by AMD, top and bottom boot sector architectures, and their sector organization is listed in Table 4.3.

TABLE 4.3 29LV010B Sector Organization

Sector Number	Top Boot Sector	Bottom Boot Sector
0	16 kB	8 kB
1	16 kB	4 kB
2	16 kB	4 kB
3	16 kB	16 kB
4	16 kB	16 kB
5	16 kB	16 kB
6	16 kB	16 kB
7	4 kB	16 kB
8	4 kB	16 kB
9	8 kB	16 kB

The reason for these mirrored architectures is that some microprocessors contain reset vectors toward the top of their address space and some toward the bottom. It is a better fit to locate the boot sectors appropriately depending on a system's CPU. As with any complex IC, there are many details relating to the operation of these flash ICs. Refer to AMD's data sheets for more information.

4.4 EEPROM

Electrically erasable programmable ROM, or EEPROM, is flash's predecessor. In fact, some people still refer to flash as "flash EEPROM," because the underlying structures are very similar. EEPROM, sometimes written as E^2PROM, is more expensive to manufacture per bit than EPROM or flash, because individual bytes may be erased randomly without affecting neighboring locations. Because of the complexity and associated cost of making each byte individually erasable, EEPROM is not commonly manufactured in large densities. Instead, it has served as a niche technology for applications that require small quantities of flexible reprogrammable ROM. Common uses for EEPROM are as program memory in small microprocessors with embedded memory and as small nonvolatile memory arrays to hold system configuration information. Serial EEPROM devices can be found in eight-pin DIP or SOIC packages and provide up to several kilobytes of memory. Their serial interface, small size, and low power consumption make them very practical as a means to hold serial numbers, manufacturing information, and configuration data.

Parallel EEPROM devices are still available from manufacturers as the 28xx family. They are pin and function compatible (for reads) with the 27xxx EPROM family that they followed. Some applications requiring reprogrammable nonvolatile memory may be more suited to EEPROM than flash, but flash is a compelling choice, because it is the more mainstream technology with the resultant benefit of further cost reduction.

Serial EEPROMs, however, are quite popular due to their very small size and low power consumption. They can be squeezed into almost any corner of a system to provide small quantities of nonvolatile storage. Microchip Technology is a major manufacturer of serial EEPROMs and offers the 24xx family. Densities range from 16 bytes to several kilobytes. Given that serial interfaces use very few pins, these EEPROMs are manufactured in packages ranging from eight-pin DIPs to five-pin SOT-23s that are smaller than a fingernail. Devices of this sort are designed to minimize system impact rather than for speed. Their power consumption is measured in nanoamps and microamps instead of milliamps, as is the case with standard flash, parallel EEPROM, and EPROM devices.

Microchip's 24LC00 is a 16-byte serial EEPROM with a two-wire serial bus. It requires only four pins: two for power and two for data communication. Like most modern flash devices, the 24LC00 is rated for one million write cycles. When not being accessed, the 24LC00 consumes about 250 nA! When active, it consumes only 500 μA. For added flexibility, the 24LC00 can operate over a variety of supply voltages from 2.5 to 6.0 V. Speed is not a concern here: writes take up to 4 ms to complete, which is not a problem when writing only a few bytes on rare occasions.

4.5 ASYNCHRONOUS SRAM

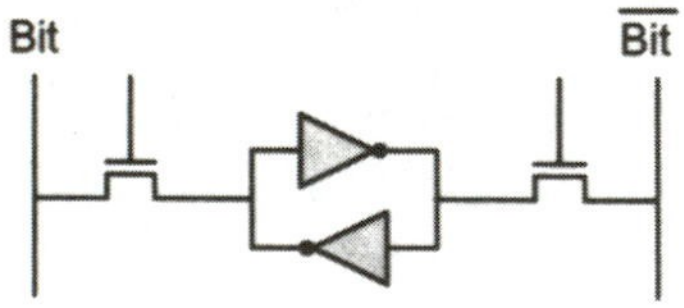

FIGURE 4.7 SRAM bit feedback latch.

Static RAM, or SRAM, is the most basic and easy to use type of volatile memory and is found in almost every computer in one form or another. An SRAM device is conceptually easy to understand, consisting of an array of latches along with control and decode logic to resolve the address that is being read or written at any given time. Each latch is a feedback circuit that traps and maintains a particular logic state. A typical SRAM bit implementation is shown in Fig. 4.7.

An SRAM latch is created by connecting two inverters in a loop. One side of the loop remains stable at the desired logic state, and the other remains stable at the opposite state. Inverters are used rather than noninverting buffers, because an inverter is the simplest logic element to construct. The two pass transistors on either side of the latch enable both writing and reading. When writing, the transistors turn on and force each half of the loop to whatever state is driven on the vertical bit lines. When reading, the transistors also turn on, but the bit lines are sensed rather than driven. Typical

SRAM implementations require six transistors per bit of memory: two transistors for each inverter and the two pass transistors. Some implementations use only a single transistor per inverter, requiring only four transistors per bit.

Discrete asynchronous SRAM devices have been around for decades. In the 1980s, the 6264 and 62256 were manufactured by multiple vendors and used in applications that required simple RAM architectures with relatively quick access times and low power consumption. The 62xxx family is numbered according to its density in kilobits. Hence, the 6264 provides 65,536 bits of RAM arranged as 8k × 8. The 62256 provides 262,144 bits of RAM arranged as 32k × 8. Being manufactured in CMOS technology and not using a clock, these devices consume very little power and draw only microamps when not being accessed.

The 62xxx family pin assignment is virtually identical to that of the 27xxx EPROM family, enabling system designs where either EPROM or SRAM can be substituted into the same location with only a couple of jumpers to set for unique signals such as the program-enable on an EPROM or write-enable on an SRAM. Like an EPROM or basic flash device, asynchronous SRAMs have a simple interface consisting of address, data, chip select, output enable, and write enable. This interface is shown in Fig. 4.8.

Writes are performed whenever the WE* signal is held low. Therefore, one must ensure that the desired address and data are stable before asserting WE* and that WE* is removed while address and data remain stable. Otherwise, the write may corrupt an undesired memory location. Unlike an EPROM, but like flash, the data bus is bidirectional during normal operation. The first two transactions shown are writes as evidenced by the separate assertions of WE* for the duration of address and data stability. As soon as the writes are completed, the microprocessor should release the data bus to the high-impedance state. When OE* is asserted, the SRAM begins driving the data bus and the output reflects the data contents at the locations specified on the address bus.

Asynchronous SRAMs are available with access times of less than 100 ns for inexpensive parts and down to 10 ns for more expensive devices. Access time measures both the maximum delay between a stable read address and its corresponding data and the minimum duration of a write cycle. Their ease of use makes them suitable for small systems where megabytes of memory are not required and where reduced complexity and power consumption are key requirements. Volatile memory doesn't get any simpler than asynchronous SRAM.

Prior to the widespread availability of flash, many computer designs in the 1980s utilized asynchronous SRAM with a battery backup as a means of implementing nonvolatile memory for storing configuration information. Because an idle SRAM draws only microamps of current, a small battery can maintain an SRAM's contents for several years while the main power is turned off. Using SRAM in this manner has two distinct advantages over other technologies: writes are quick and easy, because there are no complex EEPROM or flash programming algorithms, and there is no limit to the number of write cycles performed over the life of the product. The downsides to this approach are a lack of security for protecting valuable configuration information and the need for a battery to

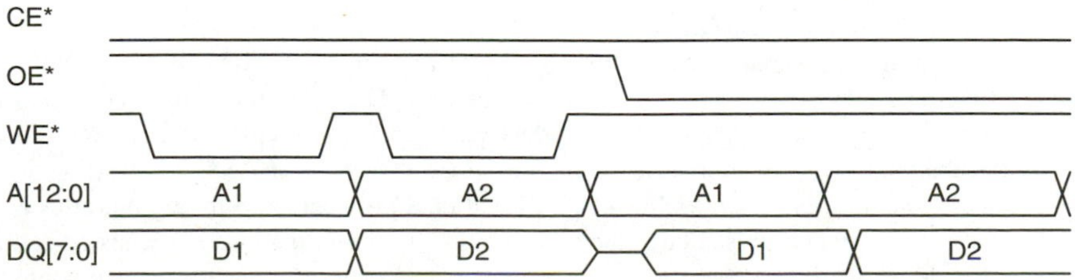

FIGURE 4.8 62xxx SRAM interface.

maintain the memory contents. Requiring a battery increases the complexity of the system and also begs the question of what happens when the battery wears out. In the 1980s, it was common for a PC's BIOS configuration to be stored in battery-backed CMOS SRAM. This is how terms like "the CMOS" and "CMOS setup" entered the lexicon of PC administration.

SRAM is implemented not only as discrete memory chips but is commonly found integrated within other types of chips, including microprocessors. Smaller microprocessors or *microcontrollers* (microprocessors integrated with memory and peripherals on a single chip) often contain a quantity of on-board SRAM. More complex microprocessors may contain on-chip data caches implemented with SRAM.

4.6 ASYNCHRONOUS DRAM

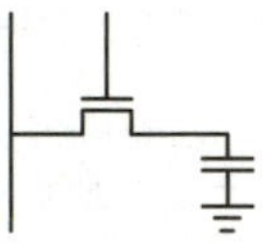

FIGURE 4.9 DRAM bit structure.

SRAM may be the easiest volatile memory to use, but it is not the least expensive in significant densities. Each bit of memory requires between four and six transistors. When millions or billions of bits are required, the complexity of all those transistors becomes substantial. Dynamic RAM, or DRAM, takes advantage of a very simple yet fragile storage component: the capacitor. A capacitor holds an electrical charge for a limited amount of time as the charge gradually drains away. As seen from EPROM and flash devices, capacitors can be made to hold charge almost indefinitely, but the penalty for doing so is significant complexity in modifying the storage element. Volatile memory must be both quick to access and not be subject to write-cycle limitations—both of which are restrictions of nonvolatile memory technologies. When a capacitor is designed to have its charge quickly and easily manipulated, the downside of rapid discharge emerges. A very efficient volatile storage element can be created with a capacitor and a single transistor as shown in Fig. 4.9, but that capacitor loses its contents soon after being charged. This is where the term *dynamic* comes from in DRAM—the memory cell is indeed dynamic under steady-state conditions. The solution to this problem of solid-state amnesia is to periodically refresh, or update, each DRAM bit before it completely loses its charge.

As with SRAM, the pass transistor enables both reading and writing the state of the storage element. However, a single capacitor takes the place of a multitransistor latch. This significant reduction in bit complexity enables much higher densities and lower per-bit costs when memory is implemented in DRAM rather than SRAM. This is why main memory in most computers is implemented using DRAM. The trade-off for cheaper DRAM is a degree of increased complexity in the memory control logic. The number one requirement when using DRAM is periodic refresh to maintain the contents of the memory.

DRAM is implemented as an array of bits with rows and columns as shown in Fig. 4.10. Unlike SRAM, EPROM, and flash, DRAM functionality from an external perspective is closely tied to its row and column organization.

SRAM is accessed by presenting the complete address simultaneously. A DRAM address is presented in two parts: a row and a column address. The row and column addresses are multiplexed onto the same set of address pins to reduce package size and cost. First the row address is loaded, or strobed, into the row address latch via *row address strobe*, or RAS*, followed by the column address with *column address strobe*, or CAS*. Read data propagates to the output after a specified access time. Write data is presented at the same time as the column address, because it is the column strobe that actually triggers the transaction, whether read or write. It is during the column address phase that WE* and OE* take effect.

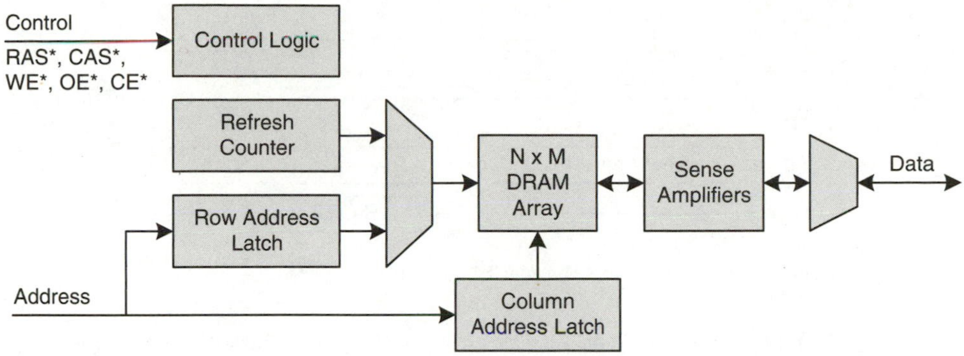

FIGURE 4.10 DRAM architecture.

Sense amplifiers on the chip are necessary to detect the minute charges that are held in the DRAM's capacitors. These amplifiers are also used to assist in refresh operations. It is the memory controller's responsibility to maintain a refresh timer and initiate refresh operations with sufficient frequency to guarantee data integrity. Rather than refreshing each bit separately, an entire row is refreshed at the same time. An internal refresh counter increments after each refresh so that all rows, and therefore all bits, will be cycled through in order. When a refresh begins, the refresh counter enables a particular memory row. The contents of the row are detected by the sense amplifiers and then driven back into the bit array to recharge all the capacitors in that row. Modern DRAMs typically require a complete refresh every 64 ms. A 64-Mb DRAM organized as 8,388,608 words × 8 bits (8 MB) with an internal array size of 4,096 × 2,048 bytes would require 4,096 refresh cycles every 64 ms. Refresh cycles need not be evenly spaced in time but are often spread uniformly for simplicity.

The complexity of performing refresh is well worth the trouble because of the substantial cost and density improvements over SRAM. One downside of DRAM that can only be partially compensated for is its slower access time. A combination of its multiplexed row and column addressing scheme plus its large memory arrays with complex sense and decode logic make DRAM significantly slower than SRAM. Mainstream computing systems deal with this speed problem by implementing SRAM-based cache mechanisms whereby small chunks of memory are prefetched into fast SRAM so that the microprocessor does not have to wait as long for new data that it requests.

Asynchronous DRAM was the prevailing DRAM technology until the late 1990s, when synchronous DRAM, or SDRAM, emerged as the dominant solution to main memory. At its heart, SDRAM works very much like DRAM but with a synchronous bus interface that enables faster memory transactions. It is useful to explore how older asynchronous DRAM works so as to understand SDRAM. SDRAM will be covered in detail later in the book.

RAS* and CAS* are the two main DRAM control signals. They not only tell the DRAM chip which address is currently being asserted, they also initiate refresh cycles and accelerate sequential transactions to increase performance. A basic DRAM read works as shown in Fig. 4.11. CE* and OE* are both assumed to be held active (low) throughout the transaction.

A transaction begins by asserting RAS* to load the row address. The strobes are falling-edge sensitive, meaning that the address is loaded on the falling edge of the strobe, sometime after which the address may change. Asynchronous DRAMs are known for their myriad detailed timing requirements. Every signal's timing relative to itself and other signals is specified in great detail, and these parameters must be obeyed for reliable operation. RAS* is kept low for the duration of the transaction. Assertion of CAS* loads the column address into the DRAM as well as the read or write status

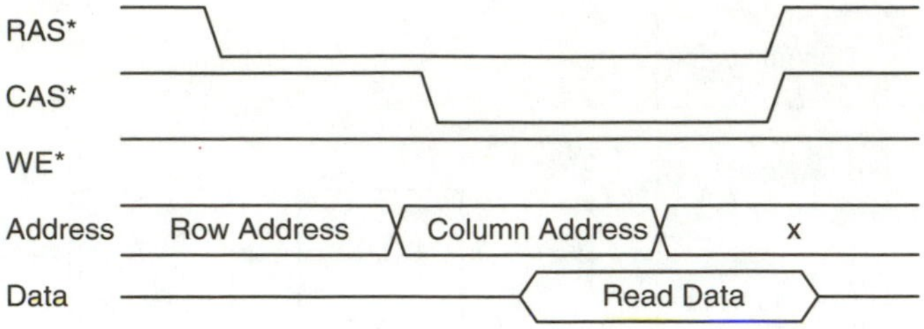

FIGURE 4.11 Basic DRAM read (CE* = 0, OE* = 0).

of the transaction. Some time later, the read data is made available on the data bus. After waiting for a sufficient time for the DRAM to return the read data, the memory controller removes RAS* and CAS* to terminate the transaction.

Basic writes are similar to single reads as shown in Fig. 4.12. Again, CE* is assumed to be held active, and, being a write, OE* is assumed to be held inactive throughout the transaction.

Like a read, the write transaction begins by loading the row address. From this it is apparent that there is no particular link between loading a row address and performing a read or a write. The identity of the transaction is linked to the falling edge of CAS*, when WE* is asserted at about the same time that the column address and write data are asserted. DRAM chips require a certain setup and hold time for these signals around the falling edge of CAS*. Once the timing requirements are met, address can be deasserted prior to the rising edge of CAS*.

A read/write hybrid transaction, called a *read-modify-write*, is also supported to improve the efficiency of the memory subsystem. In a read-modify-write, the microprocessor fetches a word from memory, performs a quick modification to it, and then writes it back as part of the same original transaction. This is an *atomic* operation, because it functions as an indivisible unit and cannot be interrupted. Figure 4.13 shows the timing for the read-modify-write. Note that CAS* is held for a longer period of time, during which the microprocessor may process the read-data before asserting WE* along with the new data to be written.

Original DRAMs were fairly slow. This was partly because older silicon processes limited the decode time of millions of internal addresses. It was also a result of the fact that accessing a single location required a time-consuming sequence of RAS* followed by CAS*. In comparison, an SRAM is quick and easy: assert the address in one step and grab the data. DRAM went through an architectural evolution that replaced the original devices with *fast-page mode* (FPM) devices that allow more efficient accesses to sequential memory locations. FPM DRAMs provide a substantial increase in usable memory bandwidth for the most common DRAM application: CPU memory. These devices take advantage of the tendency of a microprocessor's memory transactions to be sequential in nature.

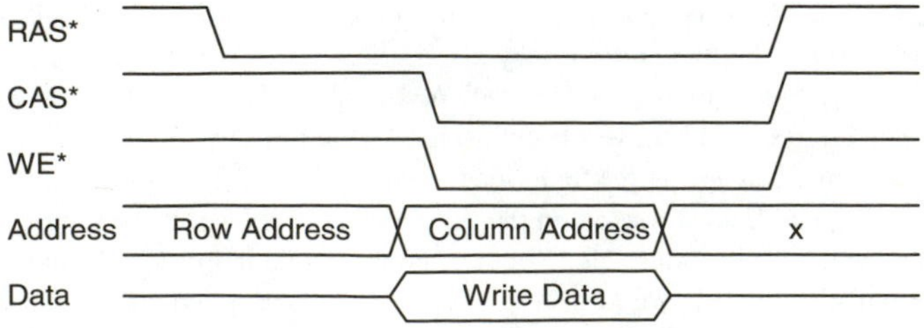

FIGURE 4.12 Basic DRAM write (CE* = 0, OE* = 1).

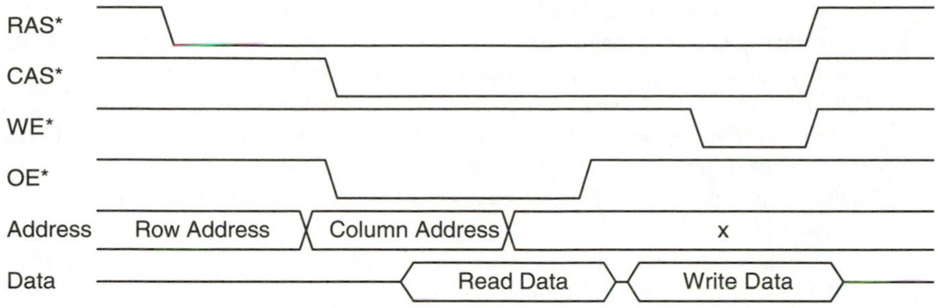

FIGURE 4.13 Read-modify-write transaction.

Software does occasionally branch back and forth in its memory space. Yet, on the whole, software moves through portions of memory in a linear fashion. FPM devices enable a DRAM controller to load a row-address in the normal manner using RAS* and then perform multiple CAS* transactions using the same row-address. Therefore, DRAMs end their transaction cycles with the rising edge of RAS*, because they cannot be sure if more reads or writes are coming until RAS* rises, indicating that the current row-address can be released.

FPM technology, in turn, gave way to *extended-data out* (EDO) devices that extend the time read data is held valid. Unlike its predecessors, an EDO DRAM does not disable the read data when CAS* rises. Instead, it waits until either the transaction is complete (RAS* rises), OE* is deasserted, or until CAS* begins a new page-mode access. While FPM and EDO DRAMs are distinct types of devices, EDO combines the page-mode features of FPM and consequently became more attractive to use. The following timing discussion uses EDO functionality as the example.

Page-mode transactions hold RAS* active and cycle CAS* multiple times to perform reads and writes as shown in Figs. 4.14 and 4.15. Each successive CAS* falling edge loads a new column address and causes either a read or write to be performed. In the read case, EDO's benefit can be properly observed. Rather than read data being removed when CAS* rises, it remains asserted until just after the next falling edge of CAS* or the rising edge of RAS* that terminates the page-mode transaction.

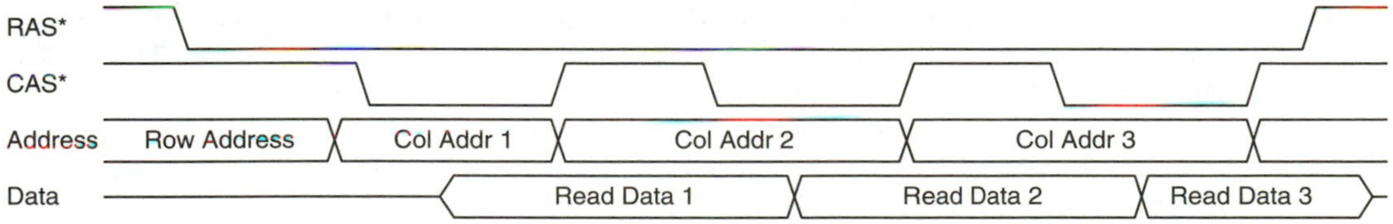

FIGURE 4.14 Page-mode reads.

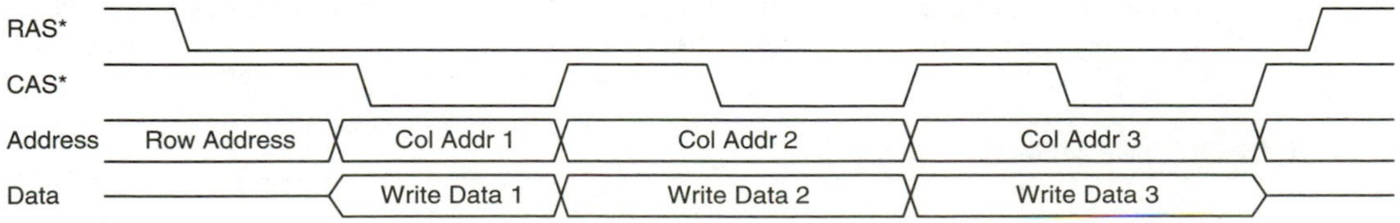

FIGURE 4.15 Page-mode writes.

There are some practical limits to the duration of a page-mode transaction. First, there is an absolute maximum time during which RAS* can remain asserted. The durations of RAS* and CAS* are closely specified to guarantee proper operation of the DRAM. Operating the DRAM with a minimum CAS* cycle time and a maximum RAS* assertion time will yield a practical limitation on the data burst that can be read or written without reloading a new row address. In reality, a common asynchronous DRAM can support over 1,000 back-to-back accesses for a given row-address. DRAM provides its best performance when operated in this manner. The longer the burst, the less overhead is experienced for each byte transferred, because the row-address setup time is amortized across each word in the burst. Cache subsystems on computers help manage the bursty nature of DRAM by swallowing a set of consecutive memory locations into a small SRAM cache where the microprocessor will then have easy access to them later without having to wait for a lengthy DRAM transaction to execute.

The second practical limitation on page-mode transactions, and all DRAM transactions in general, is refresh overhead. The DRAM controller must be smart enough to execute periodic refresh operations at the required frequency. Even if the microprocessor is requesting more data, refresh must take priority to maintain memory integrity. At any given instant in time, a scheduled refresh operation may be delayed slightly to accommodate a CPU request, but not to the point where the controller falls behind and fails to execute the required number of refresh operations. There are a variety ways to initiate a refresh operation, but most involve a so-called *CAS-before-RAS* signaling where the normal sequence of the address strobes is reversed to signal a refresh. Asserting CAS* before RAS* signals the DRAM's internal control logic to perform a row-refresh at the specific row indicated by its internal counter. Following this operation, the refresh counter is incremented in preparation for the next refresh event.

DRAM has numerous advantages over SRAM, but at the price of increased controller complexity and decreased performance in certain applications. DRAMs use multiplexed address buses, which saves pins and enables smaller, less expensive packaging and circuit board wiring. Most DRAMs are manufactured with data bus widths smaller than what is actually used in a computer to save pins. For example, when most computers used 8- or 16-bit data buses, most DRAMs were 1 bit wide. When microprocessors grew to 32 and 64 bit data buses, mainstream DRAMs grew to 4- and then 8-bit widths. This is in contrast to SRAMs, which have generally been offered with wide buses, starting out at 4 bits and then increasing to 72 bits in more modern devices. This width disparity is why most DRAM implementations in computers involve groups of four, eight, or more DRAMs on a single module. In the 1980s, eight 64k × 1 DRAMs created a 64 kB memory array. Today, eight 32M × 8 DRAMs create a 256 MB memory array that is 64 bits wide to suit the high-bandwidth 32- or 64-bit microprocessor in your desktop PC.

A key architectural attribute of DRAM is its inherent preference for sequential transactions and, accordingly, its weakness in handling random single transactions. Because of their dense silicon structures and multiplexed address architecture, DRAMs have evolved to provide low-cost bulk memory best suited to burst transactions. The overhead of starting a burst transaction can be negligible when spread across many individual memory words in a burst. However, applications that are not well suited to long bursts may not do very well with DRAM because of the constant startup penalty involved in fetching 1 word versus 1,000 words. Such applications may work better with SRAM. Planning memory architecture involves making these trade-offs between density/cost and performance.

4.7 MULTIPORT MEMORY

Most memory devices, whether volatile or nonvolatile, contain a single interface through which their contents are accessed. In the context of a basic computer system with a single microprocessor, this *single-port* architecture is well suited. There are some architectures in which multiple microprocessors or logic blocks require access to the same shared pool of memory. A shared pool of memory can be constructed in a couple of ways. First, conventional DRAM or SRAM can be combined with external logic that takes requests from separate entities (e.g., microprocessors) and arbitrates access to one requestor at a time. When the shared memory pool is large, and when simultaneous access by multiple requesters is not required, arbitration can be an efficient mechanism. However, the complexity of arbitration logic may be excessive for small shared-memory pools, and arbitration does not enable simultaneous access. A means of sharing memory without arbitration logic and with simultaneous access capability is to construct a true multiport memory element.

A multiport memory provides simultaneous access to multiple external entities. Each port may be read/write capable, read-only, or write-only depending on the implementation and application. Multiport memories are generally kept relatively small, because their complexity, and hence their cost, increases significantly as additional ports are added, each with its own decode and control logic. Most multiport memories are *dual-port* elements as shown in Fig. 4.16.

A true dual-port memory places no restrictions on either port's transactions at any given time. It is the responsibility of the engineer to ensure that one requester does not conflict with the other. Conflicts arise when one requester writes a memory location while the other is either reading or writing that same location. If a simultaneous read/write occurs, what data does the reader see? Is it the data before or after the write? Likewise, if two writes proceed at the same time, which one wins? While these riddles could be worked out for specific applications with custom logic, it is safer not to worry about such corner cases. Instead, the system design should avoid such conflicts unless there is a strong reason to the contrary.

One common application of a dual-port memory is sharing information between two microprocessors as shown in Fig. 4.17. A dual-port memory sits between the microprocessors and can be partitioned into a separate message bin, or memory area, for each side. Bin A contains messages written by CPU A and read by CPU B. Bin B contains messages written by CPU B and read by CPU A.

Notification of a waiting message is accomplished via a CPU interrupt, thereby releasing the CPUs from having to constantly poll the memory as they wait for messages to arrive. The entire process might work as follows:

1. CPU A writes a message for CPU B into Bin A.
2. CPU A asserts an interrupt to CPU B indicating the a message is waiting in Bin A.
3. CPU B reads the message in Bin A.
4. CPU B acknowledges the interrupt from CPU A.
5. CPU A releases the interrupt to CPU B.

An implementation like this prevents dual-port memory conflicts because one CPU will not read a message before it has been fully written by the other CPU and neither CPU writes to both bins.

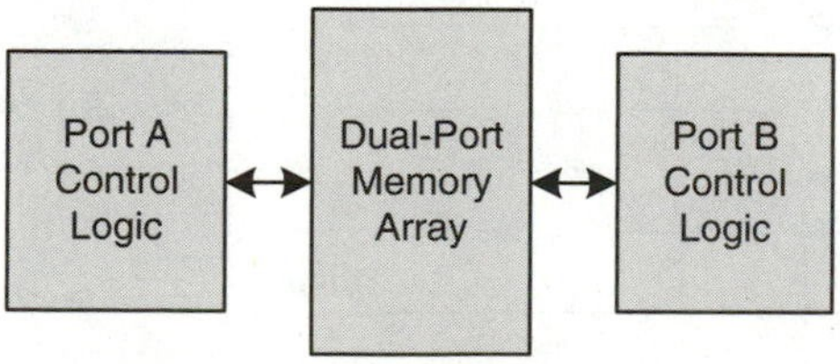

FIGURE 4.16 Dual-port memory.

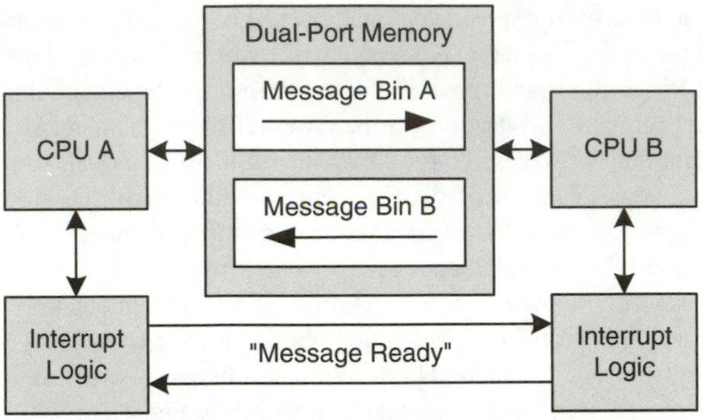

FIGURE 4.17 Dual microprocessor message passing architecture.

4.8 THE FIFO

The memory devices discussed thus far are essentially linear arrays of bits surrounded by a minimal quantity of interface logic to move bits between the port(s) and the array. *First-in-first-out* (FIFO) memories are special-purpose devices that implement a basic queue structure that has broad application in computer and communications architecture. Unlike other memory devices, a typical FIFO has two unidirectional ports without address inputs: one for writing and another for reading. As the name implies, the first data written is the first read, and the last data written is the last read. A FIFO is not a random access memory but a sequential access memory. Therefore, unlike a conventional memory, once a data element has been read once, it cannot be read again, because the next read will return the next data element written to the FIFO. By their nature, FIFOs are subject to *overflow* and *underflow* conditions. Their finite size, often referred to as *depth*, means that they can fill up if reads do not occur to empty data that has already been written. An overflow occurs when an attempt is made to write new data to a full FIFO. Similarly, an empty FIFO has no data to provide on a read request, which results in an underflow.

A FIFO is created by surrounding a dual-port memory array—generally SRAM, but DRAM could be made to work as well for certain applications—with a write pointer, a read pointer, and control logic as shown in Fig. 4.18.

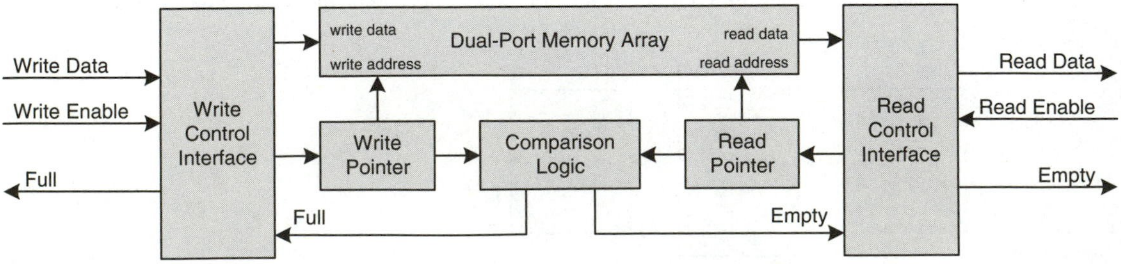

FIGURE 4.18 Basic FIFO architecture.

A FIFO is not addressed in a linear fashion; rather, it is made to form a continuous ring of memory that is addressed by the two internal pointers. The fullness of the FIFO is determined not by the absolute values of the pointers but by their relative values. An empty FIFO begins with its read and write pointers set to the same value. As entries are written, the write pointer increments. As entries are read, the read pointer increments as well. If the read pointer ever catches up to the write pointer such that the two match, the FIFO is empty again. If the read pointer fails to advance, the write pointer will eventually wrap around the end of the memory array and become equal to the read pointer. At this point, the FIFO is full and cannot accept any more data until reading resumes. Full and empty flags are generated by the FIFO to provide status to the writing and reading logic. Some FIFOs contain more detailed fullness status, such as signals that represent programmable fullness thresholds.

The interfaces of a FIFO can be asynchronous (no clock) or synchronous (with a clock). If synchronous, the two ports can be designed to operate with a common clock or different clocks. Although older asynchronous FIFOs are still manufactured, synchronous FIFOs are now more common. Synchronous FIFOs have the advantage of improved interface timing, because flops placed at a device's inputs and outputs reduce timing requirements to the familiar setup, hold, and clock-to-out specifications. Without such a registered interface, timing specifications become a function of the device's internal logic paths.

One common role that a FIFO fills is in clock domain crossing. In such an application, there is a need to communicate a series of data values from a block of logic operating on one clock to another block operating on a different clock. Exchanging data between clock domains requires special attention, because there is normally no way to perform a conventional timing analysis across two different clocks to guarantee adequate setup and hold times at the destination flops. Either an asynchronous FIFO or a dual-clock synchronous FIFO can be used to solve this problem, as shown in Fig. 4.19.

The dual-port memory at the heart of the FIFO is an asynchronous element that can be accessed by the logic operating in either clock domain. A dual-clock synchronous FIFO is designed to handle arbitrary differences in the clocks between the two halves of the device. When one or more bytes are written on clock A, the write-pointer information is carried safely across to the clock B domain within the FIFO via inter-clock domain synchronization logic. This enables the read-control interface to determine that there is data waiting to be read. Logic on clock B can read this data long after it has been safely written into the memory array and allowed to settle to a stable state.

Another common application for a FIFO is rate matching where a particular data source is bursty and the data consumer accepts data at a more regular rate. One example is a situation where a sequence of data is stored in DRAM and needs to be read out and sent over a communications interface one byte at a time. The DRAM is shared with a CPU that competes with the communications interface for memory bandwidth. It is known that DRAMs are most efficient when operated in a page-mode burst. Therefore, rather than perform a complete read-transaction each time a single byte

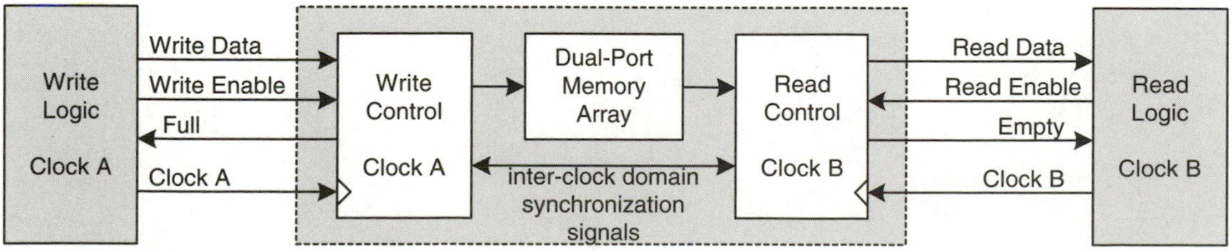

FIGURE 4.19 Clock domain crossing with synchronous FIFO.

is needed for the communications interface, a burst of data can be read and stored in a FIFO. Each time the interface is ready for a new byte, it reads it from the FIFO. In this case, only a single-clock FIFO is required, because these devices operate on a common clock domain. To keep this process running smoothly, control logic is needed to watch the state of the FIFO and perform a new burst read from DRAM when the FIFO begins to run low on data. This scheme is illustrated in Fig. 4.20.

For data-rate matching to work properly, the average bandwidth over time of the input and output ports of the FIFO must be equal, because FIFO capacity is finite. If data is continuously written faster than it can be read, the FIFO will eventually overflow and lose data. Conversely, if data is continuously read faster than it can be written, the FIFO will underflow and cause invalid bytes to be inserted into the outgoing data stream. The depth of a FIFO indicates how large a read/write rate disparity can be tolerated without data loss. This disparity is expressed as the product of rate mismatch and time. A small mismatch can be tolerated for a longer time, and a greater rate disparity can be tolerated for a shorter time.

In the rate-matching example, a large rate disparity of brief duration is balanced by a small rate disparity of longer duration. When the DRAM is read, a burst of data is suddenly written into the FIFO, creating a temporarily large rate disparity. Over time, the communications interface reads one byte at a time while no writes are taking place, thereby compensating with a small disparity over time.

DRAM reads to refill the FIFO must be carefully timed to simultaneously prevent overflow and underflow conditions. A threshold of FIFO fullness needs to be established below which a DRAM read is triggered. This threshold must guarantee that there is sufficient space available in the FIFO to accept a full DRAM burst, avoiding an overflow. It must also guarantee that under the worst-case response time of the DRAM, enough data exists in the FIFO to satisfy the communications interface, avoiding an underflow. In most systems, the time between issuing a DRAM read request and actually getting the data is variable. This variability is due to contention with other requesters (e.g., the CPU) and waiting for overhead operations (e.g., refresh) to complete.

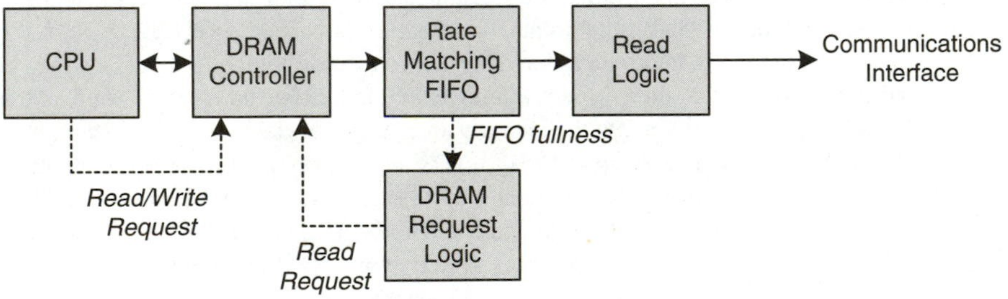

FIGURE 4.20 Synchronous FIFO application: data rate matching.

CHAPTER 5
Serial Communications

Serial communication interfaces are commonly used to exchange data with other computers. Serial interfaces are ubiquitous, because they are economical to implement over long distances as a result of their requirement of relatively few wires. Many types of serial interfaces have been developed, with speeds ranging to billions of bits per second. Regardless of the bit rate, serial communication interfaces share many common traits. This chapter introduces the fundamentals of serial communication in the context of popular data links such as RS-232 and RS-485 in which bandwidths and components lend themselves to basic circuit fabrication techniques.

The chapter first deals with the basic parallel-to-serial-to-parallel conversion process that is at the heart of all serial communication. Wide buses must be serialized at the transmitter and reconstructed at the receiver. Techniques for accomplishing this vary with the specific type of data link, but basic concepts of framing and error detection are universal.

Two widely deployed point-to-point serial communication standards, RS-232 and RS-422, are presented, along with the standard ASCII character set, to see how theory meets practice. Standards are important to communications in general because of the need to connect disparate equipment. ASCII is one of the most fundamental data representation formats with global recognition. RS-232 has traditionally been found in many digital systems, because it is a reliable standard. Understanding RS-232, its relative RS-422, and ASCII enables an engineer to design a communication interface that can work with an almost infinite range of complementary equipment ranging from computers to modems to off-the-shelf peripherals.

Systems may require more advanced communication schemes to enable data exchange between many nodes. Networks enable such communication and can range in complexity according to an application's requirements. Networking adds a new set of fundamental concepts on top of basic serial communication. Topics including network topologies and packet formats are presented to explain how networks function at a basic hardware and software level. Once networking fundamentals have been discussed, the RS-485 standard is introduced to show how a simple and fully functional network can be constructed. A complete network design example using RS-485 is offered with explanations of why various design points are included and how they contribute to the network's overall operation.

The chapter closes with a presentation of small-scale networking employed within a digital system to economically connect peripherals to a microprocessor. Interchip networks are of such narrow scope that they are usually not referred to as networks, but they can possess many fundamental properties of a larger network. Peripherals with low microprocessor bandwidth requirements can be connected using a simple serial interface consisting of just a few wires, as compared to the full complexity of a parallel bus.

5.1 SERIAL VS. PARALLEL COMMUNICATION

Most logical operations and data processing occur in parallel on multiple bits simultaneously. Microprocessors, for example, have wide data buses to increase throughput. With wide buses comes a requirement for more wires to connect the logical elements in a system. The interconnection penalty increases as distances increase. Within a chip, the penalty is small, and wide buses are common. Implementing wide buses on a circuit board is also common because of the relatively short distances involved.

The economics and technical context of interconnect changes as soon as the distances grow from centimeters to meters to kilometers. Communication is primarily concerned with transporting data from one location to another rather than processing that information as it is carried on a wire. With distance comes the expensive problem of stringing a continuous wire between two locations. Whether the wire is threaded through a conduit between floors in an office, buried under the street between buildings, or virtually constructed via radio transmission to a satellite, the cost and complexity of connecting multiple wires is many orders of magnitude greater than on a circuit board. Serial communication is well suited to long distances, because fewer wires are used as compared to a parallel bus. A serial data link implies a single-wire medium, but there can be multiwire serial links as well.

Figure 5.1 illustrates several logical components in a serial data link. At either end are the sources and consumers of the data that operate using a parallel bus. A *transceiver* converts between a parallel bus and a serial stream and handles any link-level timing necessary to properly send and receive data. A *transducer*, or *modulator* in wireless links, converts between the medium's electromagnetic signaling characteristics and the transceiver's logic-level signals. Finally, a conductive path joins the two transducers. This path can be copper wire, glass fiber optic cable, or free space. These logical components may be integrated in arbitrary physical configurations in different implementations, so not all serial links will consist of three specific discrete pieces. Simple links may have fewer pieces, and complex links may have more.

The total cost of a data link is the sum of the cost of the transceiver/transducer subsystems at each end and the cost of the physical medium itself. A serial port on a desktop computer is inexpensive because of its relatively simple electronic circuits and because the medium over which it communicates, a short copper wire, is fairly cheap. In contrast, a satellite link is very expensive as a result of the greater complexity of the ground-based transmission equipment, the high cost of the satellite itself, and the licensing costs of using the public airwaves.

If only one bit is transferred per clock cycle in a serial link, it follows that either the serial bit clock has to be substantially faster than the parallel bus, or the link's *bandwidth* will be significantly

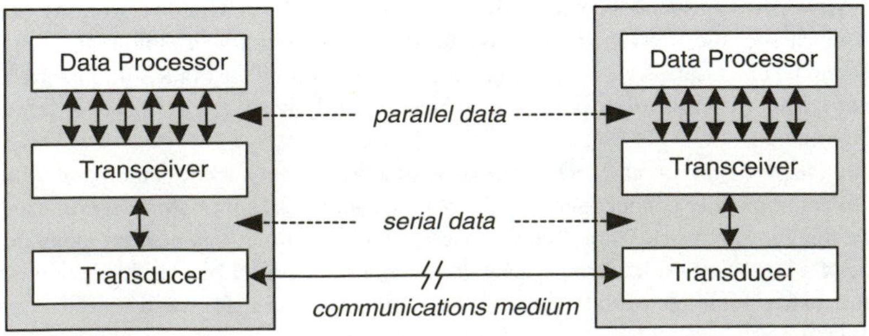

FIGURE 5.1 Components of a serial data link.

below that of the parallel bus. *Bandwidth* in a communication context refers to the capacity of the communications channel, often expressed either in bits-per-second (bps) or bytes-per-second (Bps). Serial links are available in a broad spectrum of bandwidths, from thousands of bits per second (kbps) to billions of bits per second (Gbps) and are stretching toward trillions of bits per second (Tbps)!

Most implementations in the kbps range involve applications where relatively small quantities of data are exchanged, so the cost of deploying an advanced data link is not justified. These serial links are able to run at low frequencies (several hundred kilohertz and below) and therefore do not require complex circuitry. Of course, some low-bandwidth data links can be very expensive if the medium over which they operate presents extreme technical difficulties, such as communicating across interplanetary distances. Implementations in the Gbps range serve applications such as high-end computer networks where huge volumes of data are carried. Such links are run at gigahertz frequencies and are relatively costly due to this high level of performance. Gigahertz serial transfer rates do not translate into similar logic clock frequencies. When a transceiver converts a serial data stream into a parallel bus, it contains the very high frequency complexity within itself. A 1-Gbps link requires only a 31.25 MHz clock when using a 32-bit data path.

5.2 THE UART

The *universal asynchronous receiver/transmitter* (UART) is a basic transceiver element that serializes a parallel bus when transmitting and deserializes the incoming stream when receiving. In addition to bus-width conversion, the UART also handles overhead and synchronization functions required to transport data. Data bits cannot simply be serialized onto a wire without some additional information to delineate the start and end of each unit of data. This delineation is called *framing*. The receiver must be able to recognize the start of a byte so that it can synchronize its shift registers and receive logic to properly capture the data. Basic framing is accomplished with a *start bit* that is assigned a logic state opposite to that of the transmission medium's idle state, often logic 1 for historical reasons. When no data is being sent, the transmission medium, typically a wire, may be driven to logic 1. A logic 0 start bit signals the receiver that data is on the way. The receiving UART must be configured to handle the same number of data bits sent by the transmitter. Either seven or eight data bits are supported by most UARTs. After seven or eight data bits have been captured following the start bit, the UART knows that the data unit has completed and it can resume waiting for a new start bit. One or more *stop bits* follow to provide a minimum delay between successive data units so that the receiver can complete processing of the current datum before receiving the next one.

Many UARTs also support some form of error detection in the form of a *parity bit*. The parity bit is the XOR of the data bits and is sent along with data so that it can be recalculated and verified at the receiver. Error detection is considered more important on a long-distance data link, as compared to on a circuit board, because errors are more prone over longer distances. A parity bit is added to each data unit, most often each byte, that tells the receiver if an odd or even number of 1s are in the data word. The receiver and transmitter must be configured to agree on whether even or odd parity is being implemented. Even parity is calculated by XORing all data bits, and odd parity is calculated by inverting even parity. The result is that, for even parity, the parity bit will be set if there are an odd number of 1s in the byte. Conversely, the parity bit will be cleared if there are an odd number of 1s present. Odd parity is just the opposite, as shown in Fig. 5.2.

Handshaking is another common feature of UARTs. Handshaking, also called *flow control*, is the general process whereby two ends of a data link confirm that each is ready to exchange data before the actual exchange occurs. The process can use hardware or software signaling. Hardware hand-

Data	Even Parity	Odd Parity
0xA0 = 10100000	0	1

FIGURE 5.2 Odd and even parity.

shaking involves a receiver driving a ready signal to the transmitter. The transmitter sends data only when the receiver signals that it is ready. UARTs may support hardware handshaking. Any software handshaking is the responsibility of the UART control program.

Software handshaking works by transmitting special binary codes that either pause or resume the opposite end as it sends data. *XON/XOFF* handshaking is a common means of implementing software flow control. When one end of the link is ready to accept data, it transmits a standard character called XON (0x11) to the opposite device. When the receiver has filled a buffer and is unable to accept more data, an XOFF character (0x13) is transmitted. It is by good behavior that most flow control schemes work: the device that receives an XOFF must respect the signal and pause its transmission until an XON is received. It is not uncommon to see an XON/XOFF setting in certain serial terminal configurations.

A generic UART is shown in Fig. 5.3. The UART is divided into three basic sections: CPU interface, transmitter, and receiver. The CPU interface contains various registers to configure parity, bit rate, handshaking, and interrupts. UARTs usually provide three parity options: none, even, and odd. Bit rate is selectable well by programming an internal counter to arbitrarily divide an external reference clock. The range of usable bit clocks may be from several hundred bits per second to over 100 kbps.

Interrupts are used to inform the CPU when a new byte has been received and when a new byte is ready to be transmitted. This saves the CPU from having to constantly poll the UART's status registers for this information. However, UARTs provide status bits to aid in interrupt status reporting, so a simple serial driver program could operate by polling rather than implementing an interrupt service routine. Aside from general control and status registers, the CPU interface provides access to transmit and receive buffers so that data can be queued for transmission and retrieved upon arrival. Depending on the UART, these buffers may be only one byte each, or they may be several bytes

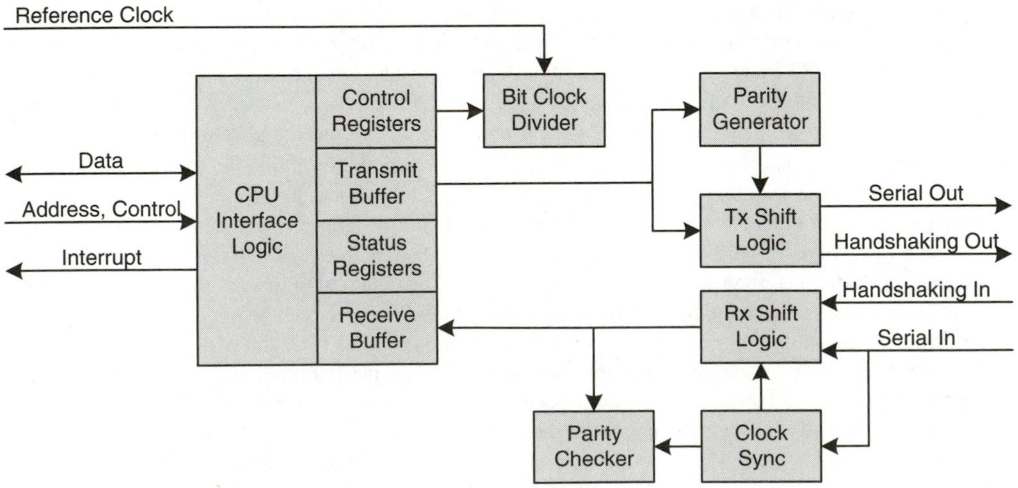

FIGURE 5.3 Generic UART block diagram.

implemented as a small FIFO. Typically, these serial ports run slow enough to not require deep buffers, because even a slow CPU can easily respond to a transmit/receive event before the data link underruns the transmit buffer or overruns the receive buffer.

The transmit section implements a parallel-to-serial shift register, parity generator, and framing logic. UARTs support framing with a start bit and one or two stop bits where the start bit is a logic 0 and stop bits are logic 1s. It is also common to transmit data LSB first. With various permutations of framing options, parity protection, and seven or eight data bits, standard configuration notation is of the form <parity:N/E/O>-<width:8/7>-<stop-bits:1/2>. For example, N-8-1 represents no parity, 8 data bits, and 1 stop bit. E-8-2 represents even parity, 8 data bits, and 2 stop bits. To help understand the format of bytes transmitted by a UART, consider Fig. 5.4. Here, two data bytes are transmitted: 0xA0 and 0x67. Keep in mind that the LSB is transmitted first.

Receiving the serial data is a bit trickier than transmitting it, because there is no clock accompanying the data with which the data can be sampled. This is where the asynchronous terminology in the UART acronym comes from. The receiver contains a clock synchronization circuit that detects the start-bit and establishes a timing reference point from which all subsequent bits in the byte will be sampled. This reference point is created using a higher-frequency receive clock. Rather than running the receiver at 1x the bit rate, it may be run at 16x the bit rate. Now the receive logic can decompose a bit into 16 time units and slide a 16-clock window according to where the start bit is observed. It is advantageous to sample each subsequent bit halfway through its validity window for maximum timing margin on either side of the sampling event. This allows maximum flexibility for settling time around the edges of the electrical pulse that defines each bit.

Consider the waveform in Fig. 5.5. When the start bit is detected, the sampling window is reset, and a sampling point halfway through is established. Subsequent bits can have degraded rising and falling edges without causing the receiver to sample an incorrect logic level.

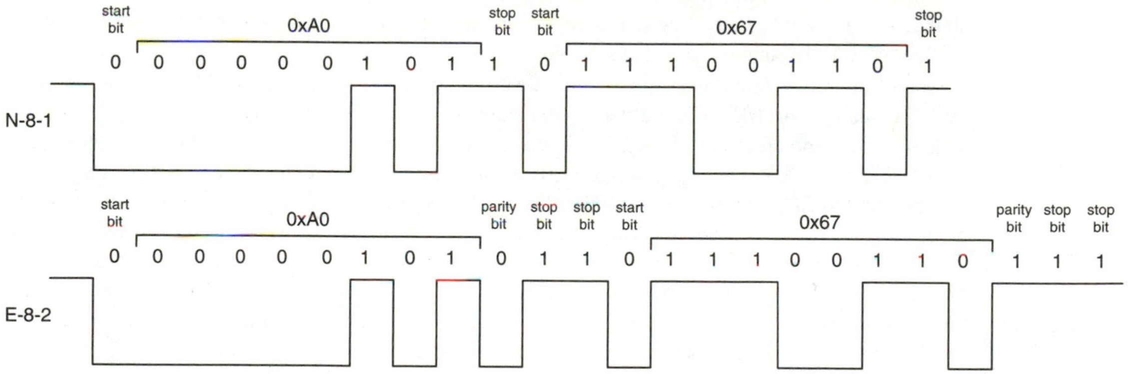

FIGURE 5.4 Common byte framing formats.

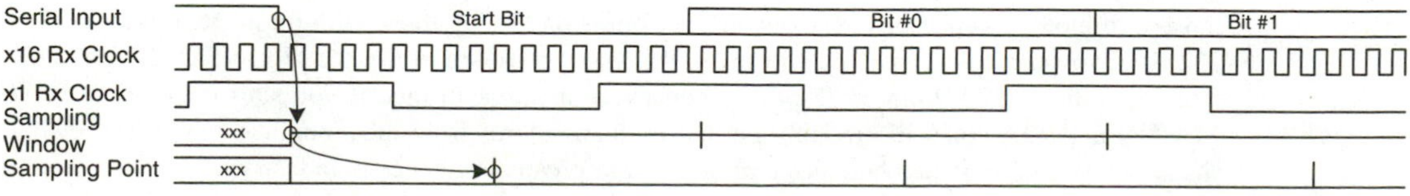

FIGURE 5.5 UART receive clock synchronization.

5.3 ASCII DATA REPRESENTATION

Successful communication requires standardized data representation so that people and computers around the world can share the same information. Alphanumeric characters are represented by a seven-bit standard representation known as the *American Standard Code for Information Interchange*, or ASCII. ASCII also includes punctuation marks and invisible control codes used to help in the display and transfer of data. ASCII was first published in 1968 by the *American National Standards Institute*, or ANSI. The original ASCII standard lacked provisions for many commonly used grammatical symbols in languages other than English. Since 1968, there have been many extensions to ASCII that have varying support throughout the world according to the prevalent language in each country. In the United States, an eight-bit ASCII variant is commonly supported that adds graphical symbols and some of the more common foreign language punctuation symbols. The original seven-bit ANSI standard ASCII mapping is shown in Table 5.1. The mappings below 0x20 are invisible control codes such as tab (0x09), carriage return (0x0D), and line-feed (0x0A). Some of the control codes are not in widespread use anymore.

5.4 RS-232

Aside from a common data representation format, communication signaling such as framing or error detection also requires standardization so that equipment manufactured by different companies can exchange information. When one begins discussing communications, an unstoppable journey into the sometimes mysterious world of industry standards begins. Navigating these standards can be tricky because of subtle differences in terminology between related standards and the everyday jargon to which the engineering community has grown accustomed. Standards are living documents that are periodically updated, revised, or replaced. This shifting base of documentation can add other challenges to fully complying with a standard.

One of the most ubiquitous serial communications schemes in use is defined by the *RS-232* family of standards. Most UARTs are designed specifically to support RS-232. Standards purists may balk at the common reference to RS-232 in the modern context, for several reasons. First, the original RS-232 document has long since been superseded by multiple revisions. Second, its name was changed first to EIA-232, then to EIA/TIA-232. And third, RS-232 is but one of a set of related standards that address asynchronous serial communications. These standards have been developed under the auspices of the Electronics Industry Alliance (formerly the Electronics Industry Association) and Telecommunications Industry Association. Technically, EIA/TIA-232 (first introduced in 1962 as RS-232) standardizes the 25-pin D-subminiature (DB25) connector and pin assignment along with an obsolete electrical specification that had limited range. EIA/TIA-423 standardizes the modern electrical characteristics that enable communication at speeds up to 100 kbps over short distances (10 m). EIA/TIA-574 standardizes the popular nine-pin DE9 connector that is used on most new "RS-232" equipped devices. These days, when most people talk about an RS-232 port, they are referring to the overall RS-232 family of related serial interfaces. In fairness to standards purists, this loose terminology is partially responsible for confusion among those who implement and use RS-232. From a practical perspective, however, it is most common to use the term RS-232 with additional qualifiers (e.g., 9-pin or 25-pin) to convey your point. In fact, if you start mentioning EIA/TIA-574 and 423, you will probably be met by blank stares from most engineers. This somewhat shady practice is continued here because of its widespread acceptance in industry.

RS-232 specifies that the least-significant bit of a byte is transmitted first and is framed by a single start bit and one or two stop bits. Common RS-232 data rates are known to many computer users.

TABLE 5.1 Seven-bit ASCII Character Mapping

Decimal	Hex	Value	Decimal	Hex	Value	Decimal	Hex	Value	Decimal	Hex	Value
0	0x00	NUL	32	0x20	SP	64	0x40	@	96	0x60	`
1	0x01	SOH	33	0x21	!	65	0x41	A	97	0x61	a
2	0x02	STX	34	0x22	"	66	0x42	B	98	0x62	b
3	0x03	ETX	35	0x23	#	67	0x43	C	99	0x63	c
4	0x04	EOT	36	0x24	$	68	0x44	D	100	0x64	d
5	0x05	ENQ	37	0x25	%	69	0x45	E	101	0x65	e
6	0x06	ACK	38	0x26	&	70	0x46	F	102	0x66	f
7	0x07	BEL	39	0x27	'	71	0x47	G	103	0x67	g
8	0x08	BS	40	0x28	(	72	0x48	H	104	0x68	h
9	0x09	HT	41	0x29	)	73	0x49	I	105	0x69	i
10	0x0A	LF	42	0x2A	*	74	0x4A	J	106	0x6A	j
11	0x0B	VT	43	0x2B	+	75	0x4B	K	107	0x6B	k
12	0x0C	FF	44	0x2C	,	76	0x4C	L	108	0x6C	l
13	0x0D	CR	45	0x2D	-	77	0x4D	M	109	0x6D	m
14	0x0E	SO	46	0x2E	.	78	0x4E	N	110	0x6E	n
15	0x0F	SI	47	0x2F	/	79	0x4F	O	111	0x6F	o
16	0x10	DLE	48	0x30	0	80	0x50	P	112	0x70	p
17	0x11	DC1/XON	49	0x31	1	81	0x51	Q	113	0x71	q
18	0x12	DC2	50	0x32	2	82	0x52	R	114	0x72	r
19	0x13	DC3/XOFF	51	0x33	3	83	0x53	S	115	0x73	s
20	0x14	DC4	52	0x34	4	84	0x54	T	116	0x74	t
21	0x15	NAK	53	0x35	5	85	0x55	U	117	0x75	u
22	0x16	SYN	54	0x36	6	86	0x56	V	118	0x76	v
23	0x17	ETB	55	0x37	7	87	0x57	W	119	0x77	w
24	0x18	CAN	56	0x38	8	88	0x58	X	120	0x78	x
25	0x19	EM	57	0x39	9	89	0x59	Y	121	0x79	y
26	0x1A	SUB	58	0x3A	:	90	0x5A	Z	122	0x7A	z
27	0x1B	ESC	59	0x3B	;	91	0x5B	[	123	0x7B	{
28	0x1C	FS	60	0x3C	<	92	0x5C	\	124	0x7C	|
29	0x1D	GS	61	0x3D	=	93	0x5D	]	125	0x7D	}
30	0x1E	RS	62	0x3E	>	94	0x5E	^	126	0x7E	~
31	0x1F	US	63	0x3F	?	95	0x5F	_	127	0x7F	DEL

Standard bit rates are 2^N multiples of 300 bps. In the 1970s, 300 bps serial links were common. During the 1980s, links went from 1,200 to 2,400, to 9,600 bps. RS-232 data links now operate at speeds from 19.2 to 153.6 kbps. Standard RS-232 bit rates are typically divided down from reference clocks such as 1.843, 3.6864, 6.144, and 11.0592 MHz. This explains why many microprocessors operate at oddball frequencies instead of even speeds such as 5, 10, or 12 MHz.

RS-232 defines signals from two different perspectives: *data communications equipment* (DCE) and *data terminal equipment* (DTE). DCE/DTE terminology evolved in the early days of computing when the common configuration was to have a dumb terminal attached to a modem of some sort to enable communication with a mainframe computer in the next room or building. A person would sit at the DTE and communicate via the DCE. Therefore, in the early 1960s, it made perfect sense to create a communication standard that specifically addressed this common configuration. By defining a set of DTE and DCE signals, not only could terminal and modem engineers design compatible systems, but cabling would be very simple: just wire each DTE signal straight through to each DCE signal. To further reduce confusion, the DTE was specified as a male DB-25 and the DCE as a female DB-25. Aside from transmit and receive data, hardware handshaking signals distinguish DCE from DTE. Some signals are specific to modems such as *carrier detect* and *ring indicator* and are still used today in many modem applications.

The principle behind RS-232 hardware handshaking is fairly simple: the DTE and DCE indicate their operational status and ability to accept data. The four main handshaking signals are *request to send* (RTS), *clear to send* (CTS), *data terminal ready* (DTR), and *data set ready* (DSR). DTR/DSR enable the DTE and DCE to signal that they are both operational. The DTE asserts DTR, which is sensed by the DCE and vice versa with DSR. RTS/CTS enable actual data transfer. RTS is asserted by the DTE to signal that the DCE can send it data. CTS is asserted by the DCE to signal the DTE that it can send data. In the case of a modem, carrier detect is asserted to signal an active connection, and ring indicator is asserted when the telephone line rings, signaling that the DTE can instruct the modem to answer the phone.

In a *null-modem* configuration, two DTEs are connected, and each considers DTR and RTS outputs and DSR and CTS inputs. This is solved by swapping DTR/DSR and RTS/CTS so that one DTE's DTR drives the other's DSR, and so on. The unidirectional carrier detect is also connected to the DTR signal at the other end (DSR at the local end) to provide positive "carrier detect" when the terminal ready signal is asserted.

Table 5.2 lists the full set of RS-232 signals with the convention that signals are named relative to the DTE. Most of the original 25 defined RS-232 signals are rarely used, as evidenced by the popularity of the smaller DE9 connector. Furthermore, a minimal RS-232 serial link can be implemented with only three wires: transmit, receive, and ground. In more recent times, the DTE/DCE distinction has created confusion in more than one engineering department, because the definitions of terminal and modem do not always hold in the more varied modern digital systems context. Often, all RS-232 ports are configured as DTE, and special crossover, or null-modem, cables are used to properly connect two DTEs. While varying subsets of the DTE pin assignment can be found in many systems, there is still a place for the original DTE/DCE configuration. It is rare, however, to find the DB25 pins that are not implemented in the DE9 actually put to use.

Not all RS-232 interfaces are configured for hardware handshaking. Some may ignore these signals entirely, and others require that these signals be tied off to the appropriate logic levels so that neither end of the link gets confused and believes that the other is preventing it from sending data. Using a software flow control mechanism can eliminate the need for the aforementioned hardware handshaking signals and reduce the RS-232 link to its three basic wires: transmit, receive, and ground. These many permutations of DTE/DCE and various degrees of handshaking are what cause substantial grief to many engineers and technicians as they build and set up RS-232 equipment. There is a healthy industry built around the common RS-232 configuration problems. *Breakout*

TABLE 5.2 RS-232 DTE Pin Assignments

DB25 DTE	DE9 DTE	Signal	Direction: DTE/DCE	Description
1	–	Shield	⇔	Shield/chassis ground
2	3	TXD	⇒	Transmit data
3	2	RXD	⇐	Receive data
4	7	RTS	⇒	Request to send
5	8	CTS	⇐	Clear to send
6	6	DSR	⇐	Data set ready
7	5	Ground	⇔	Signal ground
8	1	DCD	⇐	Data carrier detect
9	–	+V	⇔	Power
10	–	–V	⇔	Power return
11	–			Unused
12	–	SCF	⇐	Secondary line detect
13	–	SCB	⇐	Secondary CTS
14	–	SBA	⇒	Secondary TXD
15	–	DB	⇐	DCE element timing
16	–	SBB	⇐	Secondary RXD
17	–	DD	⇐	Receiver element timing
18	–			Unused/local-loopback
19	–	SCA	⇒	Secondary RTS
20	4	DTR	⇒	Data terminal ready
21	–	CQ	⇐	Signal quality detect
22	9	RI	⇐	Ring indicator
23	–	CH/CI	⇔	Data rate detect
24	–	DA	⇒	Transmitter element timing
25	–			Unused/test-mode

boxes can be purchased that consist of jumper wires, switches, and LEDs to help troubleshoot RS-232 connectivity problems by reconfiguring interfaces on the fly as the LEDs indicate which signals are active at any given moment. As a result of the male/female gender differences of various DB25/DE9 connectors, there are often cabling problems for which one needs to connect two males or two females together. Once again, the industry has responded by providing a broad array of gender-matching cables and adapters. On a conceptual level, these problems are simple; in practice, the permutations of incompatibilities are so numerous that debugging a 1960s-era RS-232 connection may not be a quick task.

Male DB25 and DE9 connectors consist of a dual row of staggered pins surrounded by a metal rim that serves as an electrical shield. The female connectors consist of matching staggered pin-sockets mounted in a solid frame whose edge forms a shield that mates with the male shield. These connectors are illustrated in Fig. 5.6.

The D-subminiature connector family uses a three-element nomenclature to specify the size of the connector housing, or shell, and the number of pins within the shell. There are five standard shell designations—A, B, C, D, E—that were originally specified with varying numbers of pins as shown in Table 5.3. DE9 connectors are commonly misrepresented as DB9, a connector configuration that is not defined. A modern D-subminiature connector that was not originally specified is the common HDE15, a high-density 15-pin connector using the E-size shell. The HDE15 is commonly used to connect monitors to desktop computers.

TABLE 5.3 Standard D-Subminiature Shell Sizes

Shell Size	Pins
A	15
B	25
C	37
D	50
E	9

Logical transceiver-level characteristics such as bit rate, error detection, and framing are accompanied by electrical transducer-level characteristics, more commonly referred to as the *physical layer* of a communications link. RS-232 refers to the logic 1 state as a *mark* and assigns it a negative potential from –3 to –25 V. The logic-0 state is a *space* and is assigned a positive potential from +3 to +25 V. Since RS-232 inverts the logic levels, an idle link is held at negative voltage, logic 1.

While RS-232 is specified with a transmitter voltage range of ±3 to ±25 V, most modern transmitters operate well below the 25-V upper bound. Many systems have been based around the ubiquitous and inexpensive 1488/1489 transmitter/receiver chipset that operate at ±12 V. These chips require an external ±12-V source for power. RS-232 circuitry was fundamentally simplified when Maxim Semiconductor created their MAX232 line of single-supply 5-V line interface ICs. These chips contain internal circuitry that generates ±8 V. Today, a variety of flexible RS-232 interface ICs are avail-

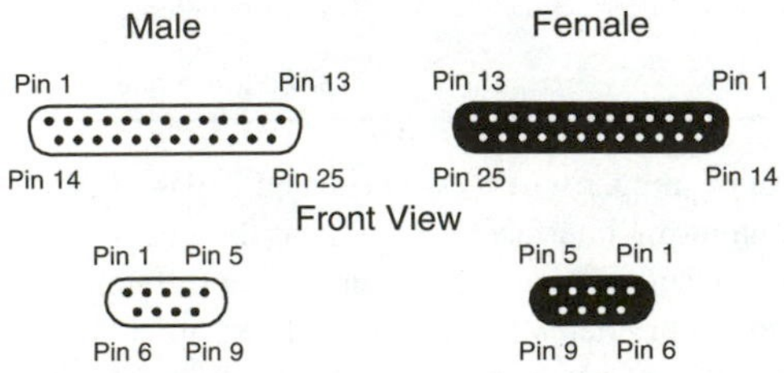

FIGURE 5.6 DB25 and DE9 connectors.

able from other manufacturers including Linear Technology, National Semiconductor, and Texas Instruments. RS-232 ports work quite well on even lower voltage ranges, because modern receivers are sensitive to smaller absolute voltages, and most RS-232 links are several meters or less in length. RS-232 was never intended to serve in truly long-distance applications.

5.5 RS-422

For crossing distances greater than several meters, RS-232 is supplemented by the *RS-422* standard. RS-422 can provide communications across more than 1.2 km at moderate bit rates such as 9.6 kbps. It is a *differential*, or *balanced*, transmission scheme whereby each logical signal is represented by two wires rather than one. RS-232 signals are *single-ended*, or *unbalanced*, signals that drive a particular voltage onto a single wire. This voltage is sensed at the receiver by measuring the signal voltage relative to the ground potential of the interface. Over long distances or at very high speeds, single-ended transmission lines are more subject to degradation resulting from ambient electrical noise. A partial explanation of this characteristic is that the electrical noise affects the active signal wire unequally with respect to ground. Differential signals, as in RS-422, drive opposing, or mirrored, voltages onto two wires simultaneously (RS-422 is specified from ±2 to ±6 V). The receiver then compares the voltages of the two wires together rather than to ground. Ambient noise tends to affect the two wires equally, because they are normally twisted together to follow the same path. Therefore, if noise causes a 1-V spike on one-half of the differential pair, it causes the same spike on the other half. When the two voltages are electrically subtracted at the receiver, the 1-V of *common-mode* noise cancels out, and the original differential voltage remains intact (subject, of course to natural attenuation over distance). The difference between RS-232 and RS-422 transmission is illustrated in Fig. 5.7.

Because of the longer distances involved in RS-422 interfaces, it is not common to employ the standard set of hardware handshaking signals that are common with RS-232. Therefore, some form of software handshaking must be implemented by the end devices to properly communicate. Some applications may not require any flow control, and some may use the XON/XOFF method. RS-422 does not specify a standard connector. It is not uncommon to see an RS-422 transmission line's bare wire ends connected to screw terminals.

Another common difference between RS-422 and RS-232 is transmission line *termination*. Transmission line theory can get rather complicated and is outside the scope of this immediate discussion.

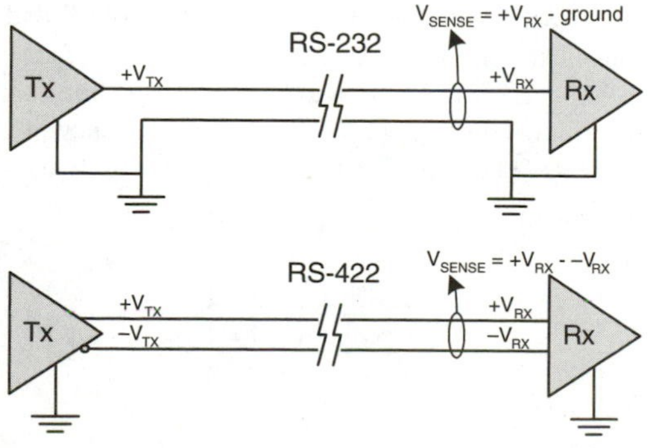

FIGURE 5.7 RS-232 vs. RS-422 signaling.

The basic practical result of transmission line theory is that, as the speed-distance product of an electrical signal increases, the signal tends to reflect off the ends of wires and bounce back and forth on the wire. When slow signals travel relatively moderate distances, the speed-distance product is not large enough to cause this phenomenon to any noticeable degree. Fast signals traveling over very short distances may also be largely immune to such reflections. However, when RS-422 signals travel over several kilometers, the speed-bandwidth product is great enough to cause previously transmitted data signals to reflect and interfere with subsequent data. This problem can be largely solved by properly *terminating* the receiving end of the transmission line with the line's *characteristic impedance*, Z_O. Typical coaxial and twisted-pair transmission lines have $Z_O = 50, 75,$ or $110\ \Omega$. Briefly put, Z_O is the impedance, or electrical resistance, that would be observed between both conductors of a balanced transmission line of infinite length. Again, there is substantial theory lurking here, but the practical result is that, by placing a resistor equal to Z_O at the far end of the line between both conductors, the transmission line will appear to be continuous and not exhibit reflections. A typical schematic diagram of a terminated RS-422 serial link is shown in Fig. 5.8.

5.6 MODEMS AND BAUD RATE

Information is conveyed by varying the electromagnetic field of a particular medium over time. The rate at which this field (e.g., voltage) changes can be represented by a certain bandwidth that characterizes the information. Transducers such as those that facilitate RS-232/RS-422 serial links place the information that is presented to them essentially unmodified onto the transmission medium. In other words, the bandwidth of the information entering the transducer is equivalent to that leaving the transducer. Such a system operates at *baseband:* the bandwidth inherent to the raw information. Baseband operation is relatively simple and works well for a transmission medium that can carry raw binary signals with minimal degradation (e.g., various types of wire, or fiber optic cable, strung directly from transmitter to receiver). However, there are many desirable communications media that are not well suited to directly carrying bits from one point to another. Two prime examples are free-space and acoustic media such as a telephone.

To launch raw information into the air or over a telephone, the bits must be superimposed upon a *carrier* that is suited to the particular medium. A carrier is a frequency that can be efficiently radiated from a transmitter and detected by a remote receiver. The process of superimposing the bits on the carrier is called *modulation*. The reverse process of detecting the bits already modulated onto the carrier is *demodulation*. For the purposes of this discussion, one of the simplest forms of modulation, binary *amplitude modulation* (AM), is presented as an example. More precisely, this type of AM is called *amplitude shift keying* (ASK). With two states, it is called 2-ASK and is illustrated in Fig. 5.9. Each time a 1 is to be transmitted, the carrier (shown as a sine wave of arbitrary frequency) is turned on with an arbitrary amplitude. Each time a 0 is to be transmitted, the carrier is turned off with an amplitude of zero. If transmitting over free space, the carrier frequency might be anywhere from hundreds of kilohertz to gigahertz. If communicating over a fiber optic cable, the carrier is

FIGURE 5.8 RS-422 transmission line termination.

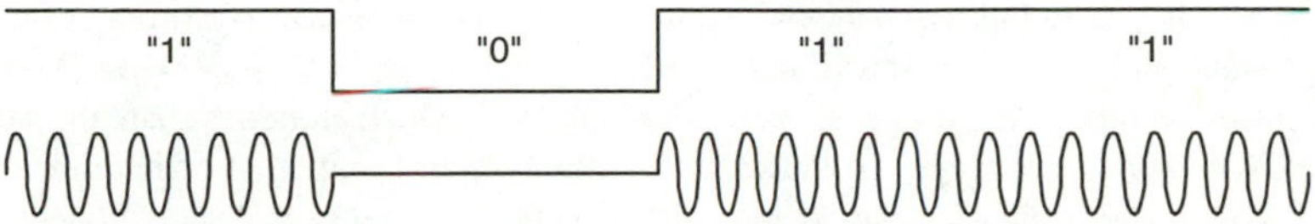

FIGURE 5.9 2-ASK modulation.

light. If an acoustic medium such as a telephone is used to send the data, the carrier is audible in the range of several kilohertz.

Frequency shift keying (FSK), a type of *frequency modulation* (FM) is a scheme that can be used to transmit multiple bits simultaneously without resorting to multiple levels of amplitude by using AM. FSK represents multiple bits by varying the frequency rather than the amplitude of the carrier. This constant amplitude approach is less susceptible to noise. Figure 5.10 shows 4-FSK modulation, in which each of the four frequency steps represents a different two-digit binary value.

A general term for a modulated data unit is a *baud*. If 2-ASK is used, each baud corresponds to one bit. Therefore, the baud rate matches the bit rate. However, the 4-FSK example shows that each baud represents two bits, making the bit rate twice that of the baud rate. This illustrates that baud rate and bit rate are related but not synonymous, despite common misuse in everyday conversation. Engineers who design modulation circuitry care about the baud rate, because it specifies how many unique data units can be transmitted each second. They also try to squeeze as many bits per baud as possible to maximize the overall bit rate of the modulator. Engineers who use modulators as black-box components do not necessarily care about the baud rate; rather, it is the system's bit rate that matters to the end application.

Enter the *modem*. A modem is simply a device that incorporates a modulator and demodulator for a particular transmission medium. The most common everyday meaning of modem is one that enables a computer to transfer bits over an analog telephone line. These modems operate using different modulation schemes depending on their bit rate. Early 300- and 1,200-bps modems operate using FSK and *phase shift keying* (PSK). Later modems, including today's 33.6- and 56-kbps models, operate using variations of *quadrature amplitude modulation* (QAM).

While *modem* often refers to telephone media, it is perfectly correct to use this term when referring to a generic modulator/demodulator circuit that operates on another medium. Digital wireless communication is increasingly common in such applications as portable cellular phones and untethered computer networking. These devices incorporate radio frequency (RF) modems in addition to digital transceivers that frame the data as it travels from one point to another.

5.7 NETWORK TOPOLOGIES

The communications schemes discussed thus far are point-to-point connections—they involve one transmitter and one receiver at either end of a given medium. Many applications require multidrop communications whereby multiple devices exchange data over the same medium. The general term

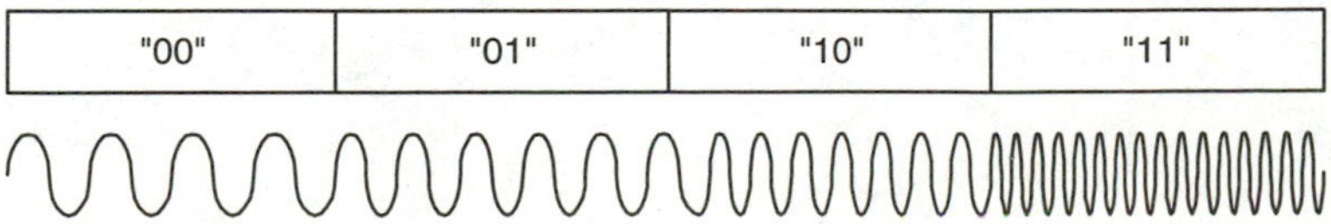

FIGURE 5.10 4-FSK modulation.

for a multidrop data link is a *network*. Networks can be constructed in a variety of topologies: buses, rings, stars, and meshes, as shown in Fig. 5.11.

A bus structure is the most basic network topology in which all nodes share the same physical medium. When one node wishes to transmit data, it must wait for another node to finish and release the bus before it can begin. The ring topology implements a daisy-chained set of connections where each node connects to its two nearest neighbors, and information usually flows in one direction (although bidirectional rings are a variation on this theme). A benefit of the ring is that a single long wire does not have to travel between all nodes. One disadvantage is that each node is burdened with the requirement of passing on information that is not destined for it to keep the message from being lost.

Mesh networks provide ultimate connectivity by connecting each node to several of its neighbors. A mesh can provide increased bandwidth as well as fault tolerance as a result of its multiple connections. Properly designed, a mesh can route traffic around a failed link, because multiple paths exist between each node in the network. The downside to these benefits is increased wiring and communications protocol complexity.

Star networks connect each node to a common central hub. The benefits of a physical star topology include ease of management, because adding or removing nodes does not affect the wiring of other nodes. A downside is that more wiring is necessary to provide a unique physical connection between each node and the central hub. A starred network may send data only to the node for which it is destined. Unlike a ring, the node does not have to pass through information that is not meant for it. And unlike a bus, the node does not have to ignore messages that are not meant for it. The requirement for a central hub increases the complexity of a star network. As more nodes are added to the network, the hub must add ports at the same rate.

A network may be wired using a physical star topology, but it may actually be a bus or ring from a logical, or electrical, perspective. Implementing differing physical and logical topologies is illustrated in Fig. 5.12. Some types of networks inherently favor bus or ring topologies, but the flexible management of star wiring is an attractive alternative to a strictly wired bus or ring. Star wiring enables nodes to be quickly added or disconnected from the central hub without disrupting other nodes. Bus and ring topologies may require the complete or partial disruption of the network medium to add or remove nodes. A star's hub typically contains electronics to include or bypass individual segments as they are added or removed from the network without disrupting other nodes.

5.8 NETWORK DATA FORMATS

Common data formats and protocols are necessary to regulate the flow of data across a network to ensure proper addressing, delivery, and access to that common resource. Several general terms for

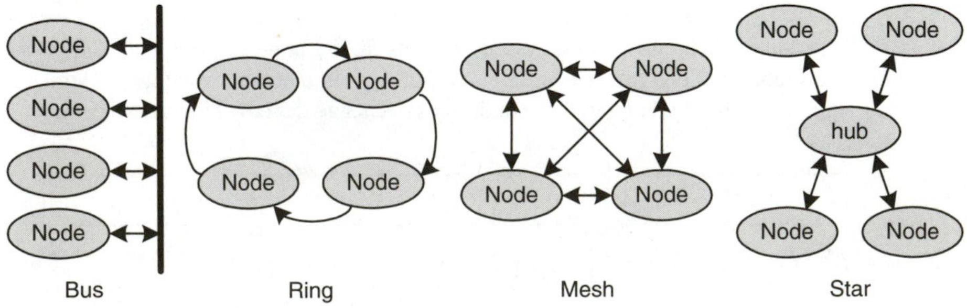

FIGURE 5.11 Basic network topologies.

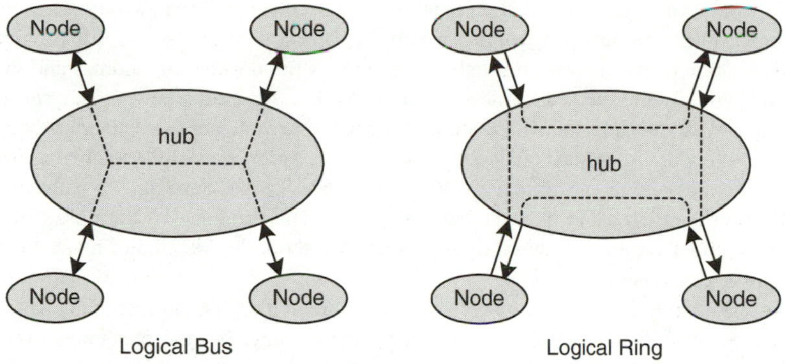

FIGURE 5.12 Physical vs. logical network topologies.

message elements on a network are *frame*, *packet*, and *cell*. Frames are sets of data that are framed at the beginning and end by special delimiters. Packets are sets of data that are not fully framed but that have some other means of determining their size, such as an embedded length field. Cells are fixed-length frames or packets. Frames and packets usually imply variable length data sets, but this is not a strict rule. As with many terms and classifications in digital systems, specific definitions are context specific and are often blurred: one system's cells may be another's frames. Frames, packets, and cells are composed of *headers*, *payloads*, and *trailers,* as shown in Fig. 5.13. The header is a collection of data fields that handle network overhead functions such as addressing and delineation. The actual data to be transmitted is placed into the payload. If present, a trailer is commonly used to implement some form of error checking and/or delineation. Not all packet formats specify the inclusion of trailers. When present, a trailer is usually substantially smaller in length than the header.

Networking is an aspect of digital systems design that directly involves hardware–software interaction at a basic level. One cannot really design networking hardware without keeping in mind the protocol, or software, support requirements. One key example is packet format. Hardware must have knowledge of the packet format so that it can properly detect a packet that is sent to it. At the same time, software must have this same knowledge so that it can properly parse received packets and generate new ones to be transmitted.

As soon as more than two nodes are connected to form a network, issues such as addressing and shared access arise. When there is only one transmitter and one receiver, it is obvious that data is intended for the only possible recipient. Likewise, the lone transmitter can begin sending data at any time it chooses, because there are no other transmitters competing for network access.

Network addressing is the mechanism by which a transmitting node indicates the destination for its packet. Each node on the network must therefore have a unique address to prevent confusion over where the packet should be delivered. In a bus topology, each node watches all the data traffic that is placed onto the network and picks out those packets that are tagged with its unique address. In a ring topology, each node passes packets on to the next node if the destination address is not matched with that node's address. If the address is matched, the node absorbs the packet and does not forward it on to the next node in the ring. Logical star and mesh topologies function a bit differently. Nodes on

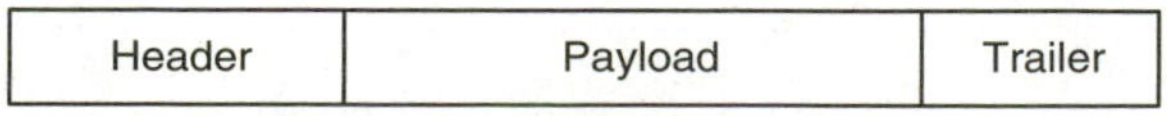

FIGURE 5.13 Generic packet structure.

these types of networks do not observe all traffic that traverses the network; rather, the network itself contains some intelligence when it comes to delivering a packet. A node in a logically starred network sends a packet to a central hub that examines the destination address and then forwards the message to only the specified node. A mesh network routes traffic partly like a ring and partly like a star; however, multiple paths between nodes exist to complicate the delivery process. Based on the destination address, the originating node sends a packet to one of its neighbors, which in turn forwards the message to one of its neighbors. This process continues until the path has been completed and the packet arrives at its intended destination. The presence of multiple valid paths between nodes requires the mesh network to use knowledge about the location of nodes to select an optimal path through the network.

Access sharing is necessary on networks to ensure that each node eventually has an opportunity to send a message. Numerous methods of access sharing have been implemented over the years. Generally speaking, the length of messages is bounded to prevent a node from transmitting an infinitely long set of data and preventing anyone else from gaining access to the shared medium. Sharing algorithms differ according to the specific network topology involved. Networks that are a collection of point-to-point links (e.g., ring, star, mesh) do not have to worry about multiple nodes fighting for access to the same physical wire, but do have to ensure that one node does not steal all the bandwidth from others. Bus networks require sharing algorithms that address both simultaneous physical contention for the same shared wire in addition to logical contention for the network's bandwidth. Arbitration schemes can be centralized (whereby a single network master provides permission to each node to transmit) or distributed (whereby each node cooperates on a peer-to-peer level to resolve simultaneous access attempts).

After deciding on a network topology, one of the first issues to resolve is the network packet format. If the network type is already established (e.g., Ethernet), the associated formats and protocols are already defined by industry and government standards committees. If an application benefits from a simple, custom network, the packet format can be tailored to suit the application's specific needs.

Delineation and addressing are the two most basic issues to resolve. Delineation can be accomplished by sending fixed-size packets, embedding a length field in the packet header, or by reserving unique data values to act as start/stop markers. Framing with unique start/stop codes places a restriction on the type of data that a packet can contain: it cannot use these unique codes without causing false start or end indications. Referring back to Table 5.1, notice that start-of-header (SOH) and end-of-transmission (EOT) are represented by 0x01 and 0x04. These (or other pairs of codes) can be used as delimiters if the packet is guaranteed to contain only alphanumeric ASCII values that do not conflict with these codes.

Addressing is normally achieved by inserting both the destination and source addresses into the header. However, some networking schemes may send only a single address. Sending both addresses enables recognition of the destination as well as a determination of which node sent the packet. Since most data exchanges are bidirectional to a certain degree, a destination node will probably need to send some form of reply to the source node of a particular packet. Many networks include a provision known as *broadcast addressing* whereby a packet is sent to all nodes on the network rather than just one. This broadcast is often indicated using a reserved *broadcast* address. In contrast to a *unicast* address that is matched by only one node, a broadcast address is matched by all nodes on the network. Some networks also have *multicast* addresses that associate multiple nodes with a single destination address.

5.9 RS-485

Whereas RS-232 and RS-422 enable point-to-point serial links, the RS-485 standard enables multiple-node networks. Like RS-422, RS-485 provides differential signaling to enable communications across spans of twisted-pair wire exceeding 1.2 km. Unlike RS-422, the RS-485 standard allows up to 32 transmit/receive nodes on a single twisted pair that is terminated at each end as shown in Fig. 5.14. Modern low-load receivers that draw very little current from the RS-485 bus can be used to increase the number of nodes on an RS-485 network well beyond the original 32-node limit to 256 nodes or more. A single pair of wires is used for both transmit and receive, meaning that the system is capable of *half-duplex* (one-way) operation rather than *full-duplex* operation (both directions at the same time). Half-duplex operation restricts the network to one-way exchange of information at any given time. When node A is sending a packet to node B, node B cannot simultaneously send a packet to node A.

RS-485 directly supports the implementation of bus networks. Bus topologies are easy to work with, because nodes can directly communicate with each other without having to pass through other nodes or semi-intelligent hubs. However, a bus network requires provisions for sharing access to be built into the network protocol. In a centralized arbitration scheme, a master node gives permission for any other node to transmit data. This permission can be a request-reply scheme whereby slave nodes do not respond unless a request for data is issued. Alternatively, slave nodes can be periodically queried by the master for transmit requests, and the master can grant permissions on an individual-node basis. There are many centralized arbitration schemes that have been worked out over the years.

A common distributed arbitration scheme on a bus network is *collision detection* with random back-off. When a node wants to transmit data, it first waits until the bus becomes idle. Once idle, the node begins transmitting data. However, when the node begins transmitting, there is a chance that one or more nodes have been waiting for an opportunity to begin transmitting and that they will begin transmitting at the same time. Collision detection circuits at each node determine that more than one node is transmitting, and this causes all active transmitters to stop. Figure 5.15 shows the imple-

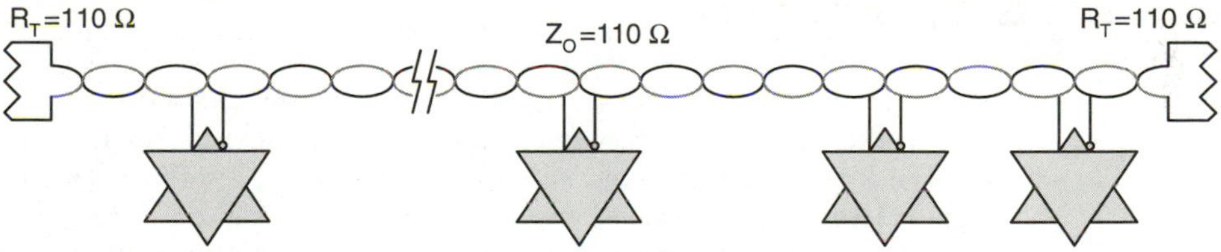

FIGURE 5.14 RS-485 bus topology.

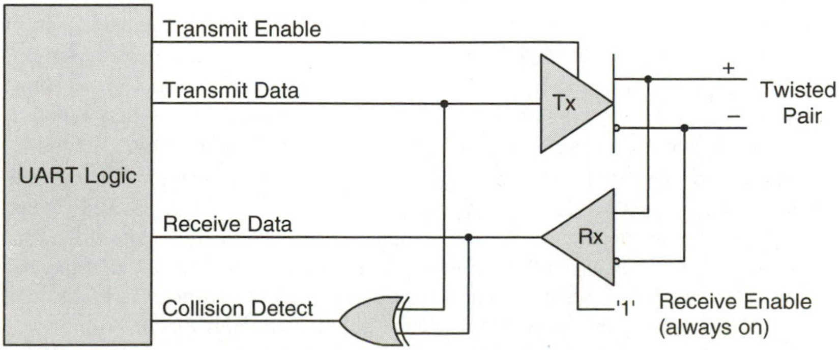

FIGURE 5.15 RS-485 collision detection transceiver.

mentation of an RS-485 transceiver with external collision detection logic. A transmit enable signal exists to turn off the transmitter when the UART is not actively sending data. Unlike an RS-422 transmitter that does not have to share access with others, the RS-485 transmitter must turn itself off when not sending data to enable others to transmit.

When transmitting, the receiver returns the logical state of the twisted-pair bus. If the bus is not at the same state as the transmitted data, a collision is most likely being caused by another transmitter trying to drive the opposite logic state. An XOR gate implements this collection detect, and the XOR output must be sampled only after allowing adequate time for the bus to settle to a stable state following the assertion of each bit from the transmitter.

Once a collision has been detected by each node and the transmitters are disabled, each node waits a different length of time before retransmitting. If all delays were equal, multiple nodes would get caught in a deadlock situation wherein each node keeps trying to transmit after the same delay interval. Random back-off delays are pseudo-random so as to not unfairly burden some nodes with consistently longer delays than other nodes. At the end of the delay, one of the nodes begins transmitting first and gains control of the bus by default. The other waiting nodes eventually exit from their delays and observe that the bus is already busy, indicating that they must wait their turn until the current packet has been completed. If, by coincidence, another node begins transmitting at the same time that the first node begins, the back-off process begins again. It is statistically possible for this process to occur several times in a row, although the probability of this being a frequent event is small in a properly designed network. A bus network constructed with too many nodes trying to send too much data at the same time can exhibit very poor performance, because it would be quite prone to collisions. In such a case, the solution may be to either reduce the network traffic or increase the network's bandwidth.

5.10 A SIMPLE RS-485 NETWORK

An example of a simple but effective network implemented with RS-485 serves as a vehicle to discuss how packet formats, protocols, and hardware converge to yield a useful communications medium. The motivation to create a custom RS-485 network often arises from a need to deploy remote actuators and data-acquisition modules in a factory or campus setting. A central computer may be located in a factory office, and it may need to periodically gather process information (e.g., temperature, pressure, fluid-flow rate) from a group of machines. Alternatively, a security control console located in one building may need to send security camera positioning commands to locations throughout the campus. Such applications may involve a collection of fairly simple and inexpensive microprocessor-based modules that contain RS-485 transceivers. Depending on the exact physical layout, it may or may not be practical to wire all remote nodes together in a single twisted-pair bus. If not, a logical bus can be formed by creating a hybrid star/bus topology as shown in Fig. 5.16. A central hub electrically connects the individual star segments so that they function electrically as a large bus but do not require a single wire to be run throughout the entire campus.

As shown, the hub does not contain any intelligent components—it is a glorified junction box. This setup is adequate if the total length of all star segments does not exceed 1.2 km, which is within the electrical limitations of the RS-485 standard. While simple, this setup suffers from a lack of fault tolerance. If one segment of the star wiring is damaged, the entire network may cease operation because, electrically, it is a single long pair of wires. Both the distance and fault-tolerance limitations can be overcome by implementing an active hub that contains *repeaters* on each star segment and smart switching logic to detect and isolate a broken segment. A repeater is an active two-port device that amplifies or regenerates the data received on one port and transmits it on the other port. An RS-

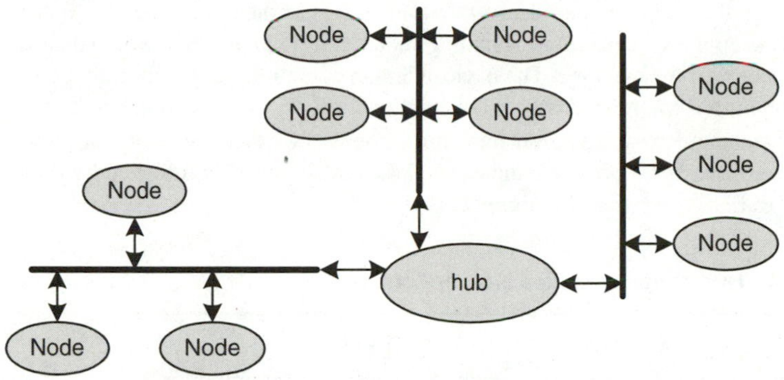

FIGURE 5.16 Hybrid star/bus network topology.

485 repeater needs a degree of intelligence, because both ports must be bidirectional. Therefore, the repeater must be able to listen for traffic on both sides, detect traffic on one side, and then transmit that traffic on the other side. A hub that detects and isolates segment failures would be well designed to report this fault information to a central control node to alert the human operator that repairs are necessary. These possible improvements in the network hub do not affect the logical operation of the network and, consequently, are not a focus of this discussion.

With a topology chosen and a general application in mind, the next step is to decide on the network's operational requirements from among the following:

1. *Support for roughly 200 nodes* provides flexibility for a variety of control applications.
2. *Central arbitration handled by master control node* for simplicity of network design. A facility control network is often a master-slave application, because all data transfers are at the request of the central controller. Central arbitration removes the need for collision-detect hardware and random back-off algorithms.
3. *Broadcast capability* enables easy distribution of network status information from the master control node.
4. *Data rate of 9600 bps* provides adequate bandwidth for small control messages without burdening the network with high frequencies that can lead to excessive noise and signal degradation.
5. *Basic error handling* prevents processing incorrect data and network lock-up conditions when occasional noise on the RS-485 twisted-pairs causes data bits to change state.

Many aspects of network functionality are directly influenced by a suitable network packet format. Other aspects are addressed by the protocol that formats data on the network, by the transceiver and UART hardware, or by a combination of these three elements.

In considering the packet format, 8-bit destination and source addresses are chosen to support more than 200 nodes on the network. A special destination address value of 0xFF represents a broadcast address, meaning that all nodes should accept the packet automatically. Such broadcast packets are useful for system-wide initialization whereby, for example, the control computer can send the current time to all nodes. This multicast address cannot be used as a normal node address, thereby limiting the network to 255 unique nodes.

It is desirable to employ variable-length packets so that a message does not have to be longer than necessary, thereby conserving network bandwidth. Variable-length packets require some mechanism to determine the length: either reserved framing codes or an explicit length count. A length count is chosen to keep the system simple. Framing codes would require that certain data values be off limits to the contents of the message. The payload length is bounded at a convenient binary threshold: 255 bytes. For simple control and data-acquisition applications, this is probably more than enough.

Based on these basic requirements and a couple of quick decisions, a packet format quickly emerges. A three-byte, fixed-length header shown in Table 5.4 is followed by a variable-length payload. No trailer is necessary in this network.

TABLE 5.4 Hypothetical Packed Header Format

Field Name	Byte	Bits	Description
DA	0	[7:0]	Destination address (0xFF = multicast)
SA	1	[7:0]	Source address
LEN	2	[7:0]	Payload length (0x0 = no payload present)

The eight-bit destination address field, DA, comes first to enable the receiving hardware to quickly determine whether the packet should be accepted by the node or ignored. A packet will be accepted if DA matches the receiver's node address, or if DA equals 0xFF, indicating a broadcast packet. At the end of the header is an eight-bit length field that indicates how many payload bytes are present after the fixed-length header. This limits the maximum packet size to 255 payload bytes plus the 3-byte header. A value of zero means that there is no payload, only a header in the packet.

Error detection can be handled by even parity. Each byte of the header and payload is sent with an accompanying parity bit. When an error is detected, the network's behavior must be clearly defined to prevent the system from either ceasing to function or acting on false data. Parity errors can manifest themselves in a variety of tricky ways. For example, if the length field has a parity error, how will the receiver know the true end of the frame? Without proper planning, a parity error on the length field can permanently knock the receivers out of sync and make automatic recovery impossible. This extreme situation can occur when an invalid length causes the receiver to either skip over the next frame header or prematurely interpret the end of the current frame as a new header. In both cases, the receiver will falsely interpret a bogus length field, and the cycle of false header detection can continue indefinitely.

If a parity error is detected on either the destination or source addresses, the receivers will not lose synchronization, but the packet should be ignored, because it cannot be known who the true recipient or sender of the packet is.

Fault tolerance in the case of an invalid payload length can be handled in a relatively simple manner. Requirements of no intrapacket gaps and a minimum interpacket gap assist in recovery from length-field parity errors. The absence of intrapacket gaps means that, once a packet has begun transmission, its bytes must be continuous without gaps. Related to this is the requirement of a minimum interpacket gap which forces a minimum idle period between the last byte of one packet and the start of the next packet. These requirements help each receiver determine when packets are starting and ending. Even if a packet has been subjected to parity errors, the receiver can wait until the current burst of traffic has ended, wait for the minimum interpacket gap, and then begin looking for the next packet to begin.

The parity error detection and accompanying recovery scheme greatly increases the probability that false data will not be acted upon as correct data and that the entire network will not stop functioning when it encounters an arbitrary parity error. However, error detection is all about probability.

A single parity bit cannot guarantee the detection of multiple errors in the same byte, because such errors can mask themselves. For example, two bit errors can flip a data bit and the parity bit itself, making it impossible for the receiver to detect the error. More complex error detection schemes are available and are more difficult to fool. Although no error detection solution is perfect, some schemes reduce the probability of undetected errors to nearly zero.

If a packet is received with an error, it cannot be acted upon normally, because its contents are suspect. For the purposes of devising a useful error-handling scheme, packet errors can be divided into two categories: those that corrupt the destination/source address information and those that do not. Parity errors that corrupt the packet's addresses must result in the packet being completely ignored, because the receiving node is unable to generate a reply message to the originator indicating that the packet was corrupted. If the source address is corrupted, the receiver does not know to whom to reply. If the destination address is corrupted, the receiver does not know whether it is the indented recipient.

In the case of an address error in which the received packet is ignored, the originator must implement some mechanism to recover from the packet loss rather than waiting indefinitely for a reply that will never arrive. A reply *timeout* can be implemented by an originator each time a packet is sent that requires a corresponding reply. A timeout is an arbitrary delay during which an originating node waits before giving up on a response from a remote node. Timeouts are common in networks because, if a packet is lost due to an error, the originator should not wait indefinitely for a response that will never come. Establishing a timeout value is a compromise between not giving up too quickly and missing a slower-than-normal reply and waiting too long and introducing unacceptable delays in system functionality when a packet is lost. Depending on the time it takes to send a packet on a network and the nodes' typical response time, timeouts can range from microseconds to minutes. Typical timeouts are often expressed in milliseconds.

When an originator times-out and concludes that its requested data somehow got lost, it can resend the request. If, for example, a security control node sends a request for a camera to pan across a room, and that request is not acknowledged within half a second, the request can be retransmitted.

In the case of a non-address error, the receiving node has enough information to send a reply back to the originator, informing it that the packet was not correctly received. Such behavior is desirable to enable the originator to retransmit the packet rather than waiting for a timeout before resending that data.

The preceding details of a hypothetical RS-485 network must be gathered into network driver software to enable proper communication across the network. While hardware controls the detection of parity errors and the flow of bits, it is usually software that generates reply messages and counts down timeouts. Figure 5.17 distills this information into a single flowchart from which software routines could be written.

As seen from this flowchart, transmit and receive processes run concurrently and are related. The transmit process does not complete until a positive acknowledgement is received from the destination node. This network control logic implemented in software is simple by mainstream networking standards, yet it is adequate for networks of limited size and complexity. Issues such as access sharing are handled inherently by the request/reply nature of this network, greatly simplifying the traffic patterns that must be handled by the software driver.

5.11 INTERCHIP SERIAL COMMUNICATIONS

Serial data links are not always restricted to long-distance communications. Within a single computer system, or even a single circuit board, serial links can provide attractive benefits as compared

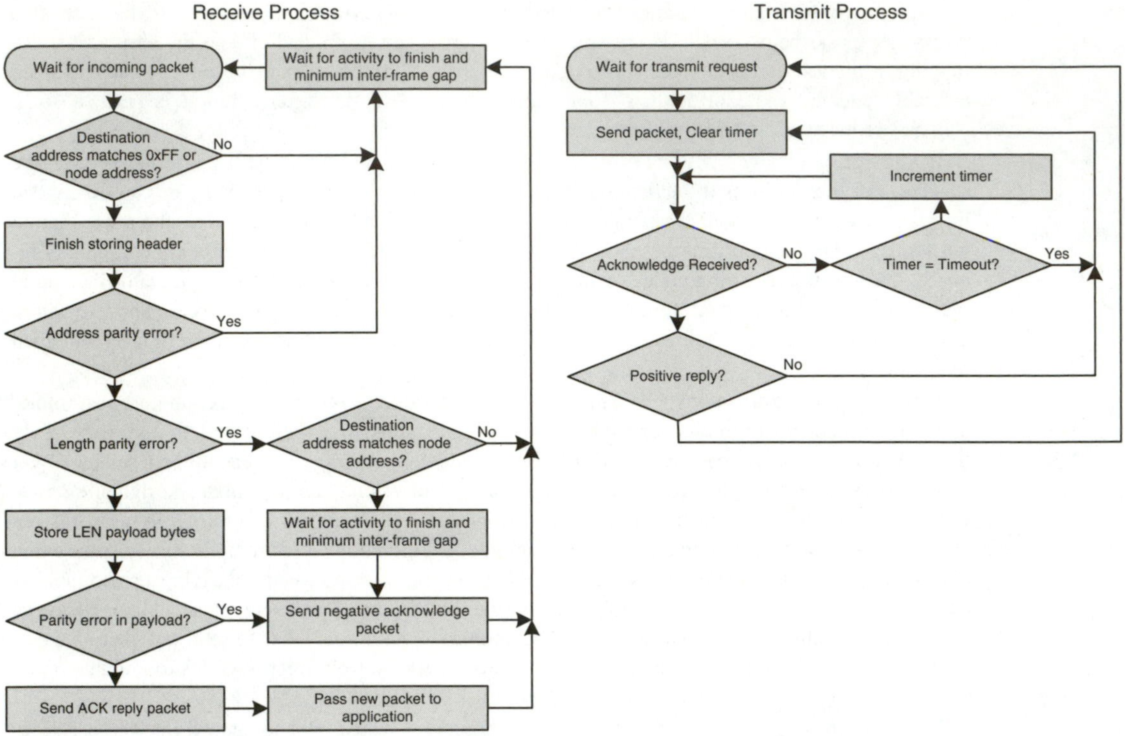

FIGURE 5.17 Hypothetical network driver flowchart.

to traditional parallel buses. Computer architectures often include a variety of microprocessor peripheral devices with differing bandwidth requirements. Main memory, both RAM and ROM, is a central part of computer architecture and is a relatively high-bandwidth element. The fact that the CPU must continually access main memory requires a simple, high-bandwidth interface—a parallel bus directly or indirectly driven by the CPU. Other devices may not be accessed as often as main memory and therefore have a substantially lower bandwidth requirement. Peripherals such as data acquisition ICs (e.g., temperature sensors), serial number EEPROMs, or liquid crystal display (LCD) controllers might be accessed only several times each second instead of millions of times per second. These peripherals can be directly mapped into the CPU's address space and occupy a spot on its parallel bus, but as the number of these low-bandwidth peripherals increases, the complexity of attaching so many devices increases.

Short-distance serial data links can reduce the cost and complexity of a computer system by reducing interchip wiring, minimizing address decoding logic, and saving pins on IC packages. In such a system, the CPU is connected to a serial controller via its parallel bus, and most other peripherals are connected to the controller via several wires in a bus topology as shown in Fig. 5.18.

Such peripherals must be specifically designed with serial interfaces, and many are. It is common for low-bandwidth peripheral ICs to be designed in both parallel and serial variants. In fact, some devices are manufactured with only serial interfaces, because their economics overwhelmingly favors the reduction in logic, wiring, and pins of a serial data link. A temperature sensor with a serial interface can be manufactured with just one or two signal pins plus power. That same sensor might

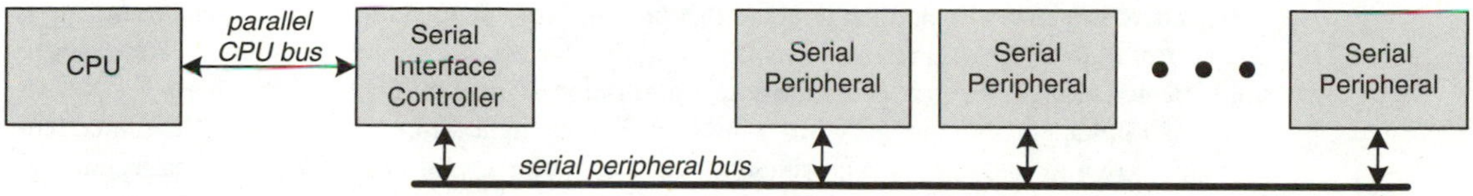

FIGURE 5.18 Generic interchip serial bus topology.

require 16 or more signal pins with a byte-wide parallel interface. Not only is the package cost reduced, its greatly reduced size enables the IC to be located in very confined spaces. Products including cell phones and handheld computers benefit tremendously from small IC packages that enable small, consumer-friendly form factors.

Interchip serial interfaces must be kept fairly simple to retain their advantages of low cost and ease of use. Industry standard interfaces exist so that semiconductor manufacturers can incorporate mainstream interfaces into their ICs, and engineers can easily connect multiple manufacturers' ICs together without redesigning the serial interface for each application. Many of these standard interfaces are actually proprietary solutions, developed by individual semiconductor manufacturers, that have gained wide acceptance. Two of the most commonly used industry standards for interchip serial communications are Philips' *inter-IC bus* (I^2C) and Motorola's *serial peripheral interface* (SPI). Both Philips and Motorola have long been leaders in the field of small, single-chip computers called *microcontrollers* that incorporate microprocessors, small amounts of memory, and basic peripherals such as UARTs. It was therefore a natural progression for these companies to add inexpensive interchip serial data links to their microcontrollers and associated peripheral products.

I^2C and SPI support moderate data rates ranging from several hundred kilobits to a few megabits per second. Because of their target applications, these networks usually involve a single CPU master connected to multiple slave peripherals. I^2C supports multiple masters and requires only two wires, as compared to SPI's four-plus wires.

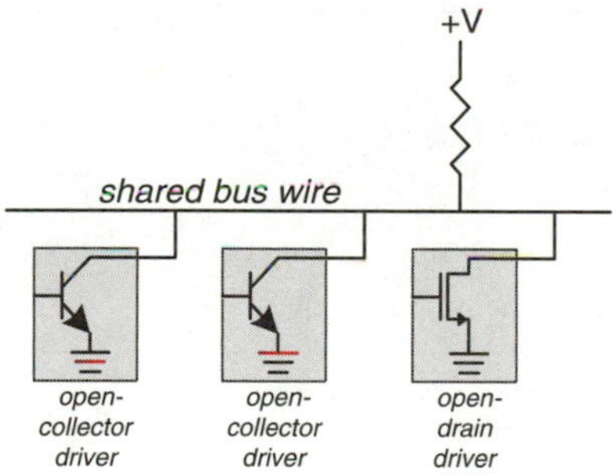

FIGURE 5.19 I^2C open-collector schematic representation.

I^2C consists of a clock signal, SCL, and a data signal, SDA. Both are *open-collector* signals, meaning that the ICs do not actively drive the signals high, only low. An open-collector driver is similar to a tri-state buffer, although no active high state is driven. Instead, the output is at either a low- or high-impedance state. The open-collector configuration is schematically illustrated in Fig. 5.19. The term *open-collector* originates from the days of bipolar logic when NPN output transistors inside the chips had no element connected to their collectors to assert a logic high. This terminology is still used for CMOS logic, although *open-drain* is the technically correct term when working with MOSFETs. A *pullup resistor* is required on each signal (e.g., SCL and SDA) to pull it to a logic 1 when the ICs release the actively driven logic 0. This open-collector arrangement enables multiple IC drivers to share the same wire without concern over electrical contention.

Under an idle condition, SCL and SDA are pulled high by their pullup resistors. When a particular IC wants to communicate, it drives a clock onto SCL and a pattern of data onto SDA. SCL may be as fast as 100 kHz for standard I^2C and up to 400 kHz for fast I^2C buses. I^2C is a real network that assigns a unique node address to each chip connected to the bus. As such, each transfer begins with a start sequence followed by seven-bit destination address. A read/write flag and data follow the address. There is a carefully defined protocol that provides for acknowledgement of write transactions and returning data for read transactions. In a multimaster configuration, collision detection can be implemented along with an appropriate access arbitration algorithm.

I^2C is implemented using only two wires, but this apparent simplicity belies its flexibility. The protocol is rich in handling special situations such as multiple masters, slow slaves that cannot respond to requests at the master's SCL frequency, and error acknowledgements. Some manufacturers that incorporate I^2C into their products pay Philips a licensing fee and are therefore able to use the trademark name in their documentation. Other manufacturers try to save some money by designing what is clearly an I^2C interface but referring to it by some generic or proprietary name such as "standard two-wire serial interface." If you come across such a product, spend a few minutes reading its documentation to make sure whether a true I^2C interface is supported.

Motorola's SPI consists of a clock signal, SCK, two unidirectional data signals, and multiple slave select signals, SS* as shown in Fig. 5.20. One data signal runs from the master to each slave and is called MOSI: master-out, slave-in. The other data signal is shared by the slaves to send data back to the master and is called MISO: master-in, slave-out. SCK is always driven by the master and can be up to several megahertz. Rather than assigning a unique address to each slave, the master must assert a unique SS* to the particular device with which it wants to exchange data. On observing SS* being asserted, a slave loads the bits on the MOSI signal into an internal shift register. If a read is being performed, the slave can reply with data shifted out onto the MISO signal. Because MISO is shared by multiple slaves, they must implement some type of contention-avoidance mechanism such as tri-state or open-collector outputs.

Each of these interchip buses proves extremely useful in simplifying many system designs. It is beyond the scope of this discussion to explain the detailed workings of either I^2C or SPI. For more information, consult the technical resources available from Philips and Motorola on their web sites or in their printed data sheets.

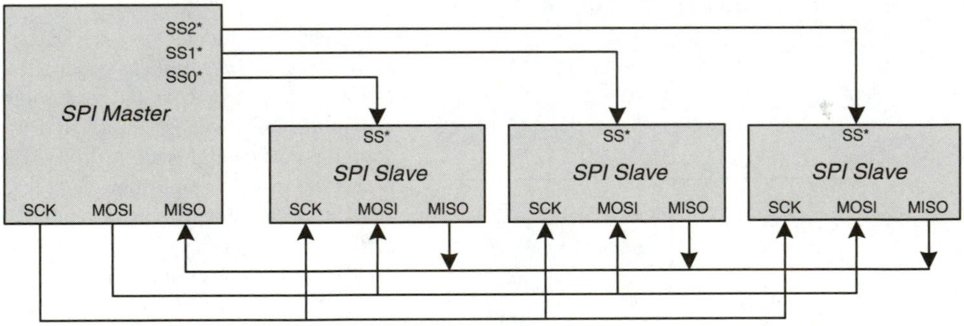

FIGURE 5.20 SPI bus organization.

CHAPTER 6

Instructive Microprocessors and Microcomputer Elements

Microprocessors, the heart of digital computers, have been in a constant state of evolution since Intel developed the first general-purpose microprocessors in the early 1970s. Intel's four-bit 4004 made history, because it was a complete microprocessor on a single chip at a time when processor modules for minicomputers filled multiple circuit boards. Over the past three decades, the complexity and throughput of microprocessors has increased dramatically as semiconductor technology has improved by leaps and bounds. Hundreds of microprocessors have come and gone over the years. There are many different architectures on the market today, each with its own claims of superior performance, lower cost, and reduced power in its intended applications.

When looking back on three decades' worth of development and the state of microprocessors today, several microprocessor families are especially worth exploring as instructional examples of basic computer architecture. Some of these families are the ancestors of very popular and widespread designs that are used to this day. Familiarity with these classic microprocessors can make it easier to learn about contemporary products that are either improved versions of the originals or members of other families that share common traits. Alternatively, some of these families are worthy of note because of their important role in permeating everyday life with microprocessors in places that most people rarely think of as computerized: cars, microwave ovens, dishwashers, and VCRs.

This chapter provides information that is both historical and directly relevant to contemporary digital systems design. Five classic microprocessor architectures are presented: Motorola 6800, Intel 8051, Microchip PIC, Intel 8086, and Motorola 68000. All of these architectures are in use today in varying forms, and each represents a different perspective on how microprocessors can accomplish similar tasks. A future design challenge may be addressed directly by one of these devices, or the solution may employ architectural concepts that they have helped to bring about.

6.1 EVOLUTION

Following the 4004's introduction in 1971, Intel enhanced the four-bit architecture by releasing the 4040 and 8008 in rapid succession. The 4040 added several instructions and internal registers, and the 8008 extended the basic architecture to eight bits. These processors ran at speeds from 100 to 200 kHz and were packaged in 16 (4004/4040) and 18 (8008) pin DIPs. While significant for their time, they had limited throughput and could address only 4 kB (4004/4040) or 16 kB (8008) of memory. In 1974, Intel made substantial improvements in microprocessor design and released the 8080, setting the stage for modern microprocessors. Whereas Intel's earlier microprocessors look like relics of a bygone era, the 8080 is architecturally not far off from many microprocessors that exist today. The 8080 was housed in a 40-pin DIP, featured a 16-bit address bus and an 8-bit data bus, and ran at 2 MHz. It also implemented a conventional stack pointer that enabled deep stacks in external memory (Intel's earlier microprocessors had internal stacks with very limited depth). The 8080 became extremely popular as a result of its performance and rich, modern instruction set. This popularity was evidenced two years later, in 1976, with Intel's enhanced 8085 and competitor Zilog's famous Z80. Designed by former Intel engineers, the Z80 was based heavily on the 8080 to the point of having a partially compatible instruction set.

Both the 8085 and Z80 were extremely popular in a variety of computing platforms from hobbyists to mainstream commercial products to video arcade games. The 8085 architecture influenced the famous 16-bit 8086 family whose strong influence continues to this day in desktop PCs. The Z80 eventually lost the mainstream microprocessor war and migrated to microcontrollers that are still available for new designs from Zilog.

As microprocessors progress, technologies that used to be leading edge first become mainstream and then appear quite pedestrian. Along the way, some microprocessor families branch into multiple product lines to suit a variety of target applications. The high-end computing market gets most of the publicity and accounts for the major technology improvements over time. Lower-end microprocessors are either made obsolete after some time or find their way into the *embedded* market. Embedded microprocessors and systems are those that may not appear to the end user as a computer, or they may not be visible at all. Instead, embedded microprocessors typically serve a control function in a machine or another piece of equipment. This is in contrast to the traditional computer with a keyboard and monitor that is clearly identified as a general-purpose computer.

Integrated microprocessor products are called *microcontrollers*, a term that has already been introduced. A microcontroller is a microprocessor integrated with a varying mix of memory and peripherals on a single chip. Microcontrollers are almost always found in embedded systems. As with many industry terms, *microcontrollers* can mean very different things to different people. In general, a microcontroller contains a relatively inexpensive microprocessor core with a complement of onboard peripherals that enable a very compact, yet complete, computing system—either on a single chip or relatively few chips. There is a vast array of single-chip microcontrollers on the market that integrate quantities of both RAM and ROM on the same chip along with basic peripherals including serial communications controllers, timers, and general I/O signal pins for controlling LEDs, relays, and so on. Some of the smallest microcontrollers can cost less than a dollar and are available in packages with as few as eight pins. Such devices can literally squeeze a complete computer into the area of a fingernail. More complex microcontrollers can cost tens of dollars and provide external microprocessor buses for memory and I/O expansion. At the very high end, there are microcontrollers available for well over $100 that include 32-bit microprocessors running at hundreds of megahertz, with integrated Ethernet controllers and DMA. Manufacturers typically refer to these high-end microcontrollers with unique, proprietary names to differentiate them from the aforementioned class of inexpensive devices.

6.2 MOTOROLA 6800 EIGHT-BIT MICROPROCESSOR FAMILY

As the microprocessor market began to take off, Motorola jumped into the fray and introduced its eight-bit 6800 in 1974, shortly after the 8080 first appeared. While no longer available as a discrete microprocessor, the 6800 is significant, because it remains in Motorola's successful 68HC05/68HC08 and 68HC11 microcontroller families and also serves as a vehicle with which to learn the basics of computer architecture. Like the 8080, the 6800 is housed in a 40-pin DIP and features a 16-bit address bus and an 8-bit data bus. All of the basic register types of a modern microprocessor are implemented in the 6800, as shown in Fig. 6.1: a program counter (PC), stack pointer (SP), index register (X), two general-purpose accumulators (ACCA and ACCB), and status flags set by the ALU in the condition code register (CCR). ACCA is the primary accumulator, and some instructions operate only on this register and not ACCB. A half-carry flag is included to enable efficient binary coded decimal (BCD) operations. After adding two BCD values with normal binary arithmetic, the half-carry is used to convert illegal results back to BCD. The 6800 provides a special instruction, decimal adjust ACCA (DAA), for this specific purpose. A somewhat out-of-place interrupt mask bit is also implemented in the CCR, because this was an architecturally convenient place to locate it. Bits in the CCR are modified through either ALU operations or directly by transferring the value in ACCA to the CCR.

The 6800 supports three interrupts: one nonmaskable, one maskable, and one software interrupt. More recent variants of the 6800 support additional interrupt sources. A software interrupt can be used by any program running on the microprocessor to immediately jump to some type of maintenance routine whose address does not have to be known by the calling program. When the software interrupt instruction is executed, the 6800 reads the appropriate interrupt vector from memory and jumps to the indicated address. The 6800's reset and interrupt vectors are located at the top of memory, as listed in Table 6.1, which generally dictates that the boot ROM be located there as well. For example, an 8-kB 27C64 EPROM (8,192 bytes = 0x2000 bytes) would occupy the address range 0xE000 through 0xFFFF. Each vector is 16 bits wide, enough to specify the full address of the associated routine. The MSB of the address, A[15:8], is located in the low, or even, byte address, and the LSB, A[7:0] is located in the high, or odd, byte address.

TABLE 6.1 6800 Reset and Interrupt Vectors

Vector Address	Purpose
0xFFFE/0xFFFF	Reset
0xFFFC/0xFFFD	Nonmaskable interrupt
0xFFFA/0xFFFB	Software interrupt
0xFFF8/0xFFF9	Maskable interrupt

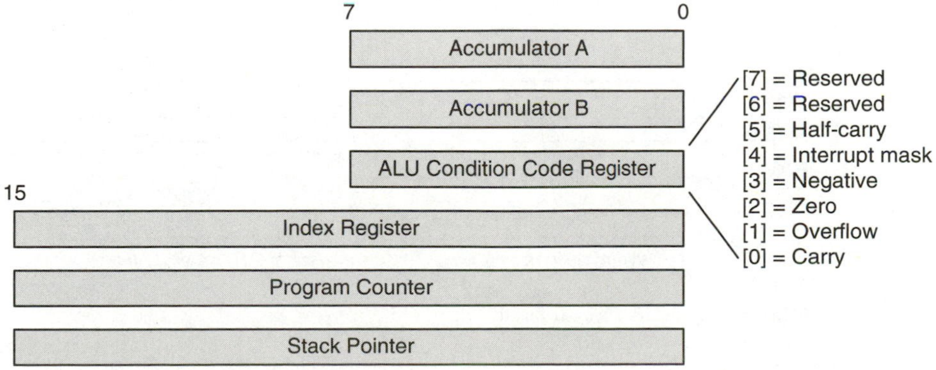

FIGURE 6.1 6800 registers.

An external clock driver circuit that provides a two-phase clock (two clock signals 180° out of phase with respect to each other) is required for the original 6800. Motorola simplified the design of 6800-based computer systems by introducing two variants, the 6802 and 6808. The 6802 includes an on-board clock driver circuit of the type that is now standard on many microprocessors available today. Such clock drivers require only an external *crystal* to create a stable, reliable oscillator with which to clock the microprocessor. A crystal is a two-leaded component that contains a specially cut quartz crystal. The quartz can be made to resonate at its natural frequency by electrical stimulus created within the microprocessor's on-board clock driver circuitry. A crystal is necessary for this purpose, because its oscillation frequency is predictable and stable. The 6802 also includes 128 bytes of on-board RAM to further simplify certain systems that have small volatile memory requirements. For customers who wanted the simplified clocking scheme of the 6802 without paying for the on-board RAM, Motorola's 6808 kept the clocking and removed the RAM.

Using a 6802 with its internal RAM, a functional computer could be constructed with only two chips: the 6802 and an EPROM. Unfortunately, such a computer would not be very useful, because it would have no I/O with which to interact with the outside world. Motorola manufactured a variety of peripheral chips intended for direct connection to the 6800 bus. Among these were the 6821 peripheral interface adapter (PIA) and the 6850 asynchronous communications interface adapter (ACIA), a type of UART. The PIA provides 20 I/O signals arranged as two 8-bit parallel ports, each with two control signals. Applications including basic pushbutton sensing and LED driving are easy with the 6821. The 6800 bus uses asynchronous control signals, meaning that memory and I/O devices do not explicitly require access to the microprocessor clock to communicate on the bus. However, many of the 6800 peripherals require their own copy of the clock to run internal logic.

As with all synchronous logic, the 6800's bus is internally controlled by the microprocessor clock, but the nature of the control signals enables asynchronous read and write transactions without referencing that clock, as shown in Fig. 6.2. An address is placed onto the bus along with the proper state of the R/W select signal (read = 1, write = 0) and a valid memory address (VMA) enable that indicates an active bus cycle. In the case of a write, the write data is driven out some time later. For reads, the data must be returned fast enough to meet the microprocessor's timing specifications. The 6802/6808 were manufactured in 1-, 1.5-, and 2-MHz speed grades. At 2 MHz, a peripheral device has to respond to a read request with valid data within 210 ns after the assertion of address, R/W, and VMA. A peripheral has up to 290 ns from the assertion of these signals to complete a write transaction.* In a real system, VMA, combined with address decoding logic, would drive the individual chip select signals to each peripheral.

In some situations, slow peripherals may be used that cannot execute a bus transaction in the time allowed by the microprocessor. The 6800 architecture deals with this by stretching the clock during

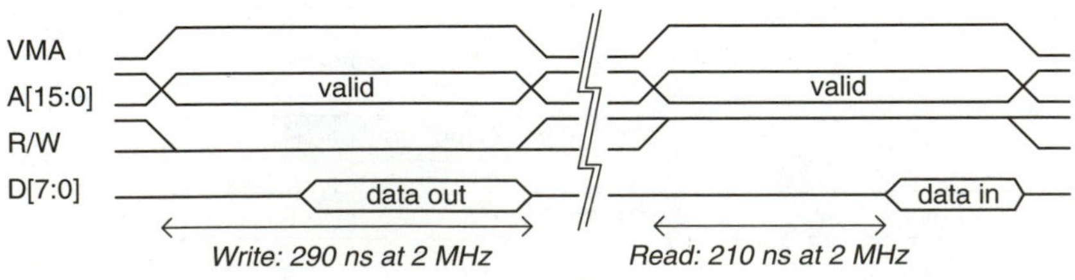

FIGURE 6.2 6802/6808 basic bus timing.

* *8-Bit Microprocessor and Peripheral Data,* Motorola, 1983, pp. 3–182.

a slow bus cycle. A clock cycle can be stretched as long as 10 μs, enabling extremely slow peripherals by delaying the next clock edge that will advance the microprocessor's internal state and terminate a pending bus cycle. This stretching is performed by an external clock circuit for a 6800, or by the internal clock of the 6802/6808. As with many modern microprocessors, the 6802/6808 provides a pin that delays the end of the current bus cycle. This memory ready (MR) signal is normally high, signaling that the addressed device is ready. When brought low, the clock is internally stretched until MR goes high again. Early microprocessors such as the 6800 used clock stretching to delay bus cycles. Most modern microprocessors maintain a constant clock frequency and, instead, insert discrete *wait states*, or extra clock cycles, into a bus transaction when a similar type of signal is asserted. This latter method is usually preferable in a synchronous system because of the desire to maintain a simple clock circuit and to not disrupt other logic that may be running on the microprocessor clock.

Motorola's success with the 6800 motivated it to introduce the upgraded 6809 in 1978. The 6809 is instruction set compatible with the 6800 but includes several new registers that enable more flexible access to memory. Two stack pointers are present: the existing hardware controlled register for subroutine calls and interrupts, and another for user control. The user stack pointer can be used to efficiently pass parameters to subroutines as they are called without conflicting with the microprocessor's push/pop operations involving the program counter and other registers. A second index register and the ability to use any of the four 16-bit pointer registers as index registers were added to enable the simultaneous handling of multiple data structure pointers without having to continually save and recall index register values. The 6809's two accumulators can be concatenated to form a 16-bit accumulator that enables 16-bit arithmetic with an enhanced ALU. This ALU is also capable of eight-bit unsigned multiplication, which made the 6809 one of the first integrated microprocessors with multiplication capability.

Other improvements in the 6809 included a direct page register (DPR) for a more flexible eight-bit direct addressing mode. The 8-bit DPR, representing A[15:8], is combined with an 8-bit direct address, representing A[7:0], to form a 16-bit direct address, thereby enabling an 8-bit direct address to reference any location in the complete 64-kB address space. The 6809 also included a more advanced bus interface with direct support for an external DMA controller. Several desktop computers, including the Tandy/Radio Shack TRS-80 Color Computer, and various platforms, including arcade games, utilized the 6809.

While still available from odd-lot retail outlets, the original 6800 family members are no longer practical to use in many computing applications. Their capabilities, once leading edge, are now available in smaller, more integrated ICs at lower cost and with lower power consumption. However, the 6800 architecture is alive and well in the 68HC05/68HC08 and 68HC11 microcontroller families that are based on the 6800/6802/6808 and 6809 architectures, respectively. These microcontrollers are available with a wide range of integrated features with on-board RAM, ROM (mask ROM, EEPROM, or EPROM), serial ports, timers, and analog-to-digital converters.

6.3 INTEL 8051 MICROCONTROLLER FAMILY

Following their success in the microprocessor market, Intel began manufacturing microcontrollers in 1976 with the introduction of the 8048 family. This early microcontroller contains 64 bytes of RAM, 1 kB of ROM, a simple 8-bit microprocessor core, and an 8-bit timer/counter as its sole on-board peripheral. (Subsequent variants, the 8049 and 8050, include double and four times the memory of the 8048, respectively.) The microprocessor consists of a 12-bit program counter, an 8-bit accumulator and ALU, and a 3-bit stack pointer. The 8048 is a complete computer on a single chip and gained a certain amount of fame in the 1980s when it was used as the standard keyboard controller on the

IBM PC because of its simplicity and low cost. The 8048 was manufactured in a 40-pin DIP and could be expanded with external memory and peripherals via an optional external address/data bus. However, when operated as a nonexpanded single-chip computer, the pins that would otherwise function as its bus were available for general I/O purposes—a practice that is fairly standard on microcontrollers.

Motivated by the popularity of the 8048, Intel introduced the 8051 microcontroller in 1980, which is substantially more powerful and flexible. The 8051's basic architecture is shown in Fig. 6.3. It contains 128 bytes of RAM, 4 kB of ROM, two 16-bit timer/counters, and a serial port. Registers within the microprocessor are 8 bits wide except for the 16-bit data pointer (DPTR) and program counter (PC). Memory is divided into mutually exclusive program and data sections that each can be expanded up to 64 kB in size via an external bus. Expansion is accomplished by borrowing pins from two of the four 8-bit I/O ports. Intel manufactured several variants of the 8051. The 8052 doubled the amount of on-chip memory to 256 bytes of RAM and 8 kB of ROM and added a third timer. The 8031/8032 are 8051/8052 chips without on-board ROM. The 8751/8752 are 8051/8052 devices with EPROM instead of mask ROM. As time went by and the popularity of the 8051 family increased, other companies licensed the core architecture and developed many variants with differing mixes of memory and peripherals.

Ports 0 through 3 are each eight-bit bidirectional I/O structures that can be used as either general-purpose signals or as dedicated interface signals according to the system configuration. In a single-chip configuration where all memory is contained on board, the four ports may be assigned freely. Some peripheral functions use these I/O pins, but if a specific function is not required, the pins may be used in a generic manner. Port 3 is the default peripheral port where pins are used for the serial port's transmit and receive, external interrupt request inputs, counter increment inputs, and external bus expansion control signals. Port 1 is a general-purpose port that is also assigned for additional peripheral support signals when an 8051 variant contains additional peripheral functions beyond what can be supported on port 3 alone.

In a multichip configuration where memory and/or additional peripherals are added externally, ports 0 and 2 are used for bus expansion. Port 0 implements a multiplexed address/data bus where the 8051 first drives the lower eight address bits and then either drives write-data or samples read-data in a conventional bidirectional data bus scheme. In this standard configuration, the lower address bits, A[7:0], are latched externally by a discrete logic chip (generally a 74LS373 or similar), and the 8051 drives an address latch enable (ALE) signal to control this latch as shown in Fig. 6.4. This multiplexed address/data scheme saves precious pins on the microcontroller that can be used

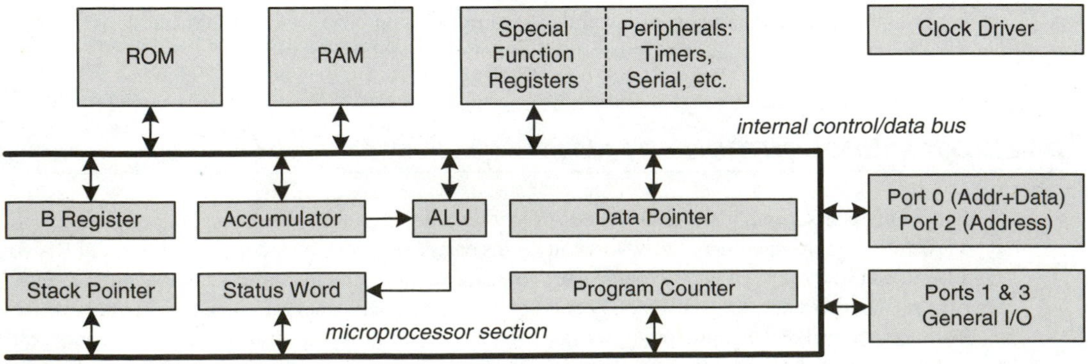

FIGURE 6.3 8051 overall architecture.

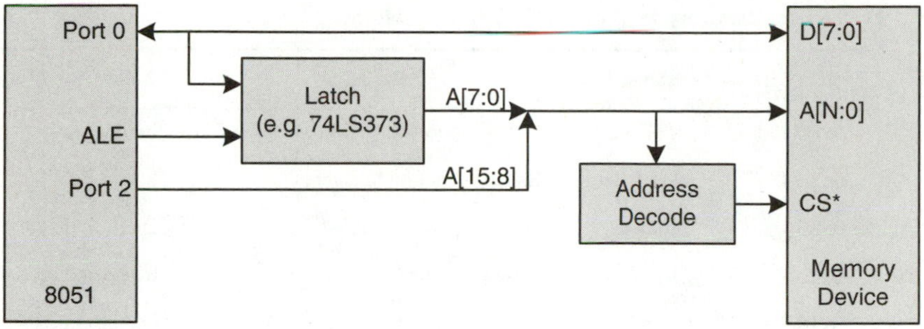

FIGURE 6.4 8051 system with external address latch.

for valuable I/O functions. Some applications may suffice with just an eight-bit external address bus. For example, if the only expansion necessary were a special purpose I/O device, 256 bytes would probably be more than enough to communicate with the device. However, some applications demand a fully functional 16-bit external address bus. In these situations, port 2 is used to drive the upper address bits, A[15:8].

The 8051's microprocessor is very capable for such an early microcontroller. It includes integer multiply and divide instructions that utilize eight-bit operands in the accumulator and B register, and it then places the result back into those registers. The stack, which grows upward in memory, is restricted to on-board RAM only (256 bytes at most), so only an eight-bit stack pointer is implemented. Aside from the general-purpose accumulator and B registers, the 8051 instruction set can directly reference 8 byte-wide general-purpose registers, numbered R0 through R7, that are mapped as 4 banks in the lower 32 bytes of on-board RAM. The active register bank can be changed at any time by modifying two bank-select bits in the status word. The map of on-board data memory is shown in Table 6.2. At reset, register bank 0 is selected, and the stack pointer is set to 0x07, meaning that the stack will actually begin at location 0x08 when the first byte is eventually pushed. Above the register banks is a 16-byte (128-bit) region of memory that is bit addressable. Microcontroller applications often involve reading status information, checking certain bits to detect particular events, and then triggering other events. Using single bits rather than whole bytes to store status information saves precious memory in a microcontroller. Therefore, the 8051's bit manipulation instructions can make efficient use of the chip's resources from both instruction execution and memory usage perspectives. The remainder of the lower 128-byte memory region contains 80 bytes of general-purpose memory.

The upper 128 bytes of data memory are split into two sections: special-function registers and RAM. Special-function registers are present in all 8051 variants, but their definitions change according to the specific mix of peripherals in each variant. Some special-function registers are standard across all 8051 variants. These registers are typically those that were implemented on the original 8051/8052 devices and include the accumulator and B registers; the stack pointer; the data pointer; and serial port, timer, and I/O port control registers. Each time a manufacturer adds an on-board peripheral to the 8051, accompanying control registers are added into the special-function memory region.

On variants that incorporate 256 bytes of on-board RAM, the upper 128 bytes are also mapped into a parallel region alongside the special-function registers. Access between RAM and special-function registers is controlled by the addressing mode used in a given instruction. Special-function registers are accessed with direct addressing only. Therefore, such an instruction must follow the opcode with an eight-bit address. The upper 128 bytes of RAM are accessed with indirect addressing only. Therefore, such an instruction must reference one of the eight general-purpose registers (R0

TABLE 6.2 Memory Map of On-Board Data Memory

Memory Range	Range Size	Purpose	Addressing Mode(s)
0x80–0xFF	128 bytes	General-purpose RAM (except 8051)	Indirect only
0x80–0xFF	128 bytes	Special-function registers	Direct only
0x30–0x7F	80 bytes	General-purpose RAM	Direct/indirect
0x20–0x2F	16 bytes	Bit-addressable RAM/general	Direct/indirect
0x18–0x1F	8 bytes	Register bank 3/general	Direct/indirect
0x10–0x17	8 bytes	Register bank 2/general	Direct/indirect
0x08–0x0F	8 bytes	Register bank 1/general	Direct/indirect
0x00–0x07	8 bytes	Register bank 0/general	Direct/indirect

through R7 in the currently selected bank) whose value is used to index into that portion of RAM. The lower 128 bytes of RAM are accessible via both direct and indirect addressing.

The 8051 is a good study in maximizing the capabilities of limited resources. Access to external memory is supported through a variety of indirect and indexed schemes that provide an option to the system designer of how extensive an external bus is implemented. Indirect access to external data memory is supported in both 8- and 16-bit address configurations. In the 8-bit mode, R0 through R7 are used as memory pointers, and the resulting address is driven only on I/O port 0, freeing port 2 for uses other than as an address bus. The DPTR functions as a pointer into data memory in 16-bit mode, enabling a full 64-kB indirect addressing range. Indexed access to external program memory is supported by both the DPTR and the PC. Being program memory (ROM), only reads are supported. Both DPTR and PC can serve as index base address registers, and the current value in the accumulator serves as an offset to calculate a final address of either DPTR+A or PC+A.

The 8051's external bus interface is asynchronous and regulated by four basic control signals: ALE, program storage enable (PSEN*), read enable (RD*), and write enable (WR*). Figure 6.5 shows the interaction of these four control signals and the two bus ports: ports 0 and 2. Recall that ALE causes an external latch to retain A[7:0] that is driven from port 0 during the first half of the access and prior to port 0 transitioning to a data bus role. The timing delays noted are for a standard 12-MHz operating frequency (the highest frequency supported by the basic 8051 devices, although certain newer devices can operate at substantially faster frequencies).*

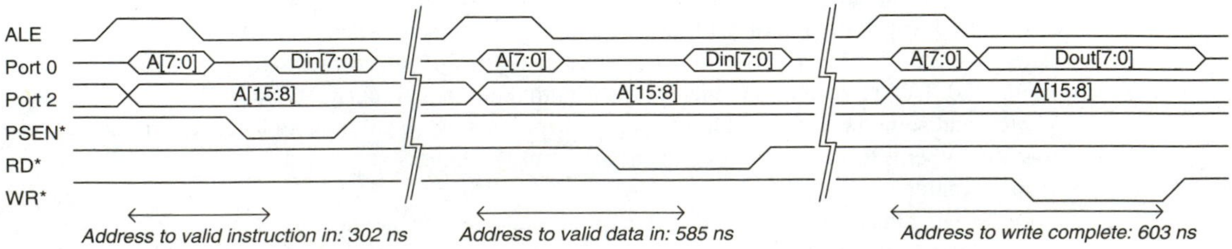

FIGURE 6.5 8051 bus interface timing.

* *Embedded Controller Handbook,* Vol. I, Intel, 1988, pp. 10-6 through 10-9.

Although the specific timing delays of program memory and data memory reads are different, they exhibit the same basic sequence of events. (More time is allowed for data reads than for instruction reads from program memory.) Therefore, if the engineer properly accounts for the timing variations by selecting memory and logic components that are fast enough to satisfy the PSEN* and RD* timing specifications simultaneously, program and data memory can actually be merged into a unified memory space external to the chip. Such unification can be performed by generating a general memory read enable, MRE*, that is the AND function of PSEN* and RD*. In doing so, whenever either read enable is driven low by the 8051, MRE* will be low. This can benefit some applications by turning the 8051 into a more general-purpose computing device that can load a program into its "data memory" and then execute that same program from "program memory." It also enables indexed addressing to operate on data memory, which normally is restricted to indirect addressing as discussed previously.

Timers such as those found in the 8051 are useful for either counting external events or triggering low-frequency events themselves. Each timer can be configured in two respects: whether it is a timer or counter, and how the count logic functions. The selection of timer versus counter is a decision between incrementing the count logic based on the microcontroller's operating frequency or on an external event sensed via an input port pin. The 8051's internal logic runs in a repetitive pattern of 12 clock cycles in which 1 machine cycle consists of 12 clock cycles. Therefore, the count logic increments once each machine cycle when in timer mode. When in counter mode, a low-to-high transition (rising edge) on a designated input pin causes the counter to increment. The counter can be configured to generate an interrupt each time it rolls over from its maximum count value back to its starting value. This interrupt can be used to either trigger a periodic maintenance routine at regular intervals (timer mode) or to take action once an external event has occurred a set number of times (counter mode). If not configured to generate an interrupt, the software can periodically poll the timer to see how many events have occurred or how much time has elapsed.

The timers inherently possess two 8-bit count registers that can be configured in a variety of ways as shown in Fig. 6.6. A timer can be configured as a conventional 16-bit counter, as two 8-bit

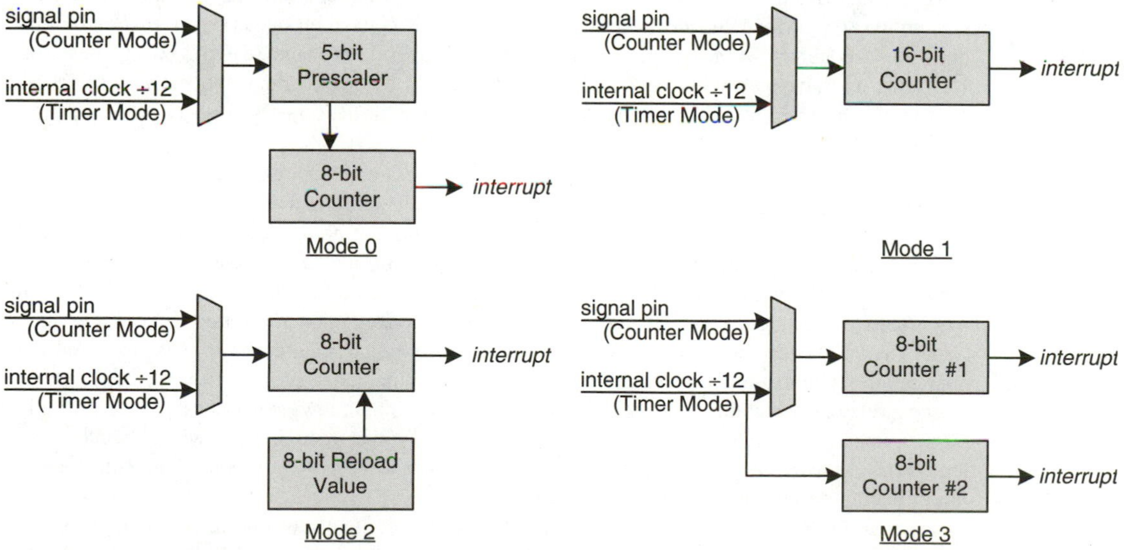

FIGURE 6.6 8051 timer configurations.

counters, as a single 8-bit counter with a 5-bit prescaler, and as a single 8-bit counter with an 8-bit reload value. The first two modes mentioned are straightforward: the timers count from 0 to either 65,535 (16-bit) or 255 (8-bit) before rolling over and perhaps generating an interrupt. The third mode is similar, but the 8-bit counter increments only once every 32 machine cycles. The 5-bit (2^5 = 32) prescaler functions as a divider ahead of the main counter. Apparently, the main reason for including this mode was to retain function compatibility with the 8048's prescaled timer. The fourth mode is interesting, because the 8-bit counter is reloaded with an arbitrary 8-bit value rather than 0 after reaching its terminal count value (255). When operated in timer mode, this feature enables the timer to synthesize a wide range of low-frequency periodic events. One very useful periodic event is an RS-232 bit-rate generator. A commonly observed 8051 operating frequency is 11.0592 MHz. When this frequency is divided by 12, a count increment rate of 921.6 kHz is obtained. Further dividing this frequency by divisors such as 96 or 384 yields the standard RS-232 bit rates 9.6 kbps and 2.4 kbps. A divisor of 384 cannot be implemented in an 8-bit counter. Instead, a selectable ÷16 or ÷32 counter is present in the serial port logic that generates the final serial bit rate.

The 8051's on-board serial port implements basic synchronous or asynchronous transmit and receive shift-register functionality but does not incorporate hardware handshaking of the type used in RS-232 communications. Serial transmission is initiated by writing the desired data to a transmit register. Incoming data is placed into a receive register, and an interrupt can be triggered to invoke a serial port ISR. The serial port can be configured in one of four modes, two of which are higher-frequency fixed bit rates, and two of which are lower-frequency variable bit rates established by the rollover characteristics of an on-board timer. Mode 0 implements a synchronous serial interface where the "receive data" pin is actually bidirectional and a transmit clock is emitted on the "transmit data" pin. This mode operates on 8-bit data and a fixed bit rate of 1/12 the operating frequency.

Mode 1 implements an asynchronous transmit/receive serial port where ten bits are exchanged for every byte: a start bit, eight data bits, and a stop bit. The bit-rate is variable according to a timer rollover rate. Mode 3 is very similar to mode 1, with the added feature that a ninth data bit is added to each byte. This extra data bit can be used for parity in an RS-232 configuration or for another application-specific purpose. These two modes can be used to implement an RS-232 serial port without hardware handshaking. Software-assisted hardware handshaking could be added using general I/O pins on the 8051. Mode 2 is identical to mode 3 except for its fixed bit rate at either 1/32 or 1/64 the operating frequency. Modes 1 and 3 can be made to operate at standard RS-232 bit rates from 19.2 kbps on downward with the aforementioned 11.0592 MHz operating frequency. A selectable ÷16 or ÷32 counter within the serial port logic combines with the timer rollover to achieve the desired serial bit rate.

Intel's 8051 architecture has been designed into countless applications in which a small, embedded computer is necessary to regulate a particular process. The original 40-pin devices are still commonly used and found in distributors' warehouses, but a host of newer devices are popular as well. Some of these variants are larger and more capable than the original and include more I/O ports, on-board peripherals, and memory. Some variants have taken the opposite direction and are available in much smaller packages (e.g., 20 pins) with low power consumption for battery-powered applications. There are even special versions of the 8051 that are radiation hardened for space and military applications. Companies that manufacture 8051 variants include Atmel, Maxim (formerly Dallas Semiconductor), and Philips. Atmel manufactures a line of small, low-power 8051 products. Maxim offers a selection of high-speed 8051 microcontrollers that run at up to 33 MHz with a 4-cycle architecture, as compared to 12 in the original 8051. Philips has a broad 8051 product line with a variety of peripherals to suit many individual applications.

The mature ROM-less 8031/8032 members of the 8051 family can be ordered through many mail order retail electronics outlets for only a few dollars apiece. The equally mature 8751/8752 EPROM devices can also be found from many of these same sources, though at a higher price as a result of the expense of the ceramic DIP in which they are most often found. More specialized 8051 variants may be available only through manufacturers' authorized distributors.

6.4 MICROCHIP PIC® MICROCONTROLLER FAMILY

By the late 1980s, microcontrollers and certain microprocessors were well established in embedded control applications. Despite advances in technology, not many devices could simultaneously address the needs for low power, moderate processing throughput, very small packages, and diverse integrated peripherals. Microchip Technology began offering a family of small peripheral interface controller (PIC®)* devices in the early 1990s that addressed all four of these needs. Microchip developed the compact PIC architecture based on a *reduced instruction set core* (RISC) microprocessor. The chips commonly run at up to 20 MHz and execute one instruction every machine cycle (four clock cycles)—except branches that consume two cycles. The key concept behind the PIC family is simplicity. The original 16C5x family, shown in Fig. 6.7, implements a 33-instruction microprocessor core with a single working register (accumulator), W, and only a two-entry subroutine stack. These devices contain as little as 25 bytes of RAM and 512 bytes of ROM, and some are housed in an 18-pin package that can be smaller than a fingernail. The PIC devices are not expandable via an external bus, further saving logic. This minimal architecture is what enables relatively high performance processing with low power consumption in a tiny package. Low-power operation is also coupled with a wide operating voltage range (2 to 6.25 V), further simplifying certain systems by not always requiring voltage regulation circuits.

No interrupt feature is included, which is a common criticism of the architecture; this was fixed in subsequent PIC microcontroller variants. PIC devices are, in general, fully static, meaning that they can operate at an arbitrarily low frequency; 32 kHz is sometimes used in very power-sensitive appli-

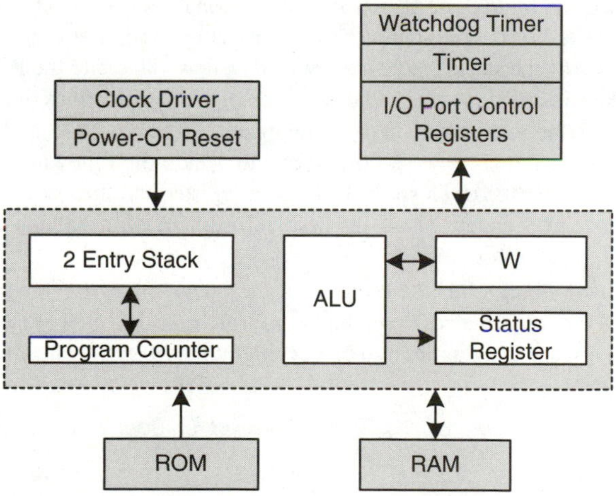

FIGURE 6.7 PIC microcontroller 16C5x architecture.

---

* The Microchip name, PIC, and PICmicro are registered trademarks of Microchip Technology Inc. in the U.S.A. and other countries.

cations in which only microamps of current are consumed. To further reduce cost and complexity, the microcontrollers contain on-board clock drivers that work with a variety of external frequency-reference components. Quartz crystals are supported, as they are very accurate references. In very small systems wherein cost and size are absolutely paramount concerns, and absolute frequency accuracy is not a concern, less-expensive and smaller frequency references can be used with a PIC microcontroller. One step down from a crystal is a *ceramic resonator,* which functions on a similar principle but with lower accuracy and cost. Finally, if the operating frequency can be allowed to vary more substantially with temperature, voltage, and time, a resistor/capacitor (RC) oscillator, the cheapest option, is supported. Tiny surface mount RC components take up very little circuit board area and cost pennies.

The original 16C5x family incorporates only the most basic of peripherals: power-on-reset, an eight-bit timer/counter, and a *watchdog timer.* A power-on reset circuit ensures that the microcontroller reliably begins operation when power is applied by automatically controlling an internal reset signal. On most microprocessors, reset is purely an external function. A *watchdog* timer can be configured to automatically reset the microcontroller if the system develops an unforeseen fault that causes the software to "crash." The watchdog functions by continuously counting, and software must periodically reset the counter to prevent it from reaching its terminal count value. If this value is reached, the internal reset signal is asserted. Under normal circumstances where software is functioning properly, it resets the watchdog timer with plenty of time to spare. However, if the software crashes, it will presumably not be able to reset the watchdog, and a system reset will soon follow. The watchdog timeout period is configurable from milliseconds to seconds. When using a watchdog, the timeout period is chosen to be long enough so that software can reliably reset the counter to prevent accidental reset, yet short enough to catch a fault and reset the system before serious problems result.

The PIC microcontroller's RISC instruction set obeys the tenets of the general RISC style: accomplish the same task with more simple instructions instead of fewer complex ones. Fewer types of simple instructions require less processing logic within the chip. As an example, there are just two branch instructions: `CALL` and `GOTO`. `CALL` is an unconditional branch-to-subroutine that places the current PC onto the stack. It is the programmer's responsibility to not nest subroutines more than two deep to avoid overflowing the stack. `GOTO` simply loads a new value into the PC. To implement conditional branches, these instructions are paired with one of four instructions that perform an action and then skip the following instruction if a particular result is true. `INCFSZ` and `DECFSZ` increment or decrement a designated register, respectively, and then skip the following instruction if the result is zero. `BTFSC` and `BTFSS` test a specified bit in a register and then skip the following instruction if the bit is 0 or 1, respectively. Using the first pair of instructions, a loop could be written as shown in Fig. 6.8.

Assembly languages commonly offer the programmer a means of representing numeric values with alphanumeric labels for convenience. Here, the loop variable `COUNT` is set to address 0 with an *equate* directive that is recognized and processed by the assembler. `MOVWF` transfers the value in the

```
COUNT          EQU        0             ; define COUNT at address 0

               MOVLW      0x09          ; 9 loop iterations
               MOVW F     COUNT         ; iteration tracking register
LOOP_START     <loop instructions>      ; body of loop
               DECFSZ     COUNT,1       ; done with loop yet?
               GOTO       LOOP_START    ; non-zero, keep going...
               <more instructions>      ; zero, loop is done...
```

FIGURE 6.8 16C5x assembly language loop.

working register into a particular location in the register file. In this example, the `GOTO` instruction is executed each time through the loop until `COUNT` is decremented to 0. (The operand "1" following `COUNT` in `DECFSZ` tells the microcontroller to place the decremented result back into `COUNT` rather than into the working register.) At this point, `GOTO` is skipped, because the result is 0, causing the microcontroller to continue executing additional instructions outside of the loop.

The second pair of skip instructions, `BTFSC` and `BTFSS`, directly supports the common situation in which the microcontroller reads a set of flag bits in a single byte and then takes action based on one of those bits. Such bit-testing instructions are common in microcontrollers by virtue of their intended applications. Some generic microprocessors do not contain bit-testing instructions, requiring software to isolate the bit of interest with a logical *mask* operation. A mask operation works as follows with an AND function, assuming that we want to isolate bit 5 of a byte so as to test its state:

	1	0	1	1	0	1	1	1	Byte to test
	0	0	1	0	0	0	0	0	Mask
AND	0	0	1	0	0	0	0	0	Bit 5 isolated

Here, the mask prevents any bit other than bit 5 from achieving a 1 state in the final result. This masking operation could then be followed with a conditional branch testing whether the overall result was 0 or non-0. In the PIC architecture, and most other microcontrollers, this process is performed directly with bit-test instructions.

Masking also works to set or clear individual bits but, here again, the PIC architecture contains special instructions to optimize this common microcontroller function. Using the above example, bit 5 can be set, regardless of its current state, by ORing the data byte with the same mask.

	1	0	1	1	0	1	1	1	Starting byte
	0	0	1	0	0	0	0	0	Mask
OR	0	0	1	0	0	0	0	0	Result

The mask ensures that only bit 5 is set, regardless of its current state. All other bits propagate through the OR process without being changed. Similarly, an individual bit can be cleared, regardless of its current state, with an inverse AND mask:.

	1	0	1	1	0	1	1	1	Starting byte
	1	1	0	1	1	1	1	1	Mask
AND	1	0	0	1	0	1	1	1	Result

Here, all bits other than bit 5 are ANDed with 1, propagating them through to the result. Bit 5 is ANDed with 0, unconditionally forcing it to a 0. Rather than having to load a mask and then execute a logical instruction, the PIC architecture contains two instructions to clear and set arbitrary bits in a specified register: `BCF` and `BSF`, respectively.

Microchip extended the 16C5x's architecture and features with the 16C6x and 16C7x families. The 16C5x's advantages of low power consumption, wide operating voltage range, and small size are retained. Improvements include more memory (up to 368 bytes of RAM and 8 kB of ROM), a more versatile microprocessor core with interrupt capability, an eight-level stack, and a wider selection of on-board peripherals including additional timers, serial ports, and *analog-to-digital* (A/D) converters. (An A/D converter is a circuit that converts an analog voltage range into a range of binary values. An 8-bit A/D converter covering the range of 0 to 5 V would express voltages in that range as a byte value with a resolution of 5 V $\div$ (2^8 – 1) increments = 19.6 mV per increment.) Between four and eight A/D converters are available in 'C7x devices.

Some PIC microcontrollers contain two serial ports on the same chip: an asynchronous port suitable for RS-232 applications and a synchronous port capable of SPI or I^2C operation in conjunction with other similarly equipped ICs in a system. At the other end of the spectrum, very small PIC devices are available in eight-pin packages—small enough to fit almost anywhere.

6.5 INTEL 8086 16-BIT MICROPROCESSOR FAMILY

Intel moved up to a 16-bit microprocessor, the 8086, in 1978—just two years after introducing the 8085 as an enhancement to the 8080. The "x86" family is famous for being chosen by IBM for their original PC. As PCs developed during the past 20 years, the x86 family grew with the industry—first to 32 bits (80386, Pentium) and more recently to 64 bits (Itanium). While the 8086 was a new architecture, it retained certain architectural characteristics of the 8080/8085 such that assembly language programs written for its predecessors could be converted over to the 8086 with little or no modification. This is one of the key reasons for its initial success.

The 8086 contains various 16-bit registers as shown in Fig. 6.9, some of which can be manipulated one byte at a time. AX, BX, CX, and DX are general-purpose registers that have alternate functions and that can be treated as single 16-bit registers or as individual 8-bit registers. The accumulator, AX, and the flags register serve their familiar functions. BX can serve as a general pointer. CX is a loop iteration count register that is used inherently by certain instructions. DX is used as a companion register to AX when performing certain arithmetic operations such as integer division or handling long integers (32 bits).

The remaining registers are pointers of various types that index into the 8086's somewhat awkward segmented memory structure. Despite being a 16-bit microprocessor with no register exceeding 16 bits in size, Intel recognized the need for more than 64 kB of addressable memory in more advanced computers. One megabyte of memory space was decided upon as a sufficiently large address space in the late 1970s, but the question remained of how to access that memory with 16-bit pointers. Intel's solution was to have programmers arbitrarily break the 1 MB address space into multiple 64-kB special-purpose segments—one for instructions (code segment), two for data (primary data and "extra" data), and one for the stack. Memory operations must reference one of these defined segments, requiring only a 16-bit pointer to address any location within a given segment. Segments can be located anywhere in memory, as shown in Fig. 6.10, and can be moved at will to provide flexibility for different applications. Additionally, there is no restriction on overlapping of segments.

Each segment register represents the upper 16 bits of a 20-bit pointer (2^{20} = 1 MB) where the lower 4 bits are fixed at 0. Therefore, a segment register directly points to an arbitrary location in 1 MB of memory on a 16-byte boundary. A pointer register is then added to the 20-bit segment address to yield a final 20-bit address, the effective address, with which to fetch or store data. Algebraically, this relationship is expressed as: effective address = (segment pointer $\times$ 16) + offset pointer.

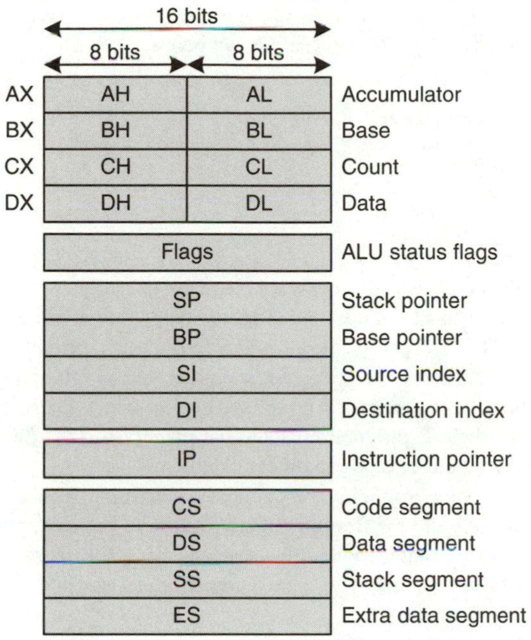

FIGURE 6.9 8086 register set.

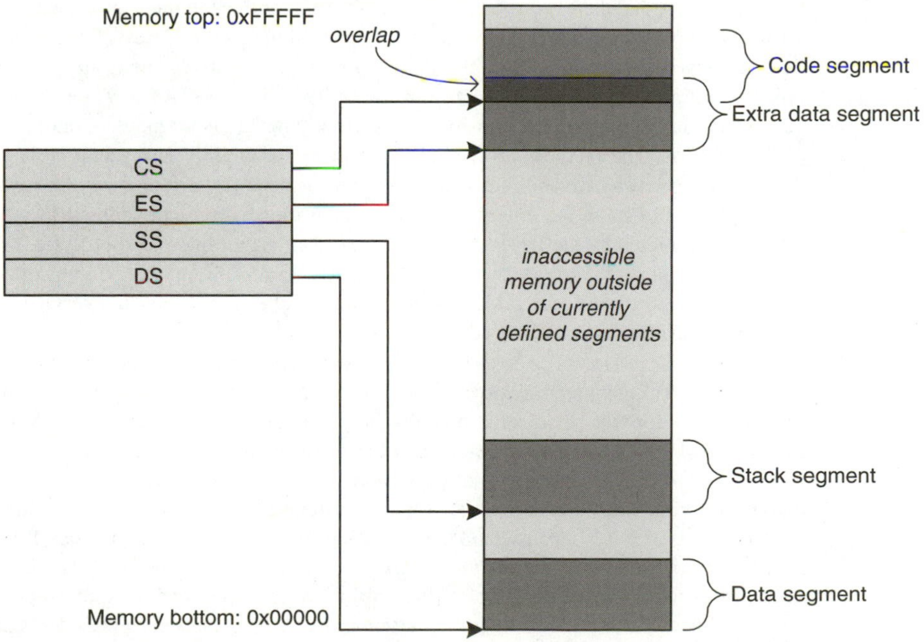

FIGURE 6.10 8086 segments.

Inside the microprocessor, this math is performed by shifting the segment pointer (0x135F) left by four bits and then adding the offset pointer (0x0102) as shown below.

	1	3	5	F	0	Segment pointer
+		0	1	0	2	Offset pointer
	1	3	6	F	2	Effective address

This segmented addressing scheme has some awkward characteristics. First, programs must organize their instructions and data into 64-kB chunks and properly keep track of which portions are being accessed. If data outside of the current segments is desired, the appropriate segment register must be updated. Second, the same memory location can be represented by multiple combinations of segment and offset values, which can cause confusion in sorting out which instruction is accessing which location in memory. Nonetheless, programmers and the manufacturers of their development tools have figured out ways to avoid these traps and others like them.

Instructions that reference memory implicitly or explicitly determine which offset pointer is added to which segment register to yield the desired effective address. For example, a push or pop instruction inherently uses the stack pointer in combination with the stack segment register. However, an instruction to move data from memory to the accumulator can use one of multiple pointer registers relative to any of the segment registers.

The 8086's reset and interrupt vectors are located at opposite ends of the memory space. On reset, the instruction pointer is set to 0xFFFF0, and the microprocessor begins executing instructions from this address. Therefore, rather than being a true vector, the 16-byte reset region contains normal executable instructions. The interrupt vectors are located at the bottom of the memory space starting from address 0, and there are 256 vectors, one for each of the 256 interrupt types. Each interrupt vector is composed of a 2-byte segment address and a 2-byte offset address, from which a 20-bit effective address is calculated. When the 8086's INTR pin is driven high, an interrupt acknowledge process begins via the INTA* output pin. The 8086 pulses INTA* low twice and, on the second pulse, the interrupting peripheral drives an interrupt type, or vector number, onto the eight lower bits of the data bus. The vector number is used to index into the interrupt vector table by multiplying it by 4 (shifting left by two bits), because each vector consists of four bytes. For example, interrupt type 0x03 would cause the microprocessor to fetch four bytes from addresses 0x0C through 0x0F. Interrupts triggered by the INTR pin are all maskable via an internal control bit. Software can also trigger interrupts of various types via the `INT` instruction. A nonmaskable interrupt can be triggered by external hardware via the NMI pin. NMI initiates the type-2 interrupt service routine at the address indicated by the vector at 0x08-0x0B.

Locating the reset boot code at the top of memory and the interrupt vectors at the bottom often leads to an 8086 computer architecture with ROM at the top and some RAM at the bottom. ROM must be at the top, for obvious reasons. Placing the interrupt vector table in RAM enables a flexible system in which software applications can install their own ISRs to perform various tasks. On the original IBM PC platform, it was not uncommon for programs to insert their own ISR addresses into certain interrupt vectors located in RAM. The system timer and keyboard interrupts were common objects of this activity. Because the PC's operating system already implemented ISRs for these interrupts, the program could redirect the interrupt vector to its own ISR and then call the system's default ISR when its own ISR completed execution. If properly done, this interrupt chaining process could add new features to a PC without harming the existing housekeeping chores performed by the standard ISRs. Chaining the keyboard interrupt could enable a program that is normally dormant to pop up each time a particular key sequence is pressed.

Despite its complexity and 16-bit processing capability, the 8086 was originally housed in a 40-pin DIP—the same package used for most 8-bit processors of the time. Intel chose to use a multiplexed address/data scheme similar to that used on the 8051 microcontroller, thereby saving 16 pins. The 8086's 20-bit address bus is shared by the data bus on the lower 16 bits and by status flags on the upper 4 bits. Combined with additional signals, these status flags control the microprocessor's bus interface. As with Intel's other microprocessors, the 8086 contains separate address spaces for memory and I/O devices. A control pin on the chip indicates whether a transaction is memory or I/O. While the memory space is 1 MB in size, the I/O space is only 64 kB. The 8086 bus interface operates in one of two modes, minimum and maximum, determined by a control pin tied either high or low, respectively. In each of these two modes, many of the control and status pins take on different functions. In minimum mode, the control signals directly drive a standard "Intel-style" bus similar to that of the 8080 and 8051, with read and write strobes and address latch enable. Other signals include a READY signal for inserting wait states for slow peripherals and a bus grant/acknowledge mechanism for supporting DMA or similar bus-sharing peripherals. Minimum mode is designed for smaller systems in which little address decoding logic is necessary to interface the 8086 to memory and peripherals devices. Maximum mode is designed for larger systems where an Intel companion chip, the 8288 bus controller, integrates more complex bus control logic onto an off-the-shelf IC. In maximum mode, certain status and control pins communicate more information about what type of transaction is being performed at any given time, enabling the 8288 to take appropriate action.

The 8086's 16-bit data bus is capable of transacting a single byte at a time for purposes of accessing byte-wide peripherals. One early advantage of the 8086 was its backward bus compatibility with the 8080/8085. In the 1970s, Intel manufactured a variety of I/O peripherals such as timers and parallel I/O devices for their eight-bit microprocessors. The 8086's ability to perform byte-wide transactions enabled easy reuse of existing eight-bit peripheral products. Two signals, byte high enable (BHE*) and address bit zero (A[0]), communicate the width and active byte of each bus transaction as shown in Table 6.3.

TABLE 6.3 8086 Bus Sizing

BHE*	A[0]	Transaction Type
0	0	16-bit transaction
0	1	8-bit transaction: high byte (odd address)
1	0	8-bit transaction: low byte (even address)
1	1	Undefined

Intel's microprocessors follow the *little-endian* byte ordering convention. *Little-endian* refers to the practice of locating the LSB of a multibyte quantity in a lower address and the MSB in a higher address. In a little-endian 16-bit microprocessor, the value 0x1234 would be stored in memory by locating 0x12 into address 1 and 0x34 into address 0. *Big-endian* is the opposite: locating the LSB in the higher address and the MSB in the lower address. Therefore, a big-endian 16-bit microprocessor would store 0x12 into address 0 and 0x34 into address 1. To clarify the difference, Table 6.4 shows little-endian versus big-endian for 16- and 32-bit quantities as viewed from a memory chip's perspective. Here, ADDR represents the base address of a multibyte data element.

Proponents of little-endian argue that it makes better sense, because the low byte goes into the low address. Proponents of big-endian argue that it makes better sense, because data is stored in

TABLE 6.4 Little-Endian vs. Big-Endian

Value	Endianness	ADDR+0	ADDR+ 1	ADDR+2	ADDR+3
0x1234	Little	0x34	0x12	X	x
0x1234	Big	0x12	0x34	X	x
0x12345678	Little	0x78	0x56	0x34	0x12
0x12345678	Big	0x12	0x34	0x56	0x78

memory as you would read and interpret it. The choice of "endianness" is rather religious and comes down to personal preference. Of course, if you are designing with a little-endian microprocessor, life will be made simpler to maintain the endianness consistently throughout the system.

At the time of the 8086's introduction, 16-bit desktop computer systems were almost unheard of and could be substantially more expensive than 8-bit systems as a result of the increased memory size required to support the larger bus. To alleviate this problem and speed market acceptance of its architecture, Intel introduced the 8088 microprocessor in 1979, which was essentially an 8086 with an eight-bit data bus. A lower-cost computer system could be built with the 8088, because fewer EPROM and RAM chips were necessary, system logic did not have to deal with two bytes at a time, and less circuit board wiring was required. A tremendous benefit to Intel in designing the 8088 was the fact that it was chosen by IBM as the low-cost 16-bit heart of the original PC/XT desktop computer, thereby locking the x86 microprocessor family into the IBM PC architecture for decades to come.

A variety of companion chips were developed by Intel to supplement the 8086/8088. Among these was the 8087 math coprocessor that enhanced the 8086's computational capabilities with *floating-point* arithmetic operations. Floating-point arithmetic refers to a computer's handling of real numbers as compared to integers. The task of adding or multiplying two real numbers of arbitrary magnitude is far more complex than similar integer operations. Certain applications such as scientific simulations and realistic games that construct a virtual reality world make significant use of floating-point operations. The 8087 is a *coprocessor* rather than a peripheral, because it sits on the microprocessor bus in parallel with the 8086 and watches for special floating-point instructions. These instructions are then executed automatically by the 8087 rather than having to wait for the 8086 to request an operation. The 8086 was designed with the 8087's existence in mind and ignores instructions destined for the 8087. Therefore, software must specifically know if a math coprocessor is installed to run correctly. Many programs that ran on older systems with or without a coprocessor would first test to see if the coprocessor was installed and then execute either an optimized set of routines for the 8087 or a slower set of routines that emulated the floating-point operations via conventional 8086 instructions.

As the x86 family developed, the optional math coprocessor was eventually integrated alongside the integer processor on the same silicon chip. The 8087 gave way to the 80287 and 80387 when the 80286 and 80386 microprocessors were produced. When Intel introduced the 80486, the coprocessor, or *floating-point unit* (FPU), was integrated on chip. This integration resulted in a somewhat more expensive product, so Intel released a lower-cost 80486SX microprocessor without the coprocessor. An 80487SX was made available to upgrade systems originally sold with the 80486SX chips, but the overall situation proved somewhat chaotic with various permutations of microprocessors and systems with and without coprocessors. Starting with the Pentium, all of Intel's high-end microprocessors contain integrated FPUs. This trend is not unique to Intel. High-performance microprocessors in general began integrating the FPU at roughly the same time because of the performance benefits and the overall simplicity of placing the microprocessor and FPU onto the same chip.

6.6 MOTOROLA 68000 16/32-BIT MICROPROCESSOR FAMILY

Motorola followed its 6800 family by leaping directly to a hybrid 16/32-bit microprocessor architecture. Introduced in 1979, the 68000 is a 16-bit microprocessor, due to its 16-bit ALU, but it contains all 32-bit registers and a linear, nonsegmented 32-bit address space. (The original 68000 did not bring out all 32 address bits as signal pins but, more importantly, there are no architectural limitations of using all 32 bits.) That the register and memory architecture is inherently 32 bits made the 68000 family easily scalable to a full 32-bit internal architecture. Motorola upgraded the 68000 family with true 32-bit devices, including the 68020, 68040, and 68060, until switching to the PowerPC architecture in the latter portion of the 1990s for new high-performance computing applications. Apple Computer used the 68000 family in their popular line of Macintosh desktop computers. Today, the 68000 family lives on primarily as a mid-level embedded-processor core product. Motorola manufacturers a variety of high-end microcontrollers that use 32-bit 68000 microprocessor cores. However, in recent years Motorola has begun migrating these products, as well as their general-purpose microprocessors, to the PowerPC architecture, reducing the number of new designs that use the 68000 family.

The 68000 inherently supports modern software *operating systems* (OSs) by recognizing two modes of operation: supervisor mode and user mode. A modern OS does not grant unlimited access to application software in using the computer's resources. Rather, the OS establishes a restricted operating environment into which a program is loaded. Depending on the specific OS, applications may not be able to access certain areas of memory or I/O devices that have been declared off limits by the OS. This can prevent a fault in one program from crashing the entire computer system. The OS *kernel*, the core low-level software that keeps the computer running properly, has special privileges that allow it unrestricted access to the computer for the purposes of establishing all of the rules and boundaries under which programs run. Hardware support for multiple privilege levels is crucial for such a scheme to prevent unauthorized programs from freely accessing restricted resources. As microprocessors developed over the last few decades, more hardware support for OS privileges was added. That the 68000 included such concepts in 1979 is a testimony to its scalable architecture.

Sixteen 32-bit general-purpose registers, one of which is a user stack pointer (USP), and an 8-bit condition code register are accessible from user mode as shown in Fig. 6.11. Additionally, a supervisor stack pointer (SSP) and eight additional status bits are accessible from supervisor mode. Computer systems do not have to implement the two modes of operation if the application does not require it. In such cases, the 68000 can be run permanently in supervisor mode to enable full access to all resources by all programs. The SSP is used for stack operations while in supervisor mode, and the USP is used for stack operations in user mode. User mode programs cannot change the USP, preventing them from relocating their stacks. Most modern operating systems are *multitasking*, meaning that they run multiple programs simultaneously. In reality, a microprocessor can only run one program at a time. A multitasking OS uses a timer to periodically interrupt the microprocessor, perhaps 20 to 100 times per second, and place it into supervisor mode. Each time supervisor mode is invoked, the kernel performs various maintenance tasks and swaps the currently running program with the next program in the list of running programs. This swap, or *context switch*, can entail substantial modifications to the microprocessor's state when it returns from the kernel timer interrupt. In the case of an original 68000 microprocessor, the kernel could change the return value of the PC, USP, the 16 general-purpose registers, and the status register. When normal execution resumes, the microprocessor is now executing a different program in exactly the same state at which it was previously interrupted, because all of its registers are in the same state in which they were left. In such a scenario, each program has its own private stack, pointed to by a kernel-designated stack pointer.

The eight data registers, D0–D7, can be used for arbitrary ALU operations. The eight address registers, A0–A7, can all be used as base addresses for indirect addressing and for certain 16- and 32-bit

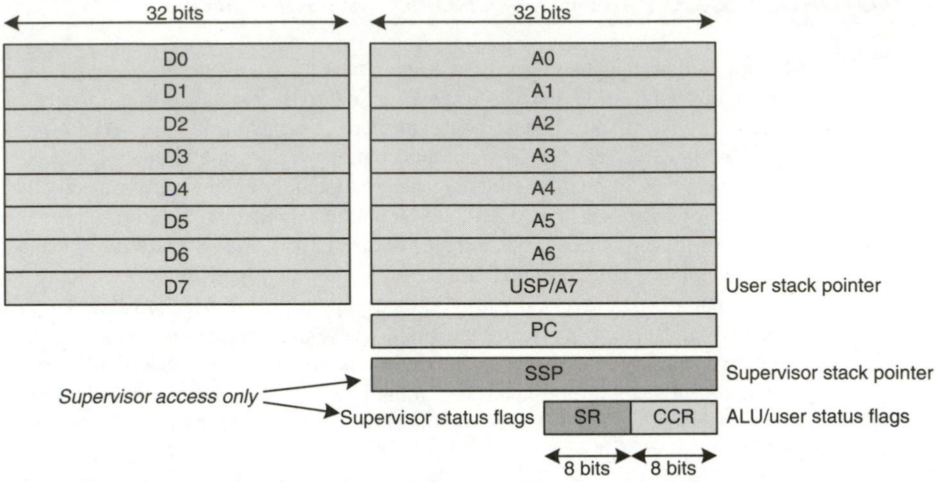

FIGURE 6.11 68000 register set.

ALU operations. All 16 registers can be used as index registers. While operating in user mode, it is illegal to access the SSP or the supervisor portion of the status register, SR. Such instructions will cause an exception, whereby a particular interrupt is asserted, which causes the 68000 to enter supervisor mode to handle the fault. (*Exception* and *interrupt* are often used synonymously in computer contexts.) Very often, the OS kernel will terminate an application that causes an exception to be generated. The registers shown above are present in all 68000 family members and, as such, are software is compatible with subsequent 68xxx microprocessors. Newer microprocessors contain additional registers that provide more advanced privilege levels and memory management. While the 68000 architecture fundamentally supports a 4-GB (32-bit) address space, early devices were limited in terms of how much physical memory could actually be addressed as a result of pin limitations in the packaging. The original 68000 was housed in a 64-pin DIP, leaving only 24 address bits usable, for a total usable memory space of 16 MB. When Motorola introduced the 68020, the first fully 32-bit 68000 microprocessor, all 32 address bits were made available. The 68000 devices are big-endian, so the MSB is stored in the lowest address of a multibyte word.

The 68000 supports a 16-MB address space, but only 23 address bits, A[23:1], are actually brought out of the chip as signal pins. A[0] is omitted and is unnecessary, because it would specify whether an even (A[0] = 0) or odd (A[0] = 1) byte is being accessed; and, because the bus is 16 bits wide, both even and odd bytes can be accessed simultaneously. However, provisions are made for byte-wide accesses in situations where the 68000 is connected to legacy eight-bit peripherals or memories. Two data strobes, upper (UDS*) and lower (LDS*), indicate which bytes are being accessed during any given bus cycle. These strobes are generated by the 68000 according to the state of the internal A0 bit and information on the size of the requested transaction. Bus transactions are triggered by the assertion of address strobe (AS*), the appropriate data strobes, and R/W* as shown in Fig. 6.12. Prior to AS*, the 68000 asserts the desired address and a three-bit function code bus, FC[2:0]. The function code bus indicates which mode the processor is in and whether the transaction is a program or data access. This information can be used by external logic to qualify transactions to certain sensitive memory spaces that may be off limits to user programs. When read data is ready, the external bus interface logic asserts data transfer acknowledge (DTACK*) to inform the microprocessor that the transaction is complete. As shown, the 68000 bus can be operated in a fully asynchronous manner. When operated asynchronously, DTACK* is removed after the strobes are

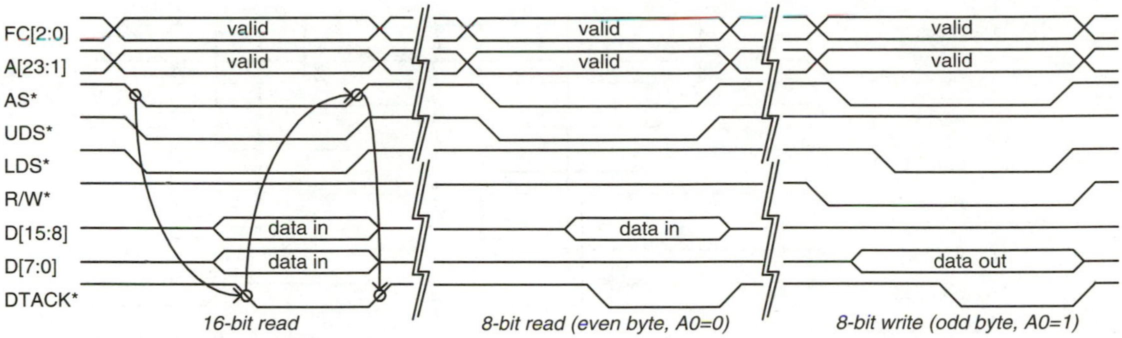

FIGURE 6.12 68000 asynchronous bus timing.

removed, ensuring that the 68000 detected the assertion of DTACK*. If DTACK* is removed prior to the strobes, there is a chance of marginal timing where the 68000 may not properly detect the acknowledge, and it may wait forever for an acknowledge that has now passed. Writes are very similar to reads, with the obvious difference that R/W* is brought low, and data is driven by the 68000. Another difference is that the data strobe assertion lags that of AS*.

Advanced microprocessors such as the 68000 are designed to recognize fault conditions wherein the requested bus transaction cannot be completed. A bus fault can be caused by a variety of problems, including unauthorized access (e.g., user mode tries to write to a protected supervisor data space) or an access to a section of memory that is not filled by a memory or peripheral device. Software should never access areas of memory that are off limits, because the results are unpredictable. Therefore, rather than simply issuing a false DTACK* and continuing with normal operation, the 68000 contains a bus error signal (BERR*) that behaves like DTACK* but triggers an exception rather than continuing normal execution. It is the responsibility of external logic to manage the DTACK* and BERR* signals according to the specific configuration and requirements of the particular system.

Operating the 68000 bus in an asynchronous manner is easy, but it reduces its bandwidth, because delays must be built into the acknowledge process to guarantee that both the 68000 and the interface logic maintain synchronization. Figure 6.12 shows read data being asserted prior to DTACK* and an arbitrary delay between the release of AS* and that of DTACK*. The data delay is necessary to guarantee that the 68000 will see valid data when it detects a valid acknowledge. The second delay is necessary to ensure that the 68000 completes the transaction, as noted previously. These delays can be eliminated if the bus is operated synchronously by distributing the microprocessor clock to the interface logic and guaranteeing that various setup and hold timing requirements are met as specified by Motorola. In such a configuration, it is known from Motorola's data sheet that the 68000 looks for DTACK* each clock cycle, starting at a fixed time after asserting the strobes, and then samples the read-data one cycle after detecting DTACK* being active. Because synchronous timing rules are obeyed, it is guaranteed that the 68000 properly detects DTACK* and, therefore, DTACK* can be removed without having to wait for the removal of the strobes. 68000 synchronous bus timing is shown in Fig. 6.13, where each transaction lasts a minimum of four clock cycles. A four-cycle transaction is a zero wait state access. Wait states can be added by simply delaying the assertion of DTACK* to the next cycle. However, to maintain proper timing, DTACK* (and BERR* and read-data) must always obey proper setup and hold requirements. As shown in the timing diagram, each signal transition, or edge, is time-bounded relative to a clock edge.

Read timing allows a single clock cycle between data strobe assertion and the return of DTACK* for a zero wait-state transaction. However, zero wait-state writes require DTACK* assertion at

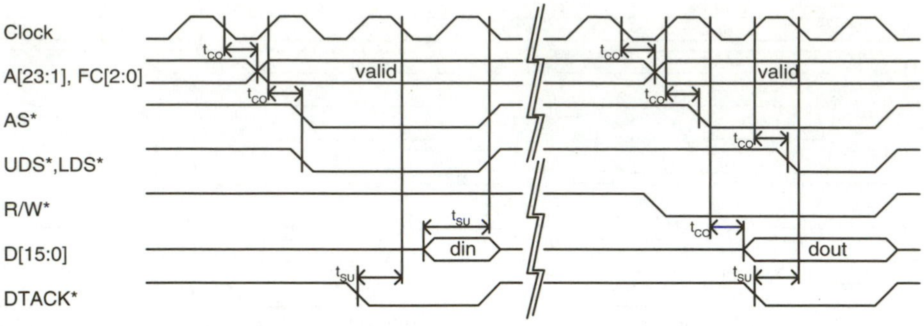

FIGURE 6.13 68000 synchronous bus timing.

roughly the same time as the data strobes. Therefore, the bus interface logic must make its decision on asserting DTACK* based on the requested address when AS* is asserted. If the requested device is operational, DTACK* can be immediately asserted for a fast transaction. Unlike reads, where the microprocessor must wait for a device to return data, writes can be acknowledged before they are actually transferred to the device. In such a scheme, writes are posted within the bus interface logic. One or two cycles later, when the device accepts the posted write data, the bus interface logic finally completes the transaction without having delayed the microprocessor. If completion of the posted-write transaction takes longer than a few cycles, it could force a subsequent access to the same device to incur wait states. Either a read or a write would be blocked until the original write was able to complete, thus freeing the device to handle the next transaction.

In addition to the basic bus interface, the 68000 supports bus arbitration to enable DMA or other logic to use the microprocessor bus for arbitrary applications. A bus request (BR*) signal is asserted by a device that wants to temporarily gain control of the bus. On the next clock cycle, when the microprocessor is not inhibited by other operations, it asserts a bus grant (BG*) signal and places its address, data, and control signals into tri-state so that they may be driven by the other device. The requesting device then asserts bus grant acknowledge (BGACK*) to signal that it is controlling the bus, and it is then free to assert its own strobes, address, and data signals.

A variety of interrupts and exceptions are supported by the 68000. Some are triggered as a result of instruction execution and some by external signals (e.g., BERR* or an interrupt request). Examples of instruction exceptions are illegal user mode register accesses or a divide-by-zero error. Most microprocessors that provide division capability contain some type of divide-by-zero error handling, because the result of such an operation is mathematically undefined and is usually the result of a fault in the program. The 68000 contains an exception vector table that is 1,024 bytes long and resides at the beginning of memory at address 0. In a multitasking system, the bus interface logic may restrict access to the vector table to supervisor mode only. In such a case, a bus error could be triggered if a user mode program, indicated by FC[2:0], tried to write the table. Each of the 256 vector entries is four bytes long and provides the starting address of the associated ISR. The one deviation from this rule is the reset vector, which actually consists of two entries at word addresses 0 and 4. Upon reset, the 68000 fetches an initial PC value from address 4 and an initial SSP value from address 0. Vectors 0 through 63 are assigned or reserved by Motorola for various hardware exceptions. Vectors 64 through 255 are assigned as user interrupt vectors. Like other microprocessors in its category, the 68000 supports bus vectoring of user interrupts where an external interrupt controller asserts an interrupt number onto the data bus during an interrupt acknowledge cycle performed by the 68000 in response to an interrupt request. This interrupt number is multiplied by four and used to index into the exception table to fetch the address of the appropriate ISR.

P · A · R · T · 2

ADVANCED DIGITAL SYSTEMS

CHAPTER 7

Advanced Microprocessor Concepts

Computer architecture is central to the design of digital systems, because most digital systems are, at their core, computers surrounded by varying mixes of interfaces to the outside world. It is difficult to know at the outset of a project how advanced architectural concepts may figure into a design, because *advanced* does not necessarily mean expensive or complex. Many technologies that were originally developed for high-end supercomputers and mainframes eventually found their way into consumer electronics and other less-expensive digital systems. This is why a digital engineer benefits from a broad understanding of advanced microprocessor and computing concepts—a wider palette of potential solutions enables a more creative and effective design process.

This chapter introduces a wide range of technologies that are alluded to in many technical specifications but are often not understood sufficiently to take full advantage of their potential. What is a 200-MHz superscalar RISC processor with a four-way set associative cache? Some people hear the term RISC and conjure up thoughts of high-performance computing. Such imagery is not incorrect, but RISC technology can also be purchased for less than one dollar. Caching is another big computer term that is more common than many people think.

1. An important theme to keep in mind is that microprocessors and the systems that they plug into are inextricably interrelated, and more so than simply by virtue of their common physical surroundings. The architecture of one directly influences the capabilities of the other. For this reason, the two need to be considered simultaneously during the design process. Among many other factors, this makes computer design an iterative process. One may begin with an assumption of the type of microprocessor required and then use this information to influence the broader system architecture. When system-level constraints and capabilities begin to come into focus, they feed back to the microprocessor requirements, possibly altering them somewhat. This cycle can continue for several iterations until a design is realized in which the microprocessor and its supporting peripherals are well matched for the application.

7.1 RISC AND CISC

One of the key features used to categorize a microprocessor is whether it supports *reduced instruction set computing* (RISC—pronounced "risk") or *complex instruction set computing* (CISC—pronounced "sisk"). The distinction is how complex individual instructions are and how many permutations exist for the same basic instruction. In practical terms, this distinction directly relates to the complexity of a microprocessor's instruction decoding logic; a more complex instruction set requires more complex decoding logic. Some engineers believe that a microprocessor should execute simple instructions at a high rate—perhaps one instruction per cycle. Others believe that a microprocessor should execute more complex instructions at a lower rate.

Operand types add complexity to an instruction set when a single general operation such as addition can be invoked with many different addressing modes. Motorola's CISC 68000 contains a basic addition instruction, among other addition operations, that can be decoded in many different ways according to the specified addressing mode. Table 7.1 shows the format of the basic ADD/ADDA/ADDX instruction word. ADD is used for operations primarily on data registers. ADDA is used for operations primarily on address registers. ADDX is used for special addition operations that incorporate the ALU extended carry bit, X, into the sum. The instruction word references Register1 directly and an effective address (EA) that can represent another register or various types of indirect and indexed addressing modes.

TABLE 7.1 68000 ADD/ADDA/ADDX Instruction Word

Bit Position	15	14	13	12	11	10	9	8	7	6	5	4	3	2	1	0
Field	Opcode = 1101				Register1			Opmode			Effective Address					
											Mode			Register2		

As listed in Table 7.2, the opmode field defines whether the operands are 8-, 16-, or 32-bit quantities and identifies the source and destination operands. In doing so, it also implies certain subclasses of instructions: ADD, ADDA, or ADDX.

TABLE 7.2 68000 ADD/ADDA/ADDX Instruction Opmode Field

Opmode Value	Operand Width	Definition of Register1	Operation	Instruction Mapping
000	8	Dn	EA + Dn ⇒ Dn	ADD
001	16	Dn	EA + Dn ⇒ Dn	ADD
010	32	Dn	EA + Dn ⇒ Dn	ADD
100	8	Dn	Dn + EA ⇒ EA	ADD/ADDX
101	16	Dn	Dn + EA ⇒ EA	ADD/ADDX
110	32	Dn	Dn + EA ⇒ EA	ADD/ADDX
011	16	An	EA + An ⇒ An	ADDA
111	32	An	EA + An ⇒ An	ADDA

The main complexity is introduced by the EA fields as defined in Table 7.3. For those modes that map to multiple functions, additional identifying fields and operands are identified by one or more extension words that follow the instruction word. One of the more complex modes involves using an address register as a base address, adding a displacement to that base to calculate a fetch address, fetching the data at that address, adding another register to the retrieved value, adding another displacement, and then using the resulting address to fetch a final operand value. `ADD/ADDA/ADDX` is a powerful instruction that requires significant decode logic behind it. Additionally, when opmode indicates an `ADD` or `ADDX` instruction, the two mode values that normally indicate simple register references now map to one of two special `ADDX` operations.

TABLE 7.3 68000 Effective Address Field Definition

Mode Field	Definition of Register2	Operand Value	Function
000	Data register N	Dn	Data register value
001	Address register N	An	Address register value
010	Address register N	(An)	Indirect address register
011	Address register N	(An)+	Indirect with post-increment
100	Address register N	–(An)	Indirect with pre-decrement
101	Address register N	(An + d_{16},)	Indirect with 16-bit displacement
110	Address register N	(An + Xn + d_8)	Indirect with index register and 8-bit displacement (extension word follows)
		(An + Xn + $d_{16,32}$)	Indirect with index register and 32- or 16-bit displacement (extension words follow)
		((An + $d_{16,32}$) + Xn + $d_{16,32}$)	Indirect with displacement to fetch pointer added to index register and displacement (extension words follow)
		((An + Xn + $d_{16,32}$) + $d_{16,32}$)	Indirect with displacement and index register to fetch pointer added to displacement (extension words follow)
111	000	d_{16}	16-bit direct address (extension word follows)
111	001	d_{32}	32-bit direct address (extension words follow)
111	100	#data	Immediate follows in extension words
111	010	(PC + d_{16})	Indirect with 16-bit displacement
111	011	<multiple>	Same as mode=110, but with PC instead of address registers

Shaded modes are invalid when EA is specified as the destination by *opmode* and change their meaning as follows:

000	Data register N	Dn	ADDX: Dregister2 + Dregister1 + X ⇒ Dregister1
001	Address register N	–(An)	ADDX: –(Aregister2) + –(Aregister1) + X ⇒ (Aregister1)

As can be readily observed, decoding an addition instruction on the 68000 is not as simple as adding two registers. For the most complex addressing modes, multiple registers must be added together to create an address from which another address is fetched that is added with an offset to yield a final address at which the true operand is located. This sounds complicated, and it is. There is really no succinct way to explain the operation of such instructions. The impact of these complex addressing modes on decoding logic is substantial, especially when it is realized that the 68000 contains dozens of instructions, each with its own permutations.

In contrast to the 68000's CISC architecture, the MIPS family of microprocessors is one of the commercial pioneers of RISC. MIPS began as a 32-bit architecture with 32-bit instruction words and 32 general-purpose registers. In the 1990s the architecture was extended to 64 bits. MIPS instruction words are classified into three basic types: immediate (I-type), jump (J-type), and register (R-type). The original MIPS architecture supports four 32-bit addition instructions without any addressing mode permutations: add signed (ADD), add unsigned (ADDU), add signed immediate (ADDI), and add unsigned immediate (ADDIU). These instructions are represented by two types of instruction words, I-type and R-type, as shown in Table 7.4.

TABLE 7.4 MIPS Addition Instruction Words

I-type bits	31:26	25:21	20:16	15:0		
Field	Opcode	Source Register	Target Register	Immediate data		
ADDI	001000	Rn	Rn	Data		
ADDIU	001001	Rn	Rn	Data		
R-type bits	31:26	25:21	20:16	15:11	10:6	5:0
Field	Opcode	Source Register	Target Register	Destination Register	Shift Amount	Function
ADD	000000	Rn	Rn	Rn	00000	100000
ADDU	000000	Rn	Rn	Rn	00000	100001

The immediate operations specify two registers and a 16-bit immediate operand: $R_T = R_S +$ Immediate. The other instructions operate on registers only and allow the programmer to specify three registers: $R_D = R_S + R_T$. If you want to add data that is in memory, that data must first be loaded into a register. Whereas a single 68000 instruction can fetch a word from memory, increment the associated pointer register, add the word to another register, and then store the result back into memory, a MIPS microprocessor would require separate instructions for each of these steps. This is in keeping with RISC concepts: use more simpler instructions to get the job done.

Instruction decode logic for a typical RISC microprocessor can be much simpler than for a CISC counterpart, because there are fewer instructions to decode and fewer operand complexities to recognize and coordinate. Generally speaking, a RISC microprocessor accesses data memory only with dedicated load/store instructions. Data manipulation instructions operate solely on internal registers and immediate operands. Under these circumstances, microprocessor engineers are able to heavily optimize their design in favor of the reduced instruction set that is supported. It turns out that not all instructions in a CISC microprocessor are used with the same frequency. Rather, there is a core set of instructions that are called most of the time, and the rest are used infrequently. Those that are used less often impose a burden on the entire system, because they increase the permutations that the decode logic must handle in any given clock cycle. By removing the operations that are not frequently used, the microprocessor's control logic is simplified and can therefore be made to run faster. The result is improved throughput for the most commonly executed operations, which translates directly into greater performance overall.

The fundamental assumption that RISC microprocessors rely on to maintain their throughput is high memory bandwidth. For a RISC microprocessor to match or outperform a CISC microprocessor, it must be able to rapidly fetch instructions, because several RISC instructions are necessary to match the capabilities of certain CISC instructions. An older computer architecture with an asynchronous memory interface may not be able to provide sufficient instruction bandwidth to make a RISC microprocessor efficient. CISC architectures dominated off-the-shelf microprocessor offerings until low-latency memory subsystems became practical at a reasonable cost. Modern computer architectures implement very fast memory interfaces that are able to provide a steady stream of instructions to RISC microprocessors.

One fundamental technique for improving the instruction fetch bandwidth is to design a microprocessor with two memory interfaces—one for instructions and one for data. This is referred to as a *Harvard* architecture, as compared to a conventional *von Neumann* architecture in which instruction and data memory are unified. Using a Harvard architecture, instruction fetches are not disrupted by load/store operations. Unfortunately, a Harvard architecture presents numerous system-level problems of how to split program and data memory and how to load programs into memory that cannot be accessed by load/store operations. Most microprocessors that implement a Harvard architecture do so with smaller on-chip memory arrays that can store segments of program and data that are fetched from and written back to a unified memory structure external to the microprocessor chip. While this may sound so complex as to only be in the realm of serious number-crunchers, the small but powerful 8-bit PIC™ RISC microcontrollers from Microchip Technology implement a Harvard architecture with mutually exclusive program and data memory structures located on chip. This illustrates the point that advanced microprocessor concepts can be applied to any level of performance if a problem needs to be solved.

The RISC concept appears to have won the day in the realm of high-performance computing. With memory bandwidth not being much of a hindrance, streamlined RISC designs can be made fast and efficient. In embedded computing applications, the victor is less clear. CISC technology is still firmly entrenched in a market where slow memory subsystems are still common and core microprocessor throughput is not always a major design issue. What is clear is that engineers and marketers will continue to debate and turn out new products and literature to convince others why their approach is the best available.

7.2 CACHE STRUCTURES

Microprocessor and memory performance have improved asymmetrically over time, leading to a well recognized performance gap. In 1980, a typical microprocessor ran at under 10 MHz, and a typical DRAM exhibited an access time of about 250 ns. Two decades later, high-end microprocessors were running at several hundred megahertz, and a typical DRAM exhibited an access time of 40 ns. Microprocessors' appetites for memory bandwidth has increased by about two orders of magnitude over 20 years while main memory technology, most often DRAM, has improved by less than an order of magnitude during that same period. To make matters worse, many microprocessors shifted from CISC to RISC architectures during this same period, thereby further increasing their demand for instruction memory bandwidth. The old model of directly connecting main memory to a microprocessor has broken down and become a performance-limiting bottleneck.

The culprits for slow main memory include the propagation delays through deep address decoding logic and the high random access latency of DRAM—the need to assert a row address, wait some time, assert a column address, and wait some more time before data is returned. These problems can be partially addressed by moving to SRAM. SRAM does not exhibit the latency penalty of DRAM, but there are still the address decoding delays to worry about. It would be nice to build main memory with SRAM, but this is prohibitively expensive, as a result of the substantially lower density of SRAM as compared to DRAM. An SRAM-based main memory requires more devices, more circuit board area, and more connecting wires—all requirements that add cost and reduce the reliability of a system. Some supercomputers have been built with main memory composed entirely of SRAM, but keep in mind that these products have minimal cost constraints, if any.

If software running on microprocessors tended to access every main memory location with equal probability, not much could be done to improve memory bandwidth without substantial increases in size and cost. Under such circumstances, a choice would have to be made between a large quantity of slow memory or a small quantity of fast memory. Fortunately, software tends to access fairly constrained sets of instructions and data in a given period of time, thereby increasing the probability of accessing sequential memory locations and decreasing the probability of truly random accesses. This property is generally referred to as *locality.* Instructions tend to be executed sequentially in the order in which they are stored in memory. When branches occur, the majority are with small displacements for purposes of forming loops and local "if…then…else" logical decisions. Data also tend to be grouped into sequential elements. For example, if a string of characters forming a person's name in a database is being processed, the characters in the string will be located in sequential memory locations. Furthermore, the entire database entry for the person will likely be stored as a unit in nearby memory locations.

Caches largely overcome main memory latency problems. A cache, pronounced "cash," is a small quantity of fast memory that is used to temporarily store portions of main memory that the microprocessor accesses often or is predicted to access in the near future. Being that cache memory is relatively small, SRAM becomes practical to use in light of its substantial benefits of fast access time and simplicity—a memory controller is not needed to perform refresh or address multiplexing operations. As shown in Fig. 7.1, a cache sits between a microprocessor and main memory and is composed of two basic elements: cache memory and a cache controller.

The cache controller watches all memory transactions initiated by the microprocessor and selects whether read data is fetched from the cache or directly from main memory and whether writes go into the cache or into main memory. Transactions to main memory will be slower than those to the cache, so the cache controller seeks to minimize the number of transactions that are handled directly by main memory.

Locality enables a cache controller to increase the probability of a *cache hit*—that data requested by the microprocessor has already been loaded into the cache. A 100 percent hit rate is impossible, because the controller cannot predict the future with certainty, resulting in a *cache miss* every so often. *Temporal* and *spatial* locality properties of instructions and data help the controller improve its hit rate. Temporal locality says that, if a memory location is accessed once, it is likely to be accessed again in the near future. This can be readily observed by considering a software loop: instructions in the body of the loop are very likely to be fetched again in the near future during the next loop itera-

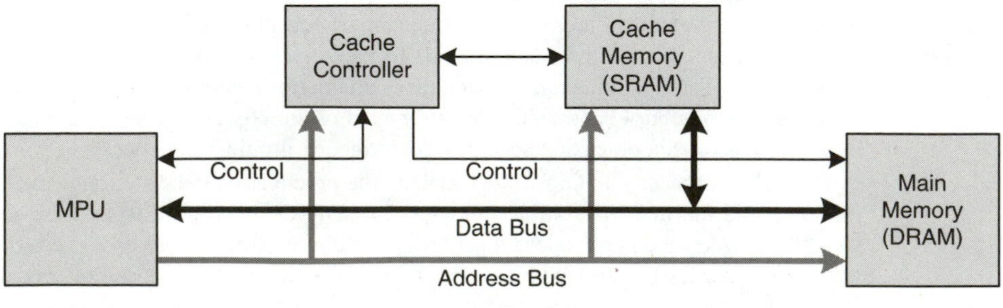

FIGURE 7.1 Computer with cache.

tion. Spatial locality says that, if a memory location is accessed, it is likely that nearby locations will be accessed in the near future. When a microprocessor fetches an instruction, there is a high probability that it will soon fetch the instructions immediately following that instruction. Practically speaking, temporal locality tells the cache controller to attempt to retain recently accessed memory locations in the expectation that they will be accessed again. Spatial locality tells the cache controller to preload additional sequential memory locations when a single location is fetched by the microprocessor, in the expectation that these locations will be soon accessed.

Given the locality properties, especially spatial locality, that need to be incorporated into the cache controller, a basic cache organization emerges in which blocks of data rather than individual bytes are managed by the controller and held in cache memory. These blocks are commonly called *lines,* and they vary in size, depending on the specific implementation. Typical cache line sizes are 16, 32, or 64 bytes. When the microprocessor reads a memory location that is not already located in the cache (a miss), the cache controller fetches an entire line from main memory and stores it as a unit. To maintain the simplicity of power-of-two logic, cache lines are typically mapped into main memory on boundaries defined by the line size. A 16-byte cache line will always hold memory locations at offsets represented by the four least-significant address bits. Main memory is therefore effectively divided into many small 16-byte lines with offsets from 0x0 to 0xF. If a microprocessor with a 32-bit address bus fetches location 0x1000800C and there is a cache miss, the controller will load locations 0x10008000 through 0x1000800F into a designated cache line. If the cache is full, and a miss occurs, the controller must *flush* a line that has a lower probability of use so as to make room for the new data. If the flushed line has been modified by writes that were not already reflected in main memory, the controller must store the line to prevent losing and corrupting the memory contents.

As more cache lines are implemented, more sections of main memory can be simultaneously held in the cache, increasing the hit rate. However, a cache's overall size must be bounded by a system's target size and cost constraints. The size of a cache line is a compromise between granularity, load/store time, and locality benefits. For a fixed overall size, larger lines reduce the granularity of unique blocks of main memory that can be simultaneously held in the cache. Larger cache lines increase the time required to load a new line and update main memory when flushing an old line. Larger cache lines also increase the probability that a subsequent access will result in a hit.

Cache behavior on reads is fairly consistent across different implementations. Writes, however, can be handled in one of three basic manners: *no-write*, *write-through*, and *write-back*. A no-write cache does not support the modification of its contents. When a write is performed to a block of memory held in a cache line, that line is flushed, and the write is performed directly into main memory. This scheme imposes two penalties on the system: writes are always slowed by the longer latency of main memory, and locality benefits are lost because the flush forces any subsequent accesses to that line to result in a miss and reload of the entire line that was already present in the cache.

Write-through caches support the modification of their contents but do not support incoherency between cache memory and main memory. Therefore, a write to a block of memory held in a cache line results in a parallel write to both the cache and main memory. This is an improvement over a no-write cache in that the cache line is not forcibly flushed, but the write is still slowed by a direct access to main memory.

A write-back cache minimizes both penalties by enabling writes to valid cache lines but not immediately causing a write to main memory. The microprocessor does not have to incur the latency penalty of main memory, because the write completes as fast as the cache can accept the new data. This scheme introduces complexity in the form of incoherency between cache and main memory: each memory structure has a different version of the same memory location. To solve the incoherency problem, a write-back cache must maintain a status bit for each line that indicates whether the line is *clean* or *dirty*. When the line is eventually flushed, dirty lines must be written back to main memory in their entirety. Clean lines can be flushed without further action. While a write-back cache cannot absolutely eliminate the longer write latency of main memory, it can reduce the overall system impact of writes, because the microprocessor can perform any number of writes to the same cache line, and only a fixed write-back penalty results upon a flush.

The central problem in designing a cache is how to effectively hold many scattered blocks from a large main memory in a small cache memory. In a standard desktop PC, main memory may consist of 256 MB of DRAM, whereas the microprocessor's cache is 256 kB—a difference of three orders of magnitude! The concept of cache lines provides a starting point with a defined granularity to minimize the problem somewhat. Deciding on a 16-byte line size, for example, indicates that a 32-bit address space needs to be handled only as 2^{28} units rather than 2^{32} units. Of course, 2^{28} is still a very large number! Each cache line must have an associated tag and/or index that identifies the higher-order address bits that its contents represent (28 bits in this example). Different cache architectures handle these tags and indices to balance cache performance with implementation expense. The three standard cache architectures are *fully associative*, *direct mapped*, and *n-way set associative*.

A fully associative cache, shown in Fig. 7.2, breaks the address bus into two sections: the lower bits index into a selected cache line to select a byte within the line, and the upper bits form a tag that is associated with each cache line. Each cache line contains a valid bit to indicate whether it contains real data. Upon reset, the valid bits for each line are cleared to 0. When a cache line is loaded with data, its tag is set to the high-order address bits that are driven by the microprocessor. On subsequent transactions, those address bits are compared in parallel against every tag in the cache. A hit occurs when one tag matches the requested address, resulting in that line's data advancing to a final multiplexer where the addressed bytes are selected by the low-order address bits. A fully associative cache is the most flexible type, because any cache line can hold any portion of main memory. The disadvantage of this scheme is its complexity of implementation. Each line requires address match-

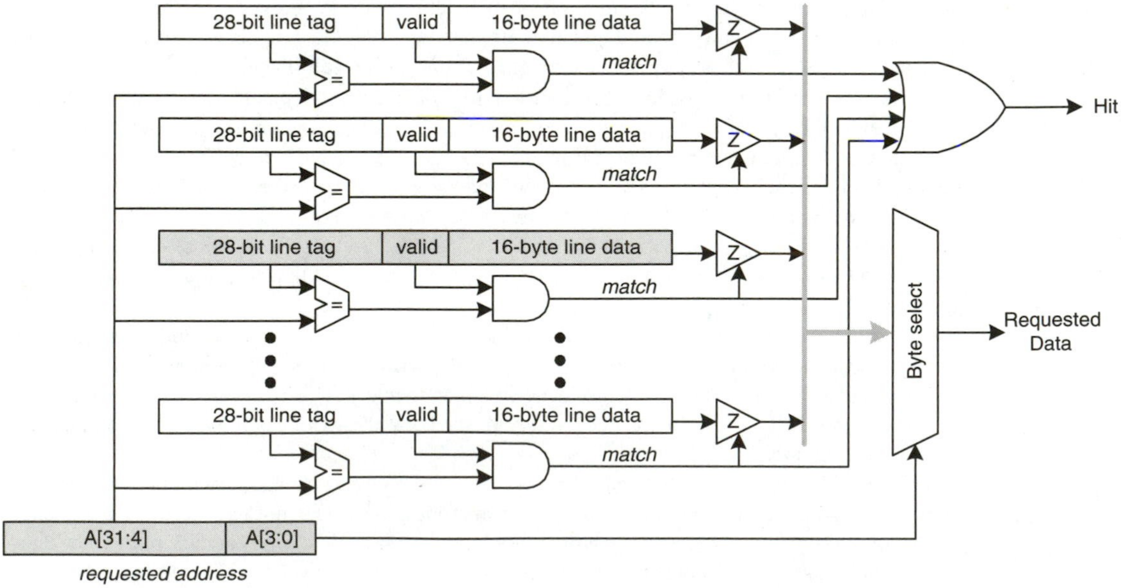

FIGURE 7.2 Fully associative cache.

ing logic, and each match signal must be logically combined in a single location to generate a final hit/miss status flag.

A direct mapped cache, shown in Fig. 7.3, breaks the address bus into three sections; the lower bits retain their index function within a selected line, the middle bits select a single line from an array of uniquely addressable lines, and the upper bits form a tag to match the selected cache line. As before, each cache line contains a valid bit. The difference here is that each block of memory can only be mapped into one cache line—the one indexed by that block's middle address bits, A[15:4] in this example (indicating a 64-kB total cache size). During a cache miss, the controller determines which line is selected by the middle address bits, loads the line, sets the valid bit, and loads the line tag with the upper address bits. On subsequent accesses, the middle address bits select a single line whose tag is compared against the upper address bits. If they match, there is a cache hit. A direct mapped cache is much easier to implement as compared to a fully associative cache, because parallel tag matching is not required. Instead, the cache can be constructed with conventional memory and logic components using off-the-shelf RAM for both the tag and line data. The control logic can index into the RAM, check the selected tag for a match, and then take appropriate action. The disadvantage to a direct mapped cache is that, because of the fixed mapping of memory blocks to cache lines, certain data access patterns can cause rapid *thrashing*. Thrashing results when the microprocessor rapidly accesses alternate memory blocks. If the alternate blocks happen to map to the same cache line, the cache will almost always miss, because each access will result in a flush of the alternate memory block.

Given the simplicity of a direct mapped cache, it would be nice to strike a compromise between an expensive fully associative cache and a thrashing-sensitive direct mapped cache. The *n-way* set associative cache is such a compromise. As shown in Fig. 7.4, a two-way set associative cache is basically two direct mapped cache elements connected in parallel to reduce the probability of thrashing. More than two sets can be implemented to further reduce thrashing potential. Four-way and two-way set associative caches are very common in modern computers. Beyond four elements, the payback of thrashing avoidance to implementation complexity declines. The term *set* refers to the number of entries in each direct mapped element, 4,096 in this example. Here, the

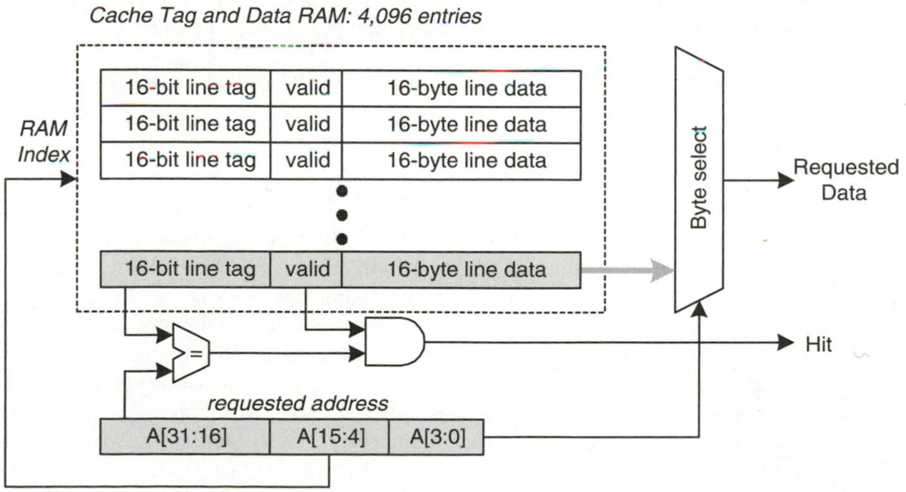

FIGURE 7.3 64-kB direct mapped cache.

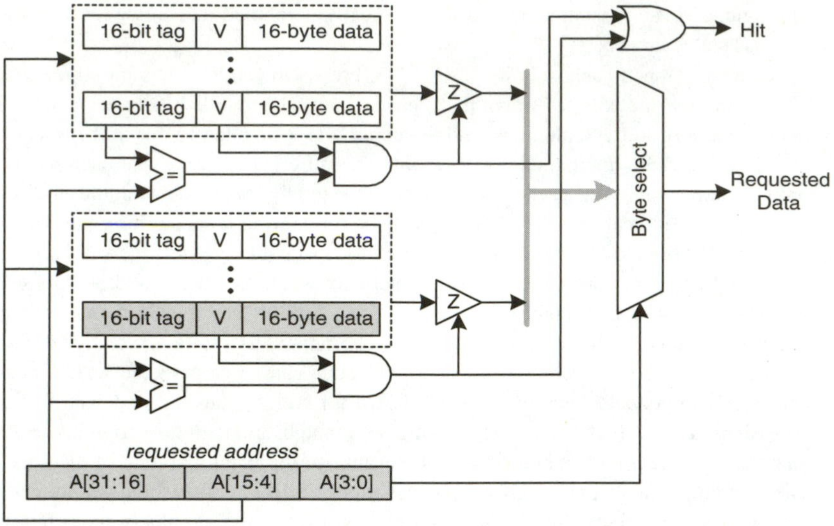

FIGURE 7.4 128-kB two-way set associative cache.

cache has expanded to 128 kB in size using two 64 kB elements. If cost constraints dictate keeping a 64 kB cache, it would be preferable to reduce the set size to 2,048 rather than halve the line size, which is already at a practical minimum of 16 bytes. Reducing the set size to 2^{11} would increase the line tag to 17 bits to maintain a 32-bit address space representation. In a cache of this type, the controller can choose which of two (or four, or *n*) cache line locations to flush when a miss is encountered.

Deciding which line to flush when a cache miss occurs can be done in a variety of ways, and different cache architectures dictate varying approaches to this problem. A fully associative cache can place any main memory block into any line, while a direct mapped cache has only one choice for any given memory block. Three basic flush, or replacement, algorithms are as follows:

- *First-in-first-out (FIFO).* Track cache line ages and replace the oldest line.
- *Least-recently-used (LRU).* Track cache line usage and replace the line that has not been accessed longest.
- *Random.* Replace a random line.

A fully associative cache has the most flexibility in selecting cache lines and therefore the most complexity in tracking line usage. To perform either a FIFO or LRU replacement algorithm on a fully associative cache, each line would need a tracking field that could be updated and checked in parallel with all other lines. N-way set associative caches are the most interesting problems from a practical perspective, because they are used most frequently. Replacement algorithms for these caches are simplified, because the number of replacement choices is restricted to N. A two-way set associative cache can implement either FIFO or LRU algorithms with a single bit per line entry. For a FIFO algorithm, the entry being loaded anew has its FIFO bit cleared, and the other entry has its FIFO bit set, indicating that the other entry was loaded first. For an LRU algorithm, the entry being accessed at any given time has its LRU bit cleared, and the other has its LRU bit set, indicating that the other entry was used least recently. These algorithms and associated hardware are only slightly more complex for a four-way set associative cache that would require two status bits per line entry.

7.3 CACHES IN PRACTICE

Basic cache structures can be applied and augmented in different ways to improve their efficacy. One common manner in which caches are implemented is in pairs: an *I-cache* to hold instructions and a *D-cache* to hold data. It is not uncommon to see high-performance RISC microprocessors with integrated I/D caches on chip. Depending on the intended application, these integrated caches can be relatively small, each perhaps 8 kB to 32 kB in size. More often than not, these are two-way or four-way set associative caches. There are two key benefits to integrating two separate caches. First, instruction and data access patterns can combine negatively to cause thrashing on a single normal cache. If a software routine operates on a set of data whose addresses happen to overlap with the I-cache's index bits, alternate instruction and data fetch operations could cause repeated thrashing on the same cache lines. Second, separate caches can effectively provide a Harvard memory architecture from the microprocessor's local perspective. While it is often not practical to provide dual instruction and data memory interfaces at the chip level, as a result of excessive pin count, such considerations are much less restrictive within a silicon die. Separate I/D caches can feed from a shared chip-level memory interface but provide independent interfaces to the microprocessor core itself. This dual-bus arrangement increases the microprocessor's load/store bandwidth by enabling it to simultaneously fetch instructions and operands without conflict.

Dual I/D caches cannot guarantee complete independence of instruction and data memory, because, ultimately, they are operating through a shared interface to a common pool of main memory. The performance boost that they provide will be dictated largely by the access patterns of the applications running on the microprocessor. Such application-dependent performance is fundamental to all types of caches, because caches rely on locality to provide their benefits. Programs that scatter instructions and data throughout a memory space and alternately access these disparate locations will show less performance improvement with the cache. However, most programs exhibit fairly beneficial locality characteristics. A system with dual I/D caches can show substantial throughput improvement when a software routine can fit its core processing instructions into the instruction cache with minimal thrashing and its data sets exhibit good locality properties. Under these circumstances, the data cache can have more time to pull in data via the common memory interface, enabling the microprocessor to simultaneously access instruction and data memory with a low miss rate.

Computer systems with caches require some assistance from the operating system and applications to maximize cache performance benefits and to prevent unexpected side effects of cached memory. It is helpful to cache certain areas of memory, but it is performance degrading or even harmful to cache other areas. Memory-mapped I/O devices are generally excluded from being cached, because I/O is a class of device that usually responds with some behavior when a control register is modified. Likewise, I/O devices frequently update their status registers to reflect conditions that they are monitoring. If a cache prevents the microprocessor from directly interacting with an I/O device, unexpected results may appear. For example, if the microprocessor wants to send data out a serial port, it might write a data value to a transmit register, expecting that the data will be sent immediately. However, a write-back cache would not actually perform the write until the associated cache line is flushed—a time delay that is unbounded. Similarly, the serial port controller could reflect handshaking status information in its status registers that the microprocessor wants to periodically read. An unknowing cache would fetch the status register memory location once and then continue to return the originally fetched value to the microprocessor until its cache line was flushed, thereby preventing the microprocessor from reading the true status of the serial port. I/O registers are unlike main memory, because memory just holds data and cannot modify it or take actions based on it.

Whereas caching an I/O region can cause system disruption, caching certain legitimate main memory regions can cause performance degradation due to thrashing. It may not be worth caching small routines that are infrequently executed, because the performance benefit of caching a quick maintenance routine may be small, and its effect on flushing more valuable cache entries may significantly slow down the application that resumes execution when the maintenance routine completes. A performance-critical application is often composed of a processing kernel along with miscellaneous initialization and maintenance routines. Most of the microprocessor time is spent executing kernel instructions, but sometimes the kernel must branch to maintenance routines for purposes such as loading or storing data. Unlike I/O regions that are inherently known to be cache averse, memory regions that should not be cached can only be known by the programmer and explicitly kept out of the cache.

Methods of excluding certain memory locations from the cache differ across system implementations. A cache controller will often contain a set of registers that enable the lockout of specific memory regions. On those integrated microprocessors that contain some address decoding logic as well as a cache controller, individual memory areas are configured into the decoding logic with programmable registers, and each is marked as cacheable or noncacheable. When the microprocessor performs a memory transaction, the address decoder sends a flag to the cache controller that tells it whether to participate in the transaction.

On the flip side of locking certain memory regions out of the cache, some applications can benefit from explicitly locking certain memory regions into the cache. Locking cache entries prevents the cache controller from flushing those entries when a miss occurs. A programmer may be able to lock a portion of the processing kernel into the cache to prevent arbitrary maintenance routines from disturbing the most frequently accessed sets of instructions and data.

Cache controllers perform burst transactions to main memory because of their multiword line architecture. Whether the cache is reading a new memory block on a cache miss or writing a dirty block back to main memory, its throughput is greatly increased by performing burst transfers rather than reading or writing a single word at a time. Normal memory transfers are executed by presenting an address and reading or writing a single unit of data. Each type of memory technology has its own associated latency between the address and data phases of a transaction. SRAM is characterized by very low access latency, whereas DRAM has a higher latency. Because main memory in most systems is composed of DRAM, single-unit memory transfers are inefficient, because each byte or word is penalized by the address phase overhead. Burst transfers, however, return multiple sequential data units while requiring only an initial address presentation, because the address specifies a starting address for a set of memory locations. Therefore, the overhead of the address phase is amortized across many data units, greatly increasing the efficiency of the memory system. Modern DRAM devices support burst transfers that nicely complement the cache controllers that often coexist in the same computer system.

As a result of cache subsystems being integrated onto the same chip along with high-performance microprocessors, the external memory interface is less a microprocessor bus and more a burst-mode cache bus. The microprocessor needs to be able to bypass the cache controller while accessing noncacheable memory locations or during boot-up when peripherals such as the cache controller have not yet been initialized. However, the external bus is often optimized for burst transfers, and absolute efficiency or simplicity when dealing with noncacheable locations may be a secondary concern to the manufacturer. If overall complexity can be reduced by giving up some performance in nonburst transfers, it may be worth the trade-off, because high performance microprocessors spend relatively little of their time accessing external memory directly. If they do, then something is wrong with the system design, because the microprocessor's throughput is sure to suffer.

Many microprocessors are designed to support multiple levels of caching to further improve performance. In this context, the cache that is closest to the microprocessor core is termed a *level-one*

(L1) cache. The L1 cache is fairly small, anywhere from 2 kB to 64 kB, with its benefit being speed. Because it is small and close to the microprocessor, it can be made to run as fast as the microprocessor can fetch instructions and data. Instruction and data caches are implemented at L1. Line sizes for L1 caches vary but are often 16 or 32 bytes. The line size needs to be kept to a practical minimum to maximize the number of unique memory blocks that can be stored in a small RAM structure.

Level two (L2) caches may reside on the same silicon chip as the microprocessor and L1 cache or externally to the chip, depending on the implementation. L2 caches are generally unified instruction and data caches to minimize the complexity of the interface between the L1 cache and the rest of the system. These caches run somewhat slower than L1 but, consequently, they can be made larger: 128 kB, 256 kB, or more. Line sizes of 64 bytes and greater are common in L2 caches to increase efficiency of main memory burst transfers. Because the L2 cache has more RAM, it can expand both the number of lines and the line size in the hope that, when the L1 cache requests a block of memory, the next sequential block will soon be requested as well. Beyond L2, some microprocessors support L3 and even L4 caches. Each successive level increases its latency of response to the cache above it but adds more cache RAM (sometimes megabytes) and sometimes larger line sizes to increase burst-mode transfer efficiency to main memory.

As core microprocessor clock frequencies commonly top several hundred megahertz, and the most advanced microprocessors exceed 1 GHz, the bandwidth disparity between the microprocessor and main memory increases. Cache misses impose severe penalties on throughput, because the effective clock speed of the microprocessor is essentially reduced to that of the memory subsystem when data is fetched directly from memory. The goal of a multilevel cache structure is to substantially reduce the probability of a cache miss that leads directly to main memory. If the L1 cache misses, hopefully, the L2 cache will be ready to supply the requested data at only a moderate throughput penalty.

Caching as a concept is not restricted to the context of microprocessors and hardware implementation. Caches are found in hard disk drives and in Internet caching products. Some high-end hard drives implement several megabytes of RAM to prefetch data beyond that which has already been requested. While the hard drive's cache, possibly implemented using DRAM, is not nearly as fast as a typical microprocessor, it is orders of magnitude faster than the drive mechanism itself. Internet caching products routinely copy commonly accessed web sites and other data onto their hard drives so that subsequent accesses do not have to go all the way out to the remote file server. Instead, the requested data is sitting locally on a cache system. Caching is the general concept of substituting a small local storage resource that is faster than the larger more remote resource. Caches can be applied in a wide variety of situations.

Caching gets somewhat more complex when the data that is being cached can be modified by another entity outside of the cache memory. This is possible in the Internet caching application mentioned above or in a multiprocessor computer. Cache coherency is the subject of many research papers and is a problem that needs to be addressed by each implementation. Simply put, how does the Internet cache know when a web site that is currently stored has been updated? A news web site, for example, may update its contents every few hours. In a multiprocessor context, multiple microprocessors may have access to the same pool of shared memory. Here, the multiple cache controllers must somehow communicate to know when memory has been modified so that the individual caches can update themselves and maintain coherency.

Determining the optimal size of a cache so that its performance improvement merits its cost has been the subject of much study. Cache performance is highly application dependent and, in general, meaningful performance improvements decline after a certain size threshold, which varies by application. A typical PC runs programs that are not very computationally intensive and that operate on limited sets of data over short time intervals. In short, they exhibit fairly good locality properties. Typical desktop PCs contain 256 kB of L2 cache and a smaller quantity of L1 cache. Computers that often operate on larger sets of data or that must run many applications simultaneously may merit larger caches to offset less optimal locality properties. Computers meant to function as computation engines and network file servers can include several megabytes of L2 cache. Smaller embedded systems may suffice with only several kilobytes of L1 cache.

7.4 VIRTUAL MEMORY AND THE MMU

Multitasking operating systems execute multiple programs at the same time by assigning each program a certain percentage of the microprocessor's time and then periodically changing which instruction sequence is being executed. This is accomplished by a periodic timer interrupt that causes the OS kernel to save the state of the microprocessor's registers and then reload the registers with preserved state from a different program. Each program runs for a while and is paused, after which execution resumes without the program having any knowledge of having been paused. In this respect, the individual programs in a multitasking environment appear to have complete control over the computer, despite sharing the resources with others. Such a perspective makes programming for a multitasking OS easier, because the programmer does not have to worry about the infinite permutations of other applications that may be running at any given time. A program can be written as if it is the only application running, and the OS kernel sorts out the run-time responsibilities of making sure that each application gets fair time to run on the microprocessor.

Aside from fair access to microprocessor time, conflicts can arise between applications that accidentally modify portions of each other's memory—either program or data. How does an application know where to locate its data so that it will not disturb that of other applications and so that it will not be overwritten? There is also the concern about system-wide fault tolerance. Even if not malicious, programs may have bugs that cause them to crash and write data to random memory locations. In such an instance, one errant application could bring down others or even crash the OS if it overwrites program and data regions that belong to the OS kernel. The first problem can be addressed with the honor system by requiring each application to dynamically request memory allocations at run time from the kernel. The kernel can then make sure that each application is granted an exclusive region of memory. However, the second problem of errant writes requires a hardware solution that can physically prevent an application from accessing portions of memory that do not belong to it.

Virtual memory is a hardware enforced and software configured mechanism that provides each application with its own private memory space that it can use arbitrarily. This virtual memory space can be as large as the microprocessor's addressing capability—a full 4 GB in the case of a 32-bit microprocessor. Because each application has its own exclusive virtual memory space, it can use any portion of that space that is not otherwise restricted by the kernel. Virtual memory frees the programmer from having to worry about where other applications may locate their instructions or data, because applications cannot access the virtual memory spaces of others. In fact, operating systems that support virtual memory may simplify the physical structure of programs by specifying a fixed starting address for instructions, the local stack, and data. UNIX is an example of an OS that does this. Each application has its instructions, stack, and data at the same virtual addresses, because they have separate virtual memory spaces that are mutually exclusive and, therefore, not subject to conflict.

Clearly, multiple programs cannot place different data at the same address or each simultaneously occupy the microprocessor's entire address space. The OS kernel configures a hardware *memory management unit* (MMU) to map each program's *virtual addresses* into unique *physical addresses* that correspond to actual main memory. Each unique virtual memory space is broken into many small *pages* that are often in the range of 2 to 16 kB in size (4 kB is a common page size). The OS and MMU refer to each virtual memory space with a *process ID* (PID) field. Virtual memory is handled on a process basis rather than an application basis, because it is possible for an application to consist of multiple semi-independent processes. The high-order address bits referenced by each instruction form the *virtual page number* (VPN). The PID and VPN are combined to uniquely map to a physical address set aside by the kernel as shown in Fig. 7.5. Low-order address bits represent offsets that directly index into mapped pages in physical memory. The mapping of virtual memory pages into physical memory is assigned arbitrarily by the OS kernel. The kernel runs in real memory rather than in virtual memory so that it can have direct access to the computer's physical resources to allocate memory as individual processes are executed and then terminated.

Despite each process having a 4-GB address space, virtual memory can work on computers with just megabytes of memory, because the huge virtual address spaces are sparsely populated. Most processes use only a few hundred kilobytes to a few megabytes of memory and, therefore, multiple processes that collectively have the potential to reference tens of gigabytes can be mapped into a much smaller quantity of real memory. If too many processes are running simultaneously, or if these processes start to consume too much memory, a computer can exhaust its physical memory resources, thereby requiring some intervention from the kernel to either suspend a process or handle the problem in some other way.

When a process is initiated, or *spawned*, it is assigned a PID and given its own virtual memory space. Some initial pages are allocated to hold its instructions and whatever data memory the process needs available when it begins. During execution, processes may request more memory from the kernel by calling predefined kernel memory management routines. The kernel will respond by allocating a page in physical memory and then returning a pointer to that page's virtual mapping. Likewise, a process can free a memory region when it no longer needs it. Under this circumstance, the kernel will remove the mapping for the particular pages, enabling them to be reallocated to another process, or the same process, at a later time. Therefore, the state of memory in a typical multitasking OS is quite dynamic, and the routines to manage memory must be implemented in software because of their complexity and variability according to the platform and the nature of processes running at any given time.

Not all mapped virtual memory pages have to be held in physical RAM at the same time. Instead, the total virtual memory allocation on a computer can spill over into a secondary storage medium such as a hard drive. The hard drive will be much slower than DRAM, but not every memory page in every process is used at the same time. When a process is first loaded, its entire instruction image is typically loaded into virtual memory. However, it will take some time for all of those instructions to reach their turn in the execution sequence. During this wait time, the majority of a process's program

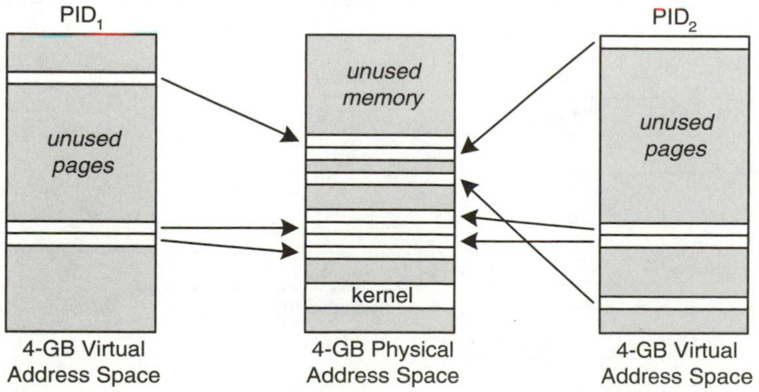

FIGURE 7.5 32-bit virtual memory mapping.

memory can be stored on the hard drive without incurring a performance penalty. When those instructions are ready to be executed, the OS kernel will have to transfer the data into physical memory. This slows the system down but makes it more flexible without requiring huge quantities of DRAM. Part of the kernel's memory management function is to decide which virtual pages should be held in DRAM and which should be *swapped* out to the disk. Pages that have not been used for a while can be swapped out to make room for new pages that are currently needed. If a process subsequently accesses a page that has been moved to the disk, that page can be swapped back into DRAM to replace another page that is not needed at the time. A computer with 256 MB of DRAM could, for example, have a 512-MB swap file on its hard drive, enabling processes to share a combined 768 MB of used virtual memory.

This scheme of expanding virtual memory onto a disk effectively turns the computer's DRAM into a large cache for an even larger disk-based memory. As with all caches, certain behavioral characteristics exist. A virtual memory page that is not present in DRAM is effectively a cache miss with a large penalty, because hard disks are much slower than DRAM. Such misses are called *page faults*. The MMU detects that the requested virtual memory address from a particular PID is not present in DRAM and causes an exception that must be handled by the OS kernel. Instead of performing a cache line fill and flush, it is the kernel's responsibility to swap pages to and from the disk. For a virtual memory system to function with reasonable performance, the *working set* of memory across all the processes running should be able to fit into the computer's physical memory. The working set includes any instructions and data that are accessed within a local time interval. This is directly analogous to a microprocessor cache's exploitation of locality. Processes with good locality characteristics will do well in a cache and in a virtual memory system. Processes with poor locality may result in thrashing as many sequential page faults are caused by random accesses throughout a large virtual memory space.

The virtual to physical address mapping process is guided by the kernel using a *page table*, which can take various forms but must somehow map each PID/VPN combination to either a physical memory page or one located on the disk drive's swap area. Virtual page mapping is illustrated in Fig. 7.6, assuming 4-kB pages, a 32-bit address space, and an 8-bit PID. In addition to basic mapping information, the page table also contains status information, including a dirty bit that indicates when a page held in memory has been modified. If modified, the page must be saved to the disk before being flushed to make room for a new virtual page. Otherwise, the page can be flushed without further action.

Given a 4-kB page size and a 32-bit address space, each process has access to $2^{20} = 1{,}048{,}576$ pages. With 256 PIDs, a brute-force page table would contain more than 268 million entries! There are a variety of schemes to reduce page table size, but there is no escaping the fact that a page table

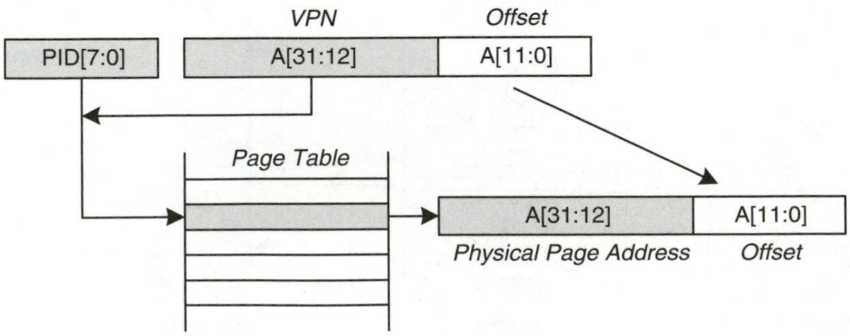

FIGURE 7.6 Virtual page mapping.

will be large. Page table management schemes are largely an issue of OS architecture and are outside the scope of this discussion. The fact that the page table is large and is parsed by software means that the mapping process will be extremely slow without hardware assistance. Every access to virtual memory, in other words almost every access performed on the computer, requires mapping, which makes hardware acceleration critical to the viability of virtual memory.

Within the MMU is a *translation lookaside buffer* (TLB), a small, fully associative cache that allows the MMU to rapidly locate recently accessed virtual page mappings. Typical sizes for a TLB are just 16 to 64 entries because of the complexity of implementing a fast fully associative cache. When a process is first spawned, it has not yet performed virtual memory accesses, so its first access will result in a TLB miss. When a TLB miss occurs, an exception is generated that invokes the kernel's memory management routine to parse the page table in search of the correct physical address mapping. The kernel routine loads a TLB entry with the mapping information and exits. On subsequent memory accesses, the TLB will hit some and miss some. It is hoped that the ratio of hits to misses will decline rapidly as the process executes. Once again, locality of reference is key to a well performing application, but the TLB and MMU are not as sensitive to locality as a normal cache, because they map multiple-kilobyte pages rather than 16 or 32 byte lines. Yet, as more processes actively vie for resources in a multitasking system, they may begin to fight each other for scarce TLB entries. The resources and architecture of a computer must be properly matched to its intended application. A typical desktop or embedded computer may get along fine with a small TLB, because it may not have many demanding processes running concurrently. A more powerful computer designed to simultaneously run many memory-intensive processes may require a larger TLB to take full advantage of its microprocessor and memory resources. The ever-present trade-off between performance and cost does not go away!

The TLB is usually located between the microprocessor and its cache subsystem as shown in Fig. 7.7, such that physical addresses are cached rather than virtual addresses. Such an arrangement adds latency to microprocessor transactions, because the virtual-to-physical mapping must take place before the L1 cache can respond. A TLB can be made very fast because of its small size, thereby limiting its time penalty on transactions. Additionally, microprocessors may implement a pipelined interface where addresses are presented each clock cycle, but their associated data are returned one or more clock cycles later, providing time for the TLB lookup.

7.5 SUPERPIPELINED AND SUPERSCALAR ARCHITECTURES

At any given time, semiconductor process technology presents an intrinsic limitation on how fast a logic gate can switch on and off and at what frequency a flip-flop can run. Other than relying on semiconductor process advances to improve microprocessor and system throughput, certain basic techniques have been devised to extract more processing power from silicon with limited switching

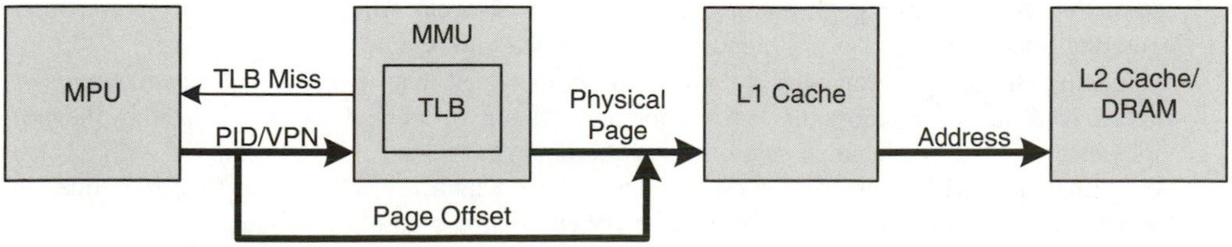

FIGURE 7.7 Location of TLB.

delays. Throughput can be enhanced in a serial manner by trying to execute a desired function faster. If each function is executed at a faster clock frequency, more functions can be executed in a given time period. An alternative parallel approach can be taken whereby multiple functions are executed simultaneously, thereby improving performance over time. These two approaches can be complementary in practice. Different logic implementations make use of serial and parallel enhancement techniques in the proportions and manners that are best suited to the application at hand.

A logic function is represented by a set of Boolean equations that are then implemented as discrete gates. During one clock cycle, the inputs to the equations are presented to a collection of gates via a set of input flops, and the results are clocked into output flops on the next rising edge. The propagation delays of the gates and their interconnecting wires largely determine the shortest clock period at which the logic function can reliably operate.

Pipelining, called *superpipelining* when taken to an extreme, is a classic serial throughput enhancement technique. Pipelining is the process of breaking a Boolean equation into several smaller equations and then calculating the partial results during sequential clock cycles. Smaller equations require fewer gates, which have a shorter total propagation delay relative to the complete equation. The shorter propagation delay enables the logic to run faster. Instead of calculating the complete result in a single 40 ns cycle, for example, the result may be calculated in four successive cycles of 10 ns each. At first glance, it may not seem that anything has been gained, because the calculation still takes 40 ns to complete. The power of pipelining is that different stages in the pipeline are operating on different calculations each cycle. Using an example of an adder that is pipelined across four cycles, partial sums are calculated at each stage and then passed to the next stage. Once a partial sum is passed to the next stage, the current stage is free to calculate the partial sum of a completely new addition operation. Therefore, a four-stage pipelined adder takes four cycles to produce a result, but it can work on four separate calculations simultaneously, yielding an average throughput of one calculation every cycle—a four-times throughput improvement.

Pipelining does not come for free, because additional logic must be created to handle the complexity of tracking partial results and merging them into successively more complete results. Pipelining a 32-bit unsigned integer adder can be done as shown in Fig. 7.8 by adding eight bits at a time and then passing the eight-bit sum and carry bit up to the next stage. From a Boolean equation perspective, each stage only incurs the complexity of an 8-bit adder instead of a 32-bit adder, enabling it to run faster. An array of pipeline registers is necessary to hold the partial sums that have been calculated by previous stages and the as-yet-to-be-calculated portions of the operands. The addition results ripple through the pipeline on each rising clock edge and are accumulated into a final 32-bit result as operand bytes are consumed by the adders. There is no feedback in this pipelined adder, meaning that, once a set of operands passes through a stage, that stage no longer has any involvement in the operation and can be reused to begin or continue a new operation.

Pipelining increases the overall throughput of a logic block but does not usually decrease the calculation latency. High-performance microprocessors often take advantage of pipelining to varying degrees. Some microprocessors implement superpipelining whereby a simple RISC instruction may have a latency of a dozen or more clock cycles. This high degree of pipelining allows the microprocessor to execute an average of one instruction each clock cycle, which becomes very powerful at operating frequencies measured in hundreds of megahertz and beyond.

Superpipelining a microprocessor introduces complexities that arise from the interactions between consecutive instructions. One instruction may contain an operand that is calculated by the previous instruction. If not handled correctly, this common circumstance can result in the wrong value being used in a subsequent instruction or a loss of performance where the pipeline is frequently stalled to allow one instruction to complete before continuing with others. Branches can also cause havoc with a superpipelined architecture, because the decision to take a conditional branch may nullify the few instructions that have already been loaded into the pipeline and partially executed. De-

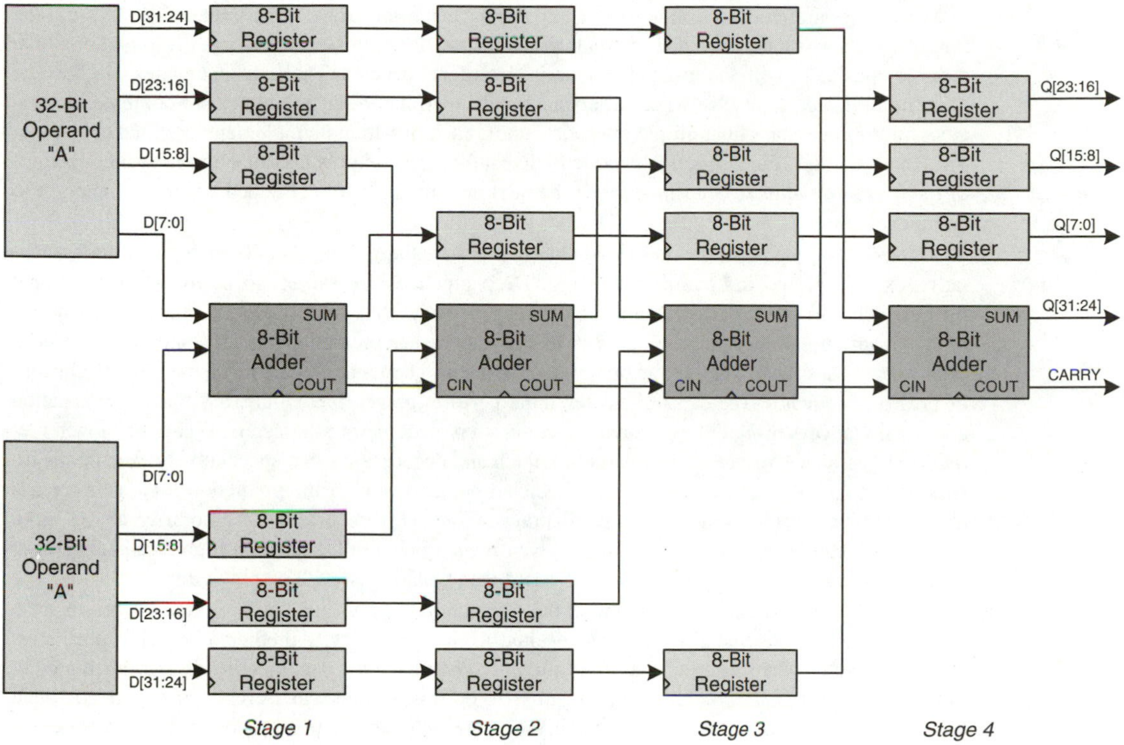

FIGURE 7.8 Four-stage pipelined adder.

pending on how the microprocessor is designed, various state information that has already been modified by these partially executed instructions may have to be rolled back as if the instructions were never fetched. Branches can therefore cause the pipeline to be flushed, reducing the throughput of the microprocessor, because there will be a gap in time during which new instructions advance through the pipeline stages and finally emerge at the output.

Traditional microprocessor architecture specifies that instructions are executed serially in the order explicitly defined by the programmer. Microprocessor designers have long observed that, within a given sequence of instructions, small sets of instructions can be executed in parallel without changing the result that would be obtained had they been executed in the traditional serial manner. *Superscalar* microprocessor architecture has emerged as a means to execute multiple instructions simultaneously within a single microprocessor that is operating on a normal sequence of instructions. A superscalar architecture contains multiple independent execution units, some of which may be identical, that are organized and replicated according to statistical studies of which instructions are executed more often and how easily they can be made parallel without excessive restrictions and dependencies. Arithmetic execution units are prime targets for replication, because calculations with floating-point numbers and large integers require substantial logic and time to fully complete. A superscalar microprocessor may contain two integer ALUs and separate FPUs for floating-point addition and multiplication operations. Floating-point operations are the most complex instructions that many microprocessors execute, and they tend to have long latencies. Most floating-point applications contain a mix of addition and multiplication operations, making them well suited to an architecture with individual FPUs that each specialize in one type of operation.

Managing parallel execution units in a superscalar microprocessor is a complex task, because the microprocessor wants to execute instructions as fast as they can be fetched—yet it must do so in a manner consistent with the instructions' serial interdependencies. These dependencies can become more complicated to resolve when superscalar and superpipelining techniques are combined to create a microprocessor with multiple execution units, each of which is implemented with a deep pipeline. In such chips, the instruction decode logic handles the complex task of examining the pipelines of the execution units to determine when the next instruction is free of dependencies, allowing it to begin execution.

Related to superpipelining and superscalar methods are the techniques of branch prediction, speculative execution, and instruction reordering. Deep pipelines are subject to performance-degrading flushes each time a branch instruction comes along. To reduce the frequency of pipeline flushes due to branch instructions, some microprocessors incorporate branch prediction logic that attempts to make a preliminary guess as to whether the branch will be taken. These guesses are made based on the history of previous branches. The exact algorithms that perform branch prediction vary by implementation and are not always disclosed by the manufacturer, to protect their trade secrets. When the branch prediction logic makes its guess, the instruction fetch and decode logic can speculatively execute the instruction stream that corresponds to the predicted branch result. If the prediction logic is correct, a costly pipeline flush is avoided. If the prediction is wrong, performance will temporarily degrade until the pipeline can be restarted. Hopefully, a given branch prediction algorithm improves performance rather than degrading it by having a worse record than would exist with no prediction at all!

The problem with branch prediction is that it is sometimes wrong, and the microprocessor must back out of any state changes that have resulted from an incorrectly predicted branch. Speculative execution can be taken a step farther in an attempt to eliminate the penalty of a wrong branch prediction by executing both possible branch results. To do this, a superscalar architecture is needed that has enough execution units to speculatively execute extra instructions whose results may not be used. It is a foregone conclusion that one of the branch results will not be valid. There is substantial complexity involved in such an approach because of the duplicate hardware that must be managed and the need to rapidly swap to the correct instruction stream that is already in progress when the result of a branch is finally known.

A superscalar microprocessor will not always be able to keep each of its execution units busy, because of dependencies across sequential instructions. In such a case, the next instruction to be pushed into the execution pipeline must be held until an in-progress instruction completes. Instruction reordering logic reduces the penalty of such instruction stalls by attempting to execute instructions outside the order in which they appear in the program. The microprocessor can prefetch a set of instructions ahead of those currently executing, enabling it to look ahead in the sequence and determine whether a later instruction can be safely executed without changing the behavior of the instruction stream. For such reordering to occur, an instruction must not have any dependencies on those that are being temporarily skipped over. Such dependencies include not only operands but branch possibilities as well. Reordering can occur in a situation in which the ALUs are busy calculating results that are to be used by the next instruction in the sequence, and their latencies are preventing the next instruction from being issued. A load operation that is immediately behind the stalled instruction can be executed out of order if it does not operate on any registers that are being used by the instructions ahead of it. Such reordering boosts throughput by taking advantage of otherwise idle execution cycles.

All of the aforementioned throughput improvement techniques come at a cost of increased design complexity and cost. However, it has been widely noted that the cost of a transistor on an IC is asymptotically approaching zero as tens of millions of transistors are squeezed onto chips that cost only several hundred dollars. Once designed, the cost of implementing deep pipelines, multiple execution units, and the complex logic that coordinates the actions of both continues to decrease over time.

7.6 FLOATING-POINT ARITHMETIC

Conventional arithmetic logic units operate on signed and unsigned integer quantities. Integers suffice for many applications, including loop count variables and memory addresses. However, our world is inherently analog and is best represented by real numbers as compared to discrete integers. Floating-point arithmetic enables the representation and manipulation of real numbers of arbitrary magnitude and precision. Historically, floating-point math was pertinent only to members of the scientific community who regularly perform calculations on large data sets to model many types of natural phenomena. Almost every area of scientific research has benefited from computational analysis, including aerodynamics, geology, medicine, and meteorology. More recently, floating-point math has become more applicable to the mainstream community in such applications as video games that render realistic three-dimensional scenes in real-time as game characters move around in virtual environments.

General mathematics represents numbers of arbitrary magnitude and precision using *scientific notation*, consisting of a signed *mantissa* multiplied by an integer power of ten. The mantissa is greater than or equal to one and less than ten. In other words, the decimal point of the mantissa is shifted left or right until a single digit remains in the 1s column. The number 456.8 would be represented as 4.568×10^2 in scientific notation. All significant digits other than the first one are located to the right of the decimal point. The number –0.000089 has only two significant digits and is represented as -8.9×10^{-5}. Scientific notation enables succinct and accurate representation of very large and very small numbers.

Floating-point arithmetic on a computer uses a format very similar to scientific notation, but binary is used in place of decimal representation. The Institute of Electrical and Electronics Engineers (IEEE) has standardized floating-point representation in several formats to express numbers of increasing magnitude and precision. These formats are used by most hardware and software implementations of floating-point arithmetic for the sake of compatibility and consistency. Figure 7.9 shows the general structure of an IEEE floating point number.

The most significant bit is defined as a sign bit where zero is positive and one is negative. The sign bit is followed by an n-bit exponent with values from 1 to $2^n - 2$ (the minimum and maximum values for the exponent field are not supported for normal numbers). The exponent represents powers of two and can represent negative exponents by means of an exponent *bias*. The bias is a fixed, standardized value that is subtracted from the actual exponent field to yield the true exponent value. It is generally $2^{(n-1)} - 1$. Following the exponent is the binary *significand*, which is a mantissa or modified mantissa. Similar to scientific notation, the mantissa is a number greater than or equal to 1 and less than the radix (2, in this case). Therefore, the whole number portion of the binary mantissa must be 1. Some IEEE floating-point formats hide this known bit and use a modified mantissa to provide an additional bit of precision in the fractional portion of the mantissa. Table 7.5 lists the basic parameters of the four commonly used floating-point formats. The IEEE-754 standard defines several formats including single and double precision. The extended and quadruple precision formats are not explicitly mentioned in the standard, but they are legal derivations from formats that provide for increased precision and exponent ranges.

It is best to use a single-precision example to see how floating-point representation actually works. The decimal number 25.25 is first converted to its binary equivalent: 11001.01. The mantissa and exponent are found by shifting the binary point four places to the left to yield 1.100101×2^4. Us-

Sign	Exponent	Modified Mantissa

FIGURE 7.9 General IEEE floating-point structure.

TABLE 7.5 IEEE/Industry Floating-Point Formats

Format	Total Bits	Exponent Bits	Exponent Bias	Smallest Exponent	Largest Exponent	Significant Bits	Mantissa MSB
Single precision	32	8	127	–126	+127	23	Hidden
Double precision	64	11	1,023	–1022	+1023	52	Hidden
Extended precision	80	15	16,383	–16382	+16383	64	Explicit
Quadruple precision	128	15	16,383	–16382	+16383	112	Hidden

ing the single-precision format, the exponent field is calculated by adding the true exponent value to the bias, 127, to get a final value of 131. Expressing these fields in a 32-bit word yields the floating point value 0x41CA0000 as shown in Fig. 7.10.

Note that the sign bit is 0 and that the mantissa's MSB has been omitted. This example is convenient, because the binary representation of 25.25 is finite. However, certain numbers that have finite representations in decimal cannot be represented as cleanly in binary, and vice versa. The number 0.23 clearly has a finite decimal representation but, when converted to binary, it must be truncated at whatever precision limitation is imposed by the floating-point format in use. The number 0.23 can be converted to a binary fraction by factoring out successive negative powers of 2 and expressing the result with 24 significant figures (leading 0s do not count), because the single precision format supports a 24-bit mantissa,

$$0.0011\_1010\_1110\_0001\_0100\_0111\_11$$

This fraction is then converted to a mantissa and power-of-two representation,

$$1.1101\_0111\_0000\_1010\_0011\_111 \times 2^{-3}$$

A single-precision floating-point exponent value is obtained by adding the bias, 127+(–3) = 124, for a final representation of

$$0011\_1110\_0110\_1011\_1000\_0101\_0001\_1111 \text{ (0x3E6B851F)}$$

These conversions are shown only to explain the IEEE formats and almost never need to be done by hand. Floating-point processing is performed either by dedicated hardware or software algorithms. Most modern high-performance microprocessors contain on-chip floating-point units (FPUs), and their performance is measured in floating-point operations per second (FLOPS). High-end microprocessors can deliver several gigaFLOPS (GFLOPS) of throughput on benchmark tests. Computers without hardware FPUs must emulate floating-point processing in software, which can be a relatively slow process. However, if a computer needs to perform only a few hundred floating-point operations per second, it may be worth saving the cost and space of a dedicated hardware FPU.

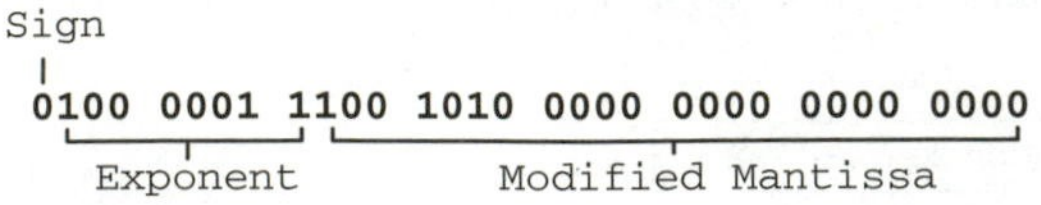

FIGURE 7.10 Single-precision floating-point expression of 25.25.

As can be readily observed from Table 7.5, very large and very small numbers can be represented because of the wide ranges of exponents provided in the various formats. However, the representation of 0 seems rather elusive with the requirement that the mantissa always have a leading 1. Values including 0 and infinity are represented by using the two-exponent values that are not supported for normal numbers: 0 and $2^n - 1$. In the case of the single-precision format, these values are 0x00 and 0xFF.

An exponent field of 0x00 is used to represent numbers of very small magnitude, where the interpreted exponent value is fixed at the minimum for that format: –126 for single precision. With a 0 exponent field, the mantissa's definition changes to a number greater than or equal to 0 and less than 1. Smaller numbers can now be represented, though with decreasing significant figures, because magnitude is now partially represented by the significand field. For example, 101×2^{-130} is expressed as 0.0101×2^{-126}. Such special-case numbers are *denormalized,* because their mantissas defy the *normalized* form of being greater than or equal to 1 and less than 2. Zero can now be expressed by setting the significand to 0 with the result that $0 \times 2^{-126} = 0$. The presence of the sign bit produces two representations of zero, positive and negative, that are mathematically identical.

Setting the exponent field to 0xFF (in single precision) is used to represent either infinity or an undefined value. Positive and negative infinity are represented by setting the significand field to 0 and using the appropriate sign bit. When the exponent field is 0xFF and the significand field is nonzero, the representation is "not a number," or *NaN*. Examples of computations that may return NaN are $0 \div 0$ and $\infty \div \infty$.

7.7 DIGITAL SIGNAL PROCESSORS

Microprocessor architectures can be optimized for increased efficiency in certain applications through the inclusion of special instructions and execution units. One major class of application-specific microprocessors is the *digital signal processor*, or DSP. DSP entails a microprocessor mathematically manipulating a sampled analog signal in a way that emulates transformation of that signal by discrete analog components such as filters or amplifiers. To operate on an analog signal digitally, the analog signal must be sampled by an analog-to-digital converter, manipulated, and then reconstructed with a digital-to-analog converter. A rough equivalency of digital signal processing versus conventional analog transformation is shown in Fig. 7.11 in the context of a simple filter.

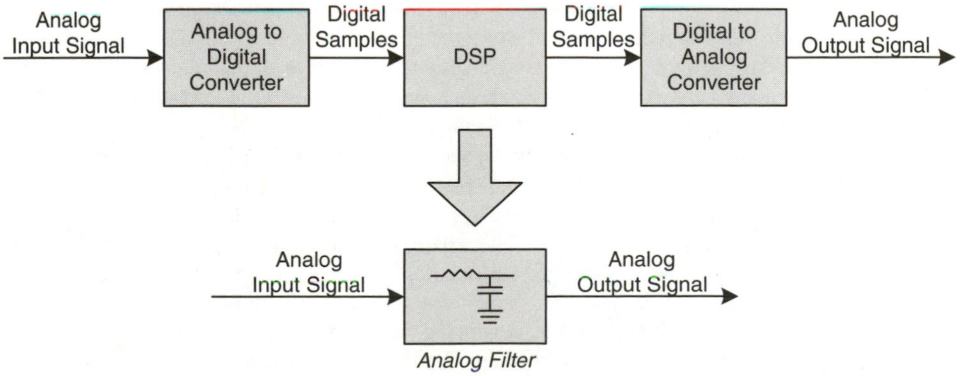

FIGURE 7.11 Digital signal processing.

In this example of a lowpass filter (the amplitude of frequencies above a certain threshold are attenuated), the complexity of digital sampling and a microprocessor appears unjustified. The power of DSP comes when much more complex analog transformations are performed that would require excessively complex analog circuit topologies. Some examples of applications in which DSPs are used include modems, cellular telephones, and radar. While sensitive analog circuits may degrade or fall out of calibration over time, digital instructions and sequences maintain their integrity indefinitely. Major manufacturers of DSPs include Analog Devices, Motorola, and Texas Instruments. Many books have been written on DSP algorithms and techniques, which are extremely diverse and challenging topics.

DSP algorithms are characterized by repetitive multiplication and addition operations carried out on the sampled data set. Multiply and addition operations are also known as *multiply and accumulate*, or MAC, operations in DSP parlance. These calculations involve the sampled data as well as coefficients that, along with the specific operations, define the transformation being performed. For DSP to be practical, it must be performed in real time, because the signals cannot be paused while waiting for the microprocessor to finish its previous operation. For DSP to be economical, this throughput must be achieved at an acceptable cost. A general-purpose microprocessor can be used to perform DSP functions, but in most cases, the solution will not be economical. This is because the microprocessor is designed to execute general programs for which there is less emphasis on specific types of calculations. A DSP is designed specifically to rapidly execute multiply and accumulate operations, and it contains additional hardware to efficiently fetch sequential operands from tables in memory. Not all of the features discussed below are implemented by all DSPs, but they are presented to provide an understanding of the overall set of characteristics that differentiates a DSP from a generic microprocessor.

At their core, DSPs contain one or more ALUs that are capable of multiplication and addition in a single cycle. This rapid calculation capability ensures that throughput can be maintained as long as operands are fed to the ALUs. DSPs are manufactured with a variety of ALU capabilities ranging from 16-bit integer to IEEE floating-point support. As with a generic microprocessor, the number of ALUs influences how many simultaneous operations can be carried out at a given time. To keep the ALUs supplied with operands, DSPs contain hardware structures called *address generators* that automatically calculate the addresses of the next operands to be used in a calculation. Sampled data is stored in a memory array, and algorithmic coefficients are stored in a separate array. Depending on the algorithm, the array entries may not be accessed sequentially. On a generic microprocessor, the software would have to add an arbitrary offset value to an index register each time a new operand was desired. Additionally, as a result of fixed array sizes, the pointer eventually wraps around from the end to the beginning, thereby requiring additional instructions to check for the wrap condition. This index register overhead slows the computation process. Address generators offload this overhead to hardware by associating additional registers with the index registers. These registers define the increment to be applied to an index register following a load or store operation and also define the start and end addresses of the memory array. Therefore, software is able to execute load/store and calculation operations without spending time on routine pointer arithmetic.

The specialized ALU and address generation hardware within the DSP core place a high demand on memory to maintain a steady flow of instructions and data. DSPs commonly implement a Harvard memory architecture in which separate buses connect to program and data memory. Most DSPs contain separate program and data memory structures integrated onto the same chip as the DSP core for minimal access latency to small repetitive DSP algorithm kernels. Program memory may be implemented as ROM or RAM, depending on whether an external interface is available from which to load programs. These on-chip memories may be as small as several kilobytes each for less expensive DSPs or hundreds of kilobytes for more powerful products. To mitigate the complexity of a Harvard architecture on the overall system design, most DSPs contain a unified external memory bus for connection to external ROM and RAM. A DSP application can boot from external ROM, then load its kernel into on-chip program memory and perform the majority of its execution without fetching additional instructions from external memory.

7.8 PERFORMANCE METRICS

Evaluating the throughput potential of a microprocessor or a complete computer is not as simple as finding out how fast the microprocessor's clock runs. System performance varies widely according to the applications being used and the computing hardware on which they run. Applications vary widely in their memory and I/O usage, both being properties whose performance is directly tied to the hardware architecture. We can consider three general sections of a computer and how each influences the speed at which an application is executed: the microprocessor, the memory system, and the I/O resources.

The usable address space of a microprocessor is an important consideration, because applications vary in their memory needs. Some embedded applications can fit into several kilobytes of memory, making an 8-bit computer with 64 kB or less of address space quite adequate. More complex embedded applications start to look like applications that run on desktop computers. If large data arrays are called for, or if a multitasking system is envisioned whereby multiple tasks each contain megabytes of program memory and high-level data structures, a 32-bit microprocessor with hundreds of megabytes of usable address space may be necessary. At the very high end, microprocessors have transitioned to 64-bit architectures with gigabytes of directly addressable memory. A microprocessor's address space can always be expanded externally by banking methods, but banking comes at a penalty of increased time to switch banks and the complexity of making an application aware of the banking scheme.

Any basic type of application can run on almost any microprocessor. The question is how fast and efficiently a particular microprocessor is able to handle the application. The instruction set is an important attribute that should be considered when designing a computer system. If a floating-point intensive application is envisioned, it should probably be run on a microprocessor that contains an FPU, and the number of floating-point execution units and their execution latencies is an important attribute to investigate. An integer-only microprocessor could most likely run the floating-point application by performing software emulation of floating-point operations, but its performance would probably be rather dismal. For smaller-scale computers and applications, these types of questions are still valid. If an application needs to perform frequent bit manipulations for testing and setting various flags, a microprocessor that directly supports bit manipulation may be better suited than a generic architecture with only logical AND/OR type instructions.

Once a suitable instruction set has been identified, a microprocessor's ability to actually fetch and execute the instructions can become an important part of system performance. On smaller systems, there are few variables in instruction fetch and execution: each instruction is fetched and executed sequentially. Superscalar microprocessors, however, must include effective instruction analysis logic to properly utilize all the extra logic that has been put onto the chip and that you are paying for. If the multiple execution units cannot be kept busy enough of the time, your application will not enjoy the benchmark performance claims of the manufacturer. Vendors of high-performance microprocessors devote much time to instruction profiling and analysis of instruction sequences. Their results improve performance on most applications, but there are always a few niche applications that have uncommon properties that can cause certain microprocessors to fall off in performance. It pays to keep in mind that common industry benchmarks of microprocessor performance do not always tell the whole story. These tests have been around for a long time, and microprocessor designers have learned how to optimize their hardware to perform well on the standard benchmarks. An application that behaves substantially differently from a benchmark test may not show the same level of performance as advertised by the manufacturer.

The microprocessor's memory interface is a critical contributor to its performance. Whether a small 8-bit microprocessor or a 64-bit behemoth, the speed with which instructions can be fetched and data can be loaded and stored affects the execution time of an application. The necessary bandwidth of a memory interface is relative and is proportional to the sum of the instruction and data bandwidths of an application. From the instruction perspective, it is clear that the microprocessor needs to keep itself busy with a steady stream of instructions. Data bandwidth, however, is very much a function of the application. Some applications may perform frequent load/store operations, whereas others may operate more on data retained within the microprocessor's register set. To the extent that load/store operations detract from the microprocessor's ability to fetch and execute new instructions, they will reduce overall throughput.

Clock frequency becomes a defining attribute of a microprocessor once its instruction set, execution capabilities, and memory interface are understood from a performance perspective. Without these supporting attributes, clock speed alone does not define the capabilities of a microprocessor. A 500-MHz single-issue, or nonsuperscalar, microprocessor could be easily outperformed by a 200-MHz four-issue superscalar design. Additionally, there may be multiple relevant clocks to consider in a complex microprocessor. Microprocessors whose internal processing cores are decoupled from the external memory bus by an integrated cache are often specified with at least two clocks: the core clock and the bus interface clock. It is necessary to understand the effect of both clocks on the processing core's throughput. A fast core can be potentially starved for instructions and data by a slow interface. Once a microprocessor's resources have been quantified, clock frequency becomes a multiplier to determine how many useful operations per second can be expected. Metrics such as instructions per second (IPS) or floating-point operations per second (FLOPS) are specified by multiplying the average number of instructions executed per cycle by how many cycles occur each second. Whereas high-end microprocessors were once measured in MIPS and MFLOPS, GIPS and GFLOPS performance levels are now attainable.

As already mentioned, memory bandwidth and, consequently, memory architecture hold key roles in determining overall system performance. Memory system architecture encompasses all memory external to the microprocessor's core, including any integrated caches that it may contain. When dealing with an older-style microprocessor with a memory interface that does not stress current memory technologies, memory architecture may not be subject to much variability and may not be a bottleneck at all. It is not hard to find flash, EPROM, and SRAM devices today with access times of 50 ns and under. A moderately sized memory array constructed from these components could provide an embedded microprocessor with full-speed random access as long as the memory transaction rate is 20 MHz or less. Many 8-, 16-, and even some 32-bit embedded microprocessors can fit comfortably within this performance window. As such, computers based on these devices can have simple memory architectures without suffering performance degradation.

Memory architecture starts to get more complicated when higher levels of performance are desired. Once the microprocessor's program and data fetch latency becomes faster than main memory's random access latency, caching and bandwidth improvement techniques become critical to sustaining system throughput. Random access latency is the main concern. A large memory array can be made to deliver adequate bandwidth given a sufficient width. As a result of the limited operating frequency of SDRAM devices, high-end workstation computers have been known to connect multiple memory chips in parallel to create 256-bit and even 512-bit wide interfaces. Using 512 Mb DDR SDRAMs, each organized as 32M × 16 and running at 167 MHz, 16 devices in parallel would yield a 1-GB memory array with a burst bandwidth of 167 MHz × 2 words/hertz × 256 bits/word = 85.5 Gbps! This is a lot of bandwidth, but relative to a microprocessor core that operates at 1 GHz or more with a 32- or 64-bit data path, such a seemingly powerful memory array may just barely be able to keep up.

While bandwidth can be increased by widening the interface, random access latency does not go away. Therefore, there is more to a memory array than its raw size. The bandwidth of the array, which is the product of its interface frequency and width, and its latency are important metrics in understanding the impact of cache misses, especially when dealing with applications that exhibit poor locality.

Caching reduces the negative effect of high random access latencies on a microprocessor's throughput. However, caches and wide arrays cannot completely balance the inequality between the bandwidth and latency that the microprocessor demands and that which is provided by SDRAM technology. Cache size, type, and latency and main memory bandwidth are therefore important metrics that contribute to overall system performance. An application's memory characteristics determine how costly a memory architecture is necessary to maintain adequate performance. Applications that operate on smaller sets of data with higher degrees of locality will be less reliant on a large cache and fast memory array, because they will have fewer cache misses. Those applications with opposite memory characteristics will increase the memory architecture's effect on the computer's overall performance. In fact, by the nature of the application being run, caching effects can become more significant than the microprocessor's core clock frequency. In some situations, a 500-MHz microprocessor with a 2-MB cache can outperform a 1-GHz microprocessor with a 256-kB cache. It is important to understand these considerations because money may be better spent on either a faster microprocessor or a larger cache according to the needs of the intended applications.

I/O performance affects system throughput in two ways: the latency of executing transactions and the degree to which such execution blocks the microprocessor from performing other work. In a computer in which the microprocessor operates with a substantially higher bandwidth than individual I/O interfaces, it is desirable to decouple the microprocessor from the slower interface as much as possible. Most I/O controllers provide a natural degree of decoupling. A typical UART, for example, absorbs one or more bytes in rapid succession from a microprocessor and then transmits them at a slower serial rate. Likewise, the UART assembles one or more whole incoming bytes that the microprocessor can read at an instantaneous bandwidth much higher than the serial rate. Network and disk adapters often contain buffers of several kilobytes that can be rapidly filled or drained by the microprocessor. The microprocessor can then continue with program execution while the adapter logic handles the data at whatever lower bandwidth is inherent to the physical interface.

Inherent decoupling provided by an I/O controller is sufficient for many applications. When dealing with very I/O-intensive applications, such as a large server, multiple I/O controllers may interact with each other and memory simultaneously in a multimaster bus configuration. In such a context, the microprocessor sets up block data transfers by programming multiple I/O and DMA controllers and then resumes work processing other tasks. Each I/O and DMA controller is a potential bus master that can arbitrate for access to the memory system and the I/O bus (if there is a separate I/O bus). As the number of simultaneous bus masters increases, contention can develop, which may cause performance degradation resulting from excessive waiting time by each potential bus master. This contention can be reduced by modifying the I/O bus architecture. A first step is to decouple the I/O bus from the memory bus into one or more segments, enabling data transfers within a given I/O segment to proceed without conflicting with a memory transfer or one contained within other I/O segments. PCI is an example of such a solution. At a more advanced level, the I/O system can be turned into a switched network in which individual I/O controllers or small segments of I/O controllers are connected to a dedicated port on an I/O switch that enables each port to communicate with any other port simultaneously insofar as multiple ports do not conflict for access to the same port. This is a fairly expensive solution that is implemented in high-end servers for which I/O performance is a key contributor to overall system throughput.

The question of how fast a computer performs does not depend solely on how many megahertz the microprocessor runs at or how much RAM it has. Performance is highly application specific and is dominated by how many cycles per second the microprocessor is kept busy with useful instructions and data.

CHAPTER 8
High-Performance Memory Technologies

Memory is an interesting and potentially challenging portion of a digital system design. One of the benefits of decades of commercial solid-state memory development is the great variety of memory products available for use. Chances are that there is an off-the-shelf memory product that fits your specific application. A downside to the modern, ever-changing memory market is rapid obsolescence of certain products. DRAM is tied closely to the personal computer market. The best DRAM values are those devices that coincide with the sweet spot in PC memory configurations. As the high-volume PC market moves on to higher-density memory ICs, that convenient DRAM that you used in your designs several years ago may be discontinued so that the manufacturer can retool the factory for parts that are in greater demand.

Rapid product development means that memory capabilities improve dramatically each year. Whether it's higher density or lower power that an application demands, steady advances in technology put more tools at an engineer's disposal. SRAM and flash EPROM devices have more stable production lives than DRAM. In part, this is because they are less dependent on the PC market, which requires ever increasing memory resources for ever more complex software applications.

Memory is a basic digital building block that is used for much more than storing programs and data for a microprocessor. Temporary holding buffers are used to store data as it is transferred from one interface to another. There are many situations in networking and communication systems where a block of data arrives and must be briefly stored in a buffer until the logic can figure out exactly what to do with it. Lookup tables are another common use for memory. A table may store precomputed terms of a complex calculation so that a result can be rapidly determined when necessary. This chapter discusses the predominant synchronous memory technologies, SDRAM and SSRAM, and closes with a presentation of CAM, a technology that is part RAM and part logic.

No book can serve as an up-to-date reference on memory technology for long, as a result of the industry's rapid pace. This chapter discusses technologies and concepts that are timeless, but specifics of densities, speeds, and interface protocols change rapidly. Once you have read and understood the basics of high-performance memory technologies, you are encouraged to browse through the latest manufacturers' data sheets to familiarize yourself with the current state of the art. Corporations such as Cypress, Hynix, Infineon, Micron, NEC, Samsung, and Toshiba provide detailed data sheets on their web sites that are extremely useful for self-education and selecting the right memory device to suit your needs.

8.1 SYNCHRONOUS DRAM

As system clock frequencies increased well beyond 50 MHz, conventional DRAM devices with asynchronous interfaces became more of a limiting factor in overall system performance. Asynchronous DRAMs have associated pulse width and signal-to-signal delay specifications that are tied closely to the characteristics of their internal memory arrays. When maximum bandwidth is desired at high clock frequencies, these specifications become difficult to meet. It is easier to design a system in which all interfaces and devices run synchronously so that interface timing becomes an issue of meeting setup and hold times, and functional timing becomes an issue of sequencing signals on discrete clock edges.

Synchronous DRAM, or SDRAM, is a twist on basic asynchronous DRAM technology that has been around for more than three decades. SDRAM can essentially be considered as an asynchronous DRAM array surrounded by a synchronous interface on the same chip, as shown in Fig. 8.1. A key architectural feature in SDRAMs is the presence of multiple independent DRAM arrays—usually either two or four banks. Multiple banks can be activated independently and their transactions interleaved with those of other banks on the IC's synchronous interface. Rather than creating a bottleneck, this functionality allows higher efficiency, and therefore higher bandwidth, across the interface. One factor that introduces latency in random accesses across all types of DRAM is the row activation time: a row must first be activated before the column address can be presented and data read or written. An SDRAM allows a row in one bank to be activated while another bank is actively engaged in a read or write, effectively hiding the row activation time in the other bank. When the current transaction completes, the previously activated row in the other bank can be called upon to perform a new transaction without delay, increasing the device's overall bandwidth.

The synchronous interface and internal state logic direct interleaved multibank operations and burst data transfers on behalf of an external memory controller. Once a transaction has been started, one data word flows into or out of the chip on every clock cycle. Therefore, an SDRAM running at 100 MHz has a theoretical peak bandwidth of 100 million words per second. In reality, of course, this number is somewhat lower because of refresh and the overhead of beginning and terminating transactions. The true available bandwidth for a given application is very much dependent on that application's data transfer patterns and the capabilities of its memory controller.

Rather than implementing a DRAM-style asynchronous interface, the SDRAM's internal state logic operates on discrete commands that are presented to it. There are still familiar sounding signals such as RAS* and CAS*, but they function synchronously as part of other control signals to form commands rather than simple strobes. Commands begin and terminate transactions, perform refresh operations, and configure the SDRAM for interface characteristics such as default burst length.

SDRAM can provide very high bandwidth in applications that exploit the technology's burst transfer capabilities. A conventional computer with a long-line cache subsystem might be able to fetch 256 words in as few as 260 cycles: 98.5 percent efficiency! Bursts amortize a fixed number of overhead cycles across the entire transaction, greatly improving bandwidth. Bandwidth can also be improved by detecting transactions to multiple banks and interleaving them. This mode of operation

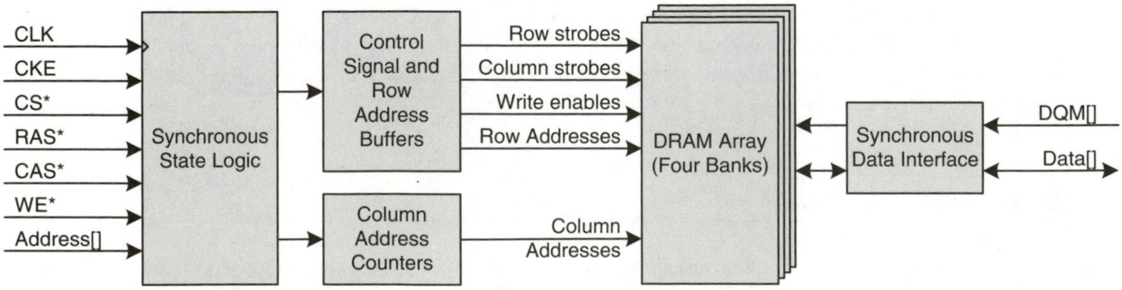

FIGURE 8.1 Basic SDRAM architecture.

allows some new burst transfers to be requested prior to the current burst ending, thereby hiding the initial startup latency of the subsequent transaction.

Most of the input signals to the state logic shown in Fig. 8.1 combine to form the discrete commands listed in Table 8.1. A clock enable, CKE, must be high for normal operation. When CKE is low, the SDRAM enters a low-power mode during which data transactions are not recognized. CKE can be tied to logic 1 for applications that are either insensitive to power savings or require continual access to the SDRAM. Interface signals are sampled on the rising clock edge. Many SDRAM devices are manufactured in multibyte data bus widths. The data mask signals, DQM[], provide a convenient way to selectively mask individual bytes from being written or being driven during reads. Each byte lane has an associated DQM signal, which must be low for the lane to be written or to enable the lane's tri-state buffers on a read.

TABLE 8.1 Basic SDRAM Command Set

Command	CS*	RAS*	CAS*	WE*	Address	AP/A10
Bank activate	L	L	H	H	Bank, row	A10
Read	L	H	L	H	Bank, column	L
Read with auto-precharge	L	H	L	H	Bank, column	H
Write	L	H	L	L	Bank, column	L
Write with auto-precharge	L	H	L	L	Bank, column	H
No operation	L	H	H	H	X	X
Burst terminate	L	H	H	L	X	X
Bank precharge	L	L	H	L	X	L
Precharge all banks	L	L	H	L	X	H
Mode register set	L	L	L	L	Configuration	Configuration
Auto refresh	L	L	L	H	X	X
Device deselect	H	X	X	X	X	X

Some common functions include activating a row for future access, performing a read, and precharging a row (deactivating a row, often in preparation for activating a new row). For complete descriptions of SDRAM interface signals and operational characteristics, SDRAM manufacturers' data sheets should be referenced directly. Figure 8.2 provides an example of how these signals are used to implement a transaction and serves as a useful vehicle for introducing the synchronous interface. CS* and CKE are assumed to be tied low and high, respectively, and are not shown for clarity.

The first requirement to read from an SDRAM is to activate the desired row in the desired bank. This is done by asserting an activate (ACTV) command, which is performed by asserting RAS* for one cycle while presenting the desired bank and row addresses. The next command issued to continue the transaction is a read (RD). However, the controller must wait a number of cycles that translates into the DRAM array's row-activate to column-strobe delay time. The timing characteristics of the underlying DRAM array is expressed in nanoseconds rather than clock cycles. Therefore, the integer number of delay cycles is different for each design, because it is a function of the clock period and the internal timing specification. If, for example, an SDRAM's RAS* to CAS* delay is 20 ns, and the clock period is 20 ns or slower, an RD command could be issued on the cycle immediately

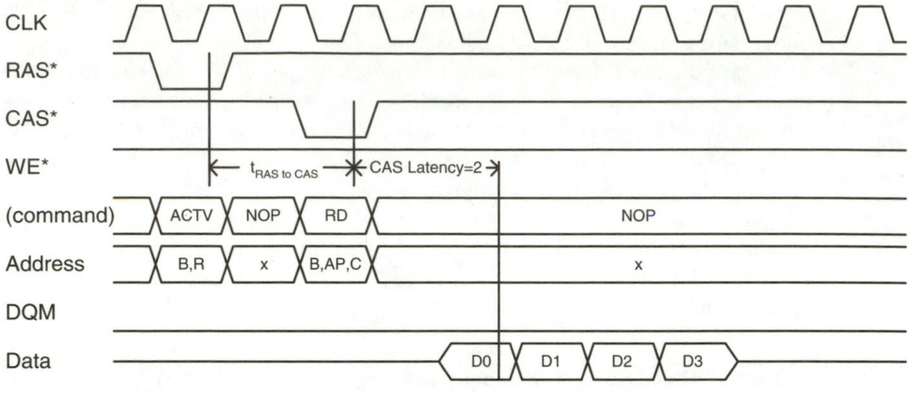

FIGURE 8.2 Four-word SDRAM burst read (CL = 2, BL = 4).

following the ACTV. Figure 8.2 shows an added cycle of delay, indicating a clock period less than 20 ns but greater than 10 ns (a 50–100 MHz frequency range). During idle cycles, a no-operation (NOP) command is indicated by leaving RAS*, CAS*, and WE* inactive.

The RD command is performed by asserting CAS* and presenting the desired bank select and column address along with the auto-precharge (AP) flag. A particular bank must be selected, because the multibank SDRAM architecture enables reads from any bank. AP is conveyed by address bit 10 during applicable commands, including reads and writes. Depending on the type of command, AP has a different meaning. In the case of a read or write, the assertion of AP tells the SDRAM to automatically precharge the activated row after the requested transaction completes. Precharging a row returns it to a quiescent state and also clears the way for another row in the same bank to be activated in the future. A single DRAM bank cannot have more than one row active at any given time. Automatically precharging a row after a transaction saves the memory controller from explicitly precharging the row after the transaction. If, however, the controller wants to take full advantage of the SDRAM's back-to-back bursting capabilities by leaving the same row activated for a subsequent transaction, it may be worthwhile to let the controller decide when to precharge a row. This way, the controller can quickly reaccess the same row without having to issue a redundant ACTV command. AP also comes into play when issuing separate precharge commands. In this context, AP determines if the SDRAM should precharge all of its banks or only the bank selected by the address bus.

Once the controller issues the RD command (it would be called RDA if AP is asserted to enable auto-precharge), it must wait a predetermined number of clock cycles before the data is returned by the SDRAM. This delay is known as *CAS latency*, or CL. SDRAMs typically implement two latency options: two and three cycles. The example in Fig. 8.2 shows a CAS latency of two cycles. It may sound best to always choose the lower latency option, but as always, nothing comes for free. The SDRAM trades off access time (effectively, t_{CO}) for CAS latency. This becomes important at higher clock frequencies where fast t_{CO} is crucial to system operation. In these circumstances, an engineer is willing to accept one cycle of added delay to achieve the highest clock frequency. For example, a Micron Technology MT48LC32M8A2-7E 256-Mb SDRAM can operate at 143 MHz with a CAS latency of three cycles, but only 133 MHz with a CAS latency of two cycles.* One cycle of additional delay will be more than balanced out by a higher burst transfer rate. At lower clock rates, it is often possible to accept the slightly increased access time in favor of a shorter CAS latency.

* 256MSDRAM_D.p65-RevD; Pub. 1/02, Micron Technologies, 2001, p. 11.

Once the CAS latency has passed, data begins to flow on every clock cycle. Data will flow for as long as the specified burst length. In Fig. 8.2, the standard burst length is four words. This parameter is configurable and adds to the flexibility of an SDRAM. The controller is able to set certain parameters at start-up, including CAS latency and burst length. The burst length then becomes the default unit of data transfer across an SDRAM interface. Longer transactions are built from multiple back-to-back bursts, and shorter transactions are achieved by terminating a burst before it has completed. SDRAMs enable the controller to configure the standard burst length as one, two, four, or eight words, or the entire row. It is also possible to configure a long burst length for reads and only single-word writes. Configuration is performed with the mode register set (MRS) command by asserting the three primary control signals and driving the desired configuration word onto the address bus.

As previously mentioned, DQM signals function as an output disable on a read. The DQM bus (a single signal for SDRAMs with data widths of eight bits or less) follows the CAS* timing and, therefore, leads read data by the number of cycles defined in the CAS latency selection. The preceding read can be modified as shown in Fig. 8.3 to disable the two middle words.

In contrast, write data does not have an associated latency with respect to CAS*. Write data begins to flow on the same cycle that the WR/WRA command is asserted, as shown in Fig. 8.4. This

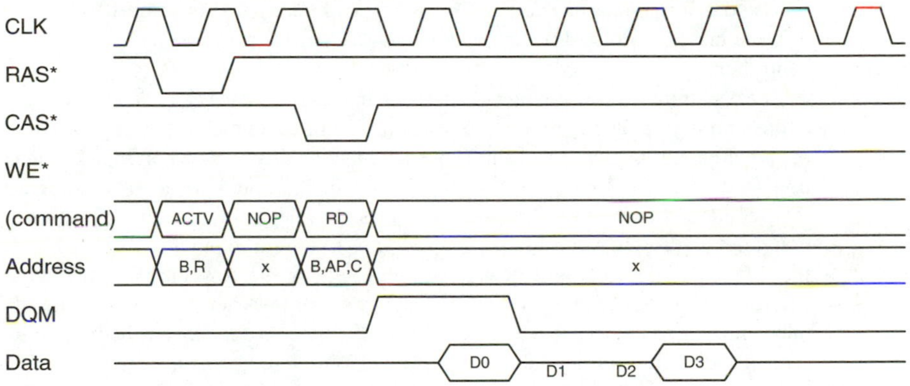

FIGURE 8.3 Four-word SDRAM burst read with DQM disable (CL = 2, BL = 4).

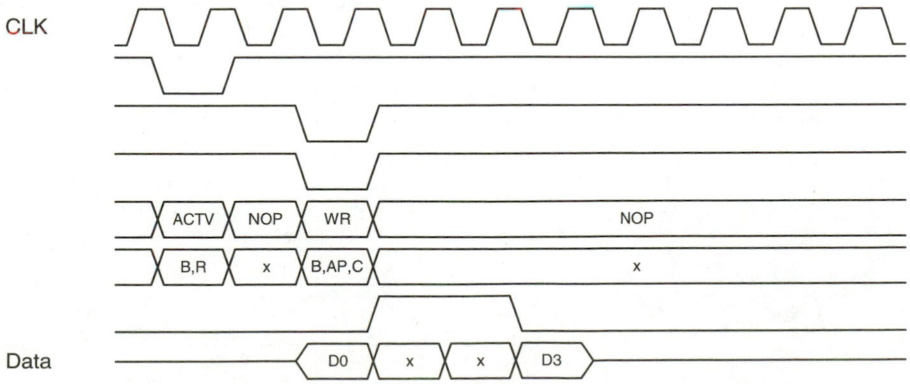

FIGURE 8.4 Four-word SDRAM burst write with DQM masking (BL = 4).

example also shows the timing of DQM to prevent writing the two middle words. Since DQM follows the CAS* timing, it is also directly in line with write data. DQM is very useful for writes, especially on multibyte SDRAM devices, because it enables the uniform execution of a burst transfer while selectively preventing the unwanted modification of certain memory locations. When working with an SDRAM array composed of byte-wide devices, it would be possible to deassert chip select to those byte lanes that you don't want written. However, there is no such option for multibyte devices other than DQM.

When the transaction completes, the row is left either activated or precharged, depending on the state of AP during the CAS* assertion. If left activated, the controller may immediately issue a new RD or WR command to the same row. Alternatively, the row may be explicitly precharged. If automatically precharged, a new row in that bank may be activated in preparation for other transactions. A new row can be activated immediately in most cases, but attention must be paid to the SDRAM's specifications for minimum times between active to precharge commands and active to active commands.

After configuring an SDRAM for a particular default burst length, it will expect all transactions to be that default length. Under certain circumstances, it may be desirable to perform a shorter transaction. Reads and writes can be terminated early by either issuing a *precharge* command to the bank that is currently being accessed or by issuing a *burst-terminate* command. There are varying restrictions and requirements on exactly how each type of transaction is terminated early. In general, a read or write must be initiated without automatic precharge for it to be terminated early by the memory controller.

The capability of performing back-to-back transactions has been already mentioned. In these situations, the startup latency of a new transaction can be accounted for during the data transfer phase of the previous transaction. An example of such functionality is shown in Fig. 8.5. This timing diagram uses a common SDRAM presentation style in which the individual control signals are replaced by their command equivalent. The control signals are idle during the data portion of the first transaction, allowing a new request to be asserted prior to the completion of that transaction. In this example, the controller asserts a new read command for the row that was previously activated. By asserting this command one cycle (CAS latency minus one) before the end of the current transaction, the controller guarantees that there will be no idle time on the data bus between transactions. If a the second transaction was a write, the assertion of WR would come the cycle after the read transaction ended to enable simultaneous presentation of write data in phase with the command. However, when following a write with a read, the read command cannot be issued until after the write data completes, causing an idle period on the data bus equivalent to the selected CAS latency.

This concept can be extended to the general case of multiple active banks. Just as the controller is able to assert a new RD in Fig. 8.5, it could also assert an ACTV to activate a different bank. Therefore, any of an SDRAM's banks can be asserted independently during the idle command time of an in-progress transaction. When these transactions end, the previously activated banks can be seamlessly read or written in the same manner as shown. This provides a substantial performance boost and can eliminate most overhead other than refresh in an SDRAM interface.

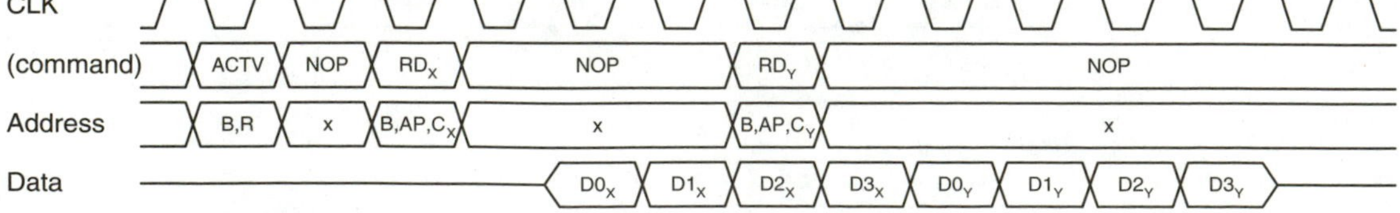

FIGURE 8.5 Back-to-back read transactions (CL = 2, BL = 4).

Periodic refresh is a universal requirement of DRAM technology, and SDRAMs are no exception. An SDRAM device may contain 4,096 rows per bank (or 8,192, depending on its overall size) with the requirement that all rows be refreshed every 64 ms. Therefore, the controller has the responsibility of ensuring that 4,096 (or 8,192) refresh operations are carried out every 64 ms. Refresh commands can be evenly spaced every 15.625 µs (or 7.8125 µs), or the controller might wait until a certain event has passed and then rapidly count out 4,096 (or 8,192) refresh commands. Different SDRAM devices have slightly differing refresh requirements, but the means of executing refresh operations is standardized. The first requirement is that all banks be precharged, because the auto-refresh (REF) command operates on all banks at once. An internal refresh counter keeps track of the next row across each bank to be refreshed when a REF command is executed by asserting RAS* and CAS* together.

It can be easy to forget the asynchronous timing requirements of the DRAM core when designing around an SDRAM's synchronous interface. After a little time spent studying state transition tables and command sets, the idea that an asynchronous element is lurking in the background can become an elusive memory. Always be sure to verify that discrete clock cycle delays conform to the nanosecond timing specifications that are included in the SDRAM data sheet. The tricky part of these timing specifications is that they affect a system differently, depending on the operating frequency. At 25 MHz, a 20-ns time delay is less than one cycle. However, at 100 MHz, that delay stretches to two cycles. Failure to recognize subtle timing differences can cause errors that may manifest themselves as intermittent data corruption problems, which can be very time consuming to track down.

SDRAM remains a mainstream memory technology for PCs and therefore is manufactured in substantial volumes by multiple manufacturers. The SDRAM market is a highly competitive one, with faster and denser products appearing regularly. SDRAMs are commonly available in densities ranging from 64 to 512 Mb in 4, 8, and 16-bit wide data buses. Older 16-Mb parts are becoming harder to find. For special applications, 32-bit wide devices are available, though sometimes at a slight premium as a result of lower overall volumes.

8.2 DOUBLE DATA RATE SDRAM

Conventional SDRAM devices transfer one word on the rising edge of each clock cycle. At any given time, there is an upper limit on the clock speed that is practical to implement for a board-level interface. When this level of performance proves insufficient, *double data rate* (DDR) SDRAM devices can nearly double the available bandwidth by transferring one word on both the rising and falling edges of each clock cycle. In doing so, the interface's clock speed remains constant, but the data bus effectively doubles in frequency. Functionally, DDR and single data rate (SDR) devices are very similar. They share many common control signals, a common command set, and a rising-edge-only control/address interface. They differ not only in the speed of the data bus but also with new DDR data control signals and internal clocking circuitry to enable reliable circuit design with very tight timing margins. Figure 8.6 shows the DDR SDRAM structure.

A DDR SDRAM contains an internal data path that is twice the width of the external data bus. This width difference allows the majority of the internal logic to run at a slower SDR frequency while delivering the desired external bandwidth with half as many data pins as would be required with a conventional SDRAM. Rather than supplying a 2× clock to the SDRAM for its DDR interface, a pair of complementary clocks, CLK and CLK*, are provided that are 180° out of phase with each other. Input and output signals are referenced to the crossings of these two clocks, during which a rising edge is always present in either clock. Commands and addresses are presented to a DDR SDRAM as they would be for an SDR device: on the rising edge of CLK. It is not necessary to dou-

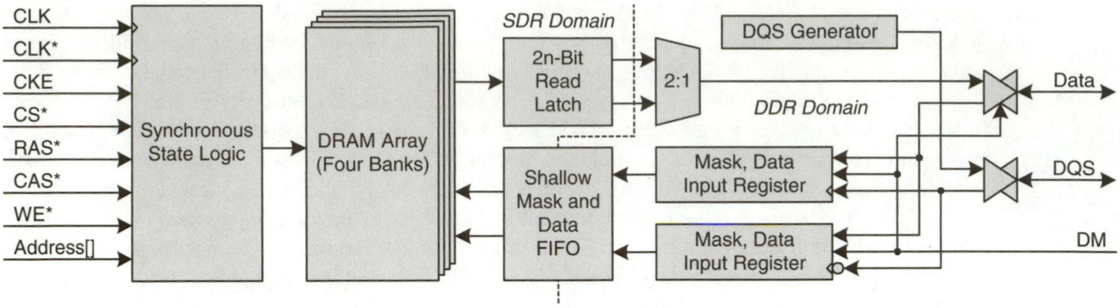

FIGURE 8.6 Basic DDR SDRAM architecture.

ble the speed of the control interface, because an SDRAM is almost always used in burst mode where the rate of commands is significantly less than the rate of data transferred.

The data interface contains a mask that has been renamed to DM and a new data strobe signal, DQS. DM functions as DQM does in an SDR device but operates at DDR to match the behavior of data. DQS is a bidirectional clock that is used to help time the data bus on both reads and writes. On writes, DQS, DM, and data are inputs and DQS serves as a clock that the SDRAM uses to sample DM and data. Setup and hold times are specified relative to both the rising and falling edges of DQS, so DQS transitions in the middle of the data valid window. DQS and data are outputs for reads and are collectively timed relative to CLK/CLK*. DQS transitions at roughly the same time as data and so it transitions at the beginning of the data valid window.

When reading, 2n bits are fetched from the DRAM array on the CLK domain and are fed into a 2:1 multiplexer that crosses the SDR/DDR clock domain. In combination with a DQS generator, the multiplexer is cycled at twice the CLK frequency to yield a double rate interface. This scheme is illustrated schematically in Fig. 8.7. Because DQS and data are specified relative to CLK/CLK* on reads, the memory controller can choose to clock its input circuitry with any of the strobe or clock signals according to the relevant timing specifications. Writes function in a reverse scheme by stacking two n-bit words together to form a 2n-bit word in the DRAM's CLK domain. Two registers are each clocked alternately on the rising and falling edges of DQS, and their contents are then transferred to a shallow write FIFO. A FIFO is necessary to cross from the DQS to CLK domains reliably as a result of skew between the two signals.

Tight timing specifications characterize DDR SDRAM because of its high-speed operation: a 333-MHz data rate with a 167-MHz clock is not an uncommon operating frequency. For reliable operation, careful planning must be done at the memory controller and in printed circuit board design to ensure that data is captured in as little as 1.5 ns (for a 333/167-MHz DDR SDRAM). These high-

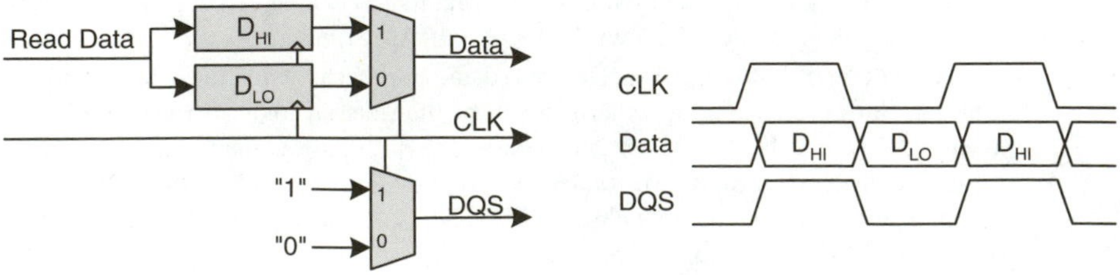

FIGURE 8.7 SDR-to-DDR data conversion scheme.

speed data buses are treated as *source-synchronous* rather than synchronous. A source-synchronous bus is one where a local clock is generated along with data and routed on the circuit board with the data signals. The clock and data signals are length-matched to a certain tolerance to greatly reduce the skew between all members of the bus. In doing so, the timing relationships between clock and data are preserved almost exactly as they are generated by the sending device. A source-synchronous bus eliminates system-level skew problems that result from clocks and data signals emanating from different sources and taking different paths to their destinations. Treating the DDR SDRAM data bus source-synchronously as shown in Fig. 8.7 guarantees that the data valid window provided by the driver will be available to the load. Likewise, because DQS is bidirectional, the SDRAM will obtain the same timing benefit when accepting write-data from the memory controller.

Methods vary across DDR SDRAM implementations. While the SDRAM requires a fixed relationship between DQS and data for writes, the memory controller may use either DQS or a source-synchronous version of CLK with which to time read data. DQS must be used for the fastest applications, because it has a closer timing relationship relative to data. The usage of DQS adds some complexity, because it is essentially a bidirectional clock. There are also multiple DQS signals in most applications, because one DQS is present for every eight bits of data.

Some applications may be able to use CLK/CLK* to register read data. The memory controller typically drives CLK/CLK* to the SDRAM along with address and control signals in a source-synchronous fashion. To achieve a source-synchronous read data bus, a skewed version of CLK/CLK* is necessary that is in phase with the returned data so that the memory controller sees timing as shown in Fig. 8.7. This skew is the propagation delay through the wires that carry the clocks from the memory controller to the SDRAM. These skews are illustrated in Fig. 8.8a, and the associated wiring implementation is shown in Fig. 8.8b. CLK′ and CLK′* are the clocks that have been skewed by propagation delay through the wiring. A source-synchronous read-data bus is achieved by generating a second pair of clocks that are identical to the main pair and then by matching their lengths to the sum of the wire lengths to and from the SDRAM. The first length component cancels out the propagation delay to the SDRAM, and the second length component maintains timing alignment, or phase, with the data bus.

With the exception of a faster data bus, a DDR SDRAM functions very much like a conventional SDRAM. Commands are issued on the rising edge of CLK and are at a single data rate. Because of the internal 2n-bit architecture, a minimum burst size of two words is supported. The other burst length options are four or eight words. To read or write a single word, DM must be used to mask or inhibit the applicable word. Two CAS latency options are supported for reads: 2 and 2.5 cycles. Two CL = 2 reads are shown in Fig. 8.9. DQS transitions from input (tri-state) to output one cycle (1.5 cycles for CL = 2.5) after the assertion of the read command. It is driven low for one full cycle (two DDR periods) and then transitions on each half of CLK for the duration of the burst, after which it

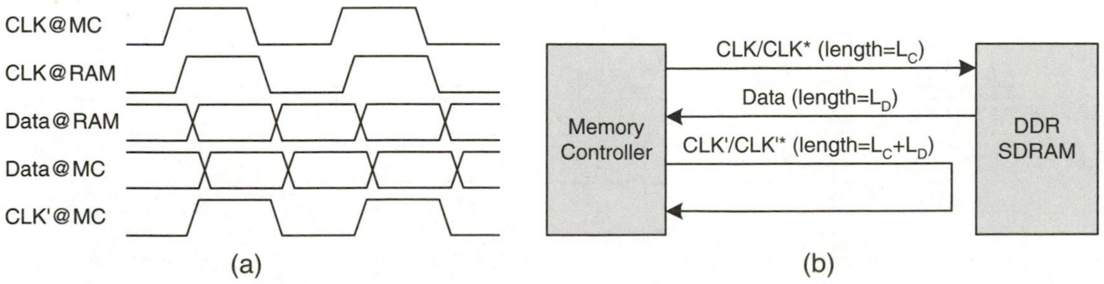

FIGURE 8.8 Source-synchronous read data with CLK/CLK*.

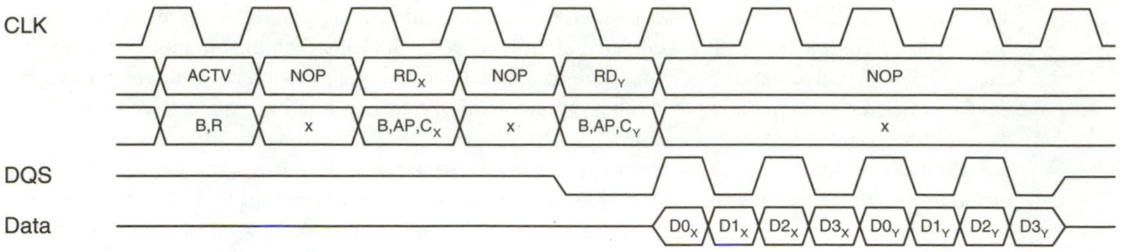

FIGURE 8.9 Consecutive DDR SDRAM reads (CL = 2, BL = 4).

returns to tri-state. Reads may be executed consecutively to achieve high bus utilization by hiding row activation and CAS latency delays, as with an SDR SDRAM.

Unlike an SDR SDRAM, writes are subject to a brief latency between assertion of the write command and delivery of write data. The first write data is presented on the first rising edge of DQS following the write command. DQS is not driven to the SDRAM until just after the write command is presented. This restriction prevents a collision between the SDRAM and the memory controller when a write follows a read by giving time for the SDRAM to turn off its DQS driver. Following the write, DQS can remain driven until a read command is asserted, at which time the SDRAM will need to drive the strobe. Write timing is shown in Fig. 8.10. Writes may also be executed consecutively to more effectively utilize the device interface.

When transitioning between reading and writing, minimum delays are introduced in a situation unlike that of a conventional SDRAM. Because write data lags the write command by a clock cycle, a cycle is lost when following a read with a write, because the write command cannot be issued until the read burst is complete (as with an SDR SDRAM). Going the other way, an explicit single-cycle delay is imposed on issuing a read command following a write burst, thereby incurring a data bus idle time equal to the selected CAS latency plus the single cycle write/read delay.

DDR SDRAM has taken the place of conventional SDRAM in many PC applications. Like SDR SDRAM, DDR devices are commonly available in densities ranging from 64 to 512 Mb in 4-, 8-, and 16-bit wide data buses. Thirty-two-bit devices are also available, although they are not the sweet spot for the industry as a whole.

8.3 SYNCHRONOUS SRAM

Like DRAM, high-performance SRAM transitioned to a synchronous interface to gain performance improvements. Several basic types of *synchronous SRAM* (SSRAM) devices appeared and became

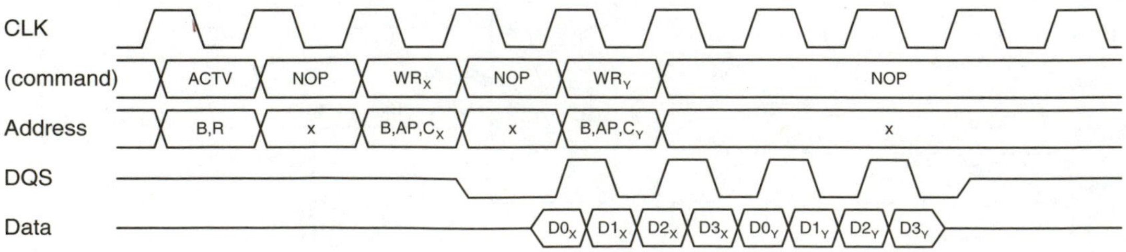

FIGURE 8.10 Consecutive DDR SDRAM writes (BL = 4).

standard offerings of numerous semiconductor vendors. SSRAMs are well suited for applications that require rapid access to random memory locations, as compared to SDRAMs that are well suited to long bursts from sequential memory locations. Many SSRAM devices can be sourced from multiple vendors with identical pinouts and functionality. An SRAM is made synchronous by registering its interface. Two basic types of SSRAMs are *flow-through* and *pipelined*. Flow-through devices register only the input signals and are therefore able to return read data immediately after the requested address is registered on a rising clock edge. Pipelined devices register both the input and output signals, incurring an added cycle of latency when returning read data. These differences are illustrated in Fig. 8.11.

As with SDRAM, there is a trade-off between access latency and clock speed. Pipelined devices can run at substantially faster clock frequencies than flow-through devices, because the SSRAM has a full clock cycle to decode the presented address once it is registered. In applications where clock speeds are under 100 MHz, flow-through SSRAMs may be preferable because of their lower latency. However, a flow-through device exhibits relatively high clock-to-data-valid timing, because the outputs are not registered. This large t_{CO} directly impacts the overall memory system design by placing tighter constraints on the interconnection delays and input register performance of the device that is reading from the SSRAM. For example, a Micron Technology MT55L512L18F-10 8-Mb flow-through SSRAM runs up to 100 MHz and exhibits a 7.5 ns access delay and a 3.0 ns data hold time after the next clock edge.* At a 10-ns clock period, there are 2.5 ns of setup budget to the next clock edge for an input register that is sampling the returned data. This 2.5-ns budget must account for interconnect delay, clock skew, and the setup time of the input flops. Alternatively, the 3 ns of hold time can help increase this timing budget, but special considerations must then be made to shift the data valid window of the input flops more in favor of hold time and less in favor of setup time. This is not always practical. In contrast, Micron's MT55L512L18P-10 8-Mb pipelined SSRAM is rated for the same 100-MHz clock but exhibits a 5.0-ns clock-to-valid delay and a 1.5-ns hold time.† For the added cycle of latency, the setup budget increases to a much more comfortable 5 ns with the same 10-ns clock period. Pipelining also allows the SSRAM to run at a much faster clock frequency: 166 MHz versus 100 MHz for the 8-Mb flow-through SSRAM. By using a pipelined SSRAM, you can choose between more favorable timing margins or increased memory bandwidth over flow-through technology.

An application in which SSRAM devices are used is a cache, which typically performs burst transactions. Caches burst data a line at a time to improve main memory bandwidth. Standard SS-RAM devices support four-word bursts by means of a loadable two-bit internal counter to assist

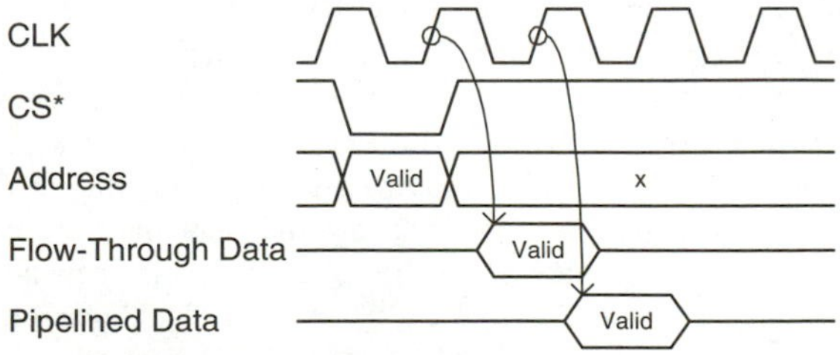

FIGURE 8.11 Flow-through vs. pipelined SSRAM reads.

---

* MT55L512L18F_2.p65-Rev. 6/01, Micron Technologies, 2001, p. 25.
† MT55L512L18P_2.p65-Rev. 6/01, Micron Technologies, 2001, p. 25.

caches and other applications that operate using bursts. An SSRAM contains one or more control signals that defines whether a memory cycle uses an externally supplied address or an internally latched address and counter. When a burst transfer is desired, the memory controller asserts a control signal to load the internal burst counter and then directs the SSRAM to use that incrementing count value for the three subsequent cycles. Bursts are supported for both reads and writes. The two-bit burst counter can be configured in one of two increment modes: linear and interleaved. Linear increment is a simple binary counter that wraps from a terminal value of 11 back to 00. Bursts can be initiated at any address, so, if the burst begins at A[1:0] = 10, the counter will count 10, 11, 00, and 01 to complete the burst. Interleaved mode forces the data access pattern into two pairs where each pair contains an odd and even address with A[1] held constant as shown in Table 8.2. Interleaving can benefit implementations that access words in specific pairs.

TABLE 8.2 SSRAM Interleaved Burst Addressing

Initial Value of A[1:0] Supplied Externally	Second Address Generated Internally	Third Address Generated Internally	Fourth Address Generated Internally
00	01	10	11
01	00	11	10
10	11	00	01
11	10	01	00

Flow-through and pipelined SSRAMs fall into two more categories: normal and *zero-bus turnaround*®(ZBT)*. Normal SSRAMs exhibit differing read and write latencies: write data can be asserted on the same cycle as the address and write enable signals, but reads have one to two cycles of latency, depending on the type of device being used. Under conditions of extended reads or writes, the SSRAM can perform a transfer each clock cycle, because the latency of sequential commands (all reads or all writes) remains constant. When transitioning from writing to reading, however, the asymmetry causes idle time on the SSRAM data bus because of the startup latency of a read command. The read command is issued in the cycle immediately following the write, and read data becomes available one or two cycles later. If an application performs few bus turnarounds because its tends to separately execute strings of reads followed by writes, the loss of a few cycles here and there is probably not a concern. However, some applications continually perform random read/write transactions to memory and may lose necessary bandwidth each time a bus turnaround is performed.

ZBT devices solve the turnaround idle problem by enforcing symmetrical delays between address and data, regardless of whether the transaction is a read or write. This fixed relationship means that any command can follow any other command without forced idle time on the data bus. Flow-through ZBT devices present data on the first clock edge following the corresponding address/command. Pipelined ZBT SSRAMs present data on the second clock edge following the corresponding address/command as shown in Fig. 8.12. As with normal SSRAMs, higher clock frequencies are possible with pipelined versus flow-through ZBT devices, albeit at the expense of additional read latency.

ZBT SSRAMs provide an advantage for applications with frequent read/write transitions. One example is a single-clock domain FIFO implemented using a discrete SSRAM and control logic. A ge-

* ZBT and Zero Bus Turnaround are trademarks of Integrated Device Technology, Inc., and the architecture is supported by Micron Technology, Inc. and Motorola Inc.

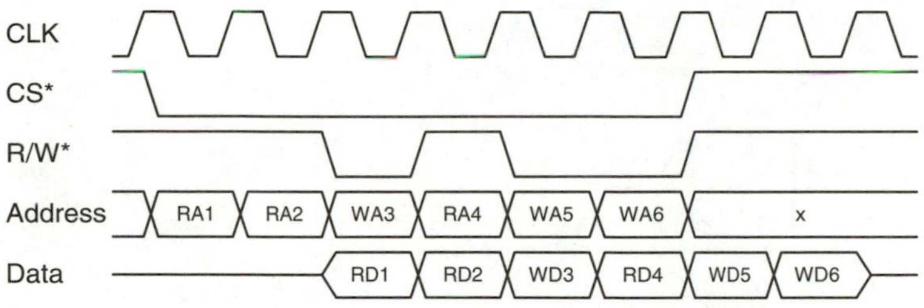

FIGURE 8.12 Pipelined ZBT SSRAM read/write timing.

neric FIFO must be capable of sustained, interleaved reads and writes, which results in frequent bus turnaround delays when using a normal SSRAM. ZBT SSRAM devices are manufactured by companies including Integrated Device Technology, Micron Technology, and Motorola. Cypress Semiconductor manufactures functionally equivalent SSRAMs under the trademark NoBL. Other manufacturers offer equivalent devices with differing naming schemes.

SSRAMs are very popular in high-performance computing and networking applications. Computers with large secondary and tertiary caches use SSRAM to hold lines of data. Networking equipment makes extensive use of SSRAMs for buffering and lookup table applications. SSRAM devices are commonly available in densities ranging from 2 to 16 Mb in 16-, 18-, 32-, and 36-bit wide data buses. The nine-bit bus multiples are used in place of eight-bit multiples for such purposes as the storage of parity and flag bits.

8.4 DDR AND QDR SRAM

SSRAM transitioned to a DDR interface to increase bandwidth in the same general manner as SDRAM. DDR SRAM devices are fully pipelined and feature fixed burst transfer lengths of two or four words to enable a less complex single-rate address/control interface. With the data bus running at twice the effective frequency of the address bus, a burst size of two guarantees that random access transfers can be issued in any order without falling behind the data interface's higher bandwidth. Burst length is fixed by the particular device being used. A burst length of four words simplifies applications such as some caches that operate using four-word transactions, although no inherent throughput advantage is gained. As with a DDR SDRAM, special clocking techniques must be employed to enable the design of reliable interfaces at effective data rates in the hundreds of megahertz. A DDR SRAM accepts a primary pair of complementary clocks, K and K*, that are each 180° out of phase with each other. Address and control signals are registered on the rising edge of K, and write-data is registered on the rising edges of both clocks. An optional secondary pair of clocks, C and C*, must be same frequency as K/K* but can be slightly out of phase to skew the timing of read-data according to an application circuit's requirements. A small degree of skewing can ease the design of the read capture logic. If such skewing is not necessary, C/C* are tied high, and all output (read) timing is referenced relative to K/K*. The SRAM automatically recognizes the inactivity on C/C* and chooses K/K* as the causal output clock. A pair of output echo clocks, CQ and CQ*, are driven by the SRAM in phase with read data such that both the echo clocks and read data are timed relative to C/C* or K/K*. These echo clocks are free running and do not stop when read activity stops. This combined clocking scheme is illustrated in Fig. 8.13. The read capture logic may choose to use the echo clocks as source-synchronous read clocks, or it may use an alternate scheme and not use the echo clocks at all. An alternative scheme could be to skew C/C* such that returning read data is in

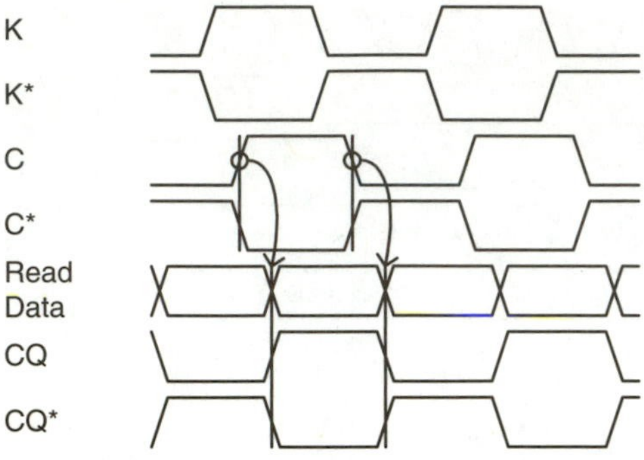

FIGURE 8.13 DDR SSRAM clocking.

proper phase with the memory controller's K/K*, thus saving it the complexity of dealing with a separate read clock domain. Such decisions are implementation specific and depend on the circuits and resources available.

Because of the high frequency of the DDR interface, bus turnaround time becomes an important design point. Idle time must be inserted onto a bidirectional data bus to enable the SRAM and memory controller time to disable their tri-state drivers when transitioning from reading to writing or vice versa. DDR SRAMs are manufactured in both *single* and *common I/O* (SIO and CIO) configurations to address turnaround timing. SIO DDR SRAMs feature two data buses—one dedicated for incoming write data and the other dedicated for outgoing read data. CIO devices feature a common bidirectional data bus. The latencies between address and data are identical between SIO and CIO devices. Write data begins on the first rising edge following the write command, and read data is returned beginning on the second falling edge following the read command. An LD* signal indicates an active read or write command. Figure 8.14 shows the timing for an SIO device in which bus turnaround is not an issue because of the dual unidirectional buses. Note that read data can overlap write data in the same clock cycle.

Commands can be issued continuously on an SIO device, because there is no possibility for data bus conflicts. A CIO device, however, requires at least one explicit idle cycle when transitioning from reading to writing, as shown in Fig. 8.15, because of the difference in data latencies for these two transactions. Without the idle cycle, write data would occur in the same cycle at the last two read words. CIO data sheets also warn that, at high frequencies, a second idle cycle may be necessary to prevent a bus conflict between the SRAM and the write data. The concern at high frequencies is that

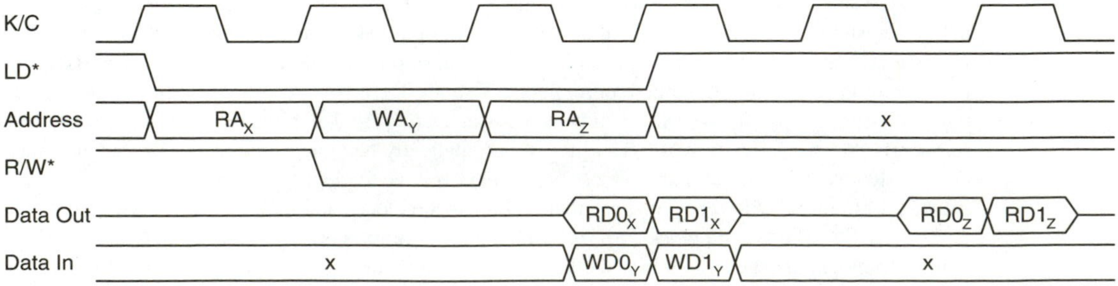

FIGURE 8.14 Separate I/O DDR SRAM read/write timing (burst length = 2).

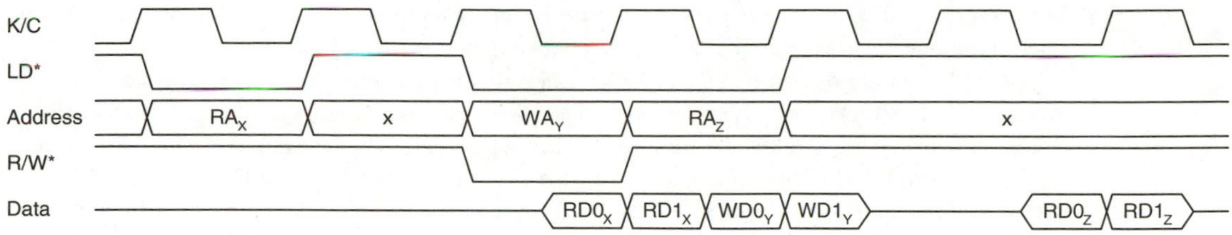

FIGURE 8.15 Common I/O DDR SRAM read/write timing (burst length = 2).

the SRAM may not be able to turn off its tri-state drivers in time for write data being driven immediately on the next cycle.

High data transfer rates are possible with CIO DDR SRAM in purely random transactions. Grouping multiple reads and writes into separate groups increases the available bandwidth by minimizing bus turnaround delays associated with read/write transitions. CIO devices have a distinct advantage in reduced signal count because of a single data bus. Balancing this out is the complexity of handling bus turnaround and somewhat reduced bandwidth in truly random transfer patterns.

SIO DDR SRAM provides a definite performance advantage in certain applications at the cost of additional signal and pin count. The concept of dual data interfaces was taken a step farther with the development of *quad data rate* ™ (QDR) SRAM technology, where the goal is to enable full utilization of the read and write data interfaces.* QDR devices are manufactured with fixed two- or four-word bursts. The address/control interface is designed so that enough commands can be issued to keep both data interfaces fully utilized. A four-word burst QDR SRAM is very similar to an SIO DDR SRAM if one were to be made with a four-word burst size. The difference is that, rather than having a R/W* select signal and an activation signal (LD*), the QDR devices implement separate read and write enables. A new command is presented during each clock cycle such that it takes two cycles to issue both a read and a write command. This frequency of commands matches perfectly with the four-word burst nature of the dual data interfaces. Each read or write command transfers four words at DDR, thereby occupying two whole clock cycles. Therefore, a read command can be issued once every two cycles, and it takes two cycles to execute. The same holds true for a write command. A two-word burst QDR SRAM differs from the four-word variety in that its address/control interface is dual rate to allow commands to be issued twice as fast to keep up with the shorter transfer duration of one cycle (two words at DDR complete in one whole cycle). Figure 8.16 shows the timing for a four-word burst QDR SRAM. If an application can make efficient use of a four-word

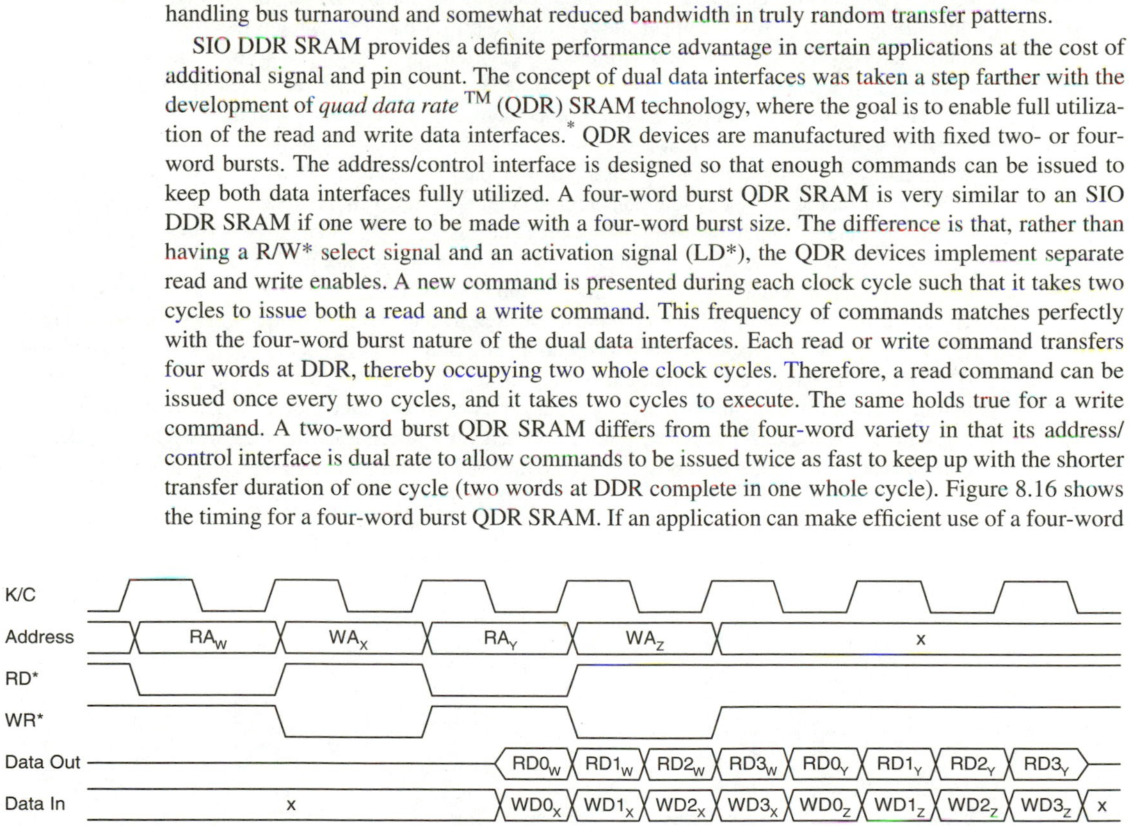

FIGURE 8.16 QDR SRAM read/write timing (burst length = 4).

* QDR is a trademark of Cypress, IDT, Micron Technology, NEC, and Samsung.

burst, the overall system design is likely to be easier, because tight DDR timing on the address/control interface is not required as it would be with a two-word burst QDR device.

As can be readily observed, a QDR device can truly provide four times the bandwidth of a conventional SDR SRAM, but only when the read and write bandwidths are symmetrical. If an application requires very high bandwidth for a long set of writes and then the same equivalent bandwidth for a long read, QDR technology will not provide any real advantage over a DDR SRAM. QDR is useful in many communications applications where it serves as an in-line buffer or FIFO between two data processing elements. Such applications exhibit symmetrical bandwidth, because they cannot store data for long and must rapidly drain data buffer as fast as data is stored to prevent an overflow or underflow.

8.5 CONTENT ADDRESSABLE MEMORY

Most types of memory are constructed from an array of data storage locations, each of which is indexed with a unique address. The set of addresses supported by the memory array is a continuous, linear range from 0 to some upper limit, usually a power of 2. For a memory array size, W, the required address bus width, N, is determined by rounding up $N = \log_2 W$ to the next whole number. Therefore, $W \leq 2^N$. Memory arrays usually store sets of data that are accessed in a sequential manner. The basic paradigm is that the microprocessor requests an arbitrary address that has no special meaning other than the fact that it is a memory index, and the appropriate data is transferred. This scheme can be modified to implement a lookup table by presenting an index that is not an arbitrary number but that actually has some inherent meaning. If, for example, a network packet arrives with an eight-bit identification tag (perhaps a source address), that tag can be used to index into a memory array to retrieve or store status information about that unique type of packet. Such status information could help implement a filter, where a flag bit indicates whether packets with certain tags should be discarded or allowed through. It could also be used to implement a unique counter for each tag to maintain statistics of how many packets with a particular tag have been observed. As shown in Fig. 8.17, when the packet arrives, its relevant eight-bit tag is used to access a single memory location that contains a filter bit and a count value that gets incremented and stored back into the memory array. This saves logic, because 256 unique counters are not required. Instead, a single +1 adder accomplishes the task with the assistance of the memory's built-in address decoding logic.

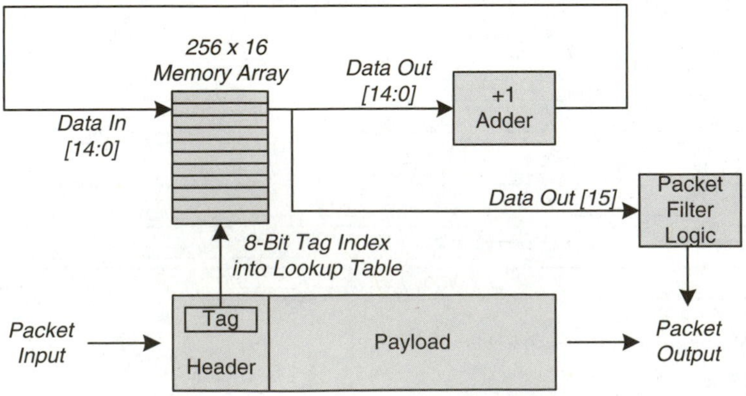

FIGURE 8.17 Using a memory array as a lookup table.

Such lookup tables are common in communication systems where decisions are made and statistics gathered according to the unique tags and network addresses present in each packet's header. When the size of a tag is bounded at a manageable width, a conventional memory array can be used to implement a lookup table. However, as tags grow to 16, 32, 64, 128, or more bits, the required memory size becomes quite impractical. The example in Fig. 8.17 would require 8 GB of memory if the tag width increased from 8 to 32 bits! If all 2^{32} tag permutations need to be accounted for independently, there would be no avoiding a large memory array. However, the majority of such lookup table applications handle a small fraction of the total set of permutations. The working set of tags sparsely populates the complete defined set of tags. So the question becomes how to rapidly index into a memory array with an N-bit tag where the array size is much less than 2^N.

A *content addressable memory* (CAM) solves this problem with an array of fully associative tags and optional corresponding data entries as shown in Fig. 8.18. Instead of decoding 2^N unique locations based on an N-bit tag, each CAM entry simultaneously matches its own tag to the one presented. The entry whose tag matches is the one that presents its associated data at the output and the one that can have its data modified as well. Alternatively, a CAM may simply return the index of the matched or winning entry in the array, if the specific device does not have any data associated with each entry. There is substantial overhead in providing each entry with a unique tag and matching logic, making CAMs substantially more expensive than conventional memories on a per-bit basis. Their increased cost is justified in those applications that require rapid searching of large yet sparsely populated index ranges.

Unlike a conventional memory, a CAM must be managed by the system's hardware and/or software to function properly. The system must load the CAM entries with relevant tags and data. Care should be taken to keep tags unique, because there is no standard means of resolving the case in which two entries' tags match the tag input. Individual CAM implementations may specify how such conflicts are resolved. Some CAMs handle read/write maintenance functions through the same interface that tags are presented and matched. Other implementations provide a separate maintenance port. Having a separate maintenance port increases the number of pins on the CAM, adding complexity to the circuit board wiring, but it may decrease overall system complexity, because maintenance logic and data logic paths do not have to be shared.

CAM tags and matching logic can be constructed in either a *binary* or *ternary* manner. A binary CAM implements a standard tag of arbitrary width and a valid bit. These CAMs are well suited to situations in which exact tag matches are desired. A ternary CAM doubles the number of tag bits to

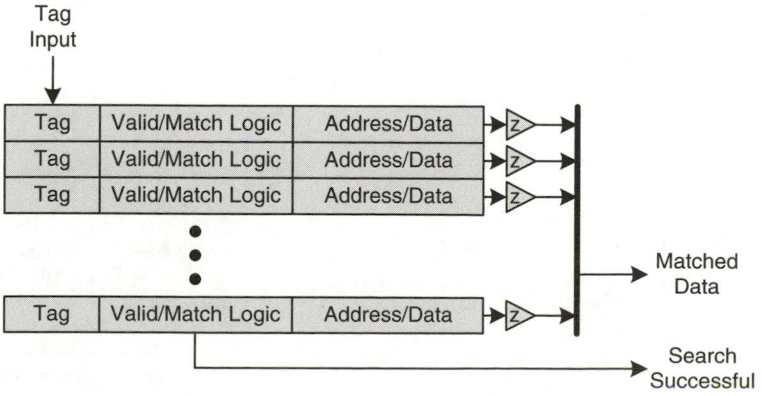

FIGURE 8.18 Basic CAM architecture.

associate a pair of tag bits with each actual tag bit. This two-bit structure allows the creation of a third "don't care" state, X. A ternary CAM is more flexible than a binary CAM, because it can match portions of a tag rather than all bits. In networking applications, this is very useful, because similar operations are often performed on groups of addresses (tags) from common destinations. It is as if the post office wanted to sort out all letters being sent to ZIP codes 11230 through 11239. A ternary CAM would be able to match the pattern 1123X with a single entry. In contrast, a binary CAM would require ten redundant entries to perform the same job.

A ternary CAM is often used to implement rather complex lookup tables with searches prioritized according to the number of X bits in each tag. Using the ZIP code example, it is possible that a post office would want to perform two overlapping searches. It may want to sort all ZIP codes from 11230 through 11239 into a particular bin, except for 11234, which should be sorted into its own bin. A ternary CAM could be setup with two overlapping entries: 11234 and 1123X. To ensure that the 11234 entry always matched ahead of the 1123X entry, it would be necessary to verify proper setup of the specific CAM being used. A ternary CAM may have a rule that the lowest or highest winning entry in the array wins. While this example is simple, the concept can be extended with many levels of overlap and priority.

Managing a ternary CAM with overlapping entries is more complex than managing a binary CAM, because the winning entry priority must be kept in sync with the application's needs, even as the CAM is updated during operation. A CAM is rarely initialized once and then left alone for the remainder of system operation. Its contents are modified periodically as network traffic changes. Let's say that the ZIP code CAM was initialized as follows in consecutive entries: 1121X, 11234, 1123X, 112XX. Where would a new special-case entry 11235 be placed? It would have to precede the 1123X entry for it to match before 1123X. Therefore, the system would have to temporarily move CAM entries to insert 11235 into the correct entry. If there is enough free space in the CAM, the system could initialize it and reserve free entries in between valid entries. But, sooner or later, the CAM will likely become congested in a local area, requiring it to be reorganized. How the data is arranged and how the CAM is reorganized will affect system performance, because it is likely that the CAM will have to be temporarily paused in its search function until the reorganization is complete. Solutions to this pause include a multibank CAM architecture whereby the system reorganizes the lookup table in an inactive bank and then quickly swaps inactive and active banks.

A CAM often does not associate general data bits with each entry, because the main purpose of a CAM is to match tags, not to store large quantities of data. It is therefore common to couple a CAM with an external SRAM that actually holds the data of interest and that can be arbitrarily expanded according to application requirements as shown in Fig. 8.19. In this example, the CAM contains

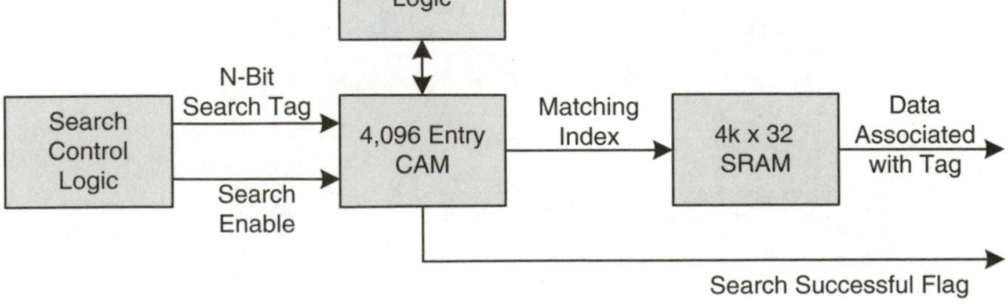

FIGURE 8.19 CAM augmentation with external SRAM.

4,096 entries and returns a 12-bit index when a tag has been successfully matched. This index serves as the address of an SRAM that has a 32-bit data path as required by the application.

When combined with conventional memory and some control logic, a CAM subsystem is sometimes referred to as a *search engine*. A search engine is differentiated from a stand-alone CAM by being capable of semi-autonomous lookups on behalf of another entity such as data processing logic in either hardware or software. A search engine's control logic can be as simple as accepting a search tag and then returning data along with a success flag. It can get more complex to include specific table maintenance functions so that CAM overhead operations are completely offloaded from the data processing logic. Search engines are especially useful when interfacing with special-purpose *network processor* devices. These processors run software to parse packets and make decisions about how each packet should be handled in the system. The tag lookup function is offloaded to a search engine when there is not enough time for a software algorithm to search a large table.

CHAPTER 9
Networking

Data communications is an essential component of every digital system. Some systems realize communications by direct interaction with the environment and some with the exchange of removable storage media such as tapes, disks, or memory modules. Many systems engage in data transfer that is more real-time in nature. When these communications begin to involve multiple end-points, high-speed transfers, and the need for reliable carriage of that data, the set of technologies that are broadly known as networking become directly relevant.

There is probably no single definition of networking that can always identify when it is or is not needed—the universe of applications is too diverse for such rigid definitions. The purpose of this chapter is to introduce mainstream networking concepts so that you can make the decision of whether a particular application demands networking or a simpler exchange of bits and bytes. Networking technologies blend hardware and software into algorithms that are implemented by either or both resources, depending on the specific context. Because of limitations of space and scope, this chapter concentrates on the hardware aspect of networking and how hardware is used to support the formats, protocols, and algorithms that make networking the flexible technology that it is.

The discussion begins with protocol layers to understand the separate logical functions that compose a network. Ethernet is frequently used as an example to further clarify networking concepts because of its ubiquity. Hardware support for networking most commonly resides at the lower layers of the protocol stack. The bulk of the chapter is concerned with transmission, recovery, and verification of data on the wire—essential tasks that serve as the foundation of data transfer. A brief presentation of Ethernet closes the chapter to provide an illustration of how networking technology functions in the real world.

9.1 PROTOCOL LAYERS ONE AND TWO

Networking systems can be highly complex and include many different hardware and software components. To facilitate the analysis and design of such systems, major functional sections are separated into *layers* whose definitions are reasonably standardized across the industry. Multiple layers are arranged from the lowest level on the bottom to the highest level on the top in a conceptual *protocol stack*. To transfer data from an application running on one computer to that on another, the data descends the stack's layers on one computer and then ascends the stack on the destination computer. The industry standard network stack definition is the *Open System Interconnection* (OSI) reference model shown in Fig. 9.1.

As with most conceptual classifications, it is important to recognize that not all networking schemes and implementations adhere strictly to the OSI seven-layer model. Some schemes may merge one or more layers together, thereby reducing the number of formally defined layers. Others

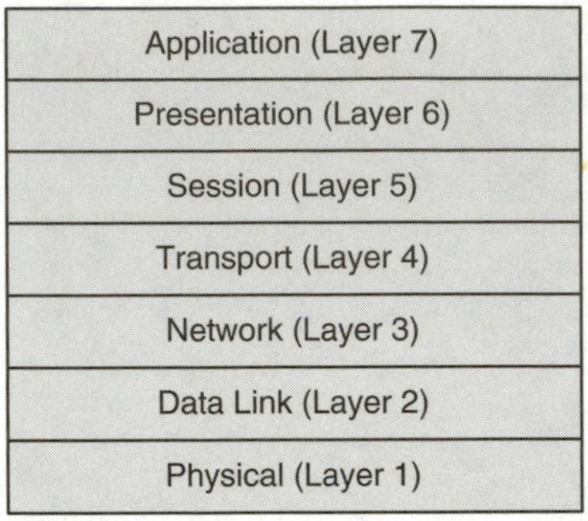

FIGURE 9.1 OSI seven-layer model.

may segment an OSI layer into multiple sublayers. The consistency of definitions decreases as one moves up the stack, because of functional protocol variations.

Layer one, the *physical layer*, comprises the electromechanical characteristics of the medium used to convey bits of information. The use of twisted pair cable, the amplitude of 1s and 0s, and associated connectors and transducers are examples of that which is specified in the physical layer. Channel coding, how the bits are represented on the physical medium, is usually classified as part of the physical layer.

The *data link layer*, layer two, encompasses the control logic and frame formatting that enables data to be injected into the network's physical layer and retrieved at the destination node. Layer-two functions are usually handled by a *media access controller* (MAC), a hardware device that contains all of the logic necessary to gain access to the network medium, properly format and transmit a frame, and properly detect and process an incoming frame. Network frame formats specify data link layer characteristics. Link level error detection mechanisms such as checksums and CRCs (more on these later) are generated and verified by the MAC. Node addresses, called *MAC addresses* in Ethernet networks, are layer-two constructs that uniquely identify individual nodes. Layer-two functions are usually handled in hardware, because they are repetitive, high-frequency, and time-critical operations. The data link layer is closely tied to the topology of the network because of its handling of access control functions and unique node addresses. Network *switches* operate at layer two by knowing which node address is connected to which port and then directing traffic to the relevant port. If port 20 of a switch is connected to node 87, all frames that enter the switch destined for node 87 will be sent out port 20. Because it is necessary to maintain unique layer-two addresses, they are generally not under the control of the user but rather are configured by the manufacturer. In the case of Ethernet, each manufacturer of equipment licenses an arbitrary range of MAC addresses from the IEEE and then assigns them one at a time as products roll off the assembly lines.

9.2 PROTOCOL LAYERS THREE AND FOUR

More flexible communications are possible when a protocol is not tied too closely to network topology or even the type of network accomplishing the exchange of information. The *network layer*, layer three, enables nodes to establish end-to-end connections without strict knowledge of the network topology. Layer-three packets are encapsulated within the payload of a layer-two frame. The packets typically contain their own header, payload, and sometimes a trailer as well. Perhaps the most common example of a layer-three protocol is *Internet Protocol* (IP). IP packets consist of a header and payload. Included within the header are 32-bit layer-three destination and source *IP addresses*. A separate set of network addresses can be implemented at layer three that is orthogonal to layer-two addresses. This gives network nodes two different addresses: one at layer three and one at layer two. For a simple network, this may appear to be redundant and inefficient. Yet modern networking protocols must support complex topologies that span buildings and continents, often with a mix of data links connecting many smaller subnetworks that may cover a single office or floor of a building. The benefit of layer-three addressing and communication is that traffic can be carried on a variety of underlying communications interfaces and not require the end points to know the exact characteristics of each interface.

Network *routers* operate at layer three by separating the many subnetworks that make up a larger network and only passing traffic that must travel between the subnetworks. Network addresses are typically broken into *subnets* that correlate to physically distinct portions of the network. A router has multiple ports, each of which is connected to a different subnetwork that is represented by a range of network addresses. A frame entering a router port will not be sent to another particular port on that router unless its network address matches a subnet configuration on that particular port. Strictly speaking, this separation could be performed by layer-two addressing, but the practical reality is that layer-two addresses are often not under the user's control (e.g., Ethernet) and therefore cannot be organized in a meaningful way. In contrast, layer-three addresses are soft properties of each network installation and are not tied to a particular type of network medium.

Layer-three functions are performed by both hardware and software according to the specific implementation and context. Layer-three packets are usually first generated by software but then manipulated by hardware as they flow through the network. A typical router processes layer-three packets in hardware so that it does not fall behind the flow of traffic and cause a bottleneck.

The bottom three layers cumulatively move data from one place to another but sometimes do not have the ability to actually guarantee that the data arrived intact. Layers one and two are collectively responsible for moving properly formatted frames onto the network medium and then recovering those in transit. The network layer adds some addressing flexibility on top of this basic function. A true end-to-end guarantee of data delivery is missing from certain lower-level protocols (e.g., Ethernet and IP) because of the complexity that this guarantee adds.

The *transport layer*, layer four, is responsible for ensuring end-to-end communication between software services running on each node. Transport layer complexity varies according to the demands of the application. Many applications are written with the simplifying assumption that once data is passed to the transport layer for transmission, it is guaranteed to arrive at the destination. *Transmission control protocol* (TCP) is one of the most common layer-four protocols, because it is used to guarantee the delivery of data across an unreliable IP network. When communicating via TCP, an application can simply transfer the desired information and then move on to new tasks. TCP is termed a *stateful* protocol, because it retains information about packets after they are sent until their successful arrival has been acknowledged. TCP operates using a sliding data transmission window shown in Fig. 9.2 and overlays a 32-bit range of indices onto the data that is being sent. Pointers are referenced into this 32-bit range to track data as it is transmitted and received.

The basic idea behind TCP is that the transmitter retains a copy of data that has already been sent until it receives an acknowledgement that the data was properly received at the other end. If an ac-

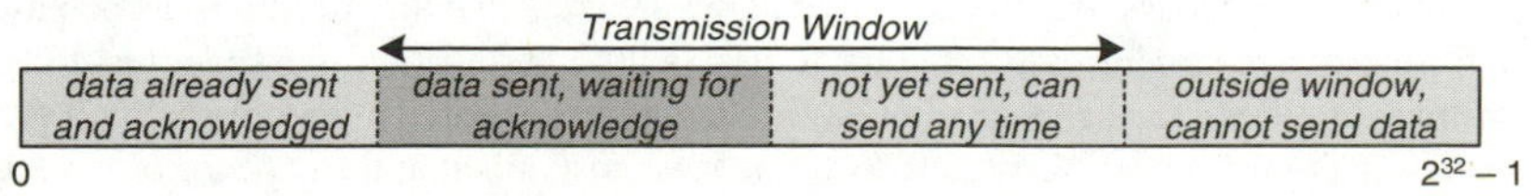

FIGURE 9.2 TCP transmission window.

knowledgement message is not received after a certain amount of time, the data is retransmitted. TCP moves the complexity of guaranteeing end-to-end data delivery into software instead of into the underlying network hardware, which is often Ethernet. When Ethernet was developed in the 1970s, the cost of logic gates was much higher than it is now, and there was a strong incentive to simplify hardware wherever possible.

The transmission window size is established by the receiver via messages that are sent to the transmitter during connection negotiation and subsequent communications. For a receiver to advertise a certain window size, it must have sufficient buffering to hold the entire contents of the window. Once the transmitter is informed of the available transmission window, it may begin sending as much data that can fit within the window. Each time the transmitter sends data, it marks that packet with a 32-bit *sequence number.* This sequence number identifies the 32-bit index that corresponds to the first data byte in the payload and enables the receiver to reconstruct the original data in its proper sequence. When the receiver has successfully received a contiguous block of data starting from the left side of the window, it sends an acknowledgement message with a 32-bit *acknowledgement number* marking the next highest expected sequence number of data. In other words, the acknowledgement number corresponds to the index of the highest byte successfully received plus 1. Upon receiving this message, the transmitter is able to slide the left side of the window up to the acknowledgment number and discard the data in its buffer that now falls outside the window on the left side. The receiver must continually extend the right side of the window to maintain data flow. If the receiver does not slide the right side of the window open, the left side will continue to advance until the window closes, preventing new data from being transmitted.

Guaranteeing end-to-end delivery of data on an inherently unreliable network adds substantial complexity to transport protocol drivers. These functions were traditionally handled by software. However, certain high-performance applications benefit from accelerating TCP in hardware—a task that is decidedly nontrivial.

There are also applications that do not require a transport protocol to guarantee delivery of data. The reason for this may be that the TCP driver is too cumbersome to implement, a proprietary mechanism is preferable, or the underlying network is, in fact, reliable. In such cases, it is unnecessary and often undesired to implement a complex protocol such as TCP. TCP's companion protocol for nonguaranteed transmission is called *user datagram protocol* (UDP). UDP is used along with IP networks to send simple messages over unreliable networks or critical data over reliable networks. It is a stateless protocol, because it simply wraps the data in a header and sends it to the network layer without retaining any information about delivery. As such, there is no sliding transmission window concept and no need for bidirectional communication at the transport layer.

Aside from guaranteeing delivery, many transport protocols implement a higher level of addressing, often referred to as *ports* or *sockets*. An individual node has a single network address associated with it. However, each application on that node can have its own associated port or socket number. These constructs allow the transport layer to direct data flows to the appropriate application on the destination node. Rather than sending each packet that arrives to each application, an application establishes a port or socket number and, henceforth, all network traffic destined for that application is marked with the correct port or socket number.

Layers five, six, and seven are more context specific and principally involve application and network driver software as part of a computer's operating system and network interface subsystem. From a design perspective, the degree of hardware responsibility decreases as one ascends the stack. Less-expensive systems will often try to use as little hardware as possible, resulting in the bare essentials of the physical and data link layers being implemented in hardware. Such systems offload as many functions as possible onto software to save cost, albeit at the expense of reducing the throughput of the network interface. As higher levels of throughput are desired, more hardware creeps into the bottom layers. On general-purpose computers, the network and transport layers are usually implemented in network driver software. However, on special-purpose platforms where high bandwidth is critical, many layer-three and layer-four functions are accelerated by hardware. How these trade-offs are made depends on the exact type of networking scheme being implemented.

9.3 PHYSICAL MEDIA

Most wired networking schemes use high-speed unidirectional serial data channels as their physical communication medium. A pair of unidirectional channels is commonly used to provide bidirectional communications between end points. Despite the fact that it is technically feasible to use a single channel in a bidirectional mode, it is easier to design electronics and associated physical apparatus that implement either a transmitter or receiver at each end of a cable, but not both. The cost of mandating a pair of cables instead of a single cable is not very burdensome, because cables are commonly manufactured as a bundle and are handled as a single unit in wiring conduits and connection points. Two ubiquitous types of media are twisted-pair wiring and fiber optic cable. It is common to find a single cable bundle containing two or more twisted pairs or a pair of fiber optic strands. Twisted-pair and fiber can often be used interchangeably by a network transceiver as long as the appropriate transducer properly converts between the transceiver's electrical signaling and the medium's signaling. In the case of twisted pair, this conversion may consist of only amplification and noise filtering. A fiber optic cable is somewhat more complex in that it requires an electro-optical conversion.

Twisted pair wiring is used in conjunction with differential signaling to provide improved noise immunity versus a single-ended, or unbalanced, transmission medium. As network data rates have increased, twisted pair wiring technology has kept pace with improved quality of manufacture to support higher bandwidths. When the majority of Ethernet connections ran at 10 Mbps (10BASE-T), *unshielded twisted pair* (UTP) *category-3* (CAT3) was a common interconnect medium. UTP wiring does not contain any surrounding grounded metal shield for added noise protection. As 100BASE-T emerged, wiring technology moved to CAT5, and this has remained the most common UTP medium for some time. CAT5 has largely replaced CAT3, because the cost differential is slim, and it exhibits better performance as a result of more twists per unit length and improved structural integrity to maintain the desired electrical characteristics over time and handling. Enhanced UTP products including CAT5e and CAT6 are emerging because of the popularity of gigabit Ethernet over twisted pair (1000BASE-T). While most twisted pair is unshielded, shielded varieties (STP) are used in specific applications. UTP is a favored wiring technology because of its relatively low cost and ease of handling: connections can be made by crimping or punching the wires onto connector terminals. The disadvantage of copper media is their susceptibility to noise and attenuation of signals over moderate distances. These characteristics limit total UTP cable length to 100 m in common Ethernet applications.

Bandwidth and distance are inversely related by the inherent characteristics of a given transmission medium. As distances increase, signal degradation increases, which reduces the available bandwidth of the channel. Fiber optic cabling is used to overcome the bandwidth and distance limitations of twisted pair wiring because of its immunity to electrical noise and very low optical attenuation over distance. Fiber optic cable is generally constructed from high-purity glass, but plastic cables have been used for special short-distance applications. Rather than being a simple extrusion of glass, a fiber optic cable contains two optical elements surrounded by a protective sheath as shown in Fig. 9.3a. The inner glass core is differentiated from the outer glass cladding by the fact that one or both have been doped with certain molecules to change their indices of refraction. The cladding has a lower index of refraction than the core, which causes the great majority of light injected into the core

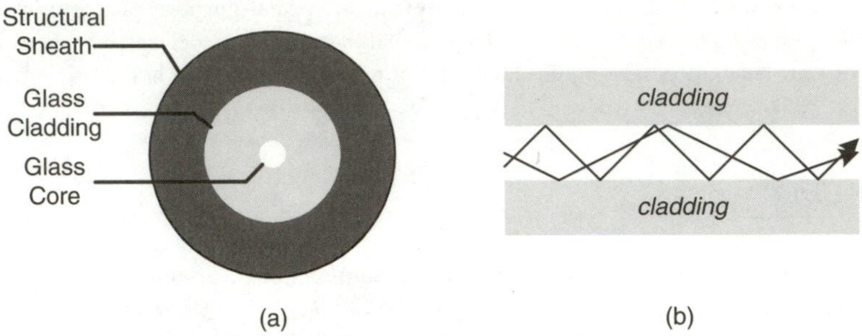

FIGURE 9.3 Fiber optic cable: (a) cross section and (b) propagation.

to bounce off the core/cladding boundary as shown in Fig. 9.3b, thereby trapping the light over very long distances.

Light is injected into the core using either an LED or laser, depending on the required quality of the signal. A laser can generate light that is coherent, meaning that its photons are at the same frequency and phase. Injecting coherent light into a fiber optic cable reduces the distortion that accumulates over distance as photons of different frequency travel through the medium at slightly different velocities. Noncoherent photons that are emitted simultaneously as part of a signal pulse will arrive at the destination spread slightly apart in time. This spreading makes reconstructing the signal more difficult at very high frequencies, because signal edges are distorted.

Even when *coherent* light is used, photons can take multiple paths in the core as they bounce off the core/cladding boundary at different angles. These multiple propagation *modes* cause distortion over distance. To deal with this phenomenon, two types of fiber optic cable are commonly used: *single-mode* and *multimode*. Single-mode fiber contains a very thin core of approximately 8 to 10 µm in diameter that constrains light to a single propagation mode, thereby reducing distortion. Multimode fiber contains a larger core, typically 62.5 µm, that allows for multiple propagation modes and hence increased distortion. Single-mode fiber is more expensive than multimode and is used in longer-distance and higher-bandwidth applications as necessary.

Fiber optic cabling is more expensive than copper wire, and the handling of optical connections is more costly and complex as compared to copper. Splicing a fiber optic cable requires special equipment to ensure a clean cut and low-loss junction between two separate cables. The best splice is obtained by actually fusing two cables together to form a seamless connection. This is substantially more involved than splicing a copper cable, which can be done with fairly simple tools. Fiber optic connectors are sensitive to dirt and other contaminants that can attenuate the photons' energy as they pass through. Additionally, fine abrasive particles can scratch the glass faces of optical interfaces, causing permanent damage. Once properly installed and sealed, however, fiber optic cable can actually be more rugged than copper cables because of its insensitivity to oxidation that degrades copper wiring over time. Aside from bandwidth issues, these environmental benefits have resulted in infrastructures such as cable TV being partially reinstalled with fiber to cut long-term maintenance costs.

9.4 CHANNEL CODING

High-speed serial data channels require the basic functionality of a UART, albeit at very high speed, to convert back and forth between serial and parallel data paths. Unlike a UART that typically functions at kilobits or a few megabits per second, specialized transceiver ICs called *serializer/deserial-*

izers, or *serdes* for short, are manufactured that handle serial rates of multiple gigabits per second. Serdes vendors include AMCC, Conexant, Intel, PMC-Sierra, Texas Instruments, and Vitesse. To simplify system design, a serdes accepts a lower-frequency reference clock that is perhaps 1/10 or 1/20 the bit frequency of the serial medium. Parallel data is usually transmitted to the serdes at this reference frequency. For example, the raw bit rate of gigabit Ethernet (IEEE 802.3z) is 1.25 Gbps, but a typical serdes accepts a 125-MHz reference clock and 10 bits per cycle of transmit data. The reference clock is internally multiplied using a phase locked loop (PLL) to achieve the final bit rate. A general serdes block diagram is shown in Fig. 9.4. An optional transmit clock is shown separately from the reference clock, because some devices support these dual clocks. The benefits of a dual-clock scheme are that a very stable reference clock can be driven by a high-accuracy source separately from a somewhat noisier source-synchronous transmit clock generated by the data processing logic. This eases the clock jitter requirements on data processing logic.

A clock recovery circuit in the receiver portion extracts a bit clock from the serial data stream that is transmitted without a separate clock. This recovery is possible, because the serial data stream is normally coded with an algorithm that guarantees a certain proportion of state transitions regardless of the actual data being transferred. Such coding can be performed within the serdes or by external data processing logic. Channel coding is necessary for more than clock recovery. Analog circuits in the signal path, notably transducer and amplifier elements, require a relatively balanced data stream to function optimally. In circuit analysis terms, they work best when the data stream has an average DC value of 0. This is achieved with a data stream that contains an equal number of 1s and 0s over short spans of time. If a 1 is represented as a positive voltage and a 0 is represented as a negative voltage of equal magnitude, equal numbers of 1s and 0s balance out to an average voltage of 0 over time.

Fortunately, it is possible to encode an arbitrary data stream such that the coded version contains an average DC value of 0, and that data can be restored to its original form with an appropriate decoding circuit. One fairly simple method of encoding data is through a scrambling polynomial im-

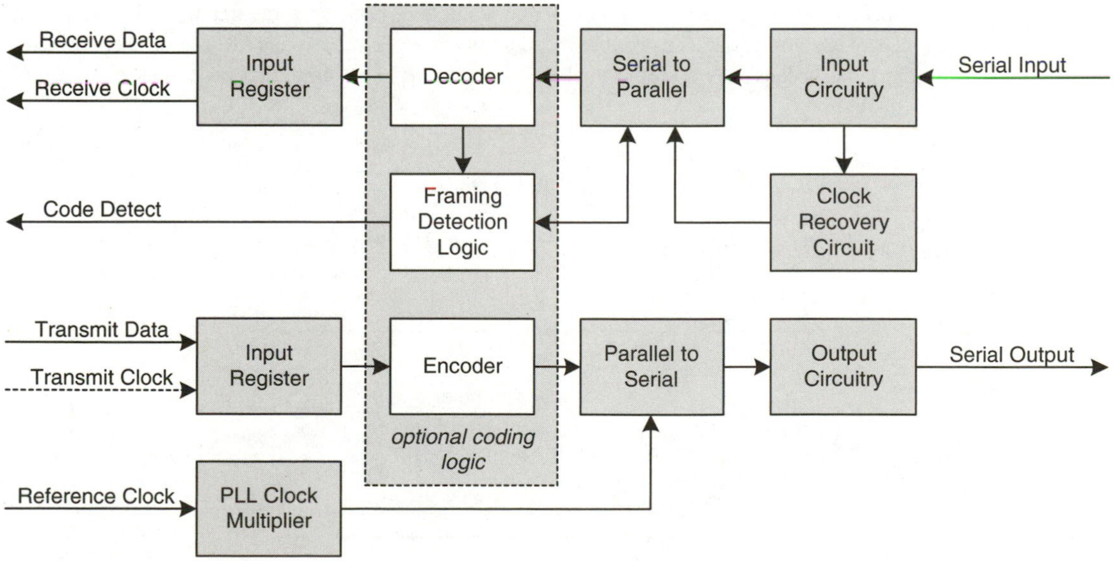

FIGURE 9.4 Serdes block diagram.

plemented with a *linear feedback shift register*—a shift register with feedback generated by a set of exclusive-OR gates. The placement of the XOR feedback terms is mathematically defined by a binary polynomial. Figure 9.5 shows scrambling logic used to encode and decode eight-bit data words using the function $F(X) = X^7 + X^4 + 1$. The mathematical theory behind such polynomials is based on *Galois fields*, discovered by Evariste Galois, a nineteenth century French mathematician. XOR gates are placed at each bit position specified by the polynomial exponents, and their outputs feed back to the shift register input to scramble and feed forward to the output to descramble.

This type of scrambling should not be confused with more sophisticated security and data protection algorithms. Data scrambled in this manner is done so for purposes of randomizing the bits on the communications channel to achieve an average DC value of 0. Polynomial scrambling works fairly well and is relatively easy to implement, but such schemes are subject to undesired cases in which the application of select repetitive data patterns can cause an imbalance in the number of 1s and 0s, thereby reducing the benefit of scrambling. The probability of settling into such cases is low, making scrambling a suitable coding mechanism for certain data links.

While shown schematically as a serial process, these algorithms can be converted to parallel logic by accumulating successive XOR operations over eight bits shifted through the polynomial register. In cases when the coding logic lies outside of the serdes in custom logic, it is necessary to convert this serial process into a parallel one, because data coming from the serdes will be in parallel form at a corresponding clock frequency. Working out the logic for eight bits at a time allows processing one byte per clock cycle. The serial to parallel algorithm conversion can be done over any number of bits that is relevant to a particular application. This process is conceptually easy, but it is rather tedious to actually work out the necessary logic.

A table can be formed to keep track of the polynomial code vector, C[6:0], and the output vector, Q[7:0], as functions of the input vector, D[7:0]. Table 9.1 shows the state of C and Q, assuming that the least-significant bit (LSB) is transmitted first, during each of eight successive cycles by listing terms that are XORed together.

The final column, D[7], and the bottom row, Q, indicate the final state of the code and output vectors, respectively. The code vector terms can be simplified, because some iterative XOR feedback terms cancel each other out as a result of the identity that $A \oplus A = 0$. Q[7:0] can be taken directly from the table because there are no duplicate XOR terms. The simplified code vector, C[6:0], is shown in Table 9.2.

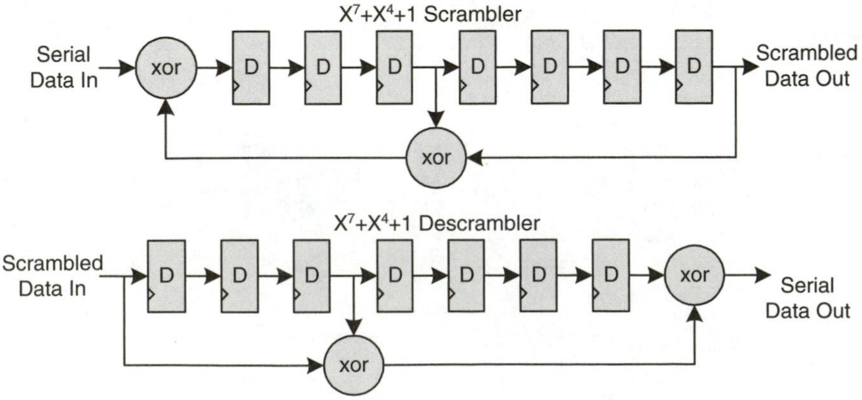

FIGURE 9.5 Eight-bit scrambling/descrambling logic.

TABLE 9.1 Scrambler Logic State Table

Code/Output Vector Bits	Input Vector Bits, One Each Clock Cycle							
	D0	D1	D2	D3	D4	D5	D6	D7
C6	C0 D0 C4	C1 D1 C5	C2 D2 C6	C3 D3 C0 D0 C4	C4 D4 C1 D1 C5	C5 D5 C2 D2 C6	C6 D6 C3 D3 C0 D0 C4	C0 D0 C4 D7 C4 D4 C1 D1 C5
C5	C6	C0 D0 C4	C1 D1 C5	C2 D2 C6	C3 D3 C0 D0 C4	C4 D4 C1 D1 C5	C5 D5 C2 D2 C6	C6 D6 C3 D3 C0 D0 C4
C4	C5	C6	C0 D0 C4	C1 D1 C5	C2 D2 C6	C3 D3 C0 D0 C4	C4 D4 C1 D1 C5	C5 D5 C2 D2 C6
C3	C4	C5	C6	C0 D0 C4	C1 D1 C5	C2 D2 C6	C3 D3 C0 D0 C4	C4 D4 C1 D1 C5
C2	C3	C4	C5	C6	C0 D0 C4	C1 D1 C5	C2 D2 C6	C3 D3 C0 D0 C4
C1	C2	C3	C4	C5	C6	C0 D0 C4	C1 D1 C5	C2 D2 C6
C0	C1	C2	C3	C4	C5	C6	C0 D0 C4	C1 D1 C5
Q	C0	C1	C2	C3	C4	C5	C6	C0 D0 C4

TABLE 9.2 Simplified Scrambling Code Vector Logic

Code Vector Bits	XOR Logic
C6	C0 D0 C1 D1 D4 C5 D7
C5	C0 D0 C3 D3 C4 C6 D6
C4	C2 D2 C5 D5 C6
C3	C1 D1 C4 D4 C5
C2	C0 D0 C3 D3 C4
C1	C2 D2 C6
C0	C1 D1 C5

Following the same process, code and output vectors for the associated descrambling logic can be derived as well, the results of which are shown in Table 9.3. In this case, the code vector, C[6:0], is easy, because it is simply the incoming scrambled data shifted one bit at a time without any XOR feedback terms. The output vector, Q[7:0], is also easier, because the XOR logic feeds forward without any iterative feedback terms.

TABLE 9.3 Descrambling Logic Code and Output Vector Logic

Code Vector Bits	Shift Logic	Output Vector Bits	XOR Logic
–	–	D7	D0 D7 D4
C6	D7	D6	C6 D6 D3
C5	D6	D5	C5 D5 D2
C4	D5	D4	C4 D4 D1
C3	D4	D3	C3 D3 D0
C2	D3	D2	C2 D2 C6
C1	D2	D1	C1 D1 C5
C0	D1	D0	C0 D0 C4

Data that has been encoded must be decoded before framing information can be extracted. Therefore, the serdes' receiving shift register must begin by performing a simple serial-to-parallel conversion until framing information can be extracted after decoding. Prior to the detection of framing information, the output of the parallel shift register will be arbitrary data at an arbitrary alignment, because there is no knowledge of where individual bytes of words begin and end in the continuous data stream. Once a framing sequence has been detected, the shift register can be "snapped" into correct alignment, and its output will be properly formatted whole bytes or words.

When reconstructing decoded data, the desired byte alignment will likely span two consecutive bytes as they come straight from the descrambling logic. The most significant bits of a descrambled byte logically follow the LSB of the next descrambled byte because of the order in which bits are shifted through the scrambler and descrambler. Therefore, when data arrives misaligned at the receiver/descrambler, bytes are reassembled by selecting the correct bits from the most significant bits (MSB) of descrambled byte N and the LSB of descrambled byte N + 1 as shown in Fig. 9.6.

Framing information is not conveyed by the scrambled coding and must therefore be extracted at a higher level. For this reason, certain serdes components that are used in scrambled coding systems

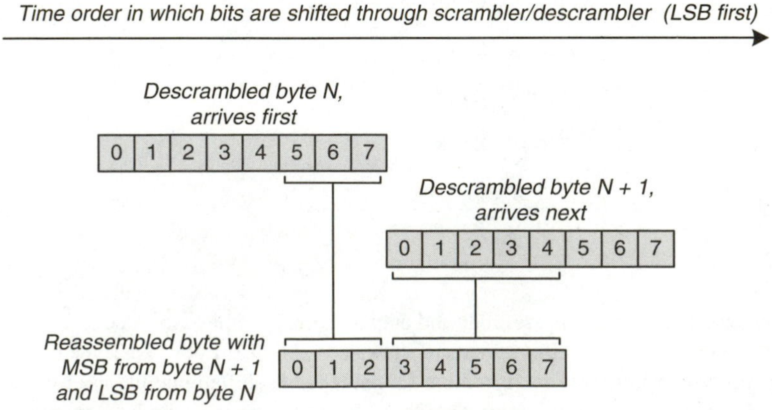

FIGURE 9.6 Reassembly of descrambled data.

do not support any descrambling or framing detection logic, because that logic is application specific. These functions must be implemented by external data processing logic. There are specific cases in which a common class of applications use standard scrambling and framing techniques. Serdes ICs designed for these applications do support the decoding and framing functions. An example of such an application is the transport of serial digital video in a TV studio setting. Companies such as Gennum Corporation manufacture serdes components with logic that can descramble 10-bit video data words and recognize framing sequences so that the resulting data stream is cleanly word aligned and directly usable without further manipulation.

9.5 8B10B CODING

An alternative, more robust coding method is *8B10B coding,* which maps byte values to 10-bit codes. Unlike polynomial coding that operates on a continuous stream of bits, 8B10B processes each byte individually. Each coded value is guaranteed to contain either five 0s and five 1s, four 0s and six 1s, or six 0s and four 1s. Therefore, a coded value contains a 0/1 imbalance of at most one bit. Some raw byte values are mapped to a perfectly balanced coded value. Those that map to imbalanced codes are given two mappings: one with four 0s and six 1s and one with six 0s and four 1s. A concept called *current running disparity* (CRD) is used to keep track of whether the last code contained a positive (1) or negative (0) imbalance. If a byte passes through the 8B10B encoder and is mapped to a code with four 0s and six 1s, a CRD state bit is set. The next time a byte passes through the encoder that has two possible mappings, the negatively imbalanced code will be chosen, and CRD will be cleared. This mechanism ensures that there are an equal number of 1s and 0s over time.

8B10B guarantees a minimum frequency of 0/1/0 transitions within the coded data stream, enabling reliable recovery of the serial bit clock at the receiver and guaranteeing an average DC value of 0 across the data stream. In mapping the 256 unique byte values to 10-bit codes, not all 1,024 code words are used. Some of these words are undesirable, because they contain greater than a single bit of 0/1 imbalance. However, some code words are left over that contain valid sequences of 0s and 1s. Rather than leave them unused, these code words are used to carry special values called *special characters* that can assist in the framing of data on a link. In particular, three special characters contain a unique *comma pattern*, where the first seven bits of a word contain two bits of one polarity followed by five bits of the opposite polarity. This comma pattern is guaranteed not to occur within any other data words, making it ideal as a marker with which data can be aligned within a serdes. A serdes that is 8B10B coding aware can search for comma patterns in the data stream and then realign the data stream such that comma patterns show up in the seven most significant bits of an incoming word. If the data link is properly encoded and is sufficiently free of disruptive noise, this alignment process should have to occur only once and thereafter will be transparent, because the comma patterns will already be in the proper bit positions. The benefits of 8B10B coding come with a 20 percent overhead penalty, because every eight data bits require ten coded bits. The cost is justified in many communications systems, because the electrical benefits of a balanced coding scheme enable more usable bandwidth to be extracted from a medium with a corresponding low bit error rate.

Serdes devices that are 8B10B coding aware contain a framing detection logic block that implements the 8B10B comma alignment function. The actual encoding and decoding may or may not be implemented in the serdes according to the parameters of the specific device. The comma detection and alignment function does not require the more complex encoding/decoding functions, because only a simple pattern match is required. A benefit of 8B10B coding is that low-level alignment operations can be performed by generic serdes logic without regard to the actual type of network traffic passing through the device. For example, Gigabit Ethernet and Fibre Channel (a storage area networking technology) both utilize 8B10B coding and can therefore take advantage of the same serdes devices on the market, despite the fact that their frame structures are completely different.

8B10B codes are broken into two sub-blocks to simplify the encoding/decoding process. Rather than directly mapping 256 data values into 1,024 code words, the data byte is separated into its three most-significant bits (designated Y) and its five least-significant bits (designated XX). The data byte is combined with a flag bit to represent its status as either a normal data character or as a special character. In notational form, a data character is represented as D and a special character as K. As shown in Fig. 9.7, the eight data bits are assigned letters from H down to A and the sub-blocks are swapped to yield a final notation of D/KXX.Y. For example, the ASCII character Z (0x5A) is split into two sub-blocks, 010 (Y) and 11010 (XX), yielding a notation of D26.2.

Once the data or special character has been split into sub-blocks, each sub-block is encoded using separate lookup tables whose inputs are the sub-blocks, CRD state information, and the special character flag. Separating the sub-blocks enables smaller lookup tables, which is advantageous from complexity and timing perspectives. The five-bit (XX) sub-block is converted to a six-bit code, and the three-bit sub-block (Y) is converted to a four-bit code, making ten bits in total. When encoding the XX sub-block, the CRD bit reflects the running disparity remaining from the previously encoded character. In the case of the Y sub-block, the CRD bit reflects the running disparity remaining from the 5B6B encoding of the current character. Tables 9.4 and 9.5 list the 5B6B and 3B4B lookup functions. Note that not all special characters are valid and that the CRD bit inverts the code when necessary to maintain a balance of 0s and 1s. The encoded bits are referred to by lower-case letters and, when sub-blocks are combined, form a string of bits: a, b, c, d, e, i, f, g, h, j. Encoded words are transmitted starting with bit "a" and ending with bit "j".

A special case exists in the 3B4B lookup table when encoding Dxx.7 characters to prevent long strings of consecutive zeroes or ones. When the CRD is negative and D17.7, D18.7, or D20.7 are being encoded, the alternate 0111 encoding is used instead of 1110. Likewise, when the CRD is positive and D11.7, D13.7, or D14.7 are being encoded, the alternate 1000 encoding is used instead of 0001. These exception cases present some additional complexity to the 3B4B translation.

In addition to the 256 possible data characters, 12 special characters are supported. All possible K28.y characters are supported along with K23.7, K27.7, K29.7 and K30.7. These special characters are identified by either a unique 5B6B code or a unique 3B4B code. The K28 is the only special character type with a unique 5B6B encoding. The other special characters are identified, because Kxx.7 has a unique 3B4B encoding. Recall that not all ten-bit code word permutations can be used because of 0/1 disparity rules inherent in the 8B10B coding algorithm.

There are numerous possible implementations of 8B10B encoder and decoder circuits, and they can vary by how many bytes are simultaneously processed, the type of lookup table resources available, and the degree of pipelining or other optimization required to achieve the desired throughput. A high-level block diagram of an 8B10B encoder is shown in Fig. 9.8. The larger 5B6B lookup table computes the intermediate CRD´ state that results from mapping the running disparity of the previ-

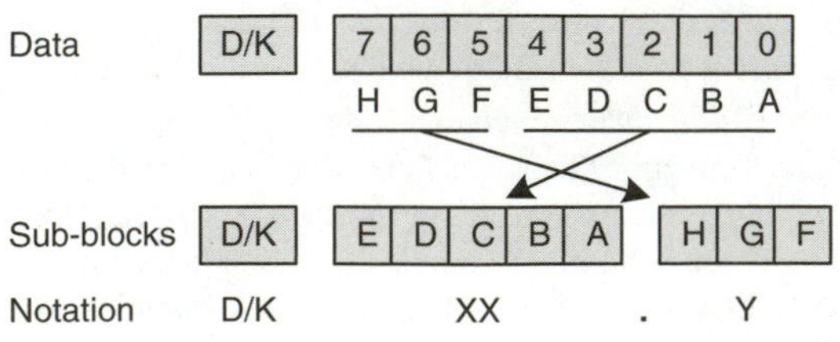

FIGURE 9.7 8B10B coding notation.

TABLE 9.4 5B6B Sub-block Encoding

Input Character	Binary Value EDCBA	Encoded Value abcdei	
		Positive Disparity	Negative Disparity
D00.y	00000	100111	011000
D01.y	00001	011101	100010
D02.y	00010	101101	010010
D03.y	00011	110001	
D04.y	00100	110101	001010
D05.y	00101	101001	
D06.y	00110	011001	
D07.y	00111	111000	000111
D08.y	01000	111001	000110
D09.y	01001	100101	
D10.y	01010	010101	
D11.y	01011	110100	
D12.y	01100	001101	
D13.y	01101	101100	
D14.y	01110	011100	
D15.y	01111	010111	101000
D16.y	10000	011011	100100
D17.y	10001	100011	
D18.y	10010	010011	
D19.y	10011	110010	
D20.y	10100	001011	
D21.y	10101	101010	
D22.y	10110	011010	
D/K23.y	10111	111010	000101
D24.y	11000	110011	001100
D25.y	11001	100110	
D26.y	11010	010110	
D/K27.y	11011	110110	001001
D28.y	11100	001110	
K28.y	11100	001111	110000
D/K29.y	11101	101110	010001
D/K30.y	11110	011110	100001
D31.y	11111	101011	010100

TABLE 9.5 5B6B Sub-block Encoding

Input Character	Binary Value HGF	Encoded Value fghj — Positive Disparity	Encoded Value fghj — Negative Disparity
D/Kxx.0	000	1011	0100
Dxx.1	001	1001	
Kxx.1	001	0110	1001
Dxx.2	010	0101	
Kxx.2	010	1010	0101
D/Kxx.3	011	1100	0011
D/Kxx.4	100	1101	0010
Dxx.5	101	1010	
Kxx.5	101	0101	1010
Dxx.6	110	0110	
Kxx.6	110	1001	0110
Dxx.7	111	1110 (0111)	0001 (1000)
Kxx.7	111	0111	1000

ous character to the current 5B6B code. If the 5B6B code is neutral, CRD´ will reflect the disparity of the generated 5B6B code. Otherwise, the lookup table will attempt to choose a code that balances out the CRD. If the character maps to a neutral code, the CRD will be passed through. The 3B4B table not only performs a simple mapping of the Y sub-block, but it also handles the alternate encoding of the special cases mentioned previously. The final CRD from this table is stored for use in the next character encoding.

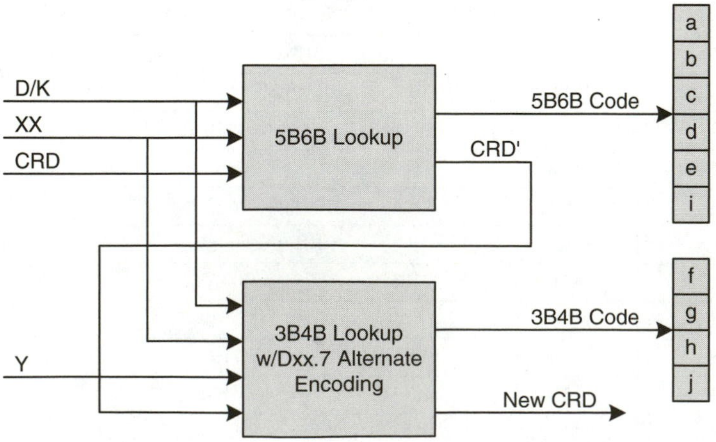

FIGURE 9.8 8B10B encoding logic.

The 8B10B decoding process requires similar lookup tables that perform the reverse operation. Decoding should also deal with CRD errors and invalid characters that indicate a bit error on the data link. However, not all bit errors on the data link will result in CRD errors or invalid characters.

9.6 ERROR DETECTION

Error detection and recovery are key requirements of data communications systems, because undesired results can occur if corrupted data is handled as if it were correct. While you might not mind an extra 0 being added to your bank account balance, you would certainly not want a 0 accidentally removed due to a data error! Transducer circuitry seeks to achieve the lowest *bit error rate* (BER) possible, but it will never be 0. Bit error rates of 10^{-10} to 10^{-12} are commonly achievable in wired (copper and fiber optic) data links. When these links carry data at 1 to 10 Gbps, errors will statistically occur every few seconds or minutes. These statistics make bit errors infrequent but recurring events that must be handled appropriately.

It is easier to detect an error than it is to correct one. Certain coding schemes, including 8B10B, provide some inherent bit error detection capability. If a bit error causes the detection of an invalid 8B10B code, or one with the wrong disparity, the receiver can detect the error. However, 8B10B coding cannot be relied upon to detect all errors, because not all errors will result in an invalid code word. Some single-bit or multibit errors will result in a different yet valid code word.

For channels with relatively low BER (e.g., high-quality wired data links), coding with a low-overhead scheme such as 8B10B is sufficient, and the responsibility for error detection and recovery can be passed up to the upper protocol levels. If errors are rare, the protocol-handling logic (both hardware and software) will not have to spend much time recovering from errors. A typical action when an error is detected at the protocol level is to request the retransmission of the affected frame. Such retransmission is expensive in terms of relative time and effort but is usually insignificant overall because of the low error rate. This situation changes when a channel has a higher BER.

A high-quality wireless data link may exhibit a BER of 10^{-6}, making errors much more frequent than in a high-quality wired channel. If all errors were handled by retransmitting frames, overall system throughput would suffer, because a much higher proportion of traffic would be devoted to error recovery. In these situations, it is worth utilizing a coding scheme with higher overhead that is capable of not only detecting errors but correcting them on the fly as well. Such schemes are called *forward error correction* (FEC). FEC codes calculate additional bits to be sent along with each data unit that have a degree of redundancy built into them so that if one or more bits get changed, a logical transformation can detect the mismatch and determine the correct values. It stands to reason that the overhead of FEC increases with the desire to correct increased numbers of bit errors within a single coded data unit. More FEC bits are required to correct two simultaneous bit errors than a single bit error. The decision on how complex a coding should be used is based on the channel's BER and on the penalty for passing error recovery functions to the protocol level. These characteristics are analyzed mathematically by considering the *coding gain* of a particular FEC code. An FEC code can be considered as its own channel that, rather than causing bit errors, resolves them. When the FEC channel is placed together with the real channel, the FEC coding gain effectively reduces the BER of the overall channel. FEC and its implementation are complex topics that are covered in specialized texts.

Regardless of whether a communications channel implements FEC, the data that is passed to higher protocol layers is subject to a net nonzero BER. The responsibility for handling these remaining errors lies at the data link layer and above. Network frame formats usually contain one or more error detection fields. These fields generally fall into one of two categories: *checksum* and *cyclic redundancy check* (CRC).

9.7 CHECKSUM

A checksum is a summation of a set of data and can be an arbitrary width, usually 8, 16, or 32 bits. Once a set of data has been summed, the checksum is sent along with the frame so that the sum can be verified at the receiver. The receiver can calculate its own checksum value by summing the relevant data and then compare its result against the frame's checksum. Alternatively, many checksum schemes enable the checksum value itself to be summed along with the data set, with a final result of zero indicating verification and nonzero indicating an error.

Perhaps the most common checksum scheme uses one's complement binary arithmetic to calculate the sum of a data set. One's complement addition involves calculating a normal two's complement sum of two values and then adding the carry bit back into the result. This constrains the running sum to the desired bit width, a necessary feature when summing hundreds or thousands of bytes where an unconstrained sum can be quite large. It is guaranteed that, when the carry bit is added back into the original result, a second carry will not be generated. For example, adding 0xFF and 0xFF yields 0x1FE. When the carry bit is added back into the eight-bit sum, 0xFF is the final one's complement result.

An interesting consequence of one's complement math is that the eight-bit values 0x00 and 0xFF (or 0xFFFF for a 16-bit value) are numerically equivalent. Consider what would happen if 1 is added to either value. In the first case, 0x00 + 0x01 = 0x01, is the obvious result. In the second case, 0xFF + 0x01 = 0x100 = 0x01, where the carry bit is added back into the eight-bit sum to yield the same final result, 0x01. A brief example of calculating a 16-bit checksum is shown in Table 9.6 to aid in understanding the one's complement checksum.

TABLE 9.6 Sixteen-Bit One's Complement Checksum

Data Value	Sum	Carry	Running Checksum
Initialize checksum to zero			0x0000
0x1020	0x1020	0	0x1020
0xFFF0	0x1010	1	0x1011
0xAD00	0xBD11	0	0xBD11
0x6098	0x1DA9	1	0x1DAA
0x701E	0x8DC8	0	0x8DC8

The logic to perform a one's complement checksum calculation can take the form of a normal two's complement adder whereby the carry bit is fed back in the next clock cycle to adjust the sum. As shown in Fig. 9.9, this forms a pipelined checksum calculator where the latency is two cycles. To begin the calculation, the accumulator and carry bit are reset to 0. Each time a word is to be summed, the multiplexer is switched from 0 to select the word. The adder has no qualifying logic and adds its two inputs, whose sum is repeatedly loaded into the accumulator and carry bits on each rising clock edge. After the last word has been summed, the control logic should wait an extra cycle, during which the multiplexer is selected to 0 to allow the most recent carry bit to be incorporated into the sum. It is guaranteed that a nonzero carry bit will not propagate into another nonzero carry bit after summing the accumulator with 0. This configuration works well in many situations, because two's complement adders are supported by many available logic implementation technologies. In high-

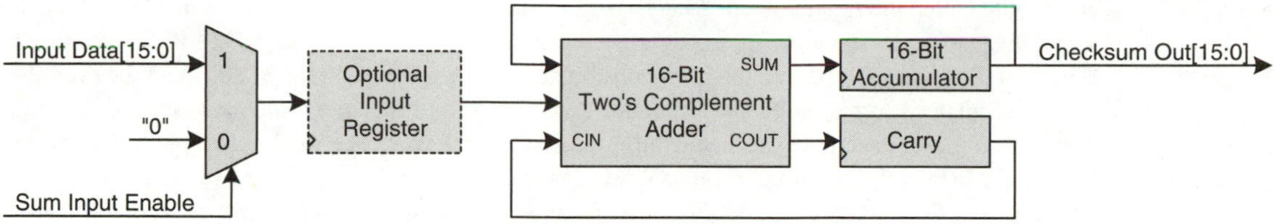

FIGURE 9.9 One's complement checksum calculator.

speed designs, logic paths with adders in them can prove difficult to meet timing. As shown in Fig. 9.9, an optional input register can be added between the multiplexer and the adder to completely isolate the adder from the control logic. This modification improves timing at the expense of an added cycle of latency to the checksum calculation.

A convenient advantage to the one's complement checksum is that, if the final result is inverted, or complemented, it can be summed along with the original data set with a final result of 0. The complement of 0x8DC8 is 0x7237, which, when added to the sum, 0x8DC8, yields 0xFFFF, the equivalent of 0x0000. Therefore, rather than placing the checksum itself into a frame, the transmitter can place the complemented value instead. The receiver's verification logic is thereby simplified by not requiring a full 16-bit comparator between the calculated checksum and the frame's checksum. It can sum all the relevant data values along with the complemented checksum and check for a result of 0x0000 or 0xFFFF.

Checksums can catch many types of errors, because most errors will cause the receiver's calculated checksum to differ from that contained in the frame. There are, however, consecutive sequences of errors, or bursts, that a checksum cannot detect. As the checksum is calculated, the first bit error will corrupt the in-progress sum. Subsequent errors can mask the previous error and return the sum to its proper value. The probability of this occurring is low, but it is not zero. Error detection is a study in probability wherein one can never attain a zero probability of undetected errors, but the probability can be brought arbitrarily close to zero as overhead is added in the form of coding and error detection fields.

9.8 CYCLIC REDUNDANCY CHECK

The CRC is a more complex calculation and one that provides a higher probability of detecting errors, especially multibit burst errors. A CRC is calculated using a linear feedback shift register, following the same basic concept behind scrambling data with a polynomial according to Galois field theory. CRCs exhibit superior error detection characteristics including the ability to detect all single-bit and double-bit errors, all odd numbers of errors, all burst errors less than or equal to the degree of the polynomial used, and most burst errors greater than the degree of the polynomial used.* The quality of CRC error detection depends on choosing the right polynomial, called a *generator polynomial*. The theory behind mathematically proving CRC validity and choosing generator polynomials is a complex set of topics about which much has been written. Different standard applications that employ CRCs have a specific polynomial associated with them.

A common CRC algorithm is the 8-bit polynomial specified by the *International Telecommunication Union* (ITU) in recommendation I.432, and it is used to protect ATM cell headers. This CRC,

* *Parallel Cyclic Redundancy Check (CRC) for HOTLink™*, Cypress Semiconductor, 1999.

commonly called the *Header Error Check* (HEC) field, is defined by the polynomial $x^8 + x^2 + x + 1$. The HEC is implemented with an eight-bit *linear feedback shift register* (LFSR) as shown in Fig. 9.10. Bytes are shifted into the LFSR one bit at a time, starting with the MSB. Input data and the last CRC bit feed to the XOR gates that are located at the bit positions indicated by the defining polynomial. After each byte has been shifted in, a CRC value can be read out in parallel with the LSB and MSB at the positions shown. When a new CRC calculation begins, this CRC algorithm specifies that the CRC register be initialized to 0x00. Not all CRC algorithms start with a 0 value; some start with each bit set to 1.

The serial LFSR can be converted into a set of parallel equations to enable practical implementation of the HEC on byte-wide interfaces. The general method of deriving the parallel equations is the same as done previously for the scrambling polynomial. Unfortunately, this is a very tedious process that is prone to human error. As CRC algorithms increase in size and complexity, the task can get lengthy. LFSRs may be converted manually or with the help of a computer program or spreadsheet. Table 9.7 lists the XOR terms for the eight-bit HEC algorithm wherein a whole byte is clocked through each cycle. Each CRC bit is referred to as Cn, where n = [7:0]. Once the equations are simplified, matching pairs of CRC and data input bits are found grouped together. Therefore, the convention Xn is adopted where Xn = Cn XOR Dn to simplify notation. Similar Boolean equations can be derived for arbitrary cases where fractions of a byte (e.g., four bits) are clocked through each cycle, or where multiple bytes are clocked through in the case of a wider data path.

TABLE 9.7 Simplified Parallel HEC Logic

CRC Bits	XOR Logic
C0	X0 X6 X7
C1	X0 X1 X6
C2	X0 X1 X2 X6
C3	X1 X2 X3 X7
C4	X2 X3 X4
C5	X3 X4 X5
C6	X4 X5 X6
C7	X5 X6 X7

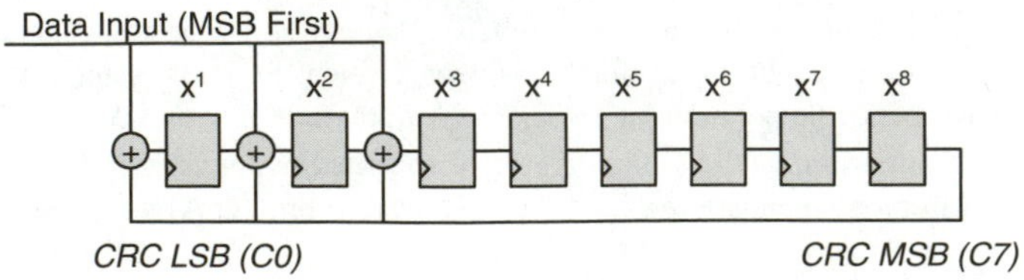

FIGURE 9.10 HEC LFSR.

When a new HEC calculation is to be started, the CRC state bits are reset to 0. Each byte is then clocked through the parallel logic at the rate of one byte per cycle. Following the final data byte, the HEC is XORed with 0x55 to yield a final result. An arbitrary number of bytes can be clocked through, and the CRC value will change each cycle. The one exception to this is the case of leading 0s. Because the HEC specifies a reset state of 0, passing 0x00 data through the CRC logic will not result in a nonzero value. However, once a nonzero value has been clocked through, the LFSR will maintain a nonzero value in the presence of a stream of 0s. This property makes the HEC nonideal for checking arbitrary strings of leading 0s, and it is a reason why other CRC schemes begin with a nonzero reset value. Table 9.8 shows an example of passing four nonzero data bytes through the parallel HEC logic and then XORing with 0x55 to determine a final CRC value.

TABLE 9.8 Examples of HEC Calculation

Data Input	HEC Value
(Initialization)	0x00
0x11	0x77
0x22	0xAC
0x33	0xD4
0x44	0xF9
XOR 0x55	0xAC

Another common CRC is the 16-bit polynomial appropriately called *CRC-16*. Its polynomial is $x^{16} + x^{15} + x^2 + 1$, and its LFSR implementation is shown in Fig. 9.11. As with the HEC, a CRC-16 can be converted to a parallel implementation. Because the CRC-16 is two bytes wide, its common implementations vary according to whether the data path is 8 or 16 bits wide. Of course, wider data paths can be implemented as well, at the expense of more complex logic. Table 9.9 lists the CRC-16 XOR terms for handling either one or two bytes per cycle.

Properly calculating a CRC-16 requires a degree of bit shuffling to conform to industry conventions. While this shuffling does not intrinsically add value to the CRC algorithm, it is important for all implementations to use the same conventions so that one circuit can properly exchange CRC codes with another. Unlike the HEC that shifts in data bytes MSB to LSB, the CRC-16 shifts in data bytes LSB to MSB. In the case of a 16-bit implementation, the high-byte, bits [15:8], of a 16-bit word is shifted in before the low-byte, bits [7:0], to match the standard order in which bytes are transmitted. What this means to the implementer is that incoming data bits must be flipped before being clocked through the parallel XOR logic. This doesn't actually add any logic to the task, and

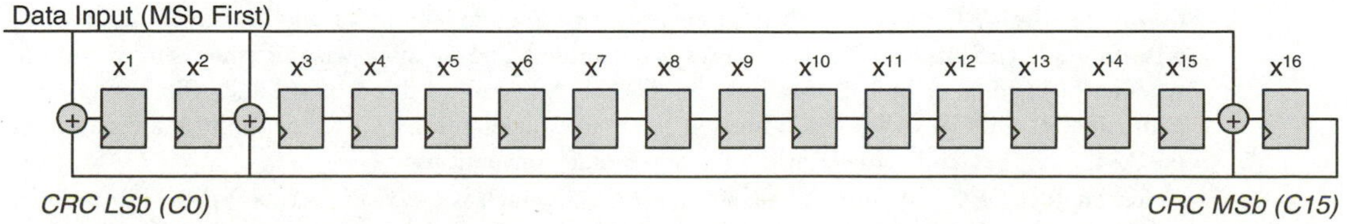

FIGURE 9.11 CRC-16 LFSR.

TABLE 9.9 CRC-16 Parallel Logic

CRC Bits	16-Bit XOR Logic	8-Bit XOR Logic
C0	X0 X1 X2 X3 X4 X5 X6 X7 X8 X9 X10 X11 X12 X13 X15	C8 C9 C10 C11 C12 C13 C14 C15 D0 D1 D2 D3 D4 D5 D6 D7
C1	X1 X2 X3 X4 X5 X6 X7 X8 X9 X10 X11 X12 X13 X14	C9 C10 C11 C12 C13 C14 C15 D1 D2 D3 D4 D5 D6 D7
C2	X0 X1 X14	C8 C9 D0 D1
C3	X1 X2 X15	C9 C10 D1 D2
C4	X2 X3	C10 C11 D2 D3
C5	X3 X4	C11 C12 D3 D4
C6	X4 X5	C12 C13 D4 D5
C7	X5 X6	C13 C14 D5 D6
C8	X6 X7	C0 C14 C15 D6 D7
C9	X7 X8	C1 C15 D7
C10	X8 X9	C2
C11	X9 X10	C3
C12	X10 X11	C4
C13	X11 X12	C5
C14	X12 X13	C6
C15	X0 X1 X2 X3 X4 X5 X6 X7 X8 X9 X10 X11 X12 X14 X15	X7 C8 C9 C10 C11 C12 C13 C14 C15 D0 D1 D2 D3 D4 D5 D6

there are a couple of ways that this can be done. One approach is to explicitly renumber the XOR terms to perform the flipping intrinsically. Another approach, the one taken here, is to maintain a consistent XOR nomenclature and simply flip the bits between the input and the XOR functions. It is convenient to adopt a common bit ordering convention across different CRC implementations, and the XOR terms shown for the CRC-16 are written with the same convention used in the HEC logic: MSB first. The actual bit-flipping in hardware is translated to a renumbering of the XOR input terms by the logic implementation software being used without any penalty of additional gates.

Once the bits have been clocked through the XOR functions in the correct order, industry convention is that the CRC register itself is flipped bit-wise and, depending on the implementation, byte-wise as well. The bit-wise flipping is always performed, and the byte-wise flipping is a function of whether big-endian or little-endian ordering is used. Since all of this talk of bit shuffling may seem confusing, Table 9.10 shows a step-by-step example of calculating a CRC-16 16 bits at a time across the 32-bit data set 0x4D41524B using the big-endian convention.

Yet another CRC is the ubiquitous 32-bit *CRC-32*, which is used in Ethernet, FDDI, Fibre Channel, and many other applications. The CRC-32 polynomial is $x^{32} + x^{26} + x^{23} + x^{22} + x^{16} + x^{12} + x^{11} + x^{10} + x^{8} + x^{7} + x^{5} + x^{4} + x^{2} + x + 1$, and its LFSR implementation is shown in Fig. 9.12. Similar to

TABLE 9.10 Step-by-Step CRC16 Calculation

Operation	Data
Initialize CRC-16 state bits	0x0000
First word to be calculated	0x4D41
Reorder bytes to end high-byte first after bit-flipping	0x414D
Flip bits for LSB-first transmission of high-byte then low-byte	0xB282
Clock word through XOR logic	0xAF06
Second word to be calculated	0x524B
Reorder bytes to end high-byte first after bit-flipping	0x4B52
Flip bits for LSB-first transmission of high byte then low byte	0x4AD2
Clock word through XOR logic	0x5CF4
Flip bits of CRC	0x2F3A
Optionally swap bytes of CRC for final result	0x3A2F

the CRC-16, parallel CRC-32 logic is commonly derived for data paths of one, two, or four bytes in width. A difference between the CRC-32 and those CRC schemes already presented is that the CRC32's state bits are initialized to 1s rather than 0s, and the final result is inverted before being used. Table 9.11 lists the CRC-32 XOR terms for handling one, two, or four bytes per cycle.

As noted, the CRC-32 state bits are initialized with 1s before calculation begins on a new data set. Words are byte-swapped and bit-flipped according to the same scheme as done for the CRC-16. When the last data word has been clocked through the parallel logic, the CRC-32 state bits are inverted to yield the final calculated value. Table 9.12 shows a step-by-step example of calculating a CRC-32 32 bits at a time using the same 32-bit data set, 0x4D41524B, as before.

CRC algorithms can be performed in software, and often are when cost savings is more important than throughput. Due to their complexity, however, the task is usually done in hardware when high-speed processing is required. Most modern networking standards place one or more CRC fields into

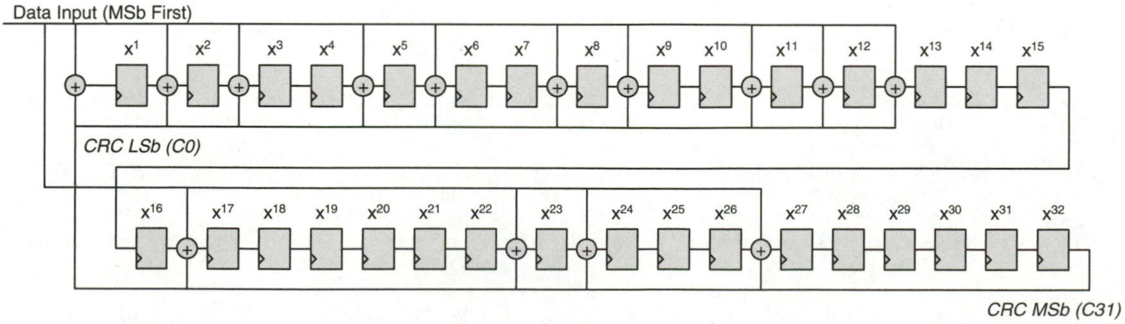

FIGURE 9.12 CRC-32 LFSR.

TABLE 9.11 CRC-32 Parallel Logic

CRC Bits	32-Bit XOR Logic	16-Bit XOR Logic	8-Bit XOR Logic
C0	X0 X6 X9 X10 X12 X16 X24 X25 X26 X28 X29 X30 X31	C16 C22 C25 C26 C28 D0 D6 D9 D10 D12	C24 C30 D0 D6
C1	X0 X1 X6 X7 X9 X11 X12 X13 X16 X17 X24 X27 X28	C16 C17 C22 C23 C25 C27 C28 C29 D0 D1 D6 D7 D9 D11 D12 D13	C24 C25 C30 C31 D0 D1 D6 D7
C2	X0 X1 X2 X6 X7 X8 X9 X13 X14 X16 X17 X18 X24 X26 X30 X31	C16 C17 C18 C22 C23 C24 C25 C29 C30 D0 D1 D2 D6 D7 D8 D9 D13 D14	C24 C25 C26 C30 C31 D0 D1 D2 D6 D7
C3	X1 X2 X3 X7 X8 X9 X10 X14 X15 X17 X18 X19 X25 X27 X31	C17 C18 C19 C23 C24 C25 C26 C30 C31 D1 D2 D3 D7 D8 D9 D10 D14 D15	C25 C26 C27 C31 D1 D2 D3 D7
C4	X0 X2 X3 X4 X6 X8 X11 X12 X15 X18 X19 X20 X24 X25 X29 X30 X31	C16 C18 C19 C20 C22 C24 C27 C28 C31 D0 D2 D3 D4 D6 D8 D11 D12 D15	C24 C26 C27 C28 C30 D0 D2 D3 D4 D6
C5	X0 X1 X3 X4 X5 X6 X7 X10 X13 X19 X20 X21 X24 X28 X29	C16 C17 C19 C20 C21 C22 C23 C26 C29 D0 D1 D3 D4 D5 D6 D7 D10 D13	C24 C25 C27 C28 C29 C30 C31 D0 D1 D3 D4 D5 D6 D7
C6	X1 X2 X4 X5 X6 X7 X8 X11 X14 X20 X21 X22 X25 X29 X30	C17 C18 C20 C21 C22 C23 C24 C27 C30 D1 D2 D4 D5 D6 D7 D8 D11 D14	C25 C26 C28 C29 C30 C31 D1 D2 D4 D5 D6 D7
C7	X0 X2 X3 X5 X7 X8 X10 X15 X16 X21 X22 X23 X24 X25 X28 X29	C16 C18 C19 C21 C23 C24 C26 C31 D0 D2 D3 D5 D7 D8 D10 D15	C24 C26 C27 C29 C31 D0 D2 D3 D5 D7
C8	X0 X1 X3 X4 X8 X10 X11 X12 X17 X22 X23 X28 X31	C16 C17 C19 C20 C24 C26 C27 C28 D0 D1 D3 D4 D8 D10 D11 D12	X0 C24 C25 C27 C28 D1 D3 D4
C9	X1 X2 X4 X5 X9 X11 X12 X13 X18 X23 X24 X29	C17 C18 C20 C21 C25 C27 C28 C29 D1 D2 D4 D5 D9 D11 D12 D13	X1 C25 C26 C28 C29 D2 D4 D5
C10	X0 X2 X3 X5 X9 X13 X14 X16 X19 X26 X28 X29 X31	C16 C18 C19 C21 C25 C29 C30 D0 D2 D3 D5 D9 D13 D14	X2 C24 C26 C27 C29 D0 D3 D5
C11	X0 X1 X3 X4 X9 X12 X14 X15 X16 X17 X20 X24 X25 X26 X27 X28 X31	C16 C17 C19 C20 C25 C28 C30 C31 D0 D1 D3 D4 D9 D12 D14 D15	X3 C24 C25 C27 C28 D0 D1 D4
C12	X0 X1 X2 X4 X5 X6 X9 X12 X13 X15 X17 X18 X21 X24 X27 X30 X31	C16 C17 C18 C20 C21 C22 C25 C28 C29 C31 D0 D1 D2 D4 D5 D6 D9 D12 D13 D15	X4 C24 C25 C26 C28 C29 C30 D0 D1 D2 D5 D6
C13	X1 X2 X3 X5 X6 X7 X10 X13 X14 X16 X18 X19 X22 X25 X28 X31	C17 C18 C19 C21 C22 C23 C26 C29 C30 D1 D2 D3 D5 D6 D7 D10 D13 D14	X5 C25 C26 C27 C29 C30 C31 D1 D2 D3 D6 D7
C14	X2 X3 X4 X6 X7 X8 X11 X14 X15 X17 X19 X20 X23 X26 X29	C18 C19 C20 C22 C23 C24 C27 C30 C31 D2 D3 D4 D6 D7 D8 D11 D14 D15	X6 C26 C27 C28 C30 C31 D2 D3 D4 D7
C15	X3 X4 X5 X7 X8 X9 X12 X15 X16 X18 X20 X21 X24 X27 X30	C19 C20 C21 C23 C24 C25 C28 C31 D3 D4 D5 D7 D8 D9 D12 D15	X7 C27 C28 C29 C31 D3 D4 D5
C16	X0 X4 X5 X8 X12 X13 X17 X19 X21 X22 X24 X26 X29 X30	X0 C16 C20 C21 C24 C28 C29 D4 D5 D8 D12 D13	C8 C24 C28 C29 D0 D4 D5
C17	X1 X5 X6 X9 X13 X14 X18 X20 X22 X23 X25 X27 X30 X31	X1 C17 C21 C22 C25 C29 C30 D5 D6 D9 D13 D14	C9 C25 C29 C30 D1 D5 D6

TABLE 9.11 CRC-32 Parallel Logic (Continued)

CRC Bits	32-Bit XOR Logic	16-Bit XOR Logic	8-Bit XOR Logic
C18	X2 X6 X7 X10 X14 X15 X19 X21 X23 X24 X26 X28 X31	X2 C18 C22 C23 C26 C30 C31 D6 D7 D10 D14 D15	C10 C26 C30 C31 D2 D6 D7
C19	X3 X7 X8 X11 X15 X16 X20 X22 X24 X25 X27 X29	X3 C19 C23 C24 C27 C31 D7 D8 D11 D15	C11 C27 C31 D3 D7
C20	X4 X8 X9 X12 X16 X17 X21 X23 X25 X26 X28 X30	X4 C20 C24 C25 C28 D8 D9 D12	C12 C28 D4
C21	X5 X9 X10 X13 X17 X18 X22 X24 X26 X27 X29 X31	X5 C21 C25 C26 C29 D9 D10 D13	C13 C29 D5
C22	X0 X9 X11 X12 X14 X16 X18 X19 X23 X24 X26 X27 X29 X31	C6 C16 C25 C27 C28 C30 D0 D9 D11 D12 D14	C14 C24 D0
C23	X0 X1 X6 X9 X13 X15 X16 X17 X19 X20 X26 X27 X29 X31	C7 C16 C17 C22 C25 C29 C31 D0 D1 D6 D9 D13 D15	C15 C24 C25 C30 D0 D1 D6
C24	X1 X2 X7 X10 X14 X16 X17 X18 X20 X21 X27 X28 X30	C8 C17 C18 C23 C26 C30 D1 D2 D7 D10 D14	C16 C25 C26 C31 D1 D2 D7
C25	X2 X3 X8 X11 X15 X17 X18 X19 X21 X22 X28 X29 X31	C9 C18 C19 C24 C27 C31 D2 D3 D8 D11 D15	C17 C26 C27 D2 D3
C26	X0 X3 X4 X6 X10 X18 X19 X20 X22 X23 X24 X25 X26 X28 X31	X10 C16 C19 C20 C22 C26 D0 D3 D4 D6	C18 C24 C27 C28 C30 D0 D3 D4 D6
C27	X1 X4 X5 X7 X11 X19 X20 X21 X23 X24 X25 X26 X27 X29	X11 C17 C20 C21 C23 C27 D1 D4 D5 D7	C19 C25 C28 C29 C31 D1 D4 D5 D7
C28	X2 X5 X6 X8 X12 X20 X21 X22 X24 X25 X26 X27 X28 X30	X12 C18 C21 C22 C24 C28 D2 D5 D6 D8	C20 C26 C29 C30 D2 D5 D6
C29	X3 X6 X7 X9 X13 X21 X22 X23 X25 X26 X27 X28 X29 X31	X13 C19 C22 C23 C25 C29 D3 D6 D7 D9	C21 C27 C30 C31 D3 D6 D7
C30	X4 X7 X8 X10 X14 X22 X23 X24 X26 X27 X28 X29 X30	X14 C20 C23 C24 C26 C30 D4 D7 D8 D10	C22 C28 C31 D4 D7
C31	X5 X8 X9 X11 X15 X23 X24 X25 X27 X28 X29 X30 X31	X15 C21 C24 C25 C27 C31 D5 D8 D9 D11	C23 C29 D5

frames (often in the trailer following the header and payload) to enable detection of infrequent bit errors. A typical system implementation contains hardware CRC generation and verification in the MAC logic to enable processing frames as fast as they can be transmitted and received.

9.9 ETHERNET

Ethernet is perhaps the most widely deployed family of networking standards in the world. It was first invented in 1973 at Xerox. There are many flavors of Ethernet (the 1973 original ran at roughly

TABLE 9.12 **Step-by-Step CRC-32 Calculation**

Operation	Data
Initialize CRC-32 state bits	0xFFFFFFFF
Word to be calculated	0x4D41524B
Reorder bytes to end high-byte first after bit-flipping	0x4B52414D
Flip bits for LSB-first transmission of high-byte then low-byte	0xB2824AD2
Clock word through XOR logic	0x5C0778F5
Flip bits of CRC	0xAF1EE03A
Optionally swap bytes of CRC for final result	0x3AE01EAF
Invert CRC state bits when input stream is completed	0xC51FE150

3 Mbps), including 10-, 100-, and 1,000-Mbps varieties. Ten-gigabit Ethernet is just now beginning to emerge. Ethernet originally ran over single shared segments of coaxial cabling, but most modern installations use twisted pair wiring in a physical star configuration. The familiar standards for Ethernet over twisted pair are 10BASE-T, 100BASE-T, and 1000BASE-T.

There is a whole family of Ethernet and related standards defined by the IEEE under the 802 LAN/MAN (local area network/metropolitan area network) Standards Committee. More specifically, the 802.3 CSMA/CD (carrier sense, multiple access, collision detect) Working Group defines Ethernet in its many forms. The 802.3 Ethernet frame format is shown in Table 9.13. A seven-byte preamble and a start of frame delimiter (collectively, a preamble) precede the main portion of the frame, which includes the header, payload, and trailer. The purpose of the preamble is to assist receivers in recognizing that a new frame is being sent so that it is ready to capture the main portion of the frame when it propagates through the wire. Not including the preamble, a traditional Ethernet frame ranges from 64 to 1,518 bytes. Two 48-bit Ethernet, or MAC, addresses are located at the start of the header: a destination address followed by a source address. The MSB of the address, bit 47, defines whether the address is unicast (0) or multicast (1). A unicast address defines a single source or destination node. A multicast address defines a group of destination nodes. The remaining address bits are broken into a 23-bit vendor block code (bits 46 through 24) and a 24-bit vendor-specific unique identifier (23 through 0). Manufacturers of Ethernet equipment license a unique vendor block code from the IEEE and then are responsible for assigning unique MAC addresses for all of their products. Each vendor block code covers 16 million (2^{24}) unique addresses.

Following the addresses is a length/type field that has two possible uses, for historical reasons. Prior to IEEE standardization, Xerox got together with Intel and Digital Equipment Corporation to agree on a standard Ethernet frame called *DIX*. DIX defines a type field that uniquely identifies the type of payload (e.g., IP) to enable easier parsing of the frame. When the IEEE first standardized Ethernet, it decided to implement a length field in place of a type field to more easily handle situations wherein payloads were less than the minimum 46 bytes allowed by the standard. This bifurcation of Ethernet caused interoperability problems. Years later, in 1997, the IEEE changed the field to be a combined length/type field. Values up to 1500 are considered lengths, and 1501 and above are considered types. Most Ethernet implementations use the original DIX-type field scheme. The IEEE has standardized a variety of type values to identify IP and certain other protocol extensions. Payloads with fewer than 46 bytes must be *padded* with extra data to meet the minimum frame size. The

TABLE 9.13 IEEE 802.3 Ethernet Frame Format

Field	Bytes	Fixed Value
Preamble	7	0x55
Start of frame delimiter	1	0xD5
Destination address	6	No
Source address	6	No
Length/type	2	No
Payload data	46–1500	No
Frame check sequence (CRC)	4	No

resolution of how many real data bytes are actually in an Ethernet frame is typically handled by higher-level protocols, such as IP, that contain their own length fields.

Modern Ethernet frames can be longer than 1,518 bytes for a couple of reasons. First, the IEEE has defined various data fields that can be thought of as extensions to the traditional Ethernet header. These include VLAN (virtual LAN) and MPLS (multiprotocol label switching) tags. Each of these extensions provides additional addressing and routing information for more advanced networking devices and adds length to the frame. Second, the industry began supporting *jumbo frames* in the late 1990s to extend an Ethernet frame to 9 kB. The advantage of a jumbo frame is that the same amount of data can be transferred with fewer individual frames, reducing overhead. Jumbo frame support is not universal, however. Older Ethernet equipment most likely will not handle such frames.

The frame check sequence is a 32-bit CRC that is computed across the entire main portion of the frame from the destination address through the last payload byte.

Because of its original topology as a shared bus, Ethernet employs a fairly simple yet effective arbitration mechanism to share access to the physical medium. This scheme is collision detection with random back-off, which was discussed earlier. Ethernet is referred to as CSMA/CD because of its access sharing mechanism. Frame size plays a role in the operation of CSMA/CD. A minimum frame size is necessary to ensure that, for a given physical network size, all nodes are capable of properly detecting a collision in time to take the correct action. Electrical signals propagate through copper wire at a finite velocity. Therefore, if two nodes at opposite ends of a bus begin transmitting at the same time, it will take a finite time for each to recognize a collision. Once a frame is successfully in progress, all other nodes must wait for that frame to end before they can transmit. A maximum frame size limits the time that a single node can occupy the shared network. Additionally, a maximum frame size limits the buffer size within Ethernet MAC logic. In the 1970s and early 1980s, the cost of memory was so high as to justify relatively small maximum frame sizes. Today, this is not a significant concern in most products, hence the emergence of jumbo frames.

Even when Ethernet networks are deployed in physical star configurations, they are often connected to hubs that electrically merge the star segments into a single, logically shared medium. A traditional Ethernet network is half-duplex, because only one frame can be in transit at any instant in time. Hubs are the least expensive way to connect several computers via Ethernet, because they do little more than merge star segments into a bus. Bus topologies present a fixed pool of bandwidth that must be shared by all nodes on that bus. As the number of nodes on a network increases, the traffic load is likely to increase as well. Therefore, there is a practical limit on the size of a bussed Ethernet network. *Bridges* were developed to mitigate Ethernet congestion problems by connecting multiple independent bus segments. A bridge operates at layer two using MAC addresses and builds a database of which addresses are on which side of the bridge. Only traffic that must cross the bridge to another segment is actually passed to the relevant segment. Otherwise, traffic can remain local on a single Ethernet segment without causing congestion on other segments. Bridging is illustrated in Fig. 9.13. Nodes 1 and 7 can simultaneously send data within their local segments. Later, node 4 can send data across the bridge to node 8, during which both network segments are burdened with the single transfer. The simplicity of this approach is that node 4 does not have any knowledge that node 8 is on a different segment. Crossing between Ethernet segments is handled transparently by the bridge.

Layer-two switches take bridging a step farther by providing many independent Ethernet ports that can exchange frames simultaneously without necessarily interfering with the traffic of other ports. As long as multiple ports are not trying to send data to the same destination port, those ports can all send data to different ports as if there existed many separate dedicated connections within the switch. This is known as *packet switching*: instantaneous connections between ports are made and then broken on a packet-by-packet basis. If two or more ports try to send data to the same port at the same time, one port will be allowed to transmit, while the others will not. Ethernet was developed to be a simple and inexpensive technology. Therefore, rather than providing special logic to handle such congestion issues, it was assumed that the network would generally have sufficient bandwidth to serve the application. During brief periods of high demand where not all data could be reliably delivered, it was assumed that higher-level protocols (e.g., TCP/IP) would handle such special cases in software, thereby saving money in reducing hardware complexity at the expense of throughput. Traditional Ethernet switches simply drop frames when congestion arises. In the case of switch congestion wherein multiple ports are sending data to a single port, all but one of those source ports may have their frames discarded. In reality, most switches contain a small amount of buffering that can temporarily hold a small number of frames that would otherwise be discarded as a result of congestion. However, these buffers do not prevent frame drops when congestion rises above a certain threshold. This behavior underscores the utility of layer-four protocols such as TCP.

Each switch port can conceivably be connected to a separate bused Ethernet segment and provide bridging functions on a broader scale than older bridges with only two ports. Switching has transformed network architecture as the cost of hardware has dropped over the years. It is common to find central computing resources such as file servers and printers with dedicated switch ports as

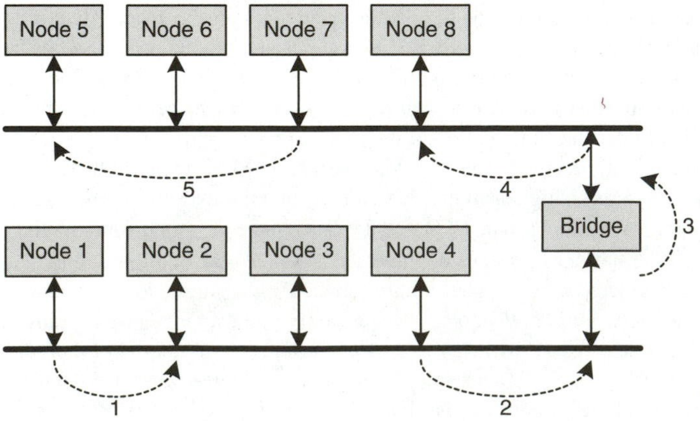

FIGURE 9.13 Ethernet bridging.

shown in Fig. 9.14. Other switch ports may connect to less-expensive hubs (some switches have built-in hubs to create a compact, integrated system). This reduces congestion by placing the most actively used nodes onto *dedicated media*, thereby eliminating collisions and increasing overall system bandwidth. Assuming 100BASE-T Ethernet segments, each file server and printer has a dedicated 100-Mbps data link into the switch. All arbitration and congestion control can be handled within the switch. If two file servers were placed onto the same *shared media* Ethernet segment, they would have to share the 100-Mbps bandwidth of a single segment and would likely experience collisions as many nodes tried to exchange data with them.

Switches and dedicated media transform Ethernet into a full-duplex-capable data link by providing separate transmit and receive signal paths between the switch and a node. Full-duplex operation is a subset of half-duplex operation, because the frame formatting is identical, but the CSMA/CD algorithm is not necessary for dedicated media applications. Some MACs may be designed for dedicated media only and can be made simpler without the necessity of media sharing logic. However, high-volume MACs may be designed into a variety of applications, requiring them to support both half- and full-duplex operation.

Ethernet ports on switches typically are capable of operating at multiple data rates to enable greater compatibility with other devices. To ease interoperability between MACs that can run at different speeds, 10/100/1000BASE-T has a mechanism called *autonegotiation* that enables two MACs to automatically determine their highest common data rate. The MACs ultimately must run at the same speed to properly exchange frames, but the initialization process of configuring the link to operate at the greatest common speed has been standardized and automated at the MAC layer. Autonegotiation works by each MAC exchanging a 16-bit message at a speed that is compatible with the slowest port type (10BASE-T). Each MAC advertises its capabilities, including speed and half/full-duplex support. The MACs may then select the greatest mutually supported link attributes. Autonegotiation is supported only for point-to-point links and is therefore most commonly observed between a switch and whatever entity is connected to it (e.g., another switch, a node, etc.).

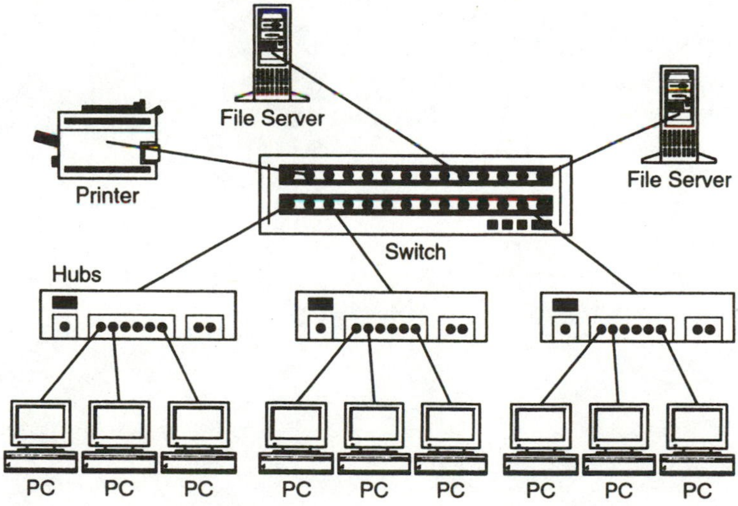

FIGURE 9.14 Switched Ethernet network.

CHAPTER 10

Logic Design and Finite State Machines

A large component of digital systems design is the implementation of application-specific algorithms. In some cases, software implements the bulk of algorithms, because microprocessors are flexible and easily programmable for a wide range of functions. Such systems may use only hardware to acquire and store data on behalf of the software. Other systems may be unable to perform their intended tasks with software alone. The reasons for this vary by application and often include throughput requirements that are not practically achievable with a microprocessor.

When algorithms are implemented in hardware, state machines are often employed to accomplish the task. A state machine can be made arbitrarily complex and can function similarly to software running on a microprocessor. Just as software moves through a sequence of tiny steps to solve a larger problem, a state machine can be designed to advance when certain conditions are satisfied. As the state machine progresses, it can activate other functions, just as software requests transactions from a microprocessor's peripherals.

This chapter focuses on higher-level logic design techniques used to implement functions ranging from basic microprocessor address decoding to clock domain crossing to state machines. Few design methodologies are mandatory, but certain techniques make the design task easier by freeing an engineer from having to worry about mundane low-level details while concentrating on the higher-level functions. Most software today is not written in assembly language, because the high-level language methodology is so productive. A software engineer does not have to worry about the accumulator and addressing modes of a microprocessor and can concentrate on the specific application being written. Likewise, more automated logic design techniques such as *hardware description languages* (HDL) handle the tasks of logic minimization and gate interconnection so that a hardware engineer can spend more time implementing the necessary algorithms for a specific application.

A great thing for the design community is that HDLs, once used by a relatively small set of companies with hefty financial resources, are becoming less expensive and more accessible with each passing year. The investment required to begin with these tools at the entry level ranges from nearly free up to a few thousand dollars, depending on the features desired and the subsidization that may be available. Because of the wide popularity of HDLs, several companies that manufacture programmable logic devices provide free or low-cost development tools to their customers. Stand-alone entry-level HDL tools are also available from companies including Model Technology and Simucad.

10.1 HARDWARE DESCRIPTION LANGUAGES

Many basic peripheral logic functions are available in off-the-shelf ICs. A variety of UART ICs are available, DMA controllers are available, and simple address decoding can be accomplished with

7400 devices such as the 74LS138. As digital systems grow more complex, the chances increase that suitable off-the-shelf logic will be either unavailable or impractical to use. The answer is to design and implement custom logic rather than relying solely on a third party to deliver a solution that does exactly what is needed.

Logic design techniques differ according to the scale of logic being implemented. If only a few gates are needed to implement a custom address decoder or timer, the most practical solution may be to write down truth tables, extract Boolean equations with Karnaugh maps, select appropriate 7400 devices, and draw a schematic diagram. This used to be the predominant means of designing logic for many applications, especially where the cost and time of building a custom IC was prohibitive. The original Apple and IBM desktop computers were designed this way, as witnessed by their rows of 7400 ICs.

When functions grow more complex, it becomes awkward and often simply impossible to implement the necessary logic using discrete 7400 devices. Reasons vary from simple density constraints—how much physical area would be consumed by dozens of 7400 ICs—to propagation delay constraints—how fast a signal can pass through multiple discrete logic gates. The answer to many of these problems is custom and semicustom logic ICs. The exact implementation technology differs according to the cost, speed, and time constraints of the application, but the underlying concept is to pack many arbitrary logic functions and flip-flops into one or more large ICs. An *application specific integrated circuit* (ASIC) is a chip that is designed with logic specific to a particular task and manufactured in a fixed configuration. A *programmable logic device* (PLD) is a chip that is manufactured with a programmable configuration, enabling it to serve in many arbitrary applications.

Once the decision is made to implement logic within custom or semicustom logic ICs, a design methodology is necessary to move ahead and solve the problem at hand. It is possible to use the same design techniques in these cases as used for discrete 7400 logic implementations. The trouble with graphical logic representations is that they are bulky and prone to human error. Hardware description languages were developed to ease the implementation of large digital designs by representing logic as Boolean equations as well as through the use of higher-level semantic constructs found in mainstream computer programming languages.

Aside from several proprietary HDLs, the major industry standard languages for logic design are *Verilog* and *VHDL* (Very high speed integrated circuits HDL). Verilog began as a proprietary product that was eventually transformed into an open standard. VHDL was developed as an open standard from the beginning. The two languages have roughly equal market presence and claim religious devotees on both sides. This book does not seek to justify one HDL over the other, nor does it seek to provide a definitive presentation of either. For the sake of practicality, Verilog is chosen to explain HDL concepts and to serve in examples of how HDLs are used in logic design and implementation.

HDLs provide logical representations that are abstracted to varying degrees. According to the engineer's choice or contextual requirements, logic can be represented at the *gate/instance* level, the *register transfer level* (RTL), or the *behavioral* level. Gate/instance-level representations involve the manual instantiation of each physical design element. An element can be an AND gate, a flop, a multiplexer, or an entire microprocessor. These decisions are left to the engineer. In a purely gate/instance-level HDL design, the HDL source code is nothing more than a glorified list of instances and connections between the input/output ports of each instance. The Verilog instance representation of $Y = A\&B + \overline{A}\&C$ is shown in Fig. 10.1. It is somewhat cumbersome but provides full control over the final implementation.

The brief listing in Fig. 10.1 incorporates many basic pieces of a generic Verilog module. First, the module is named and declared with its list of ports. Following the port list, the ports are defined as being inputs or outputs. In this case, the ports are all single net vectors, so no indices are supplied. Next is the main body that defines the function of the module. Verilog recognizes two major variable types: *wires* and *regs*. Wires simply connect two or more entities together. Regs can be assigned val-

```
module my_logic (
  A, B, C, Y
);

input A, B, C;
output Y;

wire and1_out, and2_out, notA;

and_gate u_and1 (
  .in1 (A),
  .in2 (B),
  .out (and1_out)
);

not_gate u_not (
  .in  (A),
  .out (notA)
);

and_gate u_and2 (
  .in1 (notA),
  .in2 (C),
  .out (and2_out)
);

or_gate u_or (
  .in1 (and1_out),
  .in2 (and2_out),
  .out (Y)
);

endmodule
```

FIGURE 10.1 Verilog gate/instance level design.

ues at discrete events, as will be soon discussed. When ports are defined, they are assumed to be wires unless declared otherwise. An output port can be declared as a type other than wire.

Being that this example is a gate/instance-level design, all logic is represented by instantiating other modules that have been defined elsewhere. A module is instantiated by invoking its name and following it with an instance name. Here, the common convention of preceding the name with "u_" is used, and multiple instances of the same module type are differentiated by following the name with a number. Individual ports for each module instance are explicitly connected by referencing the port name prefixed with a period and then placing the connecting variable in parentheses. Ports can be implicitly connected by listing only connecting variables in the order in which a module's ports are defined. This is generally considered poor practice, because it is prone to mistakes and is difficult to read.

HDL's textual representation of logic is converted into actual gates through a process called *logic synthesis*. A synthesis program parses the HDL code and generates a *netlist* that contains a detailed list of low-level logic gates and their interconnecting nets, or wires. Synthesis is usually done with a specific implementation target in mind, because each implementation technology differs in the logic primitives that it provides as basic building blocks. The primitive library for an ASIC will differ from that of a PLD, for example. Once synthesis is performed, the netlist can be transformed into a working chip and, hence, a working product.

A key benefit of HDL design methodology is the ability to thoroughly simulate logic before committing a netlist to a real chip. Because HDL is a programming methodology, it can be arbitrarily manipulated in a software simulation environment. The simulator allows a *test bench* to be written in either the HDL or another language (e.g., C/C++) that is responsible for creating stimulus to be applied to the logic modules. Widely used simulators include Cadence's NC-Sim, Model Technology's ModelSim, and Synopsys' VCS and Scirocco. A distinction is made between synthesizable and nonsynthesizable code when writing RTL and test benches. Synthesizable code is that which represents the logic to be implemented in some type of chip. Nonsynthesizable code is used to implement the test bench and usually contains constructs specifically designed for simulation that cannot be converted into real logic through synthesis.

An example of a test bench for the preceding Verilog module might consist of three number generators that apply pseudo-random test stimulus to the three input ports. Automatic verification of the logic would be possible by having the test bench independently compute the function $Y = A\&B + \overline{A}\&C$ and then check the result against the module's output. Such simulation, or verification, techniques can be used to root out the great majority of bugs in a complex design. This is a tremendous feature, because fixing bugs after an ASIC has been fabricated is costly and time consuming. Even in cases in which a PLD is used, it is usually faster to isolate and fix a bug in simulation than in the laboratory. In simulation, there is immediate access to all internal nodes of the design. In the lab, such access may prove quite difficult to achieve.

Verilog and VHDL both support simulation constructs that facilitate writing effective test benches. It is important to realize that these constructs are usually nonsynthesizable (e.g., a random number generator) and that they should be used only for writing test code rather than actual logic.

Gate/instance-level coding is quite useful and is used to varying degrees in almost every design, but the real power of HDL lies at the RTL and behavioral levels. Except in rare circumstances where absolute control over gates is required, instance-level coding is used mainly to connect different modules together. Most logic is written in RTL and behavioral constructs which are often treated together, hence the reason that synthesizable HDL code is often called RTL. Expressing logic in RTL frees the engineer from having to break everything down into individual gates and transfers this responsibility onto the synthesis software. The result is a dramatic increase in productivity and maintainability, because logical representations become concise. The example in Fig. 10.1 can be rewritten in Verilog RTL in multiple styles as shown in Fig. 10.2.

Each of these three styles has its advantages, each is substantially more concise and readable than the gate/instance-level version, and the styles can be freely mixed within the same module according to the engineer's preference. Style number 1 is a *continuous assignment* and makes use of the default wire data type for the output port. A wire is applicable here, because it is implicitly connecting two entities: the logic function and the output port. Continuous assignments are useful in certain cases, because they are concise, but they cannot get too complex without becoming unwieldy.

Style number 2 uses the always block, a keyword that tells the synthesis and simulation tools to perform the specified operations whenever a variable in its *sensitivity list* changes. The sensitivity list defines the variables that are relevant to the always block. If not all relevant variables are included in this list, incorrect results may occur. Always blocks are one of Verilog's fundamental constructs. A design may contain numerous always blocks, each of which contains logic functions that are activated when a variable in the sensitivity list changes state. A combinatorial always block should normally include all of its input variables in the sensitivity list. Failure to do so can lead to unexpected simulation results, because the always block will not be activated if a variable changes state and is not in the sensitivity list.

Style number 3 also uses the always block, but it uses a logical if…else construct in place of Boolean expression. Such logical representations are often preferable so that an engineer can concentrate on the functionality of the logic rather than deriving and simplifying Boolean algebra.

```
module my_logic (
  A, B, C, Y
);

input A, B, C;
output Y;

// Style #1: continuous assignment

assign Y = (A && B) || (!A && C);

// Style #2: behavioral assignment

reg Y;

always @(A or B or C)
begin
  Y = (A && B) || (!A && C);
end

// Style #3: if...then construct

reg Y;

always @(A or B or C)
begin
  if (A)
    Y = B;
  else
    Y = C;
end

endmodule
```

FIGURE 10.2 Verilog RTL-level design.

The two examples shown thus far illustrate basic Verilog HDL syntax with combinatorial logic. Clearly, synchronous logic is critical to digital systems, and it is fully supported by HDLs. D-type flip-flops are most commonly used in digital logic design, and they are directly inferred by using the correct RTL syntax. As always, gate-level instances of flops can be invoked, but this is discouraged for the reasons already discussed. Figure 10.3 shows the Verilog RTL representation of two flops, one with a synchronous reset and the other with an asynchronous reset.

The first syntactical difference to notice is the Verilog keyword *posedge*. Posedge and its complement, *negedge*, modify a sensitivity list variable to activate the always block only when it transitions. Synthesis tools are smart enough to recognize these keywords and infer a clocked flop. Clocked always blocks should not include normal regs or wires in the sensitivity list, because it is only desired to activate the block on the active clock edge or when an optional asynchronous reset transitions.

At reset, a default 0 value is assigned to Q. Constants in Verilog can be explicitly sized and referenced to a particular radix. Preceding a constant with `b denotes it as binary (`h is hex, and `d is decimal). Preceding the radix identifier with a number indicates the number of bits occupied by that constant.

Another syntactical difference to note is the use of a different type of assignment operator: <= instead of =. This is known as a non-blocking (<=) assignment as compared to a blocking (=) assignment. It is considered good practice to use non-blocking assignments when inferring flops, because

```
// synchronous reset

always @(posedge CLK)
begin
  if (RESET) // RESET evaluated only at CLK rising edge
    Q <= 1´b0;
  else
    Q <= D;
end

// asynchronous reset

always @(posedge CLK or posedge RESET)
begin
  if (RESET) // RESET evaluated whenever it goes active
    Q <= 1´b0;
  else
    Q <= D;
end
```

FIGURE 10.3 Verilog RTL flip-flop inference.

the non-blocking assignment does not take effect until after the current simulation time unit. This is analogous to the behavior of a real flop wherein the output does not transition until a finite time has elapsed from its triggering event. Under certain circumstances, either type of assignment will yield the same result in both simulation and synthesis. In other situations, the results will differ, as illustrated in Fig. 10.4.

In the first case, regs Q1 and Q2 are tracked at two different instants in time. First, their current states are maintained as they were just prior to the clock edge for the purpose of using their values in subsequent assignments. Second, their new states are assigned as dictated by the RTL. When Q2 is assigned, it takes the previous value of Q1, not the new value of Q1, which is D. Two flops are inferred.

In the second case, variables Q1 and Q2 are tracked at a single instant in time. Q1 is assigned the value of variable D, and then Q2 is assigned the new value of variable Q1. Q1 has become a temporary placeholder and has no real effect on its own. Therefore, only a single flop, Q2, is inferred.

Utilizing HDL to design logic requires software tools more complex than just pencil and paper. However, the benefits quickly accumulate for designs of even moderate complexity. The digital

```
// Non-blocking assignments: two flops inferred

always @(posedge CLK)
begin
  Q1 <= D;
  Q2 <= Q1;
end

// Blocking assignments: one flop inferred

always @(posedge CLK)
begin
  Q1 = D;
  Q2 = Q1;
end
```

FIGURE 10.4 Verilog blocking vs. non-blocking assignment.

functions and techniques discussed in the remainder of this chapter show how practical HDL design can be.

10.2 CPU SUPPORT LOGIC

Most digital systems require some quantity of miscellaneous glue logic to help tie a CPU to its memory and I/O peripherals. Some of the most common support functions are address decoding, basic I/O signals, interrupt control, and timers. Another common function is interface conversion whereby the CPU needs to talk with a peripheral that has an interface that is incompatible with that of the CPU. Interface conversion can range from simple control signal polarity adjustments to complex buffering schemes that cross clock domains with FIFOs.

Address decoding is usually a combinatorial implementation, because many CPU interfaces are nonpipelined. When performing address decoding and other bus control functions for a pipelined CPU bus, a more complex synchronous circuit is called for that can track the various pipeline stages and take the necessary actions during each stage. Basic combinatorial address decoding consists of mapping ranges of addresses to chip selects. Chip select signals are usually active-low by convention and are numbered upward from 0. For the sake of discussion, consider the 24-bit memory map in Table 10.1 to design an address decoder.

TABLE 10.1 Example Memory Map

Address Range	Qualifier	Chip Select	Function
0x000000–0x0FFFFF	RomSel=0	CS0_	1-MB default boot ROM
0x100000–0x1FFFFF	RomSel=1		
0x100000–0x1FFFFF	RomSel=0	CS1_	1-MB ROM module
0x000000–0x0FFFFF	RomSel=1		
0x200000–0x21FFFF	N/A	CS2_	128-kB SRAM
0x220000–0x2FFFFF	N/A	None	Unused
0x300000–0x30000F	N/A	CS3_	UART
0x300010–0x3FFFFF	N/A	None	Unused
0x400000–0x4FFFFF	N/A	Internal	Control/status registers
0x500000–0xFFFFFF	N/A	None	Unused

Four external chip selects are called out. Instead of using the asterisk to denote active-low signals, the underscore is used, because an asterisk is not a valid character for use in a Verilog identifier. The first two chip selects are used for ROM (e.g., flash or EPROM) and their memory ranges are swappable according to the RomSel signal. It is sometimes useful to provide an alternate boot ROM that can be installed at a later date for various purposes such as a software upgrade. When boot ROM is implemented in flash, the CPU is able to load new data into its ROM. If there is no other way to send a new software image to the system, the image can be loaded onto a ROM module that is temporarily installed into the CS1_ slot. A jumper can then be installed that causes RomSel to be asserted. When the system is turned on, RomSel=1 causes the ROM module to become the boot ROM, and new software can be loaded into CS0_ ROM.

The remainder of the address space is sparsely populated. Occupied memory regions are spread out to reduce the complexity of the decoding logic by virtue of requiring fewer address bits. If the UART were located immediately after the SRAM, the logic would have to consider the state of A[23:16] rather than just A[23:20]. The fifth and final used memory region is reserved for internal control and status registers. This decoding logic can be written in Verilog as shown in Fig. 10.5.

The address decoding logic is written here in behavioral form with a *case* construct. Case statements enable actions to be associated with individual states of a causal variable. Note that the chip select outputs are declared as regs even though they are not flops, because they are assigned in an always block instead of in a continuous assignment. Prior to the case statement, all of the always

```
module GlueLogic (
  Addr,
  RomSel,
  CS0_,
  CS1_,
  CS2_,
  CS3_
);

input  [23:20] Addr;
input          RomSel;
output         CS0_, CS1_, CS2_, CS3_;

reg            CS0_, CS1_, CS2_, CS3_;
reg            IntSel;

always @(Addr or RomSel)
begin
  CS0_   = 1´b1;  // establish default values to simplify case
  CS1_   = 1´b1;  // statement and prevent formation of latches
  CS2_   = 1´b1;
  CS3_   = 1´b1;
  IntSel = 1´b0;

  case (Addr[23:20])
    4´b0000 : begin
                CS0_ = RomSel;
                CS1_ = !RomSel;
              end
    4´b0001 : begin
                CS0_ = !RomSel;
                CS1_ = RomSel;
              End
    4´b0010 : CS2_   = 1´b0;
    4´b0011 : CS3_   = 1´b0;
    4´b0100 : IntSel = 1´b1;
  endcase
end

endmodule
```

FIGURE 10.5 Address decoding logic.

block's outputs are assigned to default inactive states. This is necessary to prevent the synthesis software from inferring an unwanted latch instead of simple combinatorial logic. If a combinatorial always block does not assign a reg value for all combinations of inputs, the synthesis tool determines that the reg should hold its previous state, thereby creating a latch. Latches are prevented by either exhaustively listing all combinations of inputs or by assigning a default value to all variables somewhere in the always block.

Unintended latches are the bane of HDL design. There are valid instances when a latch is desired, but latches are often inferred mistakenly, because the RTL code does not properly handle default cases wherein the variable is not assigned. Combinatorial logic must always assign values to variables regardless of the logic path. Otherwise, *statefullness* is implied. Verilog's blocking assignments enable multiple values to be assigned to a single variable in an always block, and the last assigned variable is the one that takes effect. Therefore, latches are avoided by assigning default values up front.

An active-high signal, IntSel, is decoded but not yet used for the purposes of selecting internal control and status registers that will be discussed shortly.

When expanding a CPU bus, an engineer must be careful that too many devices are not placed onto the bus, because output pins are rated only for certain drive strengths. As the lengths of the interconnecting wires increase and the number of loads increase, it may become necessary to extend the CPU bus using bidirectional buffers as discussed earlier. Figure 10.6 shows how our hypothetical system might use such buffers to isolate the plug-in ROM module so that the electrical impact of connectors and a separate module are minimized. For clarity, control signals are not shown. A unidirectional buffer isolates the address bus, and a bidirectional buffer isolates the data bus. No control is necessary for the address buffer, because it can be configured to always pass the address bus to the ROM module socket. However, the data buffers require control, because they must direct data out to the module for writes and in from the module for reads. Therefore, the tri-state control of the data buffer must be operated according to the address decode and read/write status.

The existing address decoding logic can be augmented to provide the necessary functionality. One additional input is required: the CPU's active-low read enable signal. An additional output is required to operate the data buffer's direction select signal. When high, the buffers will pass data from the CPU side to the ROM and, when low, the buffers will drive the CPU data bus with data presented by the ROM. A second always block can be added as shown in Fig. 10.7. New port and variable declarations are assumed.

Another common function of support logic is providing general I/O signals that the CPU can use to interact with its environment. Such interaction can include detecting an opening door and turning

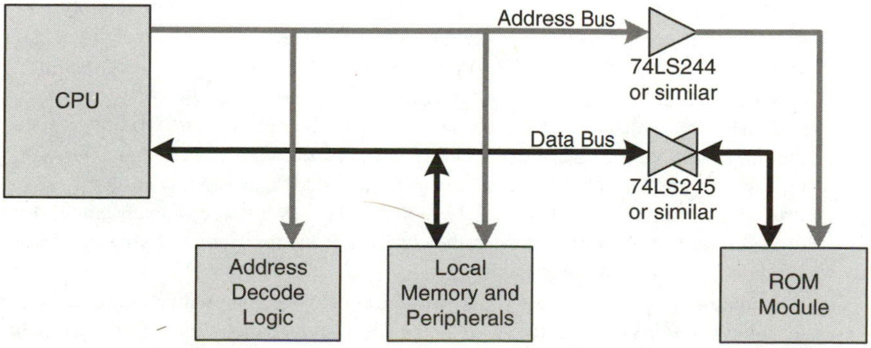

FIGURE 10.6 Bus extension buffers.

```
always @(CS1_ or Rd_)
begin
  if (!CS1_ && !Rd_)
    DataBufDir = 1'b0;  // drive CPU bus when ROM selected for read
  else
    DataBufDir = 1'b1;  // otherwise, always drive data to ROM
end
```

FIGURE 10.7 Data buffer control logic.

on an alarm. The opening door can be detected using a switch connected to an input signal. When the CPU reads the status of this signal, it can determine whether the switch is open or closed. An alarm can be turned on when the CPU sets an output signal that enables an alarm circuit. Control and status registers must be implemented to enable the CPU to read and write I/O signals. In our continuing example, we assume an eight-bit data bus coming from the CPU and the need for eight input signals and eight output signals. Implementing registers varies according to whether the CPU bus is synchronous or asynchronous. Some older microprocessors use asynchronous buses requiring latches to be formed within the support logic. Figure 10.8 shows the implementation of two registers using the previously decoded IntSel signal in both synchronous and asynchronous styles. Again, the proper declarations for ports and variables are assumed.

An added level of address decoding is required here to ensure that the two registers are not accessed simultaneously. The register logic consists of two basic sections: the write logic and read logic. The write logic (required only for the control register that drives output signals) transfers the contents of the CPU data bus to the internal register when the register is addressed and the write enable is active. The ControlRegSel signal is implemented in a case statement but can be implemented in a variety of ways. More select signals will be added in coming examples. The asynchronous write logic infers a latch, because not all permutations of input qualifiers are represented by assignments. If Reset_ is high and the control register is not being selected for a write, there is no specified action. Therefore, memory is implied and, in the absence of a causal clock, a latch is inferred. The synchronous write logic is almost identical, but it references a clock that causes a flop inference. Reset is implemented to provide a known initial state. This is a good idea so that external logic that is driven by the control register can be safely designed with the assumption that operations begin at a known state. The known state is usually inactive so that peripherals do not start operating before the CPU finishes booting and can disable them.

The read logic consists of two sections: the output multiplexer and the output buffer control. The output multiplexer simply selects one of the available registers for reading. It is not necessary to qualify the multiplexer with any other logic, because a read will not actually take place unless the output buffer control logic sends the data to the CPU. Rather than preventing a latch in ReadData by assigning it a default value before the case construct, the Verilog keyword *default* is used as the final case enumeration to specify default operation. Either solution will work—it is a matter of preference and style over which to use in a given situation. Both read-only and writable registers are included in the read multiplexer logic. Strictly speaking, it is not mandatory to have writable register contents readable by the CPU, but this is a very good practice. Years ago, when logic was very expensive, it was not uncommon to find write-only registers. However, there is a substantial drawback to this approach: you can never be sure what the contents of the register are if you fail to keep track of the exact data that has already been written!

Implementing bidirectional signals in Verilog can be done with a continuous assignment that selects between driving an active variable or a high-impedance value, Z. The asynchronous read logic is very simple: whenever the internal registers are selected and read enable is active, the tri-state buffer is enabled, and the output of the multiplexer is driven onto the CPU data bus. At all other

```
always @(Addr[3:0] or StatusInput[7:0] or ControlReg[7:0] or IntSel)
begin
  case (Addr[3:0]) // read multiplexer
    4'h0    : ReadData[7:0] = StatusInput[7:0]; // external input pins
    4'h1    : ReadData[7:0] = ControlReg[7:0];
    default : ReadData[7:0] = 8'h0; // alternate means to prevent latch
  endcase

  ControlRegSel = 1'b0;  // default inactive value

  case (Addr[3:0]) // select signal only needed for writeable registers
    4'h1 : ControlRegSel = IntSel;
  endcase
end

// Option #1A: asynchronous read logic

assign CpuData[7:0] = (IntSel && !Rd_) ? ReadData[7:0] : 8'bz;

// Option #1B: synchronous read logic

always @(posedge CpuClk)
begin
  if (!Reset_)  // synchronous reset
    CpuDataOE <= 1'b0;
    // no need to reset ReadDataReg and possibly save some logic
  else begin
    CpuDataOE         <= IntSel && !Rd_; // all outputs are registered
    ReadDataReg[7:0]  <= ReadData[7:0];
  end
end

assign CpuData[7:0] = CpuDataOE ? ReadDataReg[7:0] : 8'bz;

// Option #2A:asynchronous write logic

always @(ControlRegSel or CpuData[7:0] or Wr_ or Reset_)
begin
  if (!Reset_)
    ControlReg[7:0] = 8'h0;  // reset state is cleared
  else if (ControlRegSel && !Wr_)
    ControlReg[7:0] = CpuData[7:0];
  // missing else forces memory element: intentional latch!
end

// Option #2B: synchronous write logic

always @(posedge CpuClk)
begin
  if (!Reset_)  // synchronous reset
    ControlReg[7:0] <= 8'h0;
  else if (ControlRegSel && !Wr_)
    ControlReg[7:0] <= CpuData[7:0];
end
```

FIGURE 10.8 Control/status register logic.

times, the data bus is held in a high-impedance state. This works as expected, because the value Z can be overridden by another assignment. In simulation, the other assignment may come from a test bench that emulates the CPU's operation. In synthesis, the software properly recognizes this arrangement as inferring a tri-state bus. The continuous assignment takes advantage of Verilog's conditional operator, ? : , which serves an if...else function. When the logical expression before the question mark is true, the value before the colon is used. Otherwise, the value after the colon is used. A bidirectional port is declared using the Verilog *inout* keyword in place of *input* or *output*. The synchronous version of the read logic is very similar to the asynchronous version, except that the outputs are first registered before being used in the tri-state assignment.

Interrupt control registers are implemented by support logic when the number of total interrupt sources in the system exceeds the CPU's interrupt handling capacity. Gathering multiple interrupts and presenting them to the CPU as a single interrupt signal can be as simple as logically ORing multiple interrupt signals together. A somewhat more complex scheme, but one with value, is where interrupts can be selectively masked by the CPU, and system-wide interrupt status is accumulated in a single register. This gives the CPU more control over how it gets interrupted and, when interrupted, provides a single register that can be read to determine the source of the interrupt. Such a scheme can be implemented with two registers and associated logic: a read/write interrupt mask register and a read-only interrupt status register. Each pair of bits in the mask and status registers corresponds to a single interrupt source. When an interrupt is active, the corresponding bit in the status register is active. However, only those interrupt sources whose mask bits have been cleared will result in a CPU interrupt. At reset, the mask register defaults to 0xFF to disable all interrupts. Figure 10.9 shows a Verilog implementation of interrupt control as an extension of the previous example. Synchronous registers are assumed here, but asynchronous logic is easily adapted.

Aside from the registers themselves, the main function of the interrupt control logic is implemented by the bit-wise ORing and reduction AND of the interrupt mask bits and the external interrupt signals. Verilog's | operator is a bit-wise OR function in contrast to the || logical OR function. When two equal-size vectors are bit-wise ORed, the result is a single vector of the same size wherein each bit is the OR of the corresponding pair of bits of the operands. This first step disables any active interrupts that are masked; per DeMorgan's law, an OR acts as an active-low AND function. The second step is a reduction AND as indicated by the unary & operator. Similar to OR, & is the bit-wise version of &&. Invoking & with a single operand makes it a reduction operator that ANDs together all bits of a vector and generates a single output bit. AND is used because the interrupt polarities are active-low. Therefore, if any one interrupt is asserted (low), the reduction AND function will generate a low output. Per DeMorgan's law once again, an AND acts as an active-low OR function.

Timers are often useful structures that can periodically interrupt the CPU to invoke a time-critical interrupt service routine, enable the CPU to determine elapsed time, or trigger some other event in hardware. It is common to find timers implemented in certain CPU products such as microcontrollers. Sometimes, however, it becomes necessary to implement a custom timer in support logic. One such example is presented here with a fixed prescaler and an eight-bit counter with an eight-bit configurable terminal count value. The prescaler is used to slow down the timer so that it can interrupt the CPU over longer periods of time. For the sake of discussion, let's assume that the CPU is running at a frequency of 10 MHz and that the timer granularity should be 1 ms. Therefore, the prescaler should count from 0 to 9,999 to generate a 1-ms tick when running with a 100-ns period. A Verilog implementation of such a timer is shown in Fig. 10.10 without the associated read/write logic already presented in detail. A 14-bit prescaler is necessary to represent numbers from 0 to 9,999. It is assumed that the terminal count register, TermCount, is implemented elsewhere as a general read/write register.

When the timer rolls over, it asserts TimerRollOver for one CpuClk cycle to trigger whatever logic is desired by the application. If the trigger event is a CPU interrupt, the logic should create a

```
always @(Addr[3:0] or StatusInput[7:0] or ControlReg[7:0] or IntSel
                  or ISR[7:0] or IMR[7:0])
begin
  case (Addr[3:0]) // read multiplexer
    4'h0    : ReadData[7:0] = StatusInput[7:0]; // external input pins
    4'h1    : ReadData[7:0] = ControlReg[7:0];
    4'h2    : ReadData[7:0] = ISR[7:0];  // interrupt status
    4'h3    : ReadData[7:0] = IMR[7:0];  // interrupt mask
    default : ReadData[7:0] = 8'h0; // alternate means to prevent latch
  endcase

  ControlRegSel  = 1'b0;  // default inactive value
  IMRSel         = 1'b0;

  case (Addr[3:0]) // select signal only needed for writeable registers
    4'h1 : ControlRegSel = IntSel;
    4'h3 : IMRSel        = IntSel;
  endcase
end

// Interrupt gathering

always @(IMR[7:0] or ExtInt_[7:0])
begin
  ISR[7:0] = ExtInt_[7:0]; // reflect status of external signals
  CpuInt_  = &(IMR[7:0] | ExtInt_[7:0]); // reduction AND
end

// Write logic

always @(posedge CpuClk)
begin
  if (!Reset_)
    IMR[7:0] <= 8'hff; // mask all interrupts on reset
  else if (IMRSel && !Wr_)
    IMR[7:0] <= CpuData[7:0];
end
```

FIGURE 10.9 Interrupt control logic.

"sticky" version of TimerRollOver that does not automatically get cleared by hardware. Rather, it is sticky because, once set by hardware, the bit will retain its state until explicitly cleared by software. This provides an arbitrarily long time for software to respond to the interrupt assertion, read the interrupt status register to detect the timer roll-over event, and then clear the sticky bit. Without the sticky bit, software would have no chance of catching a pulse that is only a single cycle in width.

10.3 CLOCK DOMAIN CROSSING

Some logic design tasks involve exchanging information between logic running on unrelated clocks. When multiple independent clock domains exist in a system, there is no guaranteed skew or phase relationship between the various clocks. Synchronous timing analysis dictates that a flop's setup and hold times must be met to ensure reliable capture of the data presented to it. Yet it is impossible to guarantee proper setup and hold times when the source flop's clock has no relationship to the destination flop's clock.

```
always @(posedge CpuClk)
begin
  if (!Reset_) begin
    Prescaler[13:0] <= 14'h0;
    Timer[7:0] <= 8'h0;
    TimerRollOver <= 1'b0;
  end
  else
    if (Prescaler[13:0] == 14'd9999) begin
      Prescaler[13:0] <= 14'h0; // roll-back to zero
      if (Timer[7:0] == TermCount[7:0]) begin
        Timer[7:0] <= 8'h0; // start over
        TimerRollOver <= 1'b1; // trigger other logic
      end
    se el
        Timer[7:0] <= Timer[7:0] + 1;
    end
    else begin
      Prescaler[13:0] <= Prescaler[13:0] + 1;
      TimerRollOver <= 1'b0;
    end
end
```

FIGURE 10.10 Timer logic.

There are several scenarios that arise from clocking a flop input with unknown timing as shown in Fig. 10.11. First, there is a chance that the applied signal will be captured successfully if it happens to meet the flop's setup time specification. Second, there is a chance that the input data will be missed on the first clock cycle, because it occurs too late for the flop to detect it. If the data remains asserted through the next cycle, it will be properly captured at that time.

Finally, there is a questionable area between capturing and missing the signal. A flop is inherently an analog circuit, because it is built from transistors. Flops behave digitally when their timing specifications are adhered to. When timing violations occur, the flop circuit may behave in a nondigital manner and generate a marginal output that is somewhere between 1 and 0 before eventually settling to a valid logic state. It is not certain into which logic state the flop will settle, nor is it certain exactly how much time the flop will take to settle. This phenomenon is known as *metastability.* A flop is said to be metastable when a timing violation occurs and it takes some time for the output to stabilize. Metastability does not damage the flop, but it can cause significant trouble for a synchronous circuit that is designed with the assumption of predictable timing.

Clock domains can be reliably crossed when proper design techniques are used to accommodate likely timing violations. Some applications require that only control signals move between clock domains, and others must transport entire data paths. The simpler case of individual control signals is presented first and then used as a foundation for data paths.

Because it is impossible to avoid metastability when crossing clock domains, the phenomenon must be managed. A control signal ostensibly triggers activity in the logic that it drives, and this destination logic waits for the control signal to transition. Metastability does not prevent the signal from reaching its destination; it introduces uncertainty over exactly when the signal will stabilize in the destination clock domain. This uncertainty is dealt with by *synchronizing* the control signal using extra flops prior to treating the signal as a legal synchronous input. A two-stage synchronizing circuit is shown in Fig. 10.12. The synchronizing circuit takes advantage of the high probability that a metastable flop's output will achieve a stable digital state within a single clock cycle. If the first flop's output is stable, the second flop's output will transition perfectly and present a signal with

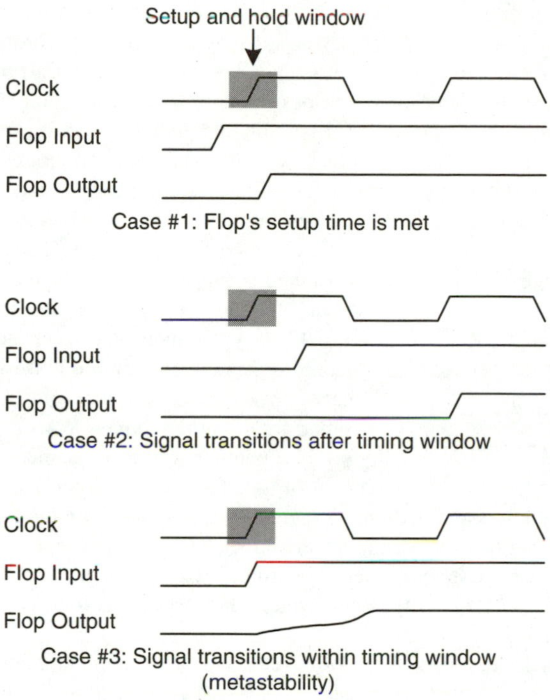

FIGURE 10.11 Flop operation with variable input timing.

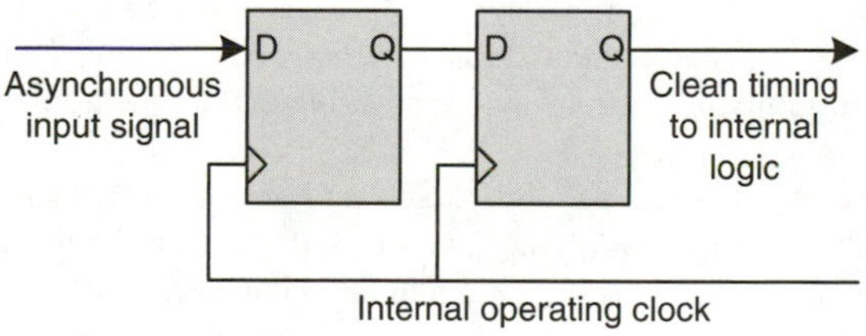

FIGURE 10.12 Synchronizing flip-flop scheme.

valid timing to the internal logic. If the first flop's output is not yet stable, chances are that it will be very close to a valid logic level, which generally should cause the second flop's output to transition cleanly. To deal with metastablility is to deal with probability. As the number of synchronizing flops is increased, the probability of a metastable state reaching the internal logic rapidly approaches zero. A general rule of thumb is that two flops reduce the probability of metastability in internal logic to practically zero.

An unavoidable downside of synchronizing flops is that they add latency to a transaction, because the internal logic will not detect an asynchronous signal's transition for two to three clock cycles after it actually transitions (when using a two-flop synchronizer). The best-case scenario of two cycles occurs when the input signal happens to transition early enough to meet the first flop's setup and hold timing. If these constraints are not met because the input signal transitions too late, the flop's output may not transition until the next clock cycle. Whether the input signal meets or misses the first flop's timing window, an extra cycle of latency is added by the second synchronizing flop.

For the synchronizing circuit to function properly, the asynchronous input must remain active for a sufficient time to be detected by the destination clock domain. If the destination clock has a 20-ns period, pulses shorter than 20 ns will have a low probability of being detected. It is best to guarantee that the pulse is active for several destination clock periods. When the two clock frequencies in a clock domain crossing are known, it is possible to determine a safe minimum pulse width. There are also situations in which one or both clock frequencies are unknown or variable. A circuit may need to run at a range of frequencies. Microprocessors are a good example of this, because they are designed to operate over a range of frequencies determined by an engineer for a specific project.

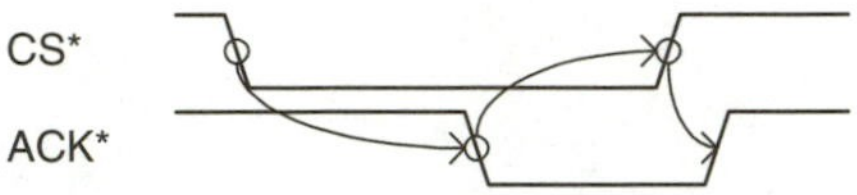

FIGURE 10.13 Four-corner microprocessor bus handshake.

In situations in which the clock frequencies are variable, asynchronous interfaces must be self-regulating to guarantee minimum pulse widths such that each clock domain can properly detect signals asserted by the other domain. *Four-corner handshaking* is a popular way to implement a self-regulated control interface. Four-corner handshaking establishes a rule that signals are not deasserted until a corresponding signal from the other entity is observed to transition, indicating that the locally generated signal has been properly detected by the remote entity. In the case of a request/acknowledge microprocessor interface, chip select is not deasserted until acknowledge is asserted, and acknowledge is not deasserted until chip select is deasserted as shown in Fig. 10.13.

Once causal control signals are properly synchronized across a clock domain boundary, arbitrary data paths can be synchronized as well. A register or data bus cannot simply be passed through a two-stage synchronizer, because there is no way of knowing if all the members of that bus passed through the flops uniformly. Even if the delays of each signal are closely matched, individual flops have minute physical differences that can cause their metastable characteristics to differ. Unlike a single control signal with just two states, an N-bit bus has 2^N states. If the control signal changes one cycle earlier or later, there is no misinterpretation of its activity. The latter case can lead to some bits transitioning before others, resulting in random values appearing at the synchronizer output for one or two clock cycles.

A basic method for synchronizing a data path is to associate it with a control signal that is passed through a synchronizer. As shown in Fig. 10.14, the data source drives a data valid signal at the same time as the data and holds the two entities stable for a minimum duration (achieved by either calculation or four-corner handshaking). The destination uses a two-flop synchronizer to move the control signal to its clock domain. The synchronization process has an inherent latency during which it is guaranteed that the data path reaches the destination logic. When the destination detects the properly synchronized control signal, it samples the data path, which by now has been stable at the destination's inputs for well over a full clock cycle and thereby meets the input flops' setup time. Metastability of the data path is avoided by an inherent synchronization time delay during which the data remains static. This scheme relies on the assumption of relatively uniform propagation delay between the data valid signal and the data path, an assumption that is achievable in most circumstances.

Synchronizing a data path with discrete control signals is a straightforward approach requiring little overhead. Its disadvantage is that multiple clock cycles are necessary to move a single unit of data. When higher transfer efficiency is necessary, a FIFO is the means of synchronizing data paths across clock domain boundaries. The FIFO must operate on two separate clocks, unlike a conventional synchronous FIFO with just a single clock. A FIFO enables data to be continuously written on one clock and read on another, subject to limitations imposed by data rate mismatches at the write

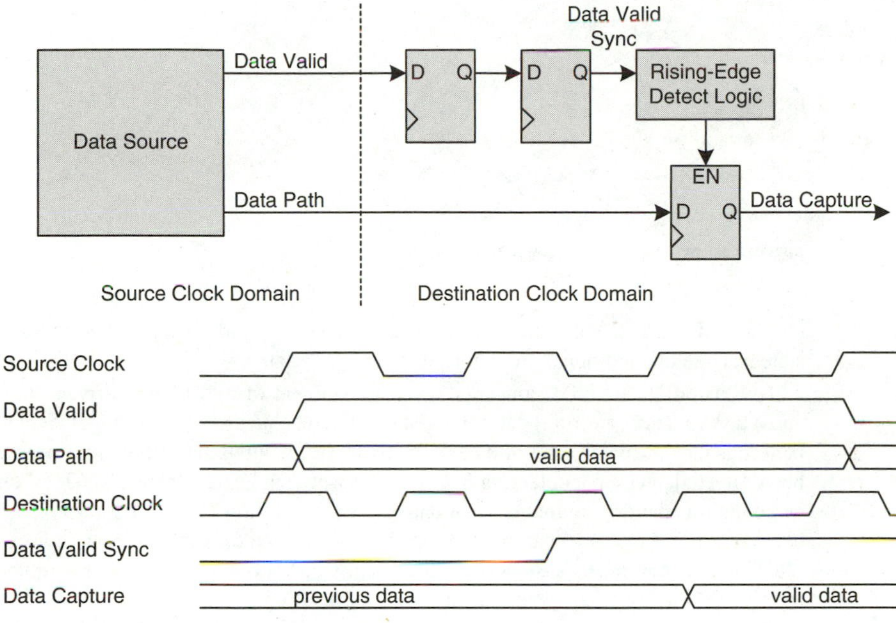

FIGURE 10.14 Data path synchronization.

and read ports. If a FIFO is used to carry 8-bit data from a 100-MHz clock to a 50-MHz clock, it is clear that the overall data rate cannot exceed the slower read rate, 50 MBps, without overflowing the FIFO and losing data. If the situation is reversed, there is still a 50-MB overall maximum data rate, and the 100-MHz read logic must inherently handle gaps in its data path, which has a 100-MB bandwidth. As long as the read and write data rates are matched over time, the dual-clock FIFO will never overflow or underflow and provides an efficient synchronization mechanism.

10.4 FINITE STATE MACHINES

Finite state machines (FSMs) are powerful design elements used to implement algorithms in hardware. A loose analogy is that an FSM is to logic design what programming is to software design. An FSM consists of a *state vector*, a set of registers, with associated logic that advances the state each clock cycle depending on external input and the current state. In this respect, an FSM is analogous to a set of software instructions that are sequenced via a microprocessor's program counter. Each state can be designed to take a different arbitrary action and branch to any other state in the same conceptual way that software branches to different program sections as its algorithm dictates. If a problem can be decomposed into a set of deterministic logical steps, that algorithm can be implemented with an FSM.

A counter is a simple example of an FSM, though its actions are very limited. Each state simply advances to the next state on each clock cycle. There is no conditional branching in a typical counter. FSMs are often represented graphically before being committed to RTL. Figure 10.15 shows a bubble diagram representation of a two-bit counter. Each state is represented by its own bubble, and arcs show the conditions that cause one state to lead to other states. An unlabeled arc is taken to mean un-

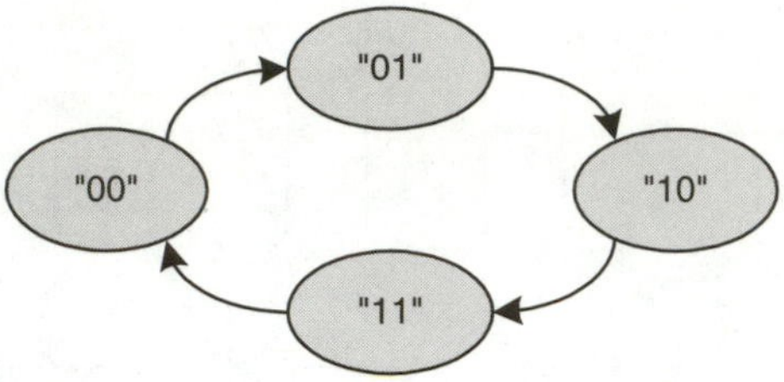

FIGURE 10.15 Two-bit counter bubble diagram.

conditional or the default if no other conditional arcs are valid. Because this is a simple counter, each state has one unconditional arc that leads to the next state.

To illustrate basic FSM concepts, consider a stream of bytes being driven onto a data interface and a task to detect a certain pattern of data within that stream when a trigger signal is asserted. That pattern is the consecutive set of data 0x01, 0x02, 0x03, and 0x04. When this consecutive pattern has been detected, an output detection flag should be set to indicate a match. The logic should never miss detecting this pattern, no matter what data precedes or follows it. After some consideration, the bubble diagram in Fig. 10.16 can be developed. It is common to name states with useful names such as "Idle" rather than binary numbers for readability. FSMs often sit in an idle or quiescent state while they wait for a triggering event that starts their execution.

The FSM waits in the Idle state until trigger is asserted. If the data value at this point is already 0x01, the FSM skips directly to the Wait02 state. Otherwise, it branches to Wait01, where it remains indefinitely until the value 0x01 is observed and it branches to Wait02. There are three possible arcs coming from Wait02. If the value 0x02 is observed, the FSM can advance to the third matching state, Wait03. If 0x01 is observed again, the FSM remains in Wait02, because 0x02 is the next value in the sequence following 0x01. If neither of these values is observed, the FSM branches back to Wait01 to start over again. Wait03 is similar, although there is no arc to remain in the same state. If the first sequence value, 0x01, is observed, the next state is Wait02. If Wait03 does succeed in immediately observing 0x03, the FSM can advance to the final matching state, Wait04. If Wait04 immediately observes 0x04, a match is declared by asserting the Match output flag, and the FSM completes its function by returning to Idle. Otherwise, like in Wait03, the FSM branches back to Wait01 or Wait02.

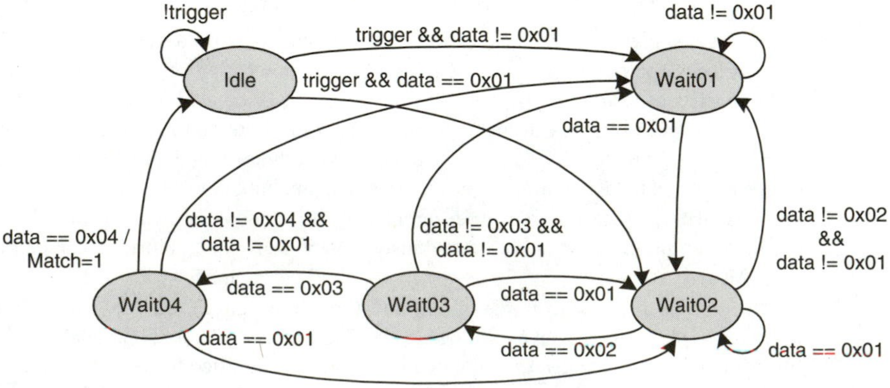

FIGURE 10.16 Pattern-matching FSM bubble diagram.

FSMs are formally classified into one of two types: *Moore* and *Mealy.* A Moore FSM is one wherein the output signals are functions of only the current state vector. A Mealy FSM is one in which the output signals are functions of both the current state as well as the inputs to the FSM. The pattern-matching FSM is a Mealy FSM, because Match is asserted when in state Wait04 and the input data is equal to 0x04. It could be converted into a Moore FSM by adding an additional state, perhaps called Matched, that would be inserted in the arc between Wait04 and Idle. The benefit of doing this would be to reduce logic complexity, because the output signals are functions of only the state vector rather than the inputs as well. However, this comes at the cost of adding an extra state. As FSMs get fairly complex, it may be too cumbersome to rigidly conform to a Moore design.

An FSM is typically coded in HDL using two blocks: a sequential (clocked) state vector block and a combinatorial state and output assignment block. The pattern-matching FSM can be written in Verilog as shown in Fig. 10.17. Rather than calling out numerical state values directly in the RTL, constants are defined to make the code more readable. Verilog supports the *`define* construct that associates a literal text string with a text identifier. The output, Match, is explicitly registered before leaving the module so that downstream logic timing will not be slowed down by incurring the penalty of the FSM logic. Registering the signal means that the output is that of a flop directly, rather than through an arbitrary number of logic gates that create additional timing delays.

In reading the FSM code, it can be seen that the state vector is three bits wide, because there are five states in total. However, 2^3 equals 8, indicating that three of the possible state vector values are invalid. Although none of these invalid values should ever arise, system reliability is increased by inserting the *default* clause into the case statement. This assignment ensures that any invalid state will result in the FSM returning to Idle. Without a default clause, any invalid state would not cause any action to be taken, and the FSM would remain in the same state indefinitely. It is good practice to insert default clauses into case statements when writing FSMs to guard against a hung condition. It is admittedly unlikely that the state vector will assume an invalid value, but if a momentary glitch were to happen, the design would benefit from a small amount of logic to restore the FSM to a valid state.

10.5 FSM BUS CONTROL

FSMs are well suited to managing arbitrarily complex bus interfaces. Microprocessors, memory devices, and I/O controllers can have interfaces with multiple states to handle arbitration between multiple requestors, wait states, and multiword transfers (e.g., a burst SDRAM). Each of these bus states can be represented by an FSM state, thereby breaking a seemingly complex problem into quite manageable pieces.

A simple example of CPU bus control is an asynchronous CPU interface with request/acknowledge signaling in which the FSM runs on a different clock from that of the CPU itself. Asynchronous CPU buses tend to implement four-corner handshaking to prevent the CPU and peripheral logic from getting out of step with one another. Two asynchronous control signals, chip select (CS_) and acknowledge (ACK_), can be managed with an FSM by breaking the distinct phases of a CPU access into separate states as shown in Fig. 10.18. The FSM waits in Idle until a synchronized CS_ is observed, at which point it executes the transaction, asserts ACK_, and then transitions to the Ack state. Once in Ack, the FSM waits until CS_ is deasserted before deasserting ACK_.

The logic is shown implemented in Verilog in Fig. 10.19. This FSM logic is simplistic in that it assumes that all transactions can be executed and acknowledged immediately. In reality, certain applications will make such rapid execution impossible. A memory controller will need to insert latency while data is fetched, for example. Such situations can be handled by adding states to the FSM that handle such cases. Rather than *Idle* immediately returning an acknowledge, it could initiate a

```
module PatternMatch (
  Clk,
  Reset_,
  Trigger,
  Data,
  Match
);

input         Clk, Reset_, Trigger;
input   [7:0] Data;
output        Match;

reg           Match, NextMatch;
reg     [2:0] State, NextState;

`define IDLE     3'h0
`define WAIT01   3'h1
`define WAIT02   3'h2
`define WAIT03   3'h3
`define WAIT04   3'h4

// sequential state vector block

always @(posedge Clk)
begin
  if (!Reset_)
    State[2:0] <= `IDLE;
  else
    State[2:0] <= NextState[2:0];
end

// register output of FSM
// could be combined with above state vector block if desired

always @(posedge Clk)
begin
  if (!Reset_)
    Match <= 1'b0;
  else
    Match <= NextMatch;
end

// FSM combinatorial logic

always @(State[2:0] or Trigger or Data[7:0])
begin
  NextState[2:0] = State[2:0]; // default values prevent latches
  NextMatch      = 1'b0;

  case (State[2:0])

    `IDLE :
       if (Trigger && (Data[7:0] == 8'h01))
         NextState[2:0] = `WAIT02;
       else if (Trigger)  // data != 0x01
         NextState[2:0] = `WAIT01;
```

FIGURE 10.17 Pattern-matching FSM logic.

```
    `WAIT01 :
      if (Data[7:0] == 8'h01)
        NextState[2:0] = `WAIT02;
      // else not required due to default assignment above

    `WAIT02 :
      if (Data[7:0] == 8'h01)
        NextState[2:0] = `WAIT02;
      else if (Data[7:0] == 8'h02)
        NextState[2:0] = `WAIT03;
      else // data != 0x01 or 0x02
        NextState[2:0] = `WAIT01;

    `WAIT03 :
      if (Data[7:0] == 8'h01)
        NextState[2:0] = `WAIT02;
      else if (Data[7:0] == 8'h03)
        NextState[2:0] = `WAIT04;
      else // data != 0x01 or 0x03
        NextState[2:0] = `WAIT01;

    `WAIT04 :
      if (Data[7:0] == 8'h01)
        NextState[2:0] = `WAIT02;
      else if (Data[7:0] == 8'h04) begin
        NextState[2:0] = `IDLE;
        NextMatch      = 1'b1;
      end
      else // data != 0x01 or 0x04
        NextState[2:0] = `WAIT01;

    default : NextState[2:0] = `IDLE; // handle unknown state vectors

  endcase
end

endmodule
```

FIGURE 10.17 (continued) Pattern-matching FSM logic.

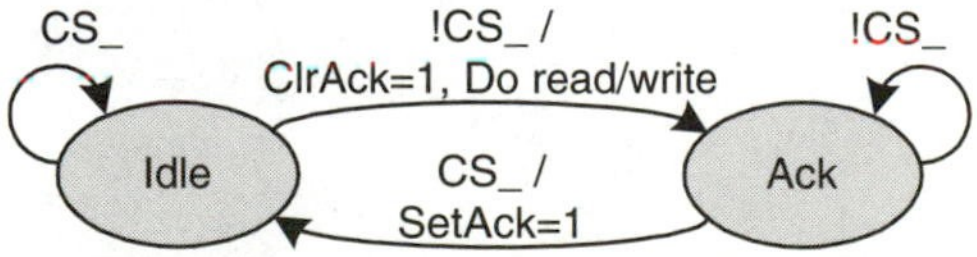

FIGURE 10.18 Asynchronous bus slave FSM bubble diagram.

secondary transaction on behalf of the microprocessor and then transition to a state that waits until the secondary transaction completes before acknowledging the microprocessor.

It may appear that writing a formal FSM when the state vector is a single flop is too cumbersome. Equivalent functionality can be obtained with a simple if…else structure instead of a case statement and separate vector assignment block. The advantages to coding the FSM in a formal manner are consistency and ease of maintenance. If the logic grows in complexity to include more states, it will be relatively easy to augment an existing FSM structure rather than rewriting one from an if…else

```
reg  State, NextState;
reg  CS_Input_, CS_Sync_, CpuAck_;

`define IDLE  1'h0
`define ACK   1'h1

always @(posedge Clk)
begin
  if (!Reset_)
    State <= `IDLE;
  else
    State <= NextState;
end

// FSM support logic: synchronization and registered output

always @(posedge Clk)
begin
  if (!Reset_) begin
    CS_Input_ <= 1'b1; // active low signals reset to high
    CS_Sync_  <= 1'b1;
    CpuAck_   <= 1'b1;
  end
  else begin
    CS_Input_ <= CS_;        // first synchronizer stage
    CS_Sync_  <= CS_Input_; // second synchronizer stage

    if (SetAck)
      CpuAck_ <= 1'b1;

    else if (ClrAck)
      CpuAck_ <= 1'b0;
  end
end

// FSM logic assumes supporting logic:
//
// CpuDataOE enables tristate output for reads
// WriteEnable enables writes to registers decoded from address inputs

always @(State or CS_Sync_ or Rd_ or Wr_)
begin
  NextState = State; // default values prevent latches

  ClrAck      = 1'b0;
  CpuDataOE   = 1'b0;
  SetAck      = 1'b0;
  WriteEnable = 1'b0;

  case (State)

    `IDLE :
       if (!CS_Sync_) begin
         NextState   = `ACK;
         ClrAck      = 1'b1;
         CpuDataOE   = !Rd_;
         WriteEnable = !Wr_;
       end

    `ACK :
       if (CS_Sync_) begin  // wait for CS_ deassertion
         NextState   = `IDLE;
         SetAck      = 1'b1;
       end

  endcase
end
```

FIGURE 10.19 Asynchronous bus slave logic.

construct. Real bus control FSMs typically requires additional states to handle more complex transaction. In the end, the decision is a matter of individual preference and style.

10.6 FSM OPTIMIZATION

FSM complexity can rapidly get out of hand when designing logic to execute a complex algorithm. Two related ways in which complexity manifests itself are excessively large FSMs and timing problems. FSMs with dozens of states can, on their own, lead to timing problems resulting from the many levels of logic necessary to map the full set of inputs and the current state vector to the full set of outputs and the next state vector. Yet, even FSMs with relatively few states can exhibit timing problems if the branch conditions get overly complex.

When complex branch conditions are combined with a very large FSM, the result can be a timing nightmare if suitable design decisions are not made from the beginning. A conceivably poor result could be logic that needs to run at 66 MHz being barely capable of 20-MHz operation. In most instances, an acceptable level of performance can be obtained by properly optimizing the FSM and its supporting logic from the start. If this is not true, chances are that a fundamental change is necessary in either the implementation technology (e.g., faster logic circuits) or the overall architectural approach to solving the problem (e.g., more parallelism in dividing a task into smaller pieces). FSM timing optimization techniques include partitioning, state vector encoding methods, and pipelining.

Proper partitioning of FSMs, and logic in general, is a major factor in successful systems development. Even the best optimization schemes can fail if the underlying FSM is improperly structured. It is usually better to design a system with multiple smaller FSMs instead of a few large ones. Smaller logic structures will tend to have fewer inputs, thereby reducing the complexity of the logic and improving its performance. When the interfaces between multiple FSMs are registered, long timing paths are broken to isolate each FSM from a timing perspective. Without registered interfaces, it is possible for multiple FSMs to form a large loop of dense logic as one feeds back to another. This would defeat a primary benefit of designing smaller FSMs.

Partitioning functionality across smaller FSMs not only makes it easier to improve their timing, it also makes it easier to design and debug the logic. Each smaller section of logic can be assigned to a different engineer for concurrent development, or the sections can be developed serially in a progressive manner by designing and testing each element sequentially. By the time the entire design has been completed, the daunting task of simulating everything at once can be substantially minimized, because each section has already been tested individually. Bugs are likely to arise when all the pieces are put together, but the overall magnitude of the debugging process should be reduced.

State-vector encoding methods are most often considered as a choice between two options: *binary encoding* and *one-hot encoding*. A binary encoded FSM is equivalent to the examples presented earlier in this chapter: a state vector is chosen with N flops such that 2^N is greater than or equal to the total number of states in the FSM. Each state is assigned a unique value in the range of 0 to $2^N - 1$. A one-hot FSM allocates one state flop for each unique state and adheres to a rule that only one flop is set (hot) on any single cycle. The benefits of a one-hot FSM include reduced complexity of output logic and reduced power consumption. Output logic complexity is often reduced by one-hot encoding, because the entire state vector does not have to be decoded. Instead, only those state flops that directly map to an output signal are included in the Boolean expression. Power consumption is reduced, because only two flops change state at a time (the old state flop transitions from high to low, and the new state flop transitions from low to high) instead of many or all state bits. The decision to implement one-hot encoding varies according to the size of the FSM and the constraints of the implementation technology. Beyond a certain size, one-hot encoding becomes too unwieldy and may actually result in more logic than a binary encoded version. Different technologies (e.g., custom versus programmable logic) provide more or less flexibility in how flops and logic gates are placed together. In many cases, custom logic (e.g., an ASIC) is more efficient with a binary encoded FSM, whereas programmable logic performs better with moderately sized one-hot FSMs.

An FSM can be explicitly written as one-hot, but most leading HDL synthesis tools (e.g., Cadence BuildGates, Exemplar Leonardo Spectrum, Synplicity Synplify, or Synopsys Design Compiler and FPGA Express) have options to automatically evaluate a binary encoded FSM in either its native encoding or a one-hot scheme and then pick the best result. These tools save the engineer the manual time of trying various encoding styles. Better yet, if an FSM works well in one-hot encoding at one phase during the design, and then better as binary encoded due to later design changes, the synthesis tool will automatically change the encoding style without manual intervention.

If a situation arises in which it is advantageous to manually write an FSM using one-hot encoding, the technique is straightforward. Once the number of required state flops has been counted, RTL can be written as shown in Fig. 10.20 using the previous example of an asynchronous bus slave FSM.

There are two significant syntactical elements that enable the one-hot FSM encoding. The first is using the case (1) construct wherein the individual test cases become the individual state vector bits as indexed by the defined state identifiers. Next is the actual NextState assignment, made by left shifting the constant 1 by the state identifiers. Left shifting by the state vector index of the desired next state causes that next state's bit, and only that bit, to be set while the other bits are cleared.

```
// Define bit positions of each state in the set of state bits

`define IDLE  0  // state = 01
`define ACK   1  // state = 10

always @(State[1:0] or CS_Sync_ or Rd_ or Wr_)
begin
  NextState[1:0] = State[1:0];

  ClrAck      = 1'b0;
  CpuDataOE   = 1'b0;
  SetAck      = 1'b0;
  WriteEnable = 1'b0;

  case (1)

    State[`IDLE] :
       if (!CS_Sync_) begin
         NextState[1:0] = 1 << `ACK;
         ClrAck         = 1'b1;
         CpuDataOE      = !Rd_;
         WriteEnable    = !Wr_;
       end

    State[`ACK] :
       if (CS_Sync_) begin
         NextState[1:0] = 1 << `IDLE;
         SetAck         = 1'b1;
       end

  endcase
end
```

FIGURE 10.20 One-hot encoded FSM.

10.7 PIPELINING

Pipelining is a significant method of improving FSM timing in a way similar to normal logic. A complex FSM contains a large set of inputs that feeds the next state logic. As the number of branch variables across the entire FSM increases, long timing paths quickly develop. It is advantageous to precompute many branch conditions and reduce a condition to a binary variable: true or false. In doing so, the branch variables are moved away from the FSM logic and behind a pipeline flop. The logic delay penalty of evaluating the branch condition is hidden by the pipeline flop. This simplifies the FSM logic because it has to evaluate only the true/false condition result, which is a function of one variable, instead of the whole branch condition, which can be a function of many variables.

An example of using pipelining to improve FSM timing is whereby an FSM is counting events and waits in a particular state until the event has occurred N times, at which point the FSM branches to a new state. In this situation, a counter lies off to the side of the FSM, and the FSM asserts a signal to increment the counter each time a particular event occurs. Without pipelining, the FSM would compare the counter bits against the constant, N, as a branch condition. If the counter is eight bits wide, there is a function of eight variables plus any other qualifiers. If the counter is 16 bits wide, at least 16 variables are present. Many such Boolean equations in the same FSM can lead to timing problems.

Pipelining can be implemented by performing the comparison outside of the FSM with the true/false result being stored in a flop that is directly evaluated by the FSM, thereby reducing the evaluation to a function of only one variable. The trick to making this work properly is that the latency of the pipelined comparison needs to be built into both the comparison logic and the FSM itself.

In a counter situation without pipelining, the FSM increments the counter when an event occurs and simultaneously evaluates the current count value to determine if N events have occurred. Therefore, the FSM would have to make its comparison against N – 1 instead of against N if it wanted to react at the proper time. When the counter is at N – 1 and the event occurs, the FSM knows that this event is the Nth event in the sequence.

Inserting a pipeline flop into the counter comparison logic adds a cycle of latency to the comparison. The FSM cannot simply look ahead an additional cycle to N – 2, because it cannot tell the future: it does not ordinarily know if an event will occur in the next cycle. This problem can be addressed in the pipelined comparison logic. The comparison logic must be triggered at N – 2 so that the FSM can observe the asserted signal on the following cycle when the counter is at N – 1, which enables the FSM to operate as before to recognize the Nth event. Because an ultimate comparison to N – 1 is desired, the pipelined logic can evaluate the expression, "If count equals N – 2 and increment is true, then count will equal N – 1 in the next cycle." In the next cycle, the FSM will have the knowledge that count equals N – 1 from a single flop, and it can qualify this in the same way it did without the pipelining with the end same result. Verilog code fragments to implement this scheme are shown in Fig. 10.21, where the FSM completes its task when 100 events have been observed.

Pipelining should be carefully considered when an FSM is evaluating long bit vectors if it is believed that proper timing will be difficult to achieve. Counter comparisons to constants are not the only candidates for pipelining. Valid pipelining candidates include counter comparisons to other registers and arithmetic embedded within branch conditions. Arithmetic expressions such as one shown in Fig. 10.22 can add long timing paths to an already complex FSM.

Adding two vectors and then comparing their sum to a third vector is a substantial quantity of logic. Such situations are not uncommon in an FSM that must negotiate complex algorithms. It would be highly advantageous to perform the arithmetic and comparison in a pipelined manner so that the FSM would have to evaluate only a single flop. The complexity involved in pipelining such an expression is highly dependent on the application's characteristics. Pipelining can get tricky when it becomes necessary to look ahead one cycle and deal with possible exceptions to the look-ahead

```
`define COMPARE_VALUE 7´d98  // N-2 due to pipeline delay

// counter logic with pipelined comparison logic

always @(posedge Clk)
begin
  if (ClrCount) begin
    CountDone  <= 1´b0;
    Count[6:0] <= 8´h1;
  end
  else if (IncCount) begin
    CountDone  <= Count[6:0] == `COMPARE_VALUE;
    Count[6:0] <= Count[6:0] + 1;
  end
end

// Partial FSM logic

...
    `WAIT_COUNT :
       if (Event && CountDone) // N-1 CountDone flag qualified with
         NextState[3:0] = `FINISH;  // Event, indicating Nth event
       else if (Event)
         IncCount = 1´b1;
         ...
       end
...
```

FIGURE 10.21 Pipelined count comparison logic.

```
    `COMPLEX_STATE :
       if ( (BytesFound[15:0] + BytesOffset[7:0]) >
                  BytesThreshold[15:0] )
         NextState[3:0] = `MORE_MATH;
         ...
       else
         ...
       end
...
```

FIGURE 10.22 Arithmetic expression in FSM branch condition.

process. Despite this complexity, there are situations in which this is the only way of obtaining the desired performance.

Pipelining is perhaps most tricky when the logic involved contains a feedback path through the FSM. This feedback path can be of the type discussed previously whereby the FSM is detecting an event, incrementing a counter, and then relying on that count value to take further action. There are other situations in which no feedback path exists. An FSM may be called on to process an incoming data stream over which it has no direct control. As in the earlier pattern matching example, the FSM may need to look for certain data values and take action if they are located. In Fig. 10.17, the FSM directly evaluated the eight-bit data stream in searching for the values 0x01, 0x02, 0x03, and 0x04. This logic can be optimized by inserting pattern-matching flops between the data input port and the FSM as shown in Fig. 10.23. In doing so, the FSM no longer needs to evaluate the entire eight-bit vector but instead can test a single flop for each pattern.

```
// pipelined pattern matching flops

always @(posedge Clk)
begin
  if (!Reset_) begin
    Data01 <= 1´b0;
    Data02 <= 1´b0;
    Data03 <= 1´b0;
    Data04 <= 1´b0;
  end
  else begin
    Data01 <= Data[7:0] == 8´h01;
    Data02 <= Data[7:0] == 8´h02;
    Data03 <= Data[7:0] == 8´h03;
    Data04 <= Data[7:0] == 8´h04;
  end
end

always @(State[2:0] or Trigger or Data01 or Data02 or Data03 or Data04)
begin
  NextState[2:0] = State[2:0];
  NextMatch      = 1´b0;

  case (State[2:0])

    `IDLE :
       if (Trigger && Data01) // Instead of Data[7:0] == 8´h01
         NextState[2:0] = `WAIT02;
...
```

FIGURE 10.23 Pattern-matching logic with pipelining.

When pipeline flops are inserted into a data path, care must be taken to apply a consistent delay to the entire data path. This example does not use the data path for any purpose other than pattern matching; therefore, there is no need to perform further operations on the data once it is checked for the relevant patterns. If, however, the logic performed pattern matching and then manipulated the data based on that matching (e.g., recognize the start of a data packet and then store the packet into a memory), it would be critical to keep the pipelined pattern-matching flops coincident with the data that they represent. Failure to do so would result in detecting the pattern late with respect to the data stream, thereby missing the packet's initial data. The data path can be delayed along with the pipelined logic by passing it through a register also. If the previous example did process the data, logic reading `DataPipe[7:0] <= Data[7:0]` could be placed into the same always block as the pattern matching logic. Subsequent references to Data[7:0] would be replaced by references to DataPipe[7:0]. This way, when Data[7:0] equals 0x01, Data01 will be set on the next cycle, and DataPipe[7:0] will be loaded with the value 0x01 on the next cycle as well.

CHAPTER 11
Programmable Logic Devices

Programmable logic is the means by which a large segment of engineers implement their custom logic, whether that logic is a simple I/O port or a complex state machine. Most programmable logic is implemented with some type of HDL that frees the engineer from having to derive and minimize Boolean expressions each time a new logical relationship is designed. The advantages of programmable logic include rapid customization with relatively limited expense invested in tools and support.

The widespread availability of flexible programmable logic products has brought custom logic design capabilities to many individuals and smaller companies that would not otherwise have the financial and staffing resources to build a fully custom IC. These devices are available in a wide range of sizes, operating voltages, and speeds, which all but guarantees that a particular application can be closely matched with a relevant device. Selecting that device requires some research, because each manufacturer has a slightly different specialty and range of products.

Programmable logic technology advances rapidly, and manufacturers are continually offering devices with increased capabilities and speeds. After completing this chapter and learning about the basic types of devices that are available, it is recommended that you to browse through the latest manufacturers' data sheets to get updated information. Companies such as Altera, Atmel, Cypress, Lattice, QuickLogic, and Xilinx provide detailed data sheets on their web sites and also tend to offer bundled development software for reasonable prices.

11.1 CUSTOM AND PROGRAMMABLE LOGIC

Beyond using discrete 7400 ICs, custom logic is implemented in larger ICs that are either manufactured with custom masks at a factory or programmed with custom data images at varying points after fabrication. Custom ICs, or *application specific integrated circuits* (ASICs), are the most flexible option because, as with anything custom, there are fewer constraints on how application specific logic is implemented. Because custom ICs are tailored for a specific application, the potential exists for high clock speeds and relatively low unit prices. The disadvantages to custom ICs are long and expensive development cycles and the inability to make quick logic changes. Custom IC development cycles are long, because a design must generally be frozen in a final state before much of the silicon layout and circuit design work can be completed. Engineering charges for designing a custom mask set (not including the logic design work) can range from $50,000 to well over $1 million, depending on the complexity. Once manufactured, the logic can't simply be altered, because the logic configuration is an inherent property of the custom design. If a bug is found, the time and money to alter the mask set can approach that of the initial design itself.

Programmable logic devices (PLDs) are an alternative to custom ASICs. A PLD consists of general-purpose logic resources that can be connected in many permutations according to an engineer's logic design. This programmable connectivity comes at the price of additional, hidden logic that makes logic connections within the chip. The main benefit of PLD technology is that a design can be rapidly loaded into a PLD, bypassing the time consuming and expensive custom IC development process. It follows that if a bug is found, a fix can be implemented very quickly and, in many cases, reprogrammed into the existing PLD chip. Some PLDs are one-time programmable, and some can be reprogrammed in circuit.

The disadvantage of PLDs is the penalty paid for the hidden logic that implements the programmable connectivity between logic gates. This penalty manifests itself in three ways: higher unit cost, slower speeds, and increased power consumption. Programmable gates cost more than custom gates, because, when a programmable gate is purchased, that gate plus additional connectivity overhead is actually being paid for. Propagation delay is an inherent attribute of all silicon structures, and the more structures that are present in a path, the slower the path will be. It follows that a programmable gate will be slower than a custom gate, because that programmable gate comes along with additional connectivity structures with their own timing penalties. The same argument holds true for power consumption.

Despite the downside of programmable logic, the technology as a whole has progressed dramatically and is extremely popular as a result of competitive pricing, high performance levels, and, especially, quick time to market. Time to market is an attribute that is difficult to quantify but one that is almost universally appreciated as critical to success. PLDs enable a shorter development cycle, because designs can be prototyped rapidly, the bugs worked out, and product shipped to a customer before some ASIC technologies would even be in fabrication. Better yet, if a bug is found in the field, it may be fixable with significantly less cost and disruption. In the early days of programmable logic, PLDs could not be reprogrammed, meaning that a bug could still force the recall of product already shipped. Many modern reprogrammable PLDs allow hardware bugs to be fixed in the field with a software upgrade consisting of a new image that can be downloaded to the PLD without having to remove the product from the customer site.

Cost and performance are probably the most debated trade-offs involved in using programmable logic. The full range of applications in which PLDs or ASICs are considered can be broadly split into three categories as shown in Fig. 11.1. At the high end of technology, there are applications in which an ASIC is the only possible solution because of leading edge timing and logic density requirements. In the mid range, clock frequencies and logic complexity are such that a PLD is capable of solving the problem, but at a higher unit cost than an ASIC. Here, the decision must be made between flexibility and time to market versus lowest unit cost. At the low end, clock frequencies and logic density requirements are far enough below the current state of silicon technology that a PLD may meet or even beat the cost of an ASIC.

It may sound strange that a PLD with its overhead can ever be less expensive than a custom chip. The reasons for this are a combination of silicon die size and volume pricing. Two of the major factors in the cost of fabricating a working silicon die are its size and manufacturing yield. As a die gets smaller, more of them can be fabricated at the same time on the same wafer using the same resources. IC manufacturing processes are subject to a certain yield, which is the percentage of working dice obtained from an overall lot of dice. Some dice develop microscopic flaws during manufacture that make them unusable. Yield is a function of many variables, including the reliability of uniformly manufacturing a certain degree of complexity given the prevailing state of technology at a point in time. From these two points, it follows that a silicon chip will be less expensive to manufacture if it is both small and uses a technology process that is mature and has a high yield.

At the low end of speed and density, a small PLD and a small ASIC may share the same mature technology process and the same yield characteristics, removing yield as a significant variable in

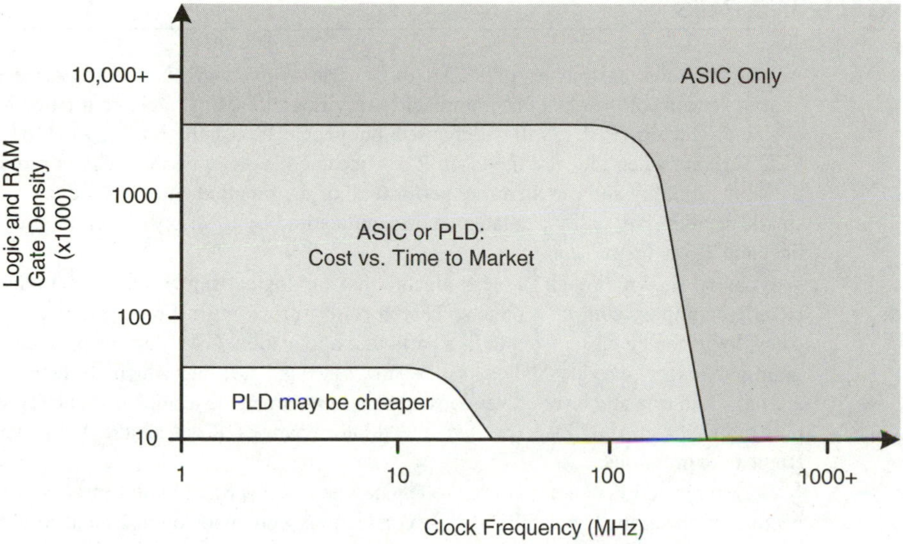

FIGURE 11.1 PLDs vs. ASICs circa 2003.

their cost differential. Likewise, raw packaging costs are likely to be comparable because of the maturation of stable packaging materials and technologies. The cost differential comes down to which solution requires the smaller die and how the overhead costs of manufacturing and distribution are amortized across the volume of chips shipped.

Die size is function of two design characteristics: how much logic is required and how many I/O pins are required. While the size of logic gates has decreased by orders of magnitude over time, the size of I/O pads, the physical structures that packaging wires connect to, has not changed by the same degree. There are nonscalable issues constraining pad size, including physical wire bonding and current drive requirements. I/O pads are often placed on the perimeter of a die. If the required number of I/O pads cannot be placed along the existing die's perimeter, the die must be enlarged even if not required by the logic. ICs can be considered as being balanced, logic limited, or pad limited. A balanced design is optimal, because silicon area is being used efficiently by the logic and pad structures. A logic-limited IC's silicon area is dominated by the internal logic requirements. At the low end being presently discussed, being logic limited is not a concern because of the current state of technology. Pad-limited designs are more of a concern at the low end, because the chip is forced to a certain minimum size to support a minimum number of pins.

Many low-end logic applications end up being pad limited as the state of silicon process technology advances and more logic gates can be squeezed into ever smaller areas. The logic shrinks, but the I/O pads do not. Once an IC is pad limited, ASIC and CPLD implementations may use the same die size, removing it as a cost variable. This brings us back to the volume pricing and distribution aspects of the semiconductor business. If two silicon manufacturers are fabricating what is essentially the same chip (same size, yield, and package), who will be able to offer the lowest final cost? The comparison is between a PLD vendor that turns out millions of the exact same chip each year versus an ASIC vendor that can manufacture only as many of your custom chips that you are willing to buy. Is an ASIC's volume 10,000 units per year? 100,000? One million? With all other factors being equal, the high-volume PLD vendor has the advantage, because the part being sold is not custom but a mass-produced generic product.

11.2 GALS AND PALS

Among the most basic types of PLDs are Generic Array Logic™ (GAL) devices.* GALs are enhanced variants of the older Programmable Array Logic™ (PAL) architecture that is now essentially obsolete. The term PAL is still widely used, but people are usually referring to GAL devices or other PLD variants when they use the term. PALs became obsolete, because GALs provide a superset of their functionality and can therefore perform all of the functions that PALs did. GALs are relatively small, inexpensive, easily available, and manufactured by a variety of vendors (e.g., Cypress, Lattice, and Texas Instruments).

It can be shown through Boolean algebra that any logical expression can be represented as an arbitrarily complex sum of products. Therefore, by providing a programmable array of AND/OR gates, logic can be customized to fit a particular application. GAL devices provide an extensive programmable array of wide AND gates, as shown in Fig. 11.2, into which all the device's input terms are fed. Both true and inverted versions of each input are made available to each AND gate. The outputs of groups of AND gates (products) feed into separate OR gates (sums) to generate user-defined Boolean expressions.

Each intersection of a horizontal AND gate line and a vertical input term is a programmable connection. In the early days of PLDs, these connections were made by fuses that would literally have to be blown with a high voltage to configure the device. Fuse-based devices were not reprogrammable;

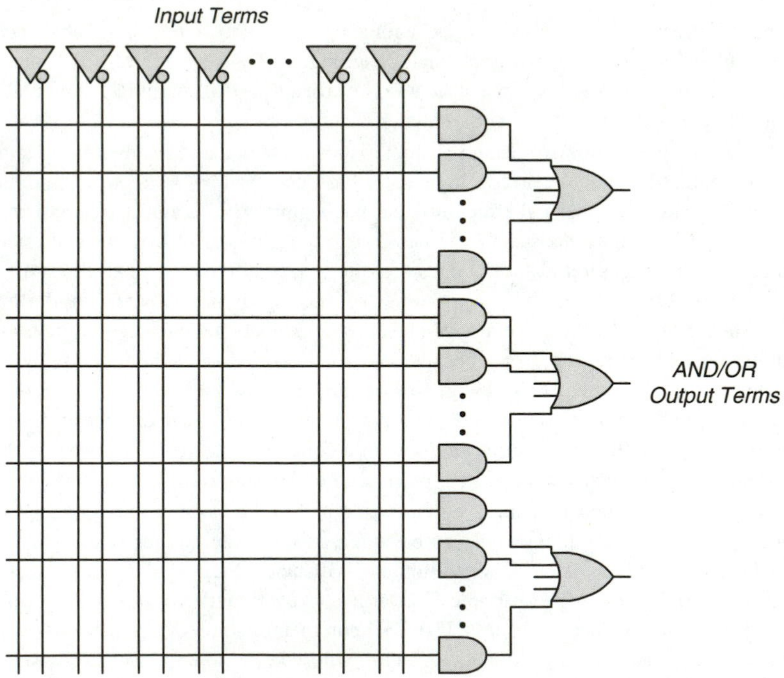

FIGURE 11.2 GAL/PAL AND/OR structure.

---

* GAL, Generic Array Logic, PAL, and Programmable Array Logic are trademarks of Lattice Semiconductor Corporation.

once a microscopic fuse is blown, it cannot be restored. Today's devices typically rely on EEPROM technology and CMOS switches to enable nonvolatile reprogrammability. However, fuse-based terminology remains in use for historical reasons. The default configuration of a connection emulates an intact fuse, thereby connecting the input term to the AND gate input. When the connection is blown, or programmed, the AND input is disconnected from the input term and pulled high to effectively remove that input from the Boolean expression. Customization of a GAL's programmable AND gate is conceptually illustrated in Fig. 11.3.

With full programmability of the AND array, the OR connections can be hard wired. Each GAL device feeds a differing number of AND terms into the OR gates. If one or more of these AND terms are not needed by a particular Boolean expression, those unneeded AND gates can be effectively disabled by forcing their outputs to 0. This is done by leaving an unneeded AND gate's inputs unprogrammed. Remember that inputs to the AND array are provided in both true and complement versions. When all AND connections are left intact, multiple expressions of the form $A\&\overline{A} = 0$ result, thereby forcing that gate's output to 0 and rendering it nonparticipatory in the OR function.

The majority of a GAL's logic customization is performed by programming the AND array. However, selecting flip-flops, OR/NOR polarities, and input/output configurations is performed by programming a configurable I/O and feedback structure called a *macrocell*. The basic concept behind a macrocell is to ultimately determine how the AND/OR Boolean expression is handled and how the macrocell's associated I/O pin operates. A schematic view of a GAL macrocell is shown in Fig. 11.4, although some GALs may contain macrocells with slightly different capabilities. Multiplexers determine the polarity of the final OR/NOR term, regardless of whether the term is registered and whether the feedback signal is taken directly at the flop's output or at the pin. Configuring the macrocell's output enable determines how the pin behaves.

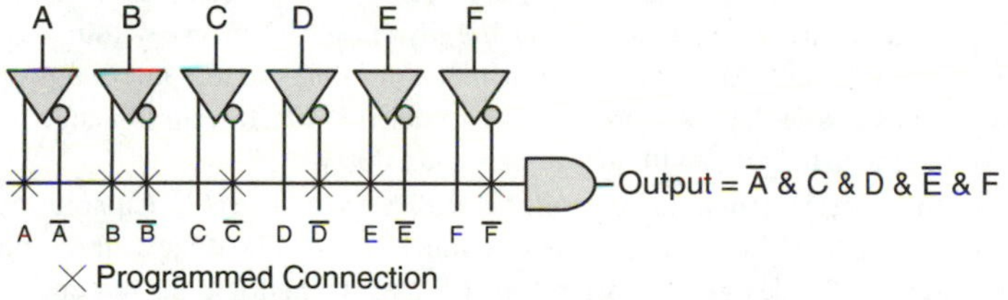

FIGURE 11.3 Programming AND input terms.

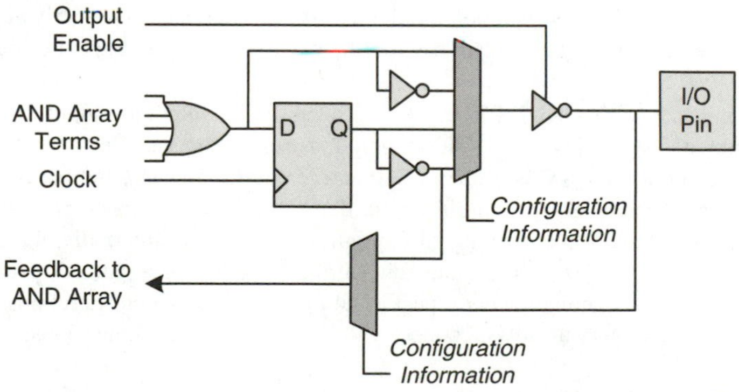

FIGURE 11.4 GAL macrocell.

There are two common GAL devices, the 16V8 and the 22V10, although other variants exist as well. They contain eight and ten macrocells each, respectively. The 16V8 provides up to 10 dedicated inputs that feed the AND array, whereas the 22V10 provides 12 dedicated inputs. One of the 22V10's dedicated inputs also serves as a global clock for any flops that are enabled in the macrocells. Output enable logic in a 22V10 is evaluated independently for each macrocell via a dedicated AND term. The 16V8 is somewhat less flexible, because it cannot arbitrarily feed back all macrocell outputs depending on the device configuration. Additionally, when configured for registered mode where macrocell flops are usable, two dedicated input pins are lost to clock and output enable functions.

GALs are fairly low-density PLDs by modern standards, but their advantages of low cost and high speed are derived from their small size. Implementing logic in a GAL follows several basic steps. First, the logic is represented in either graphical schematic diagram or textual (HDL) form. This representation is converted into a netlist using a translation or synthesis tool. Finally, the netlist is fitted into the target device by mapping individual gate functions into the programmable AND array. Given the fixed AND/OR structure of a GAL, fitting software is designed to perform logic optimizations and translations to convert arbitrary Boolean expressions into sum-of-product expressions. The result of the fitting process is a programming image, also called a *fuse map*, that defines exactly which connections, or fuses, are to be programmed and which are to be left at their default state. The programming image also contains other information such as macrocell configuration and other device-specific attributes.

Modern PLD development software allows the back-end GAL synthesis and fitting process to proceed without manual intervention in most cases. The straightforward logic flow through the programmable AND array reduces the permutations of how a given Boolean expression can be implemented and results in very predictable logic fitting. An input signal propagates through the pin and pad structure directly into the AND array, passes through just two gates, and can then either feed a macrocell flop or drive directly out through an I/O pin. Logic elements within a GAL are close to each other as a result of the GAL's small size, which contributes to low internal propagation delays. These characteristics enable the GAL architecture to deliver very fast timing specifications, because signals follow deterministic paths with low propagation delays.

GALs are a logic implementation technology with very predictable capabilities. If the desired logic cannot fit within the GAL, there may not be much that can be done without optimizing the algorithm or partitioning the design across multiple devices. If the logic fits but does not meet timing, the logic must be optimized, or a faster device must be found. Because of the GAL's basic fitting process and architecture, there isn't the same opportunity of tweaking the device as can be done with more complex PLDs. This should not be construed as a lack of flexibility on the part of the GAL. Rather, the GAL does what it says it does, and it is up to the engineer to properly apply the technology to solve the problem at hand. It is the simplicity of the GAL architecture that is its greatest strength.

Lattice Semiconductor's GAL22LV10D-4 device features a worst-case input-to-output combinatorial propagation delay of just 4 ns.* This timing makes the part suitable for address decoding on fast microprocessor interfaces. The same 22V10 part features a 3-ns t_{CO} and up to 250-MHz operation. The t_{CO} specification is a pin-to-pin metric that includes the propagation delays of the clock through the input pin and the output signal through the output pin. Internally, the actual flop itself exhibits a faster t_{CO} that becomes relevant for internal logic feedback paths. Maximum clock frequency specifications are an interesting aspect of all PLDs and some consideration. These specifications are best-case numbers usually obtained with minimal logic configurations. They may define

* GAL22LV10D, 22LV10_04, Lattice Semiconductor, 2000, p. 7.

the highest toggle rate of the device's flops, but synchronous timing analysis dictates that there is more to f_{MAX} than the flop's t_{SU} and t_{CO}. Propagation delay of logic and connectivity between flops is of prime concern. The GAL architecture's deterministic and fast logic feedback paths reduces the added penalty of internal propagation delays. Lattice's GAL22LV10D features an internal clock-to-feedback delay of 2.5 ns, which is the combination of the actual flop's t_{CO} plus the propagation delay of the signal back through the AND/OR array. This feedback delay, when combined with the flop's 3-ns t_{SU}, yields a practical f_{MAX} of 182 MHz when dealing with most normal synchronous logic that contains feedback paths (e.g., a state machine).

11.3 CPLDS

Complex PLDs, or CPLDs, are the mainstream macrocell-based PLDs in the industry today, providing logic densities and capabilities well beyond those of a GAL device. GALs are flexible for their size because of the large programmable AND matrix that defines logical connections between inputs and outputs. However, this anything-to-anything matrix makes the architecture costly to scale to higher logic densities. For each macrocell that is added, both matrix dimensions grow as well. Therefore, the AND matrix increases in a square function of the I/O terms and macrocells in the device. CPLD vendors seek to provide a more linear scaling of connectivity resources to macrocells by implementing a segmented architecture with multiple fixed-size GAL-style logic blocks that are interconnected via a central switch matrix as shown in Fig. 11.5. Like a GAL, CPLDs are typically manufactured with EEPROM configuration storage, making their function nonvolatile. After programming, a CPLD will retain its configuration and be ready for operation when power is applied to the system.

Each individual logic block is similar to a GAL and contains its own programmable AND/OR array and macrocells. This approach is scalable, because the programmable AND/OR arrays remain fixed in size and small enough to fabricate economically. As more macrocells are integrated onto the same chip, more logic blocks are placed onto the chip instead of increasing the size of individual logic blocks and bloating the AND/OR arrays. CPLDs of this type are manufactured by companies including Altera, Cypress, Lattice, and Xilinx.

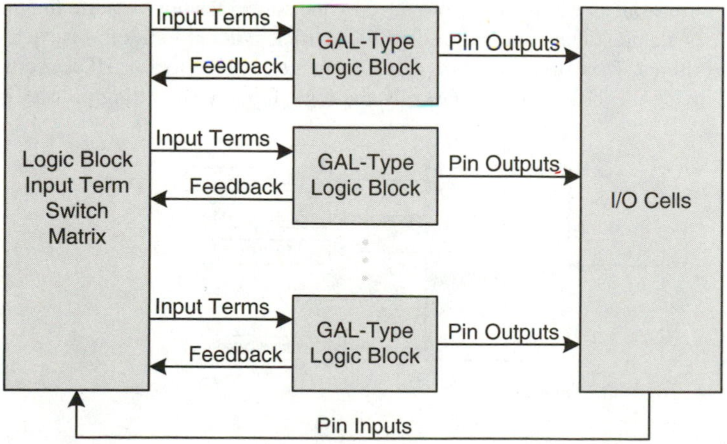

FIGURE 11.5 Typical CPLD architecture.

Generic user I/O pins are bidirectional and can be configured as inputs, outputs, or both. This is in contrast to the dedicated power and test pins that are necessary for operation. There are as many potential user I/O pins as there are macrocells, although some CPLDs may be housed in packages that do not have enough physical pins to connect to all the chip's I/O sites. Such chips are intended for applications that are logic limited rather than pad limited.

Because the size of each logic block's AND array is fixed, the block has a fixed number of possible inputs. Vendor-supplied fitting software must determine which logical functions are placed into which blocks and how the switch matrix connects feedback paths and input pins to the appropriate block. The switch matrix does not grow linearly as more logic blocks are added. However, the impact of the switch matrix's growth is less than what would result with an ever expanding AND matrix. Each CPLD family provides a different number of switched input terms to each logic block.

The logic blocks share many characteristics with a GAL, as shown in Fig. 11.6, although additional flexibility is added in the form of product term sharing. Limiting the number of product terms in each logic block reduces device complexity and cost. Some vendors provide just five product terms per macrocell. To balance this limitation, which could impact a CPLD's usefulness, product term sharing resources enable one macrocell to borrow terms from neighboring macrocells. This borrowing usually comes at a small propagation delay penalty but provides necessary flexibility in handling complex Boolean expressions with many product terms. A logic block's macrocell contains a flip-flop and various configuration options such as polarity and clock control. As a result of their higher logic density, CPLDs contain multiple global clocks that individual macrocells can choose from, as well as the ability to generate clocks from the logic array itself.

Xilinx is a vendor of CPLD products and manufactures a family known as the XC9500. Logic blocks, or function blocks in Xilinx's terminology, each contain 18 macrocells, the outputs of which feed back into the switch matrix and drive I/O pins as well. XC9500 CPLDs contain multiples of 18 macrocells in densities from 36 to 288 macrocells. Each function block gets 54 input terms from the switch matrix. These input terms can be any combination of I/O pin inputs and feedback terms from other function blocks' macrocells.

Like a GAL, CPLD timing is very predictable because of the deterministic nature of the logic blocks' AND arrays and the input term switch matrix. Xilinx's XC9536XV-3 features a maximum pin-to-pin propagation delay of 3.5 ns and a t_{CO} of 2.5 ns.* Internal logic can run as fast as 277 MHz with feedback delays included, although complex Boolean expressions likely reduce this f_{MAX} because of product term sharing and feedback delays through multiple macrocells.

CPLD fitting software is typically provided by the silicon vendor, because the underlying silicon architectures are proprietary and not disclosed in sufficient detail for a third party to design the necessary algorithms. These tools accept a netlist from a schematic tool or HDL synthesizer and automatically divide the logic across macrocells and logic blocks. The fitting process is more complex

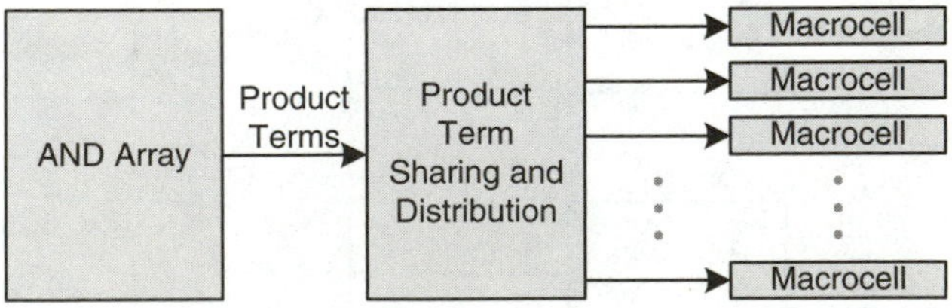

FIGURE 11.6 CPLD logic block.

* XC9536XV, DS053 (v2.2), Xilinx, August 2001, p. 4.

than for a GAL; not every term within the CPLD can be fed to each macrocell because of the segmented AND array structure. Product term sharing places restrictions on neighboring macrocells when Boolean expressions exceed the number of product terms directly connected to each macrocell. The fitting software first reduces the netlist to a set of Boolean expressions in the form that can be mapped into the CPLD and then juggles the assignment of macrocells to provide each with its required product terms. Desired operating frequency influences the placement of logic because of the delay penalties of sharing product terms across macrocells. These trade-offs occur at such a low level that human intervention is often impractical.

CPLDs have come to offer flexibility advantages beyond just logic implementation. As digital systems get more complex, logic IC supply voltages begin to proliferate. At one time, most systems ran on a single 5-V supply. This was followed by 3.3-V systems, and it is now common to find systems that operate at multiple voltages such as 3.3 V, 2.5 V, 1.8 V, and 1.5 V. CPLDs invariably find themselves designed into mixed-voltage environments for the purposes of interface conversion and bus management. To meet these needs, many CPLDs support more than one I/O voltage standard on the same chip at the same time. I/O pins are typically divided into banks, and each bank can be independently selected for a different I/O voltage.

Most CPLDs are relatively small in logic capacity because of the desire for very high-speed operation with deterministic timing and fitting characteristics at a reasonable cost. However, some CPLDs have been developed far beyond the size of typical CPLDs. Cypress Semiconductor's Delta39K200 contains 3,072 macrocells with several hundred kilobits of user-configurable RAM.* The architecture is built around clusters of 128 macrocell logic groups, each of which is similar in nature to a conventional CPLD. In a similar way that CPLDs add an additional hierarchical connectivity layer on top of multiple GAL-type logic blocks, Cypress has added a layer on top of multiple CPLD-type blocks. Such large CPLDs may have substantial benefits for certain applications. Beyond several hundred macrocells, however, engineers have tended to use larger and more scalable FPGA technologies.

11.4 FPGAS

CPLDs are well suited to applications involving control logic, basic state machines, and small groups of read/write registers. These control path applications typically require a small number of flops. Once a task requires many hundreds or thousands of flops, CPLDs rapidly become impractical to use. Complex applications that manipulate and parse streams of data often require large quantities of flops to serve as pipeline registers, temporary data storage registers, wide counters, and large state machine vectors. Integrated memory blocks are critical to applications that require multiple FIFOs and data storage buffers. *Field programmable gate arrays* (FPGAs) directly address these data path applications.

FPGAs are available in many permutations with varying feature sets. However, their common defining attribute is a fine-grained architecture consisting of an array of small logic cells, each consisting of a flop, a small lookup table (LUT), and some supporting logic to accelerate common functions such as multiplexing and arithmetic carry terms for adders and counters. Boolean expressions are evaluated by the LUTs, which are usually implemented as small SRAM arrays. Any function of four variables, for example, can be implemented in a 16 × 1 SRAM when the four variables serve as the index into the memory. There are no AND/OR arrays as in a CPLD or GAL. All Boolean functions

* Delta39K ISR CPLD Family, Document #38-03039 Rev. *.C, Cypress Semiconductor, December 2001, p. 1.

are implemented within the logic cells. The cells are arranged on a grid of routing resources that can make connections between arbitrary cells to build logic paths as shown in Fig. 11.7. Depending on the FPGA type, special-purpose structures are placed into the array. Most often, these are configurable RAM blocks and clock distribution elements. Around the periphery of the chip are the I/O cells, which commonly contain one or more flops to enable high-performance synchronous interfaces. Locating flops within I/O cells improves timing characteristics by minimizing the distance, and hence the delay between each flop and its associated pin. Unlike CPLDs, most FPGAs are based on SRAM technology, making their configurations volatile. A typical FPGA must be reprogrammed each time power is applied to a system. Major vendors of FPGAs include Actel, Altera, Atmel, Lattice, QuickLogic, and Xilinx.

Very high logic densities are achieved by scaling the size of the two-dimensional logic cell array. The primary limiting factor in FPGA performance becomes routing because of the nondeterministic nature of a multipath grid interconnect system. Paths between logic cells can take multiple routes, some of which may provide identical propagation delays. However, routing resources are finite, and conflicts quickly arise between competing paths for the same routing channels. As with CPLDs, FPGA vendors provide proprietary software tools to convert a netlist into a final programming image. Depending on the complexity of the design (e.g., speed and density), routing an FPGA can take a few minutes or many hours. Unlike a CPLD with deterministic interconnection resources, FPGA timing can vary dramatically, depending on the quality of the logic placement. Large, fast designs require iterative routing and placement algorithms.

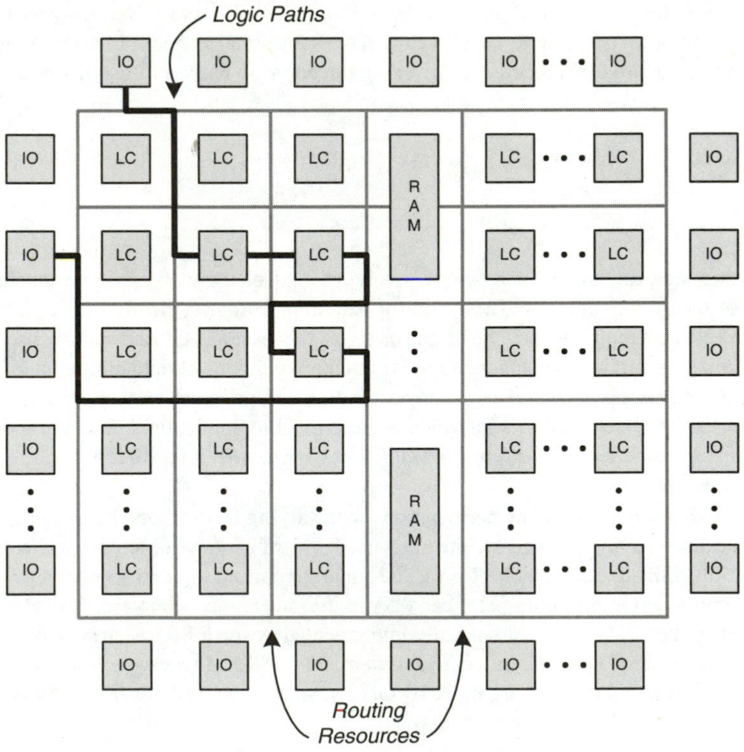

FIGURE 11.7 FPGA logic cell array.

Human intervention can be critical to the successful routing and timing of a complex FPGA design. *Floorplanning* is the process by which an engineer manually partitions logic into groups and then explicitly places these groups into sections of the logic cell array. Manually locating large portions of the logic restricts the routing software to optimizing placement of logic within those boundaries and reduces the number of permutations that it must try to achieve a successful result.

Each vendor's logic cell architecture differs somewhat, but mainly in how support functions such as multiplexing and arithmetic carry terms are implemented. For the most part, engineers do not have to consider the minute details of each logic cell structure, because the conversion of logic into the logic cell is performed by a combination of the HDL synthesis tool and the vendor's proprietary mapping software. In extreme situations, wherein a very specific logic implementation is necessary to squeeze the absolute maximum performance from a specific FPGA, optimizing logic and architecture for a given logic cell structure may have benefits. Engaging in this level of technology-specific optimization, however, can be very tricky and lead to a house-of-cards scenario in which everything is perfectly balanced for a while, and then one new feature is added that upsets the whole plan. If a design appears to be so aggressive as to require fine-tuned optimization, and faster devices cannot be obtained, it may be preferable to modify the architecture to enable more mainstream, abstracted design methodologies.

Notwithstanding the preceding comments, there are high-level feature differences among FPGAs that should be evaluated before choosing a specific device. Of course, it is necessary to pick an FPGA that has sufficient logic and I/O pins to satisfy the needs of the intended application. But not all FPGAs are created equal, despite having similar quantities of logic. While the benefits of one logic structure over another can be debated, the presence or absence of critical design resources can make implementation of a specific design possible or impossible. These resources are clock distribution elements, embedded memory, embedded third-party cores, and multifunction I/O cells.

Clock distribution across a synchronous system must be done with minimal skew to achieve acceptable timing. Each logic cell within a FPGA holds a flop that requires a clock. Therefore, an FPGA must provide at least one global clock signal distributed to each logic cell with low skew across the entire device. One clock is insufficient for most large digital systems because of the proliferation of different interfaces, microprocessors, and peripherals. Typical FPGAs provide anywhere from 4 to 16 global clocks with associated low-skew distribution resources. Most FPGAs do allow clocks to be routed using the general routing resources that normally carry logic signals. However, these paths are usually unable to achieve the low skew characteristics of the dedicated clock distribution network and, consequently, do not enable high clock speeds.

Some FPGAs support a large number of clocks, but with the restriction that not all clocks can be used simultaneously in the same portion of the chip. This type of restriction reduces the complexity of clock distribution on the part of the FPGA vendor because, while the entire chip supports a large number of clocks in total, individual sections of the chip support a smaller number. For example, an FPGA might support 16 global clocks with the restriction that any one quadrant can support only 8 clocks. This means that there are 16 clocks available, and each quadrant can select half of them for arbitrary use. Instead of providing 16 clocks to each logic cell, only 8 need be connected, thus simplifying the FPGA circuitry.

Most FPGAs provide *phase locked loops* (PLLs) or *delay locked loops* (DLLs) that enable the intentional skewing, division, and multiplication of incoming clock signals. PLLs are partially analog circuits, whereas DLLs are fully digital circuits. They have significant overlap in the functions that they can provide in an FPGA, although engineers may debate the merits of one versus the other. The fundamental advantage of a PLL or DLL within an FPGA is its ability to improve I/O timing (e.g., t_{CO}) by effectively removing the propagation delay between the clock input pin and the signal output pin, also known as *deskewing*. As shown in Fig. 11.8, the PLL or DLL aligns the incoming clock to a feedback clock with the same delay as observed at the I/O flops. In doing so, it shifts the incoming

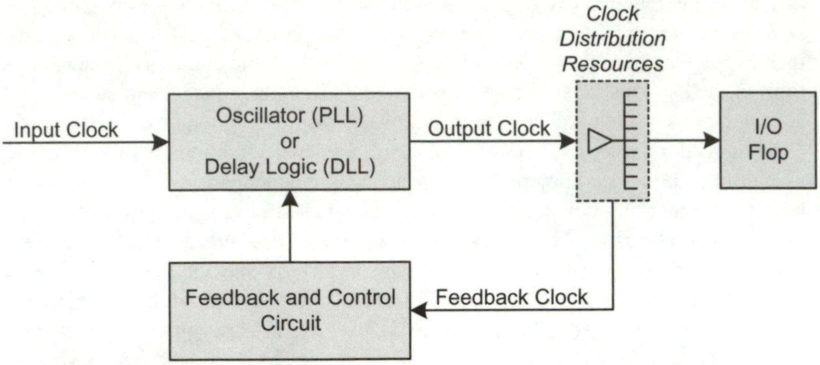

FIGURE 11.8 PLL/DLL clock deskew function within FPGA.

clock so that the causal edge observed by the I/O flops occurs at nearly the same time as when it enters the FPGA's clock input pin. PLLs and DLLs are discussed in more detail in a later chapter.

Additional circuitry enables some PLLs and DLLs to emit a clock that is related to the input frequency by a programmable ratio. The ability to multiply and divide clocks is a benefit to some system designs. An external board-level interface may run at a slower frequency to make circuit implementation easier, but it may be desired to run the internal FPGA logic as a faster multiple of that clock for processing performance reasons. Depending on the exact implementation, multiplication or division can assist with this scheme.

RAM blocks embedded within the logic cell array are a critical feature for many applications. FIFOs and small buffers figure prominently in a variety of data processing architectures. Without on-chip RAM, valuable I/O resources and speed penalties would be given up to use off-chip memory devices. To suit a wide range of applications, RAMs need to be highly configurable and flexible. A typical FPGA's RAM block is based on a certain bit density and can be used in arbitrary width/depth configurations as shown in Fig. 11.9 using the example of a 4-kb RAM block. Single- and dual-port modes are also very important. Many applications, including FIFOs, benefit from a dual-ported

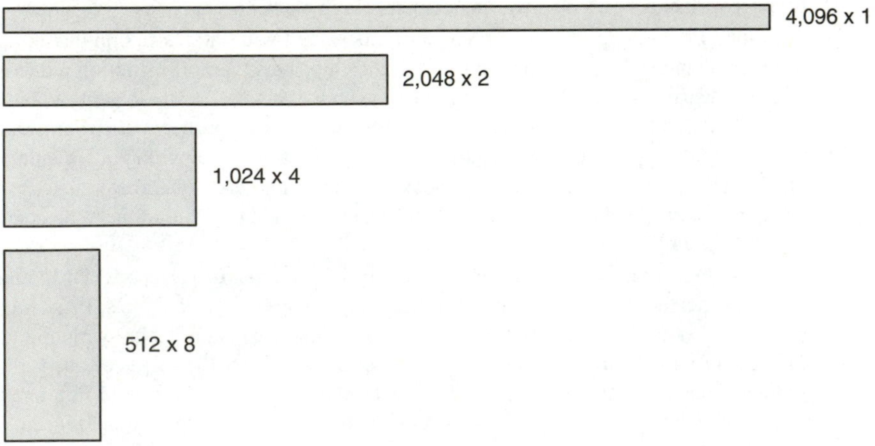

FIGURE 11.9 Configurable FPGA 4 kb RAM block.

RAM block to enable simultaneous reading and writing of the memory by different logic blocks. One state machine may be writing data into a RAM, and another may be reading other data out at the same time. RAM blocks can have synchronous or asynchronous interfaces and may support one or two clocks in synchronous modes. Supporting two clocks in synchronous modes facilitates dual-clock FIFO designs for moving data between different clock domains.

Some FPGAs also allow logic cell LUTs to be used as general RAM in certain configurations. A 16×1 four-input LUT can serve as a 16×1 RAM if supported by the FPGA architecture. It is more efficient to use RAM blocks for large memory structures, because the hardware is optimized to provide a substantial quantity of memory in a small area of silicon. However, LUT-based RAM is beneficial when a design requires many shallow memory structures (e.g., a small FIFO) and all the large RAM blocks are already used. Along with control logic, 32 four-input LUTs can be used to construct a 16×32 FIFO. If a design is memory intensive, it could be wasteful to commit one or more large RAM blocks for such a small FIFO.

Embedding third-party logic cores is a feature that can be useful for some designs, and not useful at all for others. A disadvantage of FPGAs is their higher cost per gate than custom ASIC technology. The main reason that engineers are willing to pay this cost premium is for the ability to implement custom logic in a low-risk development process. Some applications involve a mix of custom and predesigned logic that can be purchased from a third party. Examples of this include buying a microprocessor design or a standard bus controller (e.g., PCI) and integrating it with custom logic on the same chip. Ordinarily, the cost per gate of the third-party logic would be the same as that of your custom logic. On top of that cost is the licensing fee charged by the third party. Some FPGA vendors have decided that there is sufficient demand for a few standard logic cores to offer specific FPGAs that embed these cores into the silicon in a fixed configuration. The benefit of doing so is to drop the per-gate cost of the core to nearly that of a custom ASIC, because the core is hard wired and requires none of the FPGA's configuration overhead.

FPGAs with embedded logic cores may cost more to offset the licensing costs of the cores, but the idea is that the overall cost to the customer will be reduced through the efficiency of the hard-wired core implementation. Microprocessors, PCI bus controllers, and high-speed serdes components are common examples of FPGA embedded cores. Some specific applications may be well suited to this concept.

I/O cell architecture can have a significant impact on the types of board-level interfaces that the FPGA can support. The issues revolve around two variables: synchronous functionality and voltage/current levels. FPGAs support generic I/O cells that can be configured for input-only, output-only, or bidirectional operation with associated tri-state buffer output enable capability. To achieve the best I/O timing, flops for all three functions—input, output, and output-enable—should be included within the I/O cell as shown in Fig. 11.10. The timing improvement obtained by locating these three flops in the I/O cells is substantial. The alternative would be to use logic cell flops and route paths from the logic cell array directly to the I/O pin circuitry, increasing the I/O delay times. Each of the three I/O functions is provided in both registered and unregistered options using multiplexers to provide complete flexibility in logic implementation.

More advanced bus interfaces run at double data rate speeds, requiring more advanced I/O cell structures to achieve the necessary timing specifications. Newer FPGAs are available with I/O cells that specifically support DDR interfaces by incorporating two sets of flops, one for each clock edge as shown in Fig. 11.11. When configured for DDR mode, each of the three I/O functions is driven by a pair of flops, and a multiplexer selects the appropriate flop output depending on the phase of the clock. A DDR interface runs externally to the FPGA on both edges of the clock with a certain width. Internally, the interface runs at double the external width on only one edge of the same clock frequency. Therefore, the I/O cell serves as a 2:1 multiplexer for outputs and a 1:2 demultiplexer for inputs when operating in DDR mode.

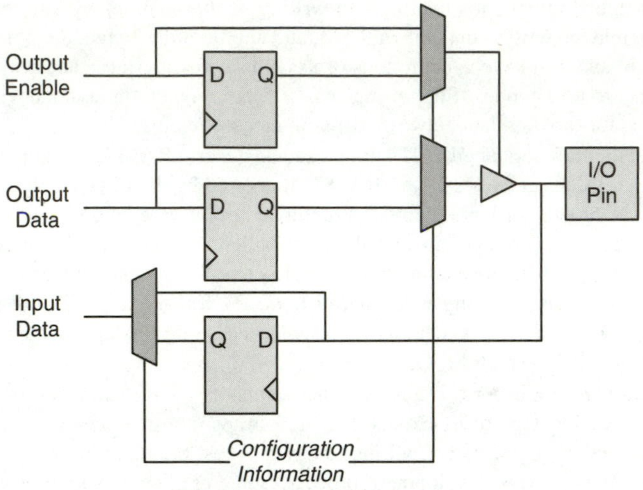

FIGURE 11.10 FPGA I/O cell structure.

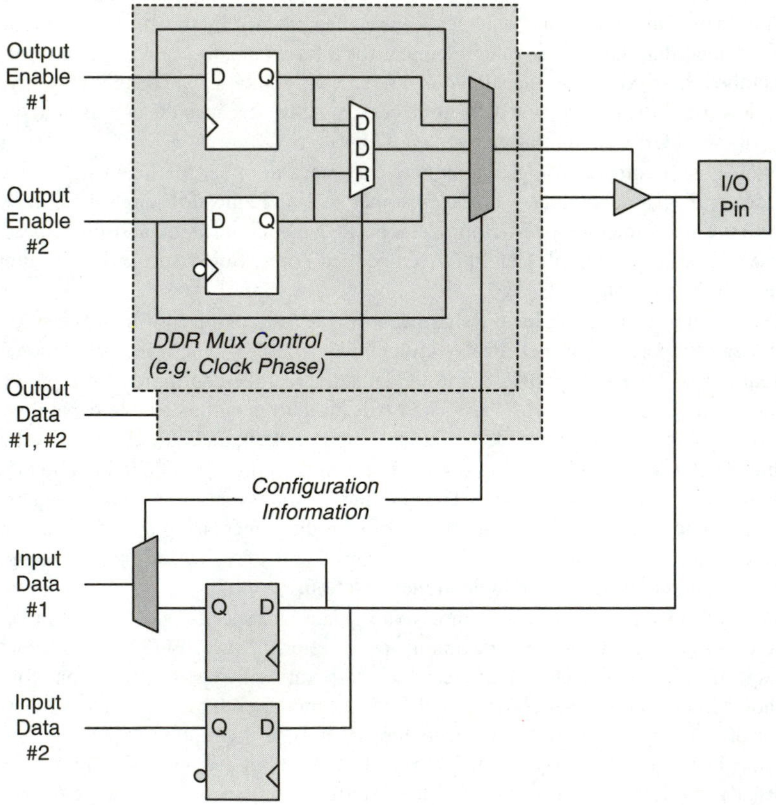

FIGURE 11.11 FPGA DDR I/O cell structure.

Aside from synchronous functionality, compliance with various I/O voltage and current drive standards is a key feature for modern, flexible FPGAs. Like CPLDs that support multiple I/O banks, each of which that can drive a different voltage level, FPGAs are usually partitioned into I/O banks as well, for the same purpose. In contrast with CPLDs, many FPGAs support a wider variety of I/O standards for greater design flexibility.

P • A • R • T • 3

ANALOG BASICS FOR DIGITAL SYSTEMS

CHAPTER 12
Electrical Fundamentals

It is easy to forget that digital systems are really large collections of analog electrical circuits being operated in a digital manner. It is easy to forget this until the digital components need to be connected to adequate power supplies, until digital circuits experience noise problems, and until digital systems need to exchange stimulus and control with our analog world. An understanding of basic circuit theory and a working knowledge of fundamental analog principles provide a digital engineer with confidence and the ability to design a system that will work reliably at the end of the day.

This chapter begins to peel back the veil of analog circuit theory by starting with the basics of Ohm's law, units of measure, and basic circuit analysis techniques. After presenting DC circuits composed of resistors and capacitors, the discussion moves on to frequency-domain analysis and AC circuits. These topics are presented with a minimum of mathematical complexity, because the goal is to support a digital system rather than focusing on analog signal processing. Filters are introduced in the second half of the chapter. Relatively simple filters are common in digital systems, partly because of the need to reduce noise so that more sensitive components such as oscillators and interface ICs can operate properly.

Constructing digital systems in a reliable manner requires a foundation in analog circuit behavior, and the remainder of this book assumes a basic knowledge of circuit theory that is covered in this chapter. Some or all of the topics presented here may be a review for some readers. A quick skim of the chapter's material will confirm whether it can be skipped.

12.1 BASIC CIRCUITS

Electric potential, called *voltage*, and current are the two basic parameters used in the analysis of electric circuits. Voltage is measured in *volts,* and current is measured in *amperes*, or *amps*. Using the analogy of a hose filled with water, voltage is akin to the water's pressure, and current is akin to the quantity of water flowing through the hose at a given pressure. Unlike water in a hose, electricity only flows when a complete loop, or *circuit*, is present. Electric charge (expressed in units of coulombs) is emitted from one terminal of a power source, flows through a circuit, and then returns in the same quantity to another terminal of that power source. Per the law of conservation of charge, the charge cannot simply disappear in the circuit; it must return to its source to complete the loop.

Georg Ohm, a nineteenth century German physicist, discovered the mathematical relationship between voltage and current that is now known as Ohm's law: $V = IR$. It states that a voltage (V) drop results by passing a current (I) through a resistance (R). Appropriately, the unit of resistance is the *ohm* and it is represented with the Greek letter omega, Ω. Consider the simple circuit in Fig. 12.1, consisting of a 10-V battery and a 50-Ω resistor. Assuming that the connecting wires have zero resistance, there is one and only one solution to Ohm's law: the current through the circuit is $I = V \div R =$

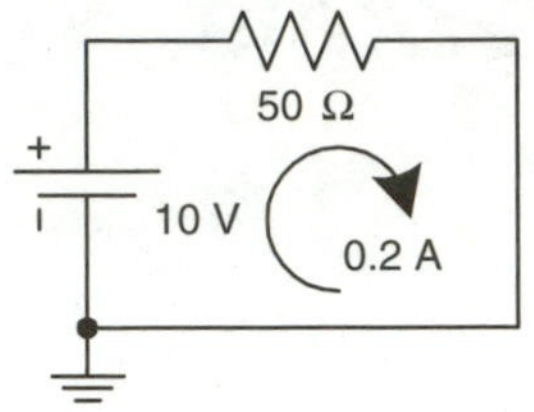

FIGURE 12.1 Simple resistive circuit.

10 ÷ 50 = 0.2 A, or 200 mA. If any of the wires are broken, the current flow instantly drops to zero, because charge cannot return to the battery. Remember that all circuits are a collection of loops. Making and breaking loops using switches (transistors) is fundamentally how digital systems operate.

Voltage is a relative measurement, and its effects can be quantified only between pairs of conductors. This is why a bird can sit on a high-voltage power line without being electrocuted and why a person can walk around on a carpet while insulated from nearby conductors and develop a high voltage static electric charge of several thousand volts. Not until the conductor (e.g., bird or person) comes into contact with another conductor (e.g., another wire or a metal door handle) does the effect of voltage become apparent. To facilitate circuit analysis, a common reference point called a *ground* node is assigned a relative voltage of zero. The ground symbol is shown attached to the battery's negative terminal in Fig. 12.1. All other nodes in the circuit can now be measured relative to ground, resulting in meaningful absolute voltage levels. This is why electrical communication signals between systems require either a common ground or a differential signaling scheme. A receiver cannot properly detect the voltage on a signal wire unless it has a valid reference to compare it against.

An important parameter derived from Ohm's law is power, expressed in *watts* (W). Power (P) is the product of current and voltage. In this simple circuit, 10 V is dropped through the circuit (all by the resistor) at a current of 0.2 A, yielding 2 W of power supplied by the battery and dissipated by the resistor. Power is an instantaneous measurement and is not energy, but the rate of flow of energy. Energy (E) is measured in *Joules* and is the product of how much power is delivered over a span of time: $E = Pt$. The resistor in this circuit converts 2 J of energy each second into heat. Per the first law of thermodynamics, energy cannot be created or destroyed. Therefore, electrical energy that is dissipated by a component is converted into thermal energy. The charge still returns to the battery, but it does so at a relative potential of 0 V. A circuit is typically characterized by its power dissipation (watts) rather than by its energy dissipation (joules) because of the time invariant nature of power.

Power can be restated depending on the unknown variable in the equation. $P = VI$ can be manipulated using Ohm's law in several common forms: $P = I^2R$ and $P = V^2 \div R$. In the first instance, the $V = IR$ definition is substituted to yield a power calculation that only considers the current passing through a resistance. Alternatively, current can be effectively removed ($I = V \div R$), and power can be restated as the square of the voltage developed across a resistance.

12.2 LOOP AND NODE ANALYSIS

The example in Fig. 12.1 is a basic application of *loop analysis*. Loop analysis is based on the rules that the sum of the voltage drops in any continuous circuit is 0 and that the instantaneous current around the loop is constant. Of course, current can change over time, and those changes apply to each component in the circuit. Based on these rules, the following general loop equation can be written for any circuit, where R_N are discrete resistors and VS_N are discrete voltage sources (e.g., a battery):

$$\sum V_N = I_{LOOP} \sum R_N + \sum VS_N = 0$$

This equation shows that a uniform loop current develops a voltage drop across one or more resistances in the circuit and that this total voltage must be offset by voltage sources. A loop equation can be written for Fig. 12.1, but special attention should be paid to the polarity of the voltage source versus the voltage drop through the resistor. The convention used to specify polarity does not change the final answer as long as the convention is applied consistently. Mistakes in loop analysis can arise from inconsistent representation of voltage polarities. In this case, the current is shown to circulate clockwise, so positive currents are clockwise currents. This means that the 50-Ω resistor will exhibit a voltage drop as the current moves through it from left to right. At the same time, the voltage source exhibits a voltage rise as the clockwise loop current passes through it. The voltage of the resistor and the voltage of the source are of opposite polarity as expressed in the following loop equation:

$$I_{LOOP} \sum R_N + \sum VS_N = I_{LOOP} \times 50\ \Omega - 10\text{ V} = 0$$

$$I_{LOOP} = \frac{10\text{ V}}{50\ \Omega} = 0.2\text{ A}$$

Keep in mind that polarity notation is a convention and not a physical rule. As long as polarities are treated consistently, the correct answer will result. Figure 12.2 shows an example of a loop circuit wherein consistent polarity notation is critical to a correct answer. Two voltage sources are present in this circuit, but they are inserted with different polarities.

Circulating around the loop clockwise starting from the ground node, there is a 10-V rise, a voltage drop through the resistor, and a 5-V drop through the voltage source. The loop equation for this circuit is written as follows, yielding $I_{LOOP} = 0.1$ A:

$$I_{LOOP} \sum R_N + \sum VS_N = I_{LOOP} \times 50\ \Omega - 10\text{ V} + 5\text{ V} = 0$$

Loop analysis in the context of a single loop circuit may not sound very different from the basics of Ohm's law. It can be truly helpful when multiple loops are present in a circuit. The double-loop circuit in Fig. 12.3 is a somewhat contrived example but one that serves as a quick illustration of the concept. There are three unknowns in this circuit: the two loop currents and the voltage at the intermediate node, V_X. Once the loop currents are known, V_X can be calculated in three different ways. The voltage drop across either R1 or R3 can be calculated. Alternatively, the current through R2 can be determined as the sum of I_{LOOP1} and I_{LOOP2}. Because these currents are both shown using the clockwise convention, they pass each other through R2 with opposite polarities and end up subtracting.

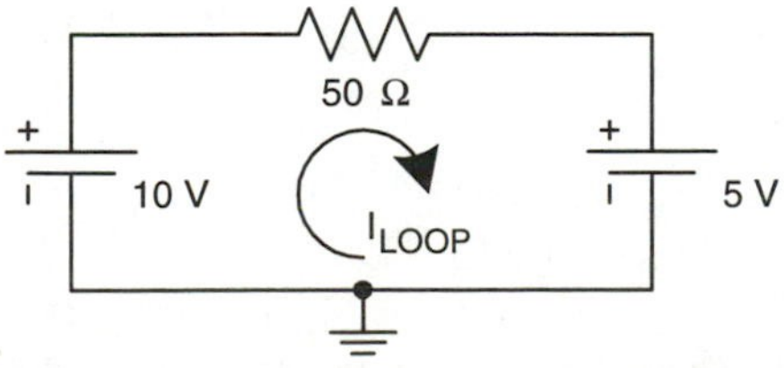

FIGURE 12.2 Circuit with two voltage sources.

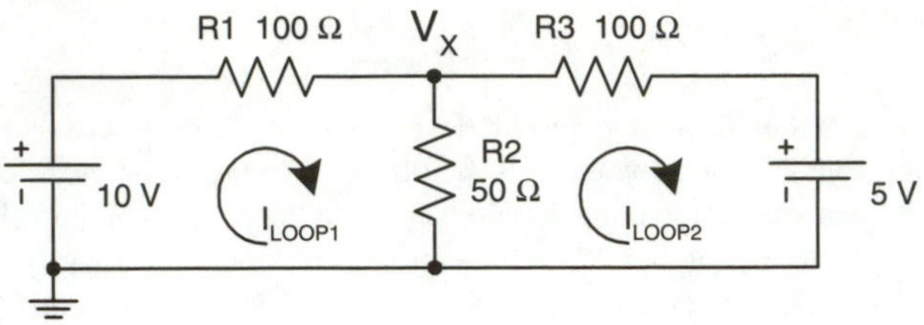

FIGURE 12.3 Double-loop circuit.

The first step is to set up the two loop equations. It is critical to adopt a consistent convention for currents and voltages. Here, positive currents flow clockwise, voltage sources are negative when they rise with the direction of the clockwise loop current, and resistors drop voltage with the same polarity as the loop current. The resulting loop equations are as follows:

$$I_{LOOP1}R1 + (I_{LOOP1} - I_{LOOP2})R2 - 10\text{ V} = 0$$

$$I_{LOOP2}R3 + (I_{LOOP2} - I_{LOOP1})R2 + 5\text{ V} = 0$$

Notice that the 10-V source appears as a negative in the first loop equation, because it represents a voltage rise, but the 5-V source appears as a positive in the second equation, because it represents a voltage drop. Our two equations with two unknowns can be solved. I_{LOOP1} can be solved as a function of I_{LOOP2} using the second equation as follows:

$$I_{LOOP1} = \frac{1}{10} + 3I_{LOOP2}$$

I_{LOOP2} is solved next by substituting the I_{LOOP1} expression into the first equation,

$$100I_{LOOP1} + 50I_{LOOP1} - 50I_{LOOP2} - 10\text{ V} = 0$$

$$150I_{LOOP1} - 50I_{LOOP2} = 10\text{ V}$$

$$\frac{150}{10} + 450I_{LOOP2} - 50I_{LOOP2} = 10\text{ V}$$

$$15 + 400I_{LOOP2} = 10\text{ V}$$

$$I_{LOOP2} = \frac{-5}{400} = -0.0125\text{ A}$$

Finally, I_{LOOP1} is solved,

$$I_{LOOP1} = 0.0625\text{ A}$$

Notice that I_{LOOP2} is a negative quantity. This is because the actual direction of current flow is out of the 5-V source rather than into it—the current is flowing counterclockwise, which is a negative clockwise current. Our results can be verified to make sure that everything adds up correctly. The voltage drops across R1 and R3 are 6.25 V and 1.25 V, respectively. Therefore, V_X = 10 V – 6.25 V = 5 V – 1.25 V = 3.75 V. This means that the current through R3 must be 3.75 V ÷ 50 Ω = 0.075 A, which is exactly equal to the difference of the two loop currents ($I_{R2} = I_{LOOP1} - I_{LOOP2}$).

Node analysis is the complement to loop analysis. Rather than dealing with a fixed current around a loop in which the voltages sum to zero, node analysis examines an individual node with a fixed voltage where the currents sum to zero. Together, these loop and node analysis methods form the basis of circuit analysis. You may find that some circuits are more easily solved with one or the other. As before, the consistent usage of a polarity convention is critical to finding the correct answer. The circuit in Fig. 12.3 can be evaluated with just one node equation to determine V_X. In the following equation, the convention is taken that positive currents flow out of the node.

$$0 = \sum I_N = \frac{V_X - 10\text{ V}}{R1} + \frac{V_X - 0\text{ V}}{R2} + \frac{V_X - 5\text{ V}}{R3}$$

This node equation does not worry about whether V_X is actually higher or lower than the voltage at the other end of each resistor. If current is actually flowing into the node because the reverse voltage relationship is true, that current will have a negative polarity. Working through the algebra shows that V_X = 3.75 V, the same answer that was obtained using loop analysis, although with only a single equation and one unknown. If the current through any resistor must be calculated, it can be done once V_X is known.

12.3 RESISTANCE COMBINATION

Circuits can be simplified for analysis purposes when multiple resistances are present in various series and parallel topologies. Multiple resistors can be combined in arbitrary configurations and represented as a single resistance. Two resistors placed in series add, while resistors placed in parallel result in a smaller overall resistance. Resistors in parallel create a combined resistance that is less than the smallest valued resistor in the parallel group. This resulting value is determined using the inverse relationship,

$$R_{TOTAL} = \frac{1}{\sum \frac{1}{R_N}}$$

Multiple resistors placed in parallel can be indicated using a parallel bar notation. $R1 \parallel R2$ indicates that two resistors, referred to as R1 and R2, are in parallel. When two resistors are placed in a parallel arrangement, the above expression can be rewritten for this special circumstance:

$$R_{TOTAL} = \frac{R1 \times R2}{R1 + R2}$$

Figure 12.4 shows resistors in both series and parallel topologies. Series resistors R1 and R2 add to form a 150-Ω resistance. Parallel resistors R3 and R4 combine to form a smaller 33.3-Ω resistance. Placing two resistors back to back increases the total resistance observed by a current. Placing two resistors in parallel provides a second path for current to flow through, thereby reducing the overall resistance. After performing these two combinations, the circuit can be simplified a third time by adding the resulting series resistances, 150 Ω + 33.3 Ω = 183.3 Ω. When the circuit has been

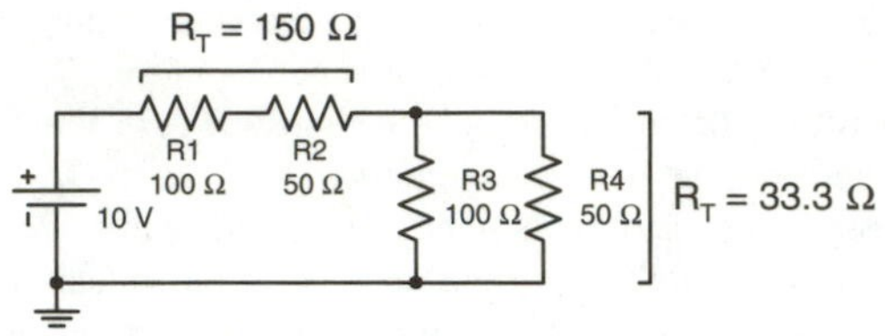

FIGURE 12.4 Combining multiple resistors.

fully simplified, its overall current flow and power dissipation can be calculated: I = 54.5 mA and P = 545 mW.

Power dissipation for the resistors in Fig. 12.4 can be determined in multiple ways: VI, I^2R, or $V^2 \div R$. Power for each resistor can be calculated individually, but care must be taken to use the true voltage drop or current through each component. R1 and R2 have the same current passing through them because both have no parallel components to divert current, but they have differing voltage drops because their resistances are not equal. In contrast, R3 and R4 have identical voltage drops because they connect the same two nodes, but differing currents because their resistances are not equal.

The two resistor pairs, R1/R2 and R3/R4, form a basic voltage divider at the intermediate node connecting R2 and R3/R4. This voltage can be calculated knowing the current through R1 and R2 (54.5 mA) by either calculating the combined voltage drop of R1 and R2 and then subtracting this from the battery voltage or by just calculating the voltage drop across R3 and R4. The parallel combination of R3 and R4 equates to 33.3 Ω, indicating a voltage drop of 1.82 V at I = 54.5 mA. This is the voltage of the intermediate node because the lower node of R3 and R4 is ground, or 0 V. The alternate approach yields the same answer.

$$V_{NODE} = V_{BATT} - I\,(\text{R1} + \text{R2}) = 10\text{ V} - 54.5\text{ mA }(150\ \Omega) = 10\text{ V} - 8.18\text{ V} = 1.82\text{ V}$$

12.4 CAPACITORS

Resistors respond to changes in current in a linear fashion according to Ohm's law by exhibiting changes in voltage drop. Similarly, changing the voltage across a resistor causes the current through that resistor to change linearly. Resistors behave this way because they do not store energy; they dissipate some energy as heat and pass the remainder through the circuit. *Capacitors* store energy, and, consequently, their voltage/current relationship is nonlinear.

A capacitor stores charge on parallel conductive plates, each of which is at a different arbitrary potential relative to the other. In this respect, a capacitor functions like a very small battery and holds a voltage according to how much charge is stored on its plates. Capacitance (C) is measured in *farads*. One farad of capacitance is relatively large. Most capacitors that are used in digital systems are measured in microfarads (μF) or picofarads (pF). As a capacitor builds up charge, its voltage increases in a linear fashion as defined by the equation, $Q = CV$, where Q is the charge expressed in coulombs.

One of the basic demonstrations of a capacitor's operation is in the common series resistor-capacitor (RC) circuit shown in Fig. 12.5, where a resistor controls the charging rate of the capacitor. The capacitor's voltage does not change in a linear fashion. From the relationship $Q = CV$, it is known that the voltage is a function of how much charge has been stored on the capacitor's plates. How much charge flows into the capacitor is a function of the current that flows around the circuit over a span of time (one amp is the flow of one coulomb per second). As the voltage across the capacitor in-

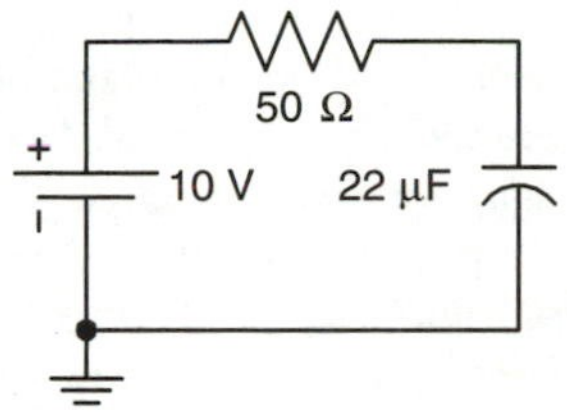

FIGURE 12.5 Simple RC circuit.

creases, the voltage drop across the resistor decreases, causing the current through the circuit to decrease as well. Therefore, the capacitor begins charging at a high rate when its voltage is 0 and the circuit's current is limited only by the resistor. The charging rate decreases as the charge on the capacitor builds up.

Normalized to 1 V, the voltage across a capacitor in an RC circuit is defined as

$$V_C = 1 - e^{-\frac{t}{RC}}$$

where e is the base of the natural logarithm, an irrational mathematical constant roughly equivalent to 2.718. Starting from the initial condition when the capacitor is fully discharged, $t = 0$ and $V_C = 0$. The *RC time constant*, expressed in seconds, is simply the product of R and C and is a measure of how fast the capacitor charges. Every RC seconds, the voltage across the capacitor's terminals changes by 63.2 percent of the remaining voltage differential between the initial capacitor voltage and the applied voltage to the circuit, in this case the 10-V battery. By rule of thumb, a capacitor is often considered fully charged after 5 RC seconds, at which point it achieves more than 99 percent of its full charge. In this example, RC = 1,100 μs = 1.1 ms. Therefore, the capacitor would be at nearly 10 V after 5.5 ms of connecting the battery to the circuit.

RC circuits are used in timer applications where low cost is paramount. The accuracy of an RC timer is relatively poor, because capacitors exhibit significant capacitance variation, thereby altering the time constant. A simple oscillator can be constructed using an inverter (e.g., 74LS04) and an RC as shown in Fig. 12.6. When the inverter's input is below its switching threshold, the output is high, causing the capacitor to charge through the resistor. At some point, the capacitor voltage rises above the switching threshold and causes the inverter's output to go low. This, in turn, causes the capacitor to begin discharging through the resistor. When the capacitor's voltage declines past the switching threshold, the process begins again.

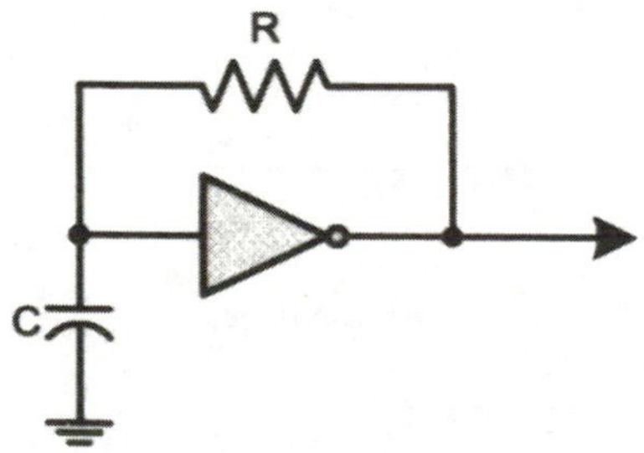

FIGURE 12.6 RC oscillator.

In reality, the inverter does not switch sharply as the input voltage slowly passes out of the valid logic-0 and logic-1 voltage thresholds. This creates a separate switching threshold when going from low to high versus high to low. The voltage difference between these thresholds depends on temperature, the operating voltage, and the individual chip's characteristics. Therefore, a simple inverter-based oscillator does not provide a very stable oscillation period over time without additional circuitry. The waveform in Fig. 12.7 shows how this RC oscillator behaves.

This waveform shows the nonlinear charging characteristics of an RC circuit. The frequency at which this oscillator runs depends on the differential between the two switching thresholds, the input current of the inverter, and the RC values. The ideal RC charge/discharge curve is modified by the inverter's input current. When the capacitor is at logic-1, the positive input (sink) current of the inverter works with the resistor to more rapidly discharge the capacitor. Similarly, when the capacitor is at logic-0, the negative input (source) current of the inverter works with the resistor to more rapidly charge the capacitor. These source and sink currents are unequal in a bipolar logic IC; therefore, the charge/discharge effects of the inverter will be unequal as well.

12.5 CAPACITORS AS AC ELEMENTS

Capacitors have a far broader range of applications than just RC oscillators. The circuits presented up to this point are *direct current* (DC) circuits. In our context, DC refers to steady-state signals with no frequency content being applied to a circuit. While the RC oscillator example certainly varies its voltages, it is a piecewise DC circuit that flips its applied charging voltage at regular intervals. The world is not static, however, and signals are characterized by direct and *alternating current* (AC) components. AC circuits deal with the time-varying properties of signals, their frequency content.

Just as resistors exhibit resistance, capacitors exhibit *reactance*. Both measures are expressed in ohms, but resistance is constant across frequency, whereas reactance (X) varies with frequency. The two terms are combined into a single *impedance* expression, $Z = R + jX$. Impedance is the overall resistance of an element that includes both DC and AC components. The imaginary (AC) component, a function of frequency, is marked by the constant j, the imaginary number $\sqrt{-1}$.

AC circuit analysis expresses frequencies in radians per second rather than in hertz and uses the lower-case Greek letter omega, ω, to denote radian frequency. There are 2π radians in a 360° circle,

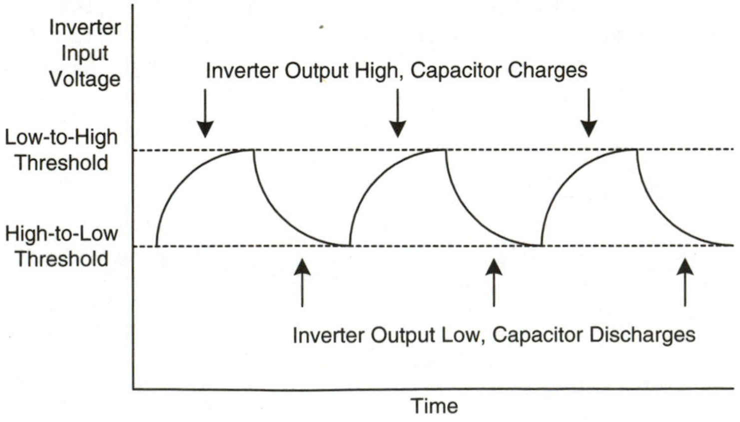

FIGURE 12.7 RC oscillator waveform.

or cycle. Therefore, there are 2π radians per hertz, or ω = 2πf. The 2πf equivalent is primarily used in this book to illustrate how various circuits interact with digital systems running at arbitrary frequencies.

The impedance of an ideal capacitor is inversely related to the frequency applied to it and is expressed as

$$Z_C = \frac{1}{2\pi f C}$$

At low frequencies, the capacitor exhibits very high impedance and appears essentially as an open circuit. Considering a basic RC circuit, once the capacitor fully charges, the circuit transitions to a steady-state DC circuit without any AC component. Therefore, no more current flows through the circuit, because the capacitor has achieved the same voltage as the battery, and there is no voltage drop across the charging resistor. In AC analysis terms, the frequency of the circuit is zero, and the impedance of the capacitor is infinite. Conversely, a capacitor's impedance asymptotically approaches zero at very high frequencies and becomes a short circuit.

A capacitor can be used to reduce noise in a system, because its impedance is a function of frequency. *Decoupling*, or *bypass*, capacitors are arranged in shunt (parallel) configurations across power supplies that may contain noise, as shown in Fig. 12.8. Power distribution wires can have noise injected back into them by the high-speed on/off switching of a digital circuit. Each time a logic gate or flop transitions, a small surge of current is created to establish the new voltage level. When hundreds or thousands of signals within and external to an IC switch on and off, noise can become a substantial problem.

Decoupling capacitors can be chosen to present a low impedance at the noise frequencies of interest in a digital system. Lower-frequency systems often use 0.1-μF capacitors that exhibit impedance of under 1 Ω at over 1 MHz. Higher-frequency systems may use 0.01-μF or 0.001-μF capacitors for the same purpose. The capacitor functions as a selective resistor that only kicks in at certain frequencies of interest and leaves the DC signal, in this case power, undisturbed. Figure 12.9 illustrates the effect of a decoupling capacitor in shunting the majority of the high-frequency AC component (noise) to ground. Most of the noise is removed, but some remains, as will be explained later.

Multiple capacitors arranged together follow the same series and parallel impedance calculation rule as resistors. However, because of the capacitor's inversely proportional impedance characteris-

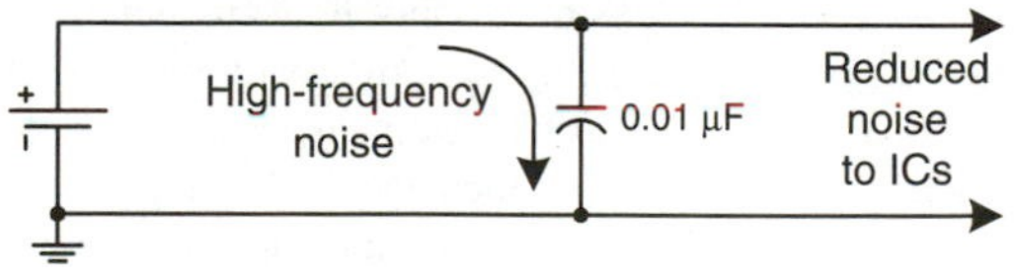

FIGURE 12.8 Noise filtering with shunt capacitance.

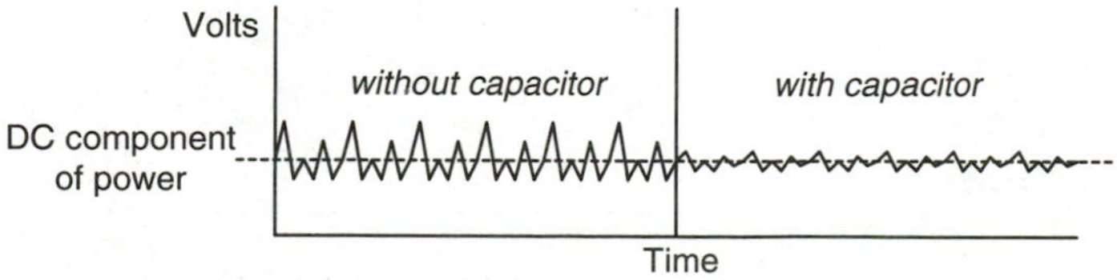

FIGURE 12.9 AC noise removed from DC power.

tic, total capacitance is increased when multiple capacitors are placed in parallel and decreased when placed in series. Consider the following calculation for two equal values capacitors in parallel:

$$Z_{TOTAL} = \frac{1}{\sum \frac{1}{Z_N}} = \frac{1}{2\pi fC + 2\pi fC} = \frac{1}{2\pi f(2C)}$$

The resulting impedance is what would be obtained with a capacitor twice as large as each individual capacitor. Similarly, capacitance is reduced as follows when two capacitors are in series:

$$Z_{TOTAL} = \sum Z_N = \frac{1}{2\pi fC} + \frac{1}{2\pi fC} = \frac{2}{2\pi fC} = \frac{1}{2\pi f(0.5C)}$$

The series impedance is what would be obtained with a capacitor half the size of each individual capacitor.

12.6 INDUCTORS

An *inductor*, also called a *choke* or *coil*, is the other basic circuit element whose impedance changes as a function of frequency. An inductor is a coil of wire that stores energy as a magnetic field. The strength of a magnetic field in a coil is proportional to the current flowing through the coil. If the current flowing through an inductor changes suddenly, the existing magnetic field tends to resist that change. Therefore, inductors present low impedances at low frequencies and high impedances at higher frequencies, the opposite the behavior of a capacitor. Inductance (L) is measured in *henries*. Most inductors encountered in digital systems are measured in microhenries (μH) or nanohenries (nH).

The impedance of an ideal inductor is expressed as $Z_L = 2\pi fL$, clearly showing the proportional relationship of frequency and impedance. In digital systems, inductance is used mainly for filtering purposes and for analyzing the (usually unwanted) inductance of other components. All conductors exhibit some inductance, because a magnetic field naturally develops in proportion to current flow. A straight wire may have much less inductance than a coil, but its inductance is not zero. Such attributes become very important at high frequencies because of the relationship between frequency and impedance.

Just as a capacitor can be used as a parallel element to shunt high-frequency noise, an inductor can be used as a series element to block that noise. Because of their varying impedance, DC signals such as power are passed without attenuation, whereas high frequencies are met with a large impedance. Effective noise filters are made with both series (inductance) and shunt (capacitance) elements as shown in Fig. 12.10. The inductance and capacitance values are chosen to be effective at the desired noise frequency, resulting in a large series resistance followed by a very small shunt resistance. This creates a basic voltage divider arrangement, effective only at high frequencies as intended.

Because of their proportional impedance versus frequency characteristic, inductance combines in the same way as resistance: decreasing with parallel inductors and increasing with series inductors.

12.7 NONIDEAL RLC MODELS

Having presented the three basic passive circuit elements in idealized form, the ways in which resistance, capacitance, and inductance combine in real-world components can be discussed. All conduc-

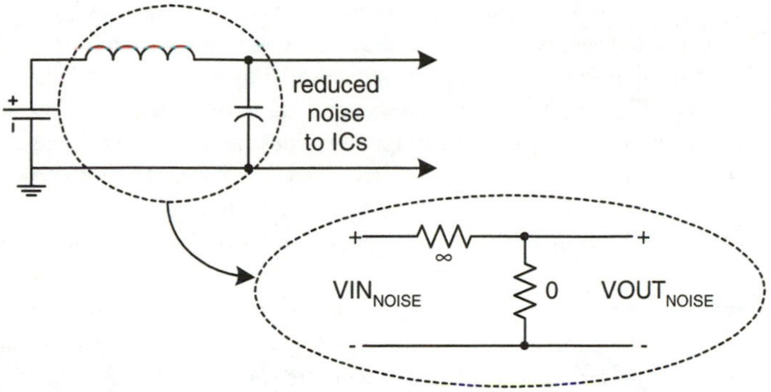

FIGURE 12.10 Noise filtering with LC network.

tors exhibit some series resistance and inductance, and all nearby pairs of conductors exhibit some mutual capacitance. A resistor consists of a resistive element encapsulated in some packaging material with a relatively small conductor at each end to connect the resistive element to an external circuit. Depending on the type of resistor, the conductors may be wires (leaded resistor) or small pieces of metal foil (surface mount). The resistive element itself will vary in size according to its power rating, material, and desired resistance. The finite lengths of the connecting leads and the resistive element each contribute a small quantity of inductance. There is also a small capacitance between the resistor's leads. These unwanted extras are called *parasitic* properties, because they usually detract from the performance of a system rather than improving it. A resistor's function is to provide a certain resistance value in a circuit, but its physical construction results in finite parasitic inductance and capacitance. Figure 12.11 shows a model of a nonideal resistor that enables analysis of its parasitic properties.

Each type of resistor exhibits different magnitudes of parasitic properties. Applications at lower frequencies often ignore these properties, because the parasitic inductance and capacitance is negligible as a result of the frequency/impedance relationships of inductors and capacitors. As the frequencies involved increase, series inductance is generally the first problem that is encountered. Inductance is minimized in resistors that have small leads or, better yet, no leads at all, as is the case with surface mount resistors. Inter-lead capacitance does not become a problem until frequencies get significantly higher.

Similarly, a capacitor exhibits parasitic resistance and inductance. The conductors that form the capacitor have finite resistance and inductance associated with them. A nonideal model of a capacitor is shown in Fig. 12.12. Inductance figures into a capacitor in much the same way that it does a resistor. Smaller leads and components result in reduced parasitic inductance.

At high frequencies, however, the capacitor's parasitic inductance has noticeable effects. The earlier example of using a capacitor to filter high-frequency noise showed that the capacitor removed most of the noise, but not all of it. As the frequency rises, the capacitor's impedance steadily decreases as expected. At a certain point, however, the frequency becomes high enough to cause no-

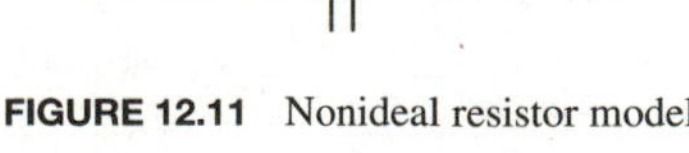

FIGURE 12.11 Nonideal resistor model.

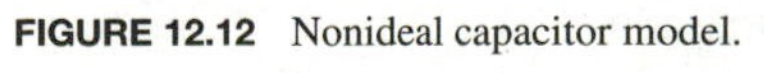

FIGURE 12.12 Nonideal capacitor model.

ticeable impedance resulting from parasitic inductance. As the frequency continues to rise, the impedance of the capacitor begins to be more affected by the inductance. Because inductors resist high-frequency signals by increasing their impedance, the filtering capacitor loses its effect above a certain frequency limit. Figure 12.13 shows the general curve of impedance versus frequency for a capacitor. The curve shows that, above a certain frequency, a capacitor no longer behaves as expected from an ideal perspective. This threshold frequency is different for each type of capacitor and is determined by its physical construction. This is why power filter (e.g., decoupling or bypass) capacitors are ideally chosen based on the expected frequencies of noise that they are expected to attenuate.

As with parasitic inductance, a capacitor's parasitic resistance is a function of its physical construction. The industry-standard term that specifies this attribute is *equivalent series resistance* (ESR). Certain applications for capacitors tend to be more sensitive to ESR than others. ESR is generally not a major concern in high-frequency decoupling applications. However, when power-supply ripple needs to be attenuated, such as in a switching power supply, low-ESR capacitors may be critical to a successful circuit.

Inductors are subject to parasitic properties as well, in the form of series resistance in the wire coil and capacitance between individual coil windings and between the terminals. The nonideal inductor looks very much like the nonideal resistor in Fig. 12.11. Inductors that are used to filter power in a series topology must have a low enough series resistance to handle the current that is passed through them. Inductors are available in a wide variety of sizes, partly because of the need to handle the spectrum of low- and high-power applications.

A major concern in operating inductors at high frequencies is the detrimental effects of their parasitic capacitance. Just as a capacitor's parasitic inductor reduces its effectiveness above a certain frequency, a similar effect is observed in a real inductor as shown in Fig. 12.14. Placing an inductor and capacitor in parallel creates a circuit that resonates at a certain frequency,

$$f_{RES} = \frac{1}{2\pi\sqrt{LC}}$$

A resonant LC circuit can be useful in many applications, including radio tuners. When an engineer specifically wants a circuit to resonate at a particular frequency, such as when trying to pick up radio waves with an antenna, a so-called *LC tank* is desired. However, an LC tank is generally not

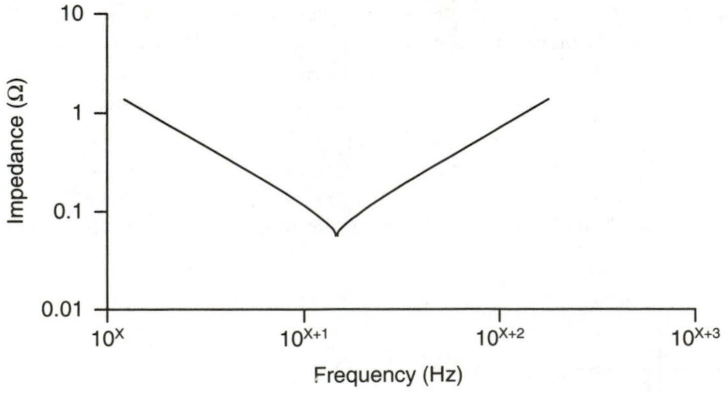

FIGURE 12.13 Nonideal capacitor impedance vs. frequency curve.

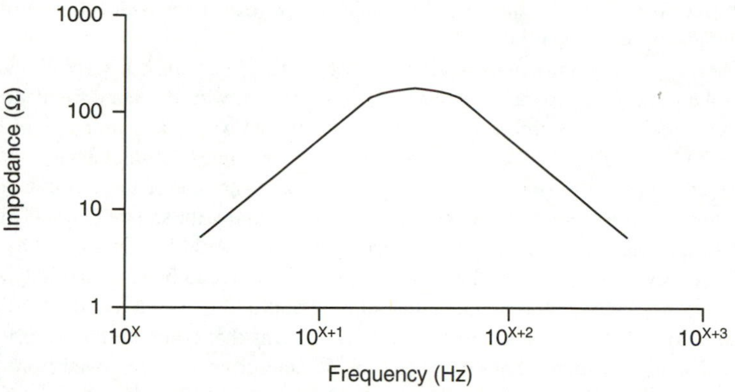

FIGURE 12.14 Nonideal inductor impedance vs. frequency curve.

desired in digital system applications that use inductors. Inductor manufacturers specify an inductor with a certain *self-resonant frequency* (SRF) to characterize the detrimental effects of the parasitic capacitance. Above the SRF, the inductor's impedance declines with increasing frequency. Therefore, if an inductor is operated near its SRF, its parasitic properties should be investigated to ensure that unexpected behavior does not result.

Many filtering applications in digital systems benefit from surface mount *ferrite* beads or chips. Ferrite is a magnetic ceramic material that behaves like an inductor: its impedance rises with frequency. A ferrite bead's parasitic capacitance is lower than that of a standard inductor, because there are no wire coils to capacitively couple with one another. Ferrites are suitable for attenuating high-frequency noise on power supplies and other signals, because they typically have high SRFs. A variety of ferrite materials exist with peak impedances at different frequencies to suit specific applications.

12.8 FREQUENCY DOMAIN ANALYSIS

Electrical signals on a wire can be viewed with an oscilloscope as a plot of voltage (or current) versus time. This is a *time-domain* view of the signals and it provides much useful information for a digital system designer. Using an oscilloscope, an engineer can verify the proper timing of a clock signal and its associated data and control signals. However, time-domain analysis is not very good at determining the frequency content of complex electrical signals. AC components are selected based on their impedances at certain frequencies. Therefore, a method is needed of evaluating a signal's frequency content, thereby knowing the frequencies of interest that the components must handle and enabling selection of suitable values.

Frequency-domain analysis enables an understanding of exactly how an overall AC circuit and its individual components respond to various frequencies that are presented to them. A frequency-domain view of a complex signal allows an engineer to tailor a circuit precisely to the application by relating frequencies and amplitudes rather than time and amplitudes in a conventional time-domain view. Pure sine waves are a convenient representation for signals, because they are easy to manipulate mathematically. While most real-world signals are not sine waves, Joseph Fourier, an eighteenth century French mathematician, demonstrated that an arbitrary signal (e.g., a microprocessor's square wave clock signal) can be expressed as the sum of many sine waves. Frequency-domain analysis is often referred to as *Fourier analysis*. Likewise, a *Fourier transform* is the process of decomposing a signal into a sum of sine waves.

Fourier analysis is represented graphically in Fig. 12.15, where a 1-MHz digital clock signal is shown broken into four separate sine waves. The sine wave with the same frequency as the clock is referred to as the *fundamental frequency.* Subsequent sine waves are multiples of the fundamental frequency and are called *harmonics*. If a sine wave is three times the fundamental frequency, it is the third harmonic. It is clear that, as higher-order harmonics are added to the fundamental frequency, the resulting signal looks more and more like a clock signal with square edges. This example stops after the seventh harmonic, but a more perfect clock signal could be constructed by continuing with higher-frequency harmonics. From a practical perspective, it can be seen that only a few harmonics are necessary to obtain a representation that closely approximates the real digital signal. Therefore, it is often convenient to consider the few sine wave terms that compose the majority of the signal's energy. This simplification allows many less-significant terms to be removed from the relevant calculations. Depending on how accurately the digital signal really needs to be represented, it may be possible to make a gross simplification and consider just the fundamental frequency and one or two subsequent harmonics.

Each harmonic in a Fourier analysis has varying amplitude, frequency, and phase relationships such that their sum yields the desired complex signal. The energy in harmonics generally decreases as their frequency increases. The clock signal shown in this example has energy only at the odd harmonics, because it is a symmetrical signal. Furthermore, the harmonics are in phase with each other. More complex real-world asymmetrical signals can also be handled with Fourier analysis but with more variation across the harmonics, including even harmonics and phase differences.

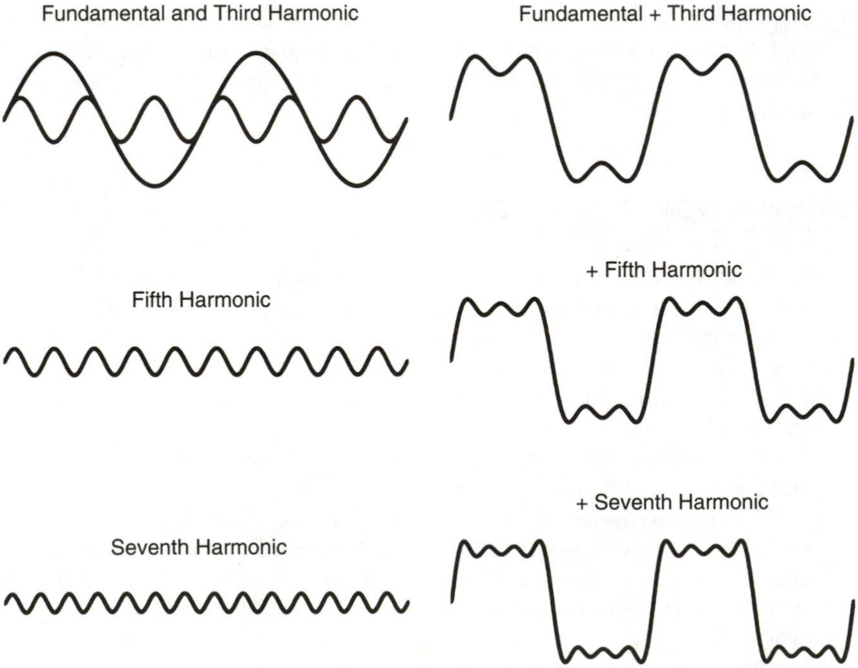

FIGURE 12.15 Digital clock composed of sine waves.

Oscilloscopes are used to view signals in the time domain, whereas *spectrum analyzers* are used to view signals in the frequency domain. Figure 12.16 shows an example of a frequency-domain view of an electrical signal as observed on a spectrum analyzer (courtesy of Agilent Technologies). Rather than viewing voltage versus time, amplitude versus frequency is shown. Time is not shown, because the signals are assumed to be repetitive. Clearly, AC circuits operate on both repetitive and nonrepetitive signals. The analysis assumes repetitive signals, because an AC circuit's response is continuous. It does not have the ability to recognize sequences of signals in a digital sense and modify its behavior accordingly. Pure sine waves are represented by a vertical line on a frequency domain plot to indicate their amplitude at a single specific frequency. Since most real-world signals are not perfect sine waves, it is common to observe a frequency distribution around a single central frequency of interest.

While not strictly necessary, frequency-domain plots are usually drawn with *decibel* (dB) scales that are inherently logarithmic. The decibel is a relative unit of measurement that enables the comparison of power levels *(P)* entering and exiting a circuit. On its own, a decibel value does not indicate any absolute power level or measurement. A decibel is defined as a ratio of power entering and leaving a circuit:

$$\text{dB} = 10\log_{10}\frac{P_{OUT}}{P_{IN}}$$

When the input and output power are identical, a level of 0 dB is achieved. Negative decibel levels indicate attenuation of power through the circuit, and positive decibel levels indicate amplification.

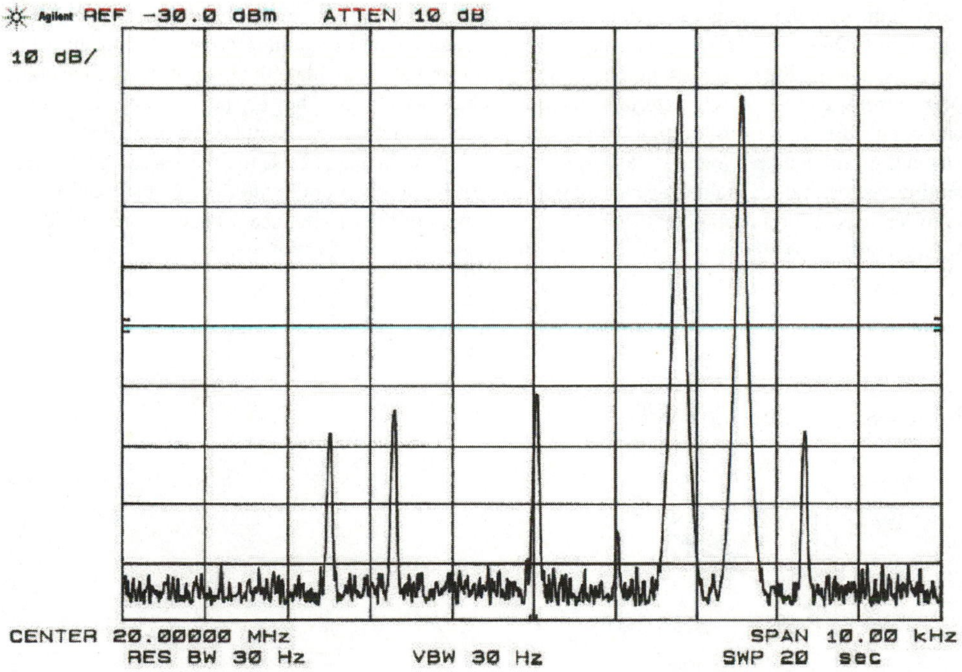

FIGURE 12.16 Spectrum analyzer frequency/amplitude plot. *(Reprinted with permission from Agilent Technologies.)*

Decibel measurements can also be calculated using voltage instead of power assuming that voltages are applied across a constant resistance. Substituting Ohm's law for the power terms yields the following definition:

$$\mathrm{dB} = 10\log_{10}\frac{\frac{V_{OUT}^2}{R}}{\frac{V_{IN}^2}{R}} = \frac{V_{OUT}^2}{V_{IN}^2} = 20\log_{10}\frac{V_{OUT}}{V_{IN}}$$

There are some common decibel levels with inherent reference points used to indicate the strength of an audio or radio signal that transform the decibel into an absolute measurement. The unit dBm is commonly seen in audio applications, where 0 dBm is one milliwatt. Therefore, 8 dBm is 6.3 mW. Radio applications sometimes use the absolute unit dBμV, where 0 dBμV is 1 μV.

Decibel units enable the analysis of signals with very low and very high amplitudes. A typical radio receiver may be sensitive enough to detect signals with –90 to –110 dBm of strength. Trying to work with such small numbers without a logarithmic scale is rather awkward.

A common decibel value that is used in frequency domain analysis is –3 dB, which corresponds to a roughly 50 percent reduction in power through a circuit element ($10 \log_{10} 0.5 \approx -3$). Because decibels are a logarithmic function, the addition of decibel measurements corresponds to a multiplication of the underlying absolute values. Therefore, –6 dB corresponds to quartering the power through a circuit (0.5×0.5 = –3 dB + –3 dB = –6 dB), and –9 dB corresponds to approximately 12.5 percent of the power passing through.

Frequency domain analysis takes into account the real and imaginary components of impedance to form the complex number expression for impedance, $Z = R + jX$, as already discussed. When combined with Ohm's law, currents and voltages with both real and imaginary components result from the impedance being a complex number. It is often desirable to calculate the magnitude of such complex currents and voltages to determine the peak values in a circuit. They are peak values, rather than static, because AC signals are time varying. If the real and imaginary components of an impedance, current, or voltage are plotted on a Cartesian grid as done in Fig. 12.17, the magnitude of their resulting vector can be obtained according to the Pythagorean theorem's statement of the relationship between legs of a right triangle: the square of the length of the hypotenuse (c) equals the sum of the squares of the two other legs (a and b). Therefore,

$$c = \sqrt{a^2 + b^2}$$

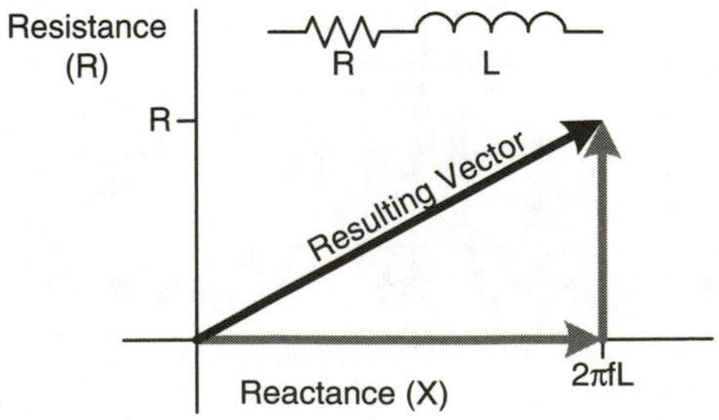

FIGURE 12.17 Finding the magnitude of an impedance.

In this example, a resistor and inductor are placed in series with a resulting impedance, $Z = R + j2\pi fL$. The magnitude is expressed as follows:

$$|Z| = \sqrt{R^2 + (2\pi fL)^2}$$

12.9 LOWPASS AND HIGHPASS FILTERS

Filtering is the general process of attenuating the energy of a certain range of frequencies while passing a desired range with little or no attenuation. The noise filtering examples discussed previously are designated *lowpass* filters, because they are designed to pass lower frequencies while attenuating higher frequencies above a certain threshold. They are also *passive* filters, because the circuits are constructed entirely of passive components (resistors, inductors, and capacitors) without any active components (e.g., transistors) to provide amplification. Passive filters are very practical because of their simplicity and small size. They are suitable for applications wherein the signal that is desired to pass through the filter has sufficient amplitude to be used after the filter. In situations in which very weak signals are involved (e.g., tuning a radio signal), amplification may be necessary before, after, or within the filter. Such filters are termed *active* filters. Whether a filter is active or passive, the underlying analysis of how desired frequencies are passed and undesired frequencies are attenuated remains the same.

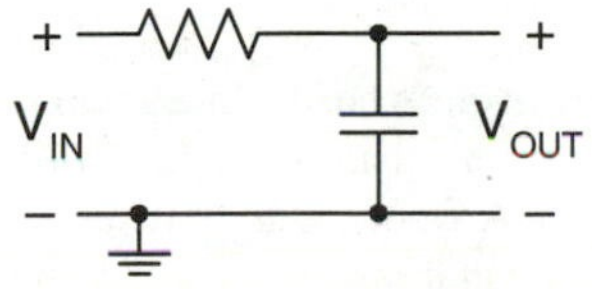

FIGURE 12.18 First-order RC lowpass filter.

A filter's frequency response can be determined by combining basic circuit principles with the frequency-dependent impedance characteristics of capacitors and inductors. The most basic lowpass filter is a series resistor and shunt capacitor as shown in Fig. 12.18. This is referred to as a first-order lowpass filter, because the circuit contains a single AC element, a capacitor. Since a capacitor's impedance drops as the frequency increases, higher frequencies are short-circuited to ground.

The filter's output voltage can be calculated by combining the impedances of each element into a single voltage-divider expression,

$$V_{OUT} = V_{IN}\frac{Z_C}{Z_C + Z_R} = V_{IN}\frac{\frac{1}{2\pi fC}}{\frac{1}{\frac{1}{2\pi fC} + R}} = V_{IN}\frac{1}{1 + 2\pi fRC}$$

Rather than continuing to explicitly reference the input and output voltages, it is common to refer to the filter's *transfer function*, or *gain* (A), which is the relationship between input and output voltage. The gain is obtained by simply dividing both sides of the equation by the input voltage.

$$A_{FILTER} = \frac{V_{OUT}}{V_{IN}} = \frac{1}{1 + 2\pi fRC}$$

A lowpass filter is considered to be effective starting at a certain *cutoff frequency* (f_C), the frequency at which its power gain in halved. This is also called the *half-power point*. Recall that a decibel level is calculated as $10 \log_{10} A_{POWER}$. Therefore, the filter's gain declines by $10 \log_{10} 0.5 = -3$

dB at f_C. Because a passive filter has an ideal gain of zero at DC (no amplification), its gain at f_C is –3 dB.

However, the gain expression for this filter is in terms of voltage, not power. The relationship between voltage gain and decibel level is a multiplier of 20 rather than 10. Therefore, rather than looking for a gain of one-half at f_C, we must find a gain equal to the square root of one half because

$$-3\text{ dB} = 20\log_{10}\frac{1}{\sqrt{2}}$$

The filter's gain expression is complex and must be converted into a real magnitude. Conveniently, the magnitude representation looks fairly close to what we are looking for.

$$|A_{FILTER}| = \frac{1}{\sqrt{1 + (2\pi fRC)^2}}$$

By setting the denominator equal to the square root of 2, it can be observed that a gain of –3 dB is achieved when $2\pi f = 1 \div RC$. Furthermore, it can be shown that the lowpass filter's gain declines at the rate of 20 dB per radian frequency decade because, for large *f*, the constant 1 in the denominator has an insignificant affect on the magnitude of the overall expression,

$$20\log_{10}\frac{1}{2\pi fRC}$$

Rather than trying to keep all of these numbers in one's head while attempting to solve a problem, it is common to plot the transfer function of a filter on a diagram called a *Bode plot*, named after the twentieth century American engineer H. W. Bode. A Bode magnitude plot for this lowpass filter is shown in Fig. 12.19. Filters and AC circuits in general also modify the phase of signals that pass through them. More complete AC analyses plot a circuit's phase transfer function using a similar Bode phase diagram. Phase diagrams are not shown, because these topics are outside the scope of this discussion.

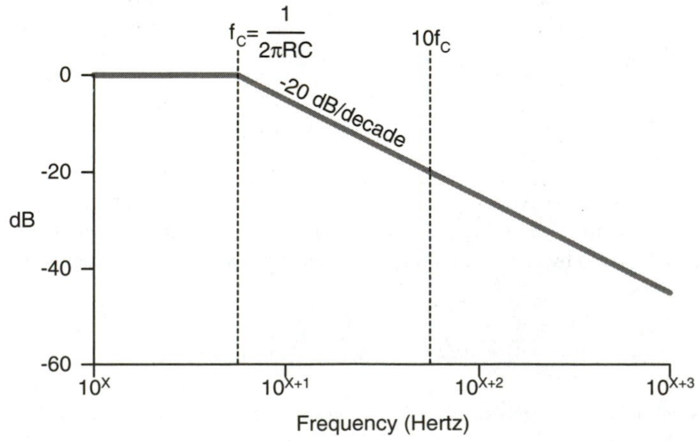

FIGURE 12.19 Bode magnitude plot for RC lowpass filter.

This Bode plot is idealized, because it shows a constant gain up to f_C and then an abrupt roll-off in filter gain. A real magnitude calculation would show a smooth curve that conforms roughly, but not exactly, to the idealized plot. For first-order evaluations of a filter's frequency response, this idealized form is usually adequate.

A filter's *passband* is the range of frequencies that are passed by a filter with little or no attenuation. Conversely, the *stopband* is the range of frequencies that are attenuated. The trick in fitting a filter to a particular application is in designing one that has a sufficiently sharp roll-off so that unwanted frequencies are attenuated to the required levels, and the desired frequencies are passed with little attenuation. If the passband and desired stopband are sufficiently far apart, a simple first-order filter will suffice, as shown in Fig. 12.20. Here, the undesired noise is almost three decades beyond the signal of interest. The RC filter's 20-dB/decade roll-off provides an attenuation of up to 60 dB, or 99.9 percent, at these frequencies.

As the passbands and stopbands get closer together, more complex filters with sharper roll-offs are necessary to sufficiently attenuate the undesired frequencies while not disturbing those of interest. Incorporating additional AC elements into the filter design can increase the slope of the gain curve beyond f_C. A second-order lowpass filter can be constructed by substituting an inductor in place of the series resistor in a standard RC circuit as shown in Fig. 12.21.

The LC filter's gain can be calculated as follows:

$$A = \frac{Z_C}{Z_C + Z_L} = \frac{\frac{1}{2\pi fC}}{\frac{1}{2\pi fC} + 2\pi fL} = \frac{1}{1 + (2\pi f)^2 LC}$$

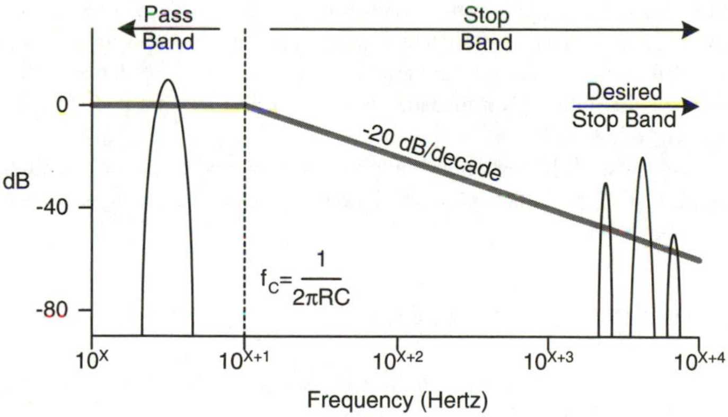

FIGURE 12.20 Widely separated pass and stopbands.

RC cutoff freq (low pass) = $\frac{1}{2\pi RC}$

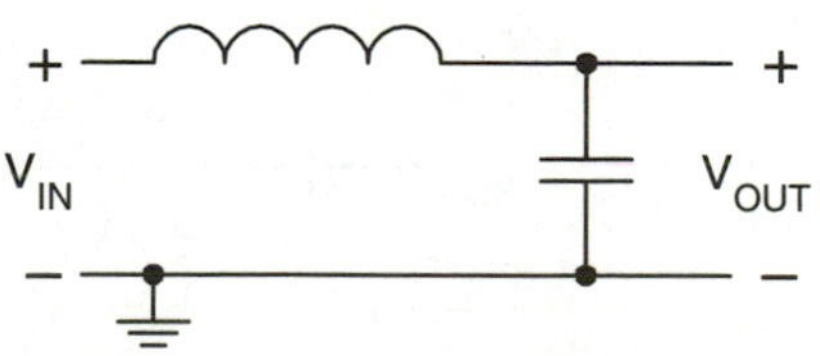

FIGURE 12.21 Second-order LC lowpass filter.

The cutoff frequency is determined as was done previously for the first-order filter, for which the magnitude of the transfer function's denominator is the square root of 2. To meet this criterion, $2\pi f = 1/\sqrt{LC}$. The additional AC element in the filter introduces a frequency-squared term that doubles the slope of the gain curve beyond f_C. Therefore, this second-order lowpass filter declines at 40 dB per decade instead of just 20 dB per decade.

Higher-order filters can be created by adding LC segments to the basic second-order circuit to achieve steeper gain curves as shown in Fig. 12.22. These basic topologies are commonly referred to as *T* and *pi* due to their graphical resemblance to the two characters.

Lowpass filters are probably the most common class of filters used in purely digital systems for purposes of noise reduction. However, when analog circuits are mixed in, typically for interface applications including audio and radio frequencies, other types of filters become useful. The inverse of a lowpass filter is a *highpass* filter, which attenuates lower frequencies and passes higher frequencies. A first-order RC highpass filter is very similar to the lowpass version except that the topology is reversed as shown in Fig. 12.23. Here, the capacitor blocks lower frequency signals but allows higher frequencies to pass as its impedance drops with increasing frequency.

As done previously, the transfer function can be calculated by combining the impedances of each element into a single expression:

$$A = \frac{Z_R}{Z_R + Z_C} = \frac{R}{R + \frac{1}{2\pi fC}} = \frac{2\pi fRC}{2\pi fRC + 1}$$

It can be observed from the transfer function that, as the frequency approaches DC, the filter's gain approaches 0. At higher frequencies, the filter's gain approaches 1, or 0 dB. First- and second-order highpass filters have gain slopes of 20 and 40 dB per decade, respectively, as mirror images of the lowpass filter frequency response. A second-order highpass filter can be created by substituting an appropriate inductor in place of the shunt resistor. The inductor appears as a short circuit to the negative voltage rail at low frequencies and gradually increases its impedance as the frequency rises. At high frequencies, the inductor's impedance is sufficiently high that it is shunting almost no current to the negative voltage rail.

Just as for lowpass filters, higher-order highpass filters can be created by adding LC segments in T and pi topologies, albeit with the locations of the inductors and capacitors swapped to achieve the highpass transfer function.

12.10 BANDPASS AND BAND-REJECT FILTERS

Some signal manipulation applications require the passing or rejection of a selective range of frequencies that do not begin at DC (a lowpass filter) nor end at an upper limit that is conceptually in-

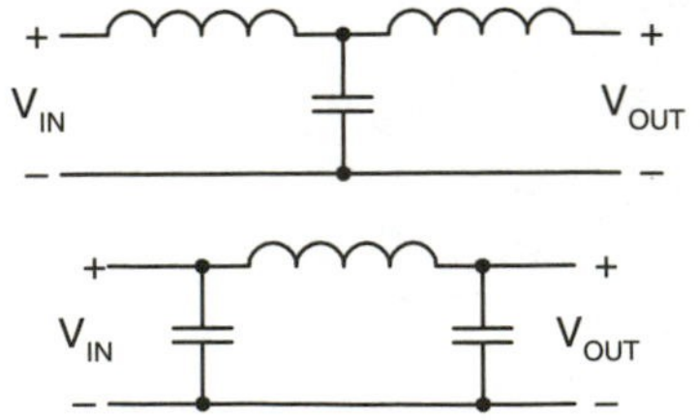

FIGURE 12.22 Third-order LC lowpass filters.

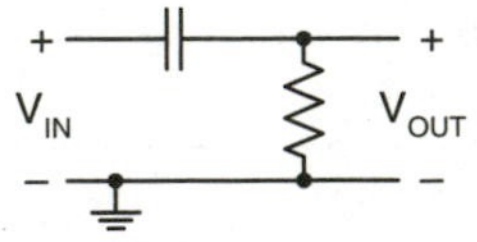

FIGURE 12.23 First-order RC highpass filter.

finity (a highpass filter). When a radio or television is tuned to a certain channel, a *bandpass* filter selects a certain narrow range of frequencies to pass while attenuating frequencies above and below that range. Bandpass filters may have limited utility in a typical digital system, but certain interface circuitry may require bandpass filtering to attenuate low-frequency and DC content while also reducing high frequency noise.

Bandpass filter design can get fairly complex, depending on the required AC specifications. Design issues include the width of the passband and the slope of the gain curve on either side of the passband. Two basic bandpass topologies are shown in Fig. 12.24. These are second-order circuits that pair an inductor and capacitor together in either a series or shunt configuration. The series topology operates by blocking the high-frequency stopband with the inductor and the low-frequency stopband with the capacitor. In the middle is a range of frequencies that are passed by both elements. The shunt topology operates by diverting the high-frequency stopband to ground with the capacitor and the low-frequency stopband with the inductor. This topology takes special advantage of the parallel LC resonant circuit mentioned previously.

Each of these topologies can be designed with passbands of arbitrary width. The bandwidth of the filter has a direct impact on the shape of its transfer function. However, for a second-order bandpass filter, each AC element contributes a –20 dB per decade roll-off on either side of the center frequency. Therefore, a narrow ideal second-order bandpass filter has the transfer function shown in Fig. 12.25, and its center frequency is

$$f_C = \frac{1}{2\pi\sqrt{LC}}$$

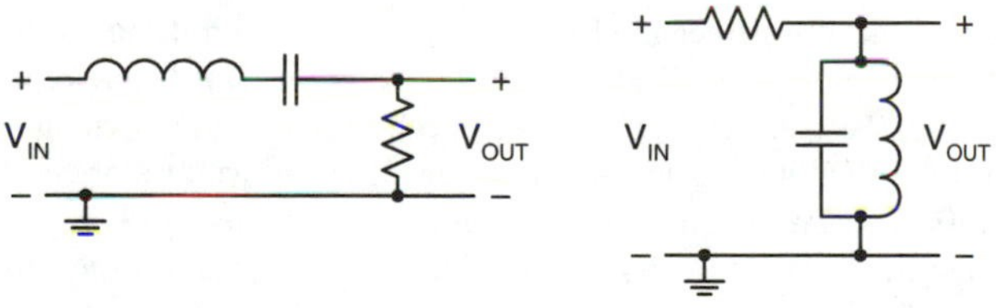

FIGURE 12.24 Second-order bandpass filter topologies.

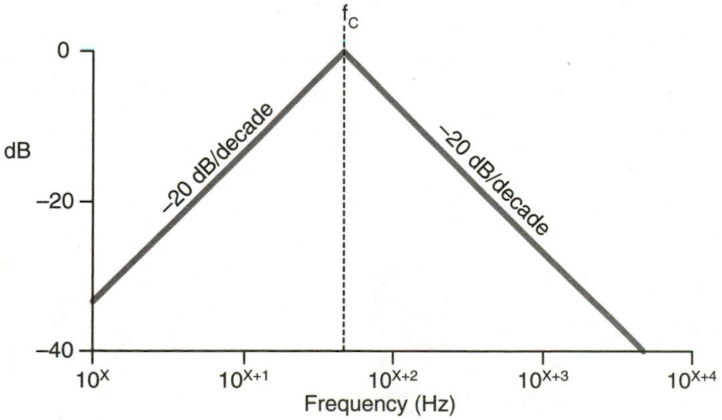

FIGURE 12.25 Idealized bandpass filter frequency response.

By now, the symmetry of filtering leaves no surprise that a bandpass topology can be rearranged to form a *band-reject* filter, often called a *notch* filter. Whereas the bandpass filter passes a narrow range of frequencies, a band-reject filter attenuates a narrow range of frequencies. The two basic second-order band-reject filter topologies are shown in Fig. 12.26. In the reverse of a bandpass filter, the series topology takes advantage of the LC circuit's high impedance at its resonant frequency to block the incoming signal. Similarly, the shunt topology's low combined impedance at a narrow range of frequencies diverts that energy to ground.

Digital systems often require only that a filter attenuate unwanted frequencies, typically noise, to the point at which the digital circuitry is not adversely impacted by that noise. Clock generators and clock circuits are usually the digital components that are most sensitive to noise. Alternatively, digital systems that incorporate sensitive analog interface components may require filters to separate and "clean up" the power, data, and control signals that pass between the digital and analog circuitry. Such applications often require only a first-order of approximation to determine cutoff or center frequencies for filters.

Designing filters for analog applications where specific characteristics of transfer functions have a major impact on circuit performance requires far more detailing and circuit analysis theory than has been presented here. If your filter is more analog than digital in its application, you are strongly advised to spend time in obtaining a thorough understanding of AC circuit analysis and filter design.

12.11 TRANSFORMERS

When two inductors are placed together in close proximity, they exhibit mutual inductance where one coil's magnetic field couples onto the other coil and vice-versa. This behavior may be undesired in many situations and can be largely avoided by physically separating individual inductors. However, the phenomenon of mutual inductance has great benefit when two or more coils are assembled together to form a *transformer*. Mutual inductance is enhanced not only by close physical proximity, but also by winding the transformer's coils around common cores made from ferrous metals that conduct magnetic fields. Transformers have many uses, but their basic function is to transfer AC energy from one coil to another without having a DC connection between the two ends. A transformer does not pass DC because there is no direct connection between the two coils.

Neglecting losses resulting from finite resistance and less-than-ideal efficiency in magnetic field coupling, the voltage excited in one coil winding is related to that created in another winding by the proportion of the number of winding turns (N) in each coil. A transformer with two windings is said to have a *primary* and a *secondary* winding. The voltage relationship of each winding in an ideal transformer is as follows:

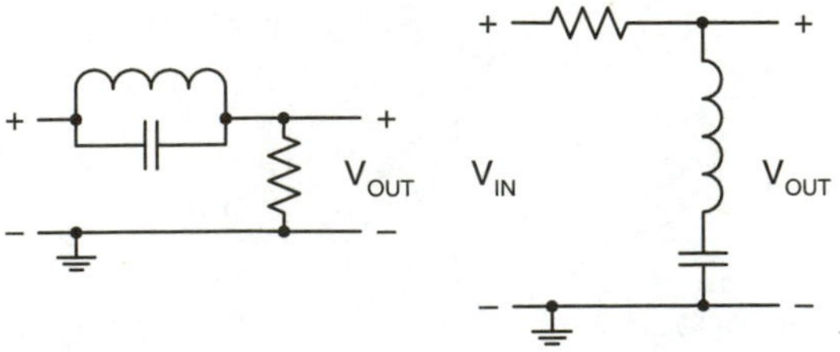

FIGURE 12.26 Second-order band-reject filter topologies.

$$\frac{V_{PRIMARY}}{V_{SECONDARY}} = \frac{N_{PRIMARY}}{N_{SECONDARY}}$$

A transformer is a passive component; it has no capability of amplifying a signal. Consequently, an ideal transformer passes 100 percent of the power applied to it and satisfies the equation $V_S I_S = V_P I_P$. If the primary coil has more windings than the secondary coil, an AC signal of lesser magnitude will be induced on the secondary coil when an AC signal is applied to the primary coil. The current flowing through the secondary coil will be higher than that in the primary so that conservation of energy is preserved. In reality, of course, a transformer has less than 100 percent efficiency as a result of parasitic properties, including finite resistance of the coils and less-than-perfect magnetic coupling.

One of the most common uses of a transformer is in power distribution in which AC power is either stepped up or stepped down, depending on the application. Figure 12.27 shows a basic transformer with a 120 VAC signal injected into the primary coil and a load resistor on the secondary coil. The ratio of the primary to secondary windings is 10:1; perhaps the primary coil has 1,000 windings whereas the secondary has 100. The result is a step-down of high-voltage power from a wall outlet to a more manageable 12 VAC. This illustrates why AC power distribution is so convenient: voltages can be arbitrarily transformed without any complicated electronic circuits. It is advantageous to distribute power at a higher voltage to reduce the current draw for a given power level. Lower current means lower I^2R power losses in the distribution wiring. The 12 VAC transformer output may power the voltage regulator of a digital circuit. If the digital circuit draws 10 A at 12 VAC, it will draw only 1 A at 120 VAC.

Transformers are critical to power distribution both at the system level and at the generation and utility levels. Power is stepped up at generating plants with transformers to as high as 765,000 V for efficient long-distance distribution. As the power gets closer to your home or office, it is stepped down to intermediate levels and finally enters the premises at 120 and 240 VAC.

Aside from power supply applications, transformers are also used for filtering and impedance matching of interface signals. The use of a transformer as a filter requires knowledge of the physical orientations of the primary and secondary windings with respect to one another. Each winding in a transformer has two terminals. When a signal is applied to the primary coil, a decision is made as to which terminal is connected to the positive portion of the circuit and which is connected to the negative portion (ground in some applications). That signal will induce a signal of equivalent polarity on the secondary coil when the appropriate choice of positive and negative terminals is made at the other end of the transformer. Alternatively, a signal of opposite polarity will be induced if the secondary coil's positive and negative connections are swapped. The graphical convention of distinguishing the matching terminals of the primary and secondary coils is by placing matching reference dots next to one terminal of each coil as shown in Fig. 12.28. The coils may be drawn with their dots on the same side or on opposite ends of the transformer.

The relative polarity of the primary and secondary windings is important for filtering applications to ensure that magnetic fields in each coil either add to or cancel each other as appropriate. Trans-

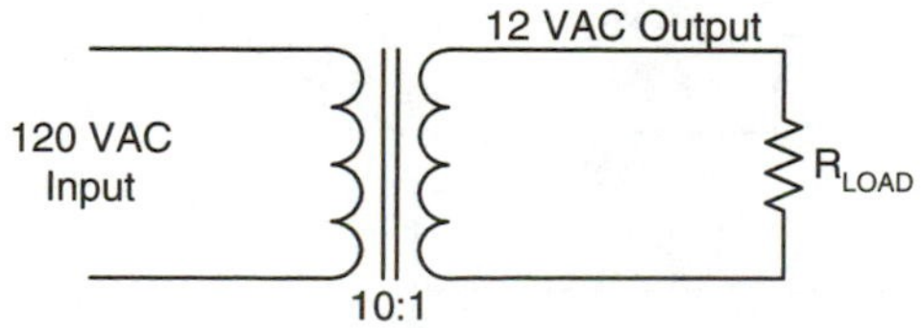

FIGURE 12.27 Basic transformer operation.

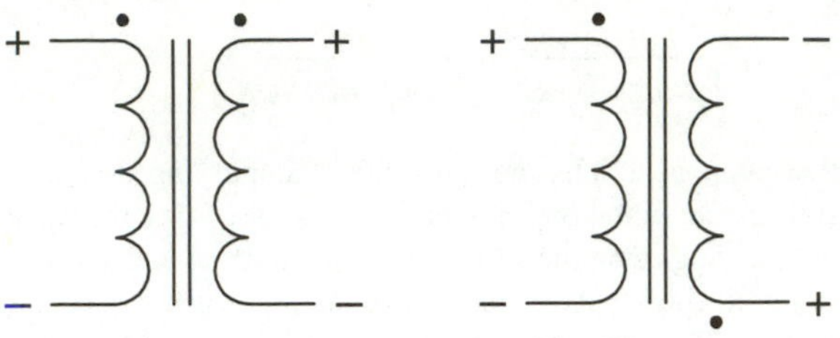

FIGURE 12.28 Transformer coil polarity graphical representation.

formers are sometimes used to attenuate *common mode noise*. A common-mode signal is one wherein current flows in the same direction on both halves of a circuit. This is in contrast to a differential-mode signal wherein current flows in opposite directions. As shown in Fig. 12.29, common-mode current can flow when a return path exists, usually ground.

To attenuate common-mode noise, the transformer is inserted into the circuit as shown in Fig. 12.30 such that each circuit half passes through the coils oriented in the same direction. Each coil's magnetic field has the same magnitude and phase as the other, because each coil has the same common-mode current passing through it. The magnetic fields do not cancel each other out, causing a voltage drop across each coil that attenuates the noise. In other words, the coils present high impedance to the common-mode portion of the signal passing through them. The differential signal, typically a desired signal carrying meaningful data, passes through the common-mode filter, because its currents are flowing in opposite directions. The magnetic fields in the coils cancel each other out, because they are of equal magnitude but opposite phase. Without a magnetic field built up, the coils do not develop a voltage drop and thereby present low impedance to the desired signal.

Impedance matching between a driver and a load is another common use of a transformer. It can be shown mathematically that maximum power is delivered to a load when the source's internal im-

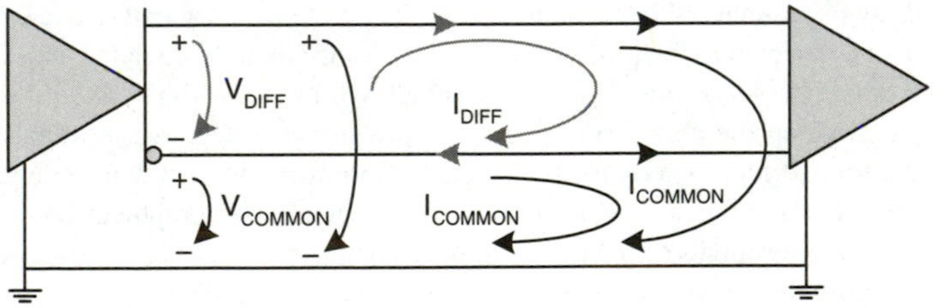

FIGURE 12.29 Common-mode current flow.

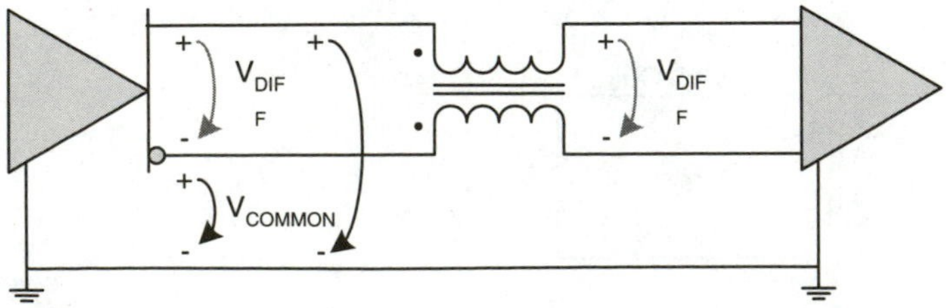

FIGURE 12.30 Common-mode filter transformer.

pedance equals that of the load. Under these conditions, only half of the total power is delivered to the load, the remaining half being dissipated by the source impedance. For unequal load and source impedances, more or less voltage and current will be delivered to the load, but always in proportions that result in less than 50 percent of the total power. It may be desirable to deliver maximum power to a load in situations where the impedance of a driver (e.g., audio amplifier) does not match that of its load (e.g., speaker). A transformer can be used to isolate the source and load from each other and present each with the necessary impedance for maximum power transfer.

If a load resistor is connected to a transformer's secondary coil as shown in Fig. 12.31, Ohm's law can be used to relate V_P and I_P by R_{LOAD}. To simplify the analysis, the secondary coil is said to have one winding and the primary N windings, for a simple N:1 ratio. Therefore,

$$\frac{V_P}{V_S} = \frac{N}{1} \text{ and } \frac{I_P}{I_S} = -\frac{1}{N}$$

More primary windings increase the voltage ratio between the primary and secondary but decrease the current because a transformer cannot amplify a signal and, hence, $V_P I_P = -V_S I_S$. The current in the secondary winding flows in the direction opposite of that in the primary, because the primary is feeding current into the transformer and the load is pulling current out.

It is known that the relationship between the voltages on either side of the transformer is given by $V_P = NV_S$ and that, maintaining proper current polarity, $V_S = -I_S R_{LOAD}$. Therefore, $V_P = -NI_S R_{LOAD}$. The secondary current term can be substituted out of the equation by using the relationship of currents to coil windings, $I_S = -NI_P$, to yield $V_P = N^2 I_P R_{LOAD}$. Finally, Ohm's law is used to remove the primary voltage and current terms: $R_{SOURCE} = N^2 R_{LOAD}$. Put in more general terms, $R_P = N^2 R_S$.

A high ratio of primary to secondary windings will present the signal source with an apparently higher impedance than is actually connected to the secondary. Similarly, the load will be presented with an apparently lower impedance than is actually present in the source. Considering the circuit in Fig. 12.32, an 8-Ω speaker can be matched to an amplifier with a 100-Ω output impedance using a transformer with a 3.5:1 winding ratio.

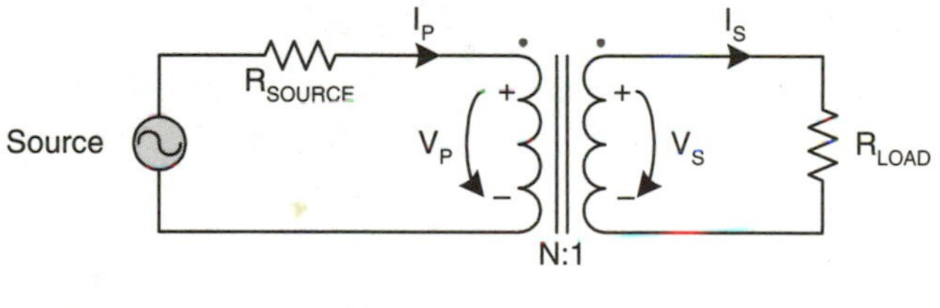

FIGURE 12.31 Impedance transformation.

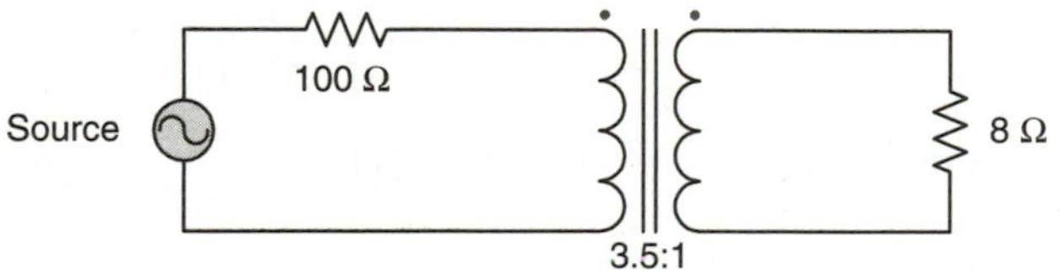

FIGURE 12.32 Amplifier/speaker impedance matching.

CHAPTER 13
Diodes and Transistors

Most of the semiconductors that a digital system employs are fabricated as part of integrated circuits. Yet there are numerous instances in which discrete semiconductors, most notably diodes and transistors, are required to complete a system. Diodes are found in power supplies, where they serve as rectifiers and voltage references. It is difficult these days *not* to find a light emitting diode, or LED, in one's immediate vicinity, on some appliance or piece of electronic equipment. Discrete transistors are present in switching power supplies and in circuits wherein a digital IC must drive a heavy load. There are many other uses for diodes and transistors in analog circuit design, most notably in signal amplification. These more analog topics are not discussed here.

Diodes and transistors are explained in this chapter from the perspective of how they are applied in the majority of digital systems. As such, the level of theory and mathematics used to explain their operation is limited. The first portion of the chapter introduces diodes and provides examples of how they are used in common power and digital applications. Bipolar junction and field-effect transistors are discussed in the remainder of the chapter. The intent of this chapter is to show how diodes and transistors can be put to immediate and practical use in a digital systems context. As such, useful example circuits are presented whenever possible.

13.1 DIODES

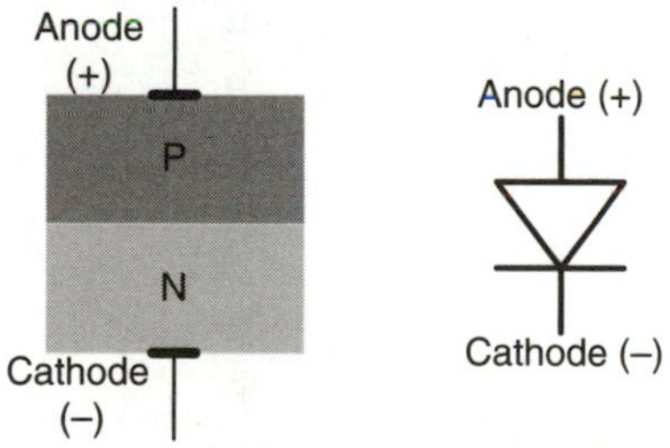

FIGURE 13.1 Diode structure and graphical representation.

An ideal diode is a nonlinear circuit element that conducts current only when the device is *forward biased*, i.e., when the voltage applied across its terminals is positive. It thereby behaves as a one-way electrical valve that prevents current from flowing under conditions of *reverse bias*. A diode has two terminals: the *anode* and *cathode*. For the diode to be forward biased, the anode must be at a more positive voltage than the cathode. Diodes are the most basic semiconductor structures and are formed by the junctions of two semiconductor materials of slightly differing properties. In the case of a silicon diode, the anode is formed from positively doped silicon, and the cathode is formed from negatively doped silicon. Along this *pn junction* is where the physical phenomenon occurs that creates a diode. Figure 13.1 shows the general silicon structure of a diode and its associated symbolic representation.

Real diodes differ from the ideal concept in several ways. Most significantly, a real diode must be forward biased beyond a certain threshold before the device will conduct. This threshold is called the *forward voltage*, V_F. A diode conducts very little current below V_F, measured in micro- or nanoamps. The relationship between a diode's current and voltage is exponential and therefore increases rapidly around V_F. Above V_F, the diode presents very low impedance and appears almost like a short circuit, or a piece of wire. A typical silicon diode's I-V characteristic is shown in Fig. 13.2 with V_F of approximately 0.7 V. For many applications, especially those in digital systems, a diode's I-V characteristic can be simplified to a step function from zero to infinite current at the forward voltage. Real diodes also have a reverse breakdown voltage, called the *Zener* voltage, V_Z, at which point they will conduct under reverse-biased conditions. Normal diodes are usually not subjected to reverse-bias voltages sufficient to achieve significant conduction. Special-purpose Zener diodes are designed specifically to operate under reverse-bias conditions and are commonly used in voltage reference and regulation applications.

Saying that a diode conducts "infinite" current beyond V_F really means that its impedance becomes so low that it no longer becomes the limiting factor in a circuit. If a typical diode is connected directly to a battery such that it is forward biased, the diode will form nearly a short circuit, which will cause a large current flow. Fairly soon afterward, the diode will likely fail due to thermal overload. Diodes are specified with maximum *forward currents*, I_F. Exceeding I_F causes the diode to dissipate more power than it is designed for, usually with destructive consequences.

When a diode is used in an application wherein it may be forward biased and driven with excessive current, a current-limiting resistor is inserted into the circuit to keep the diode within its specified operating limits. Diodes are useful for providing a fixed voltage reference regardless of a circuit's operating voltage. The circuit shown in Fig. 13.3 takes advantage of a diode's relatively static forward voltage with respect to current. A loosely regulated 12-V supply may have a tolerance of ±20 percent—a range of 9.6 to 14.4 V. If a resistor divider is used to generate a reference voltage, its accuracy could be no better than that of the 12-V supply. Some applications require a more accurate voltage reference with which to sense an incoming signal. A 1N4148 exhibits V_F = 0.7 V at

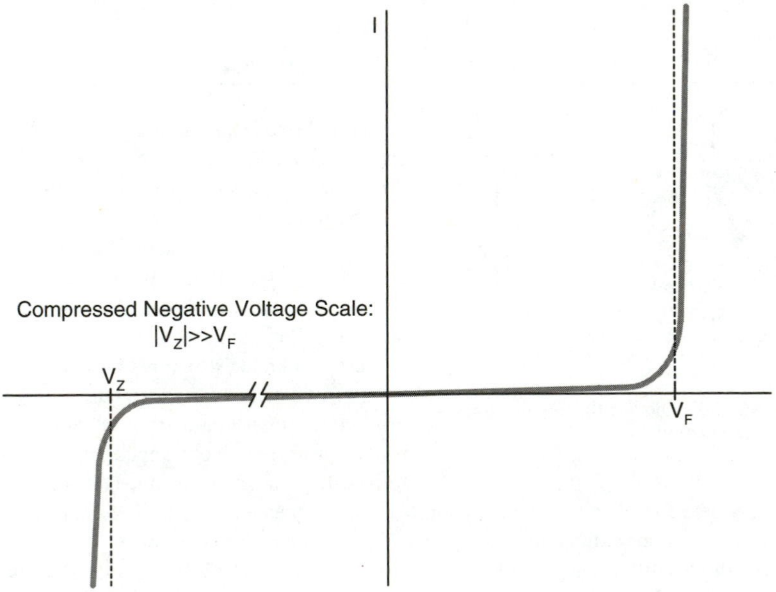

FIGURE 13.2 Silicon diode I-V characteristic.

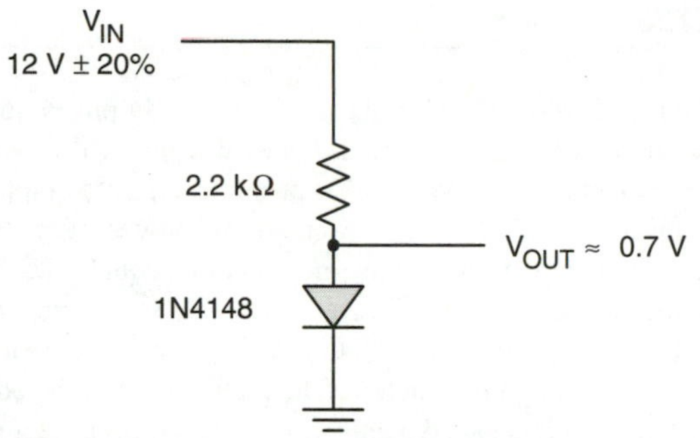

FIGURE 13.3 Diode-based voltage reference.

5 mA under typical conditions.* The 2.2-kΩ resistor limits the current through the diode to approximately 5 mA when V_{IN} = 12 V. If the input changes by 20 percent and causes a corresponding change in the current, the diode voltage changes by a small fraction. Using a basic small signal diode in this manner is an effective scheme for many applications. If tighter tolerance is desired, more stable voltage reference diodes are available. Thermal overload is not a problem for this diode, because its power dissipation is relatively constant at 0.7 V × 5 mA = 3.5 mW.

Diodes are available with a broad spectrum of characteristics. Aside from silicon diodes, there are *Schottky* diodes that exhibit lower forward voltages of under 0.5 V. Lower forward voltages provide benefits for high-power applications in which heat and power dissipation are prime concerns. Reduced V_F means reduced power. Diodes are manufactured in a variety of packages according to the amount of power that they are designed to handle. Small-signal diodes are not intended to handle much power and are available in small, surface mount packages. At the other extreme, diodes can be as large as hockey pucks for very high-power applications. Small-signal diodes are manufactured with varying response times to changes in voltage. A diode can be used to *clip* a signal to prevent it from exceeding a certain absolute voltage, as shown in Fig. 13.4. As the signal's edge rate increases, a slower diode may not respond fast enough to be effective. If a single diode's forward voltage is insufficient, multiple diodes can be placed in series to increase the clipping threshold. Some of the more common small-signal diodes used in digital circuits include the leaded 1N914 and 1N4148 devices, and their surface mount equivalents, the SOT-23 MMBD914 and MMBD4148.

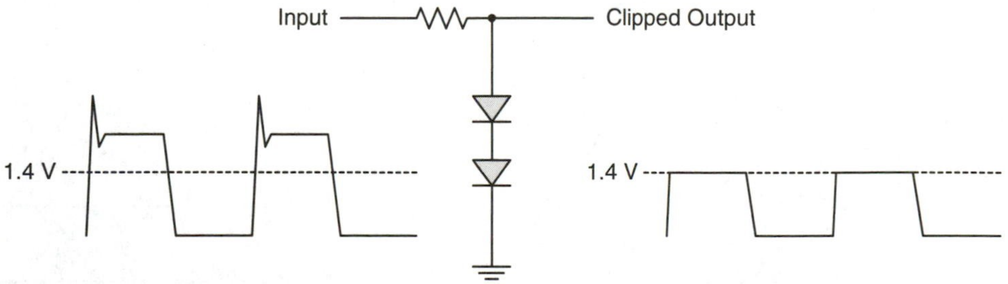

FIGURE 13.4 Clipping a signal with a diode.

* 1N4148, Fairchild Semiconductor Corporation, 2002, p. 2.

13.2 POWER CIRCUITS WITH DIODES

A major use for diodes is in the *rectification* of AC signals, specifically in power supplies in which the conversion from AC to DC is required. Small-signal diodes can be used as rectifiers in non-power or low-power applications. Larger diodes with higher power ratings are employed when constructing power supply circuits meant to provide more power. An AC power signal is a sine wave of arbitrary amplitude that is centered about 0 V. Its voltage peaks are of equal magnitude above and below 0 V. A digital circuit requires a steady DC power supply. The first step in creating a steady DC power supply is to rectify the AC input such that the negative AC sine wave excursions are blocked. Figure 13.5 shows a single diode performing this function. The rectified output is reduced in voltage by the diode's forward voltage. This circuit is called a *half-wave* rectifier, because it passes only half of the incoming power signal. Once rectified, capacitors and inductors can smooth out (lowpass filter) the rectified AC signal to create a steady DC output.

The single-diode half-wave rectifier does the job, but does not take advantage of the negative portion of the AC input. Four diodes can be assembled into a *full-wave bridge rectifier* that passes the positive portion of the sine wave and inverts the negative portion relative to the DC ground. This circuit is shown in Fig. 13.6. The bridge rectifier works by providing a current conduction path through the resistor to ground regardless of the polarity of the incoming AC signal. When the AC input is positive with respect to the polarity markings shown in the diagram, diodes D1 and D3 are forward biased, conducting current from D1 through the resistor, then through D3 to the negative AC input wire. When the AC input is negative during the next half of the sinusoid, D2 and D4 are forward biased and allow current to flow in the same direction through the resistor. The result is that a positive voltage is always developed across the load with respect to ground. Note that, because of the two diodes in series with the load, the rectified output voltage is reduced by twice the diodes' forward voltage.

Power rectifier circuits are generally found in systems wherein a high-voltage input (e.g., 120 VAC) must be converted into a low-voltage output such as +5 VDC to power a digital logic circuit. Transformers are used in conjunction with bridge rectifiers to step down the high-voltage AC input to a more appropriate intermediate level that is much closer to the final voltage level required by the system. A power filter circuit can then be used to smooth the heavily rippled rectified signal into a more stable DC input. Finally, a voltage regulator performs the final adjustments to convert the

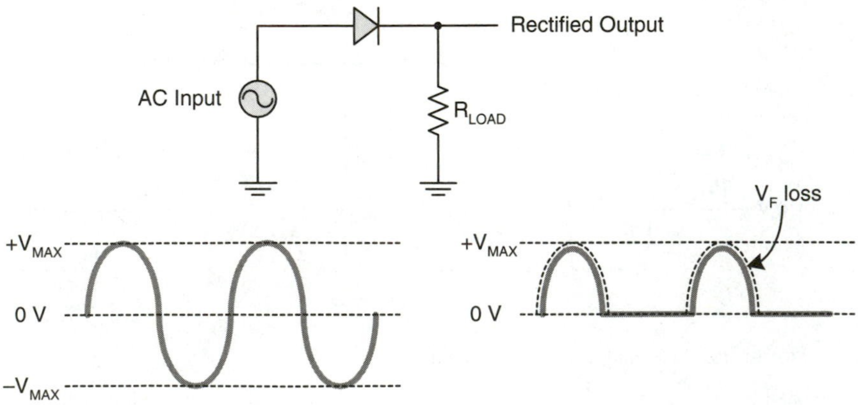

FIGURE 13.5 Half-wave rectifier circuit.

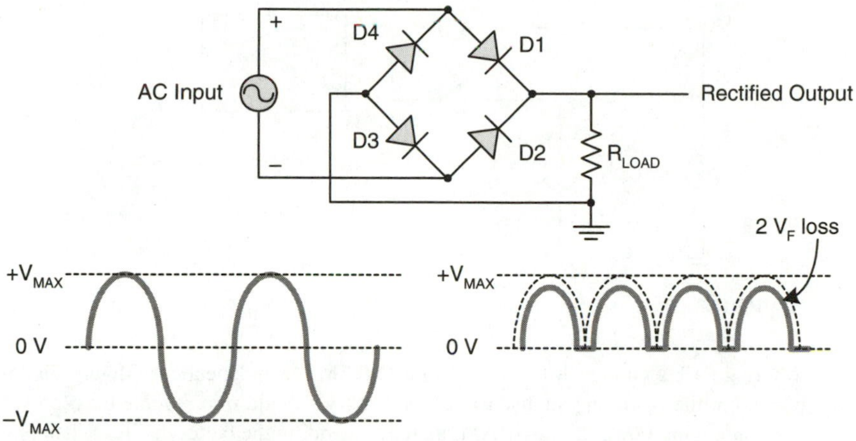

FIGURE 13.6 Bridge rectifier circuit.

intermediate voltage into a more accurate digital supply voltage. This common AC-to-DC power supply configuration is illustrated in Fig. 13.7.

Another power application of diodes is in combining multiple power supplies to feed a single component or group of components while ensuring that the failure or disappearance of one supply does not cause that component to lose power and cease operation. This concept relies on the fact that a standard diode will not conduct under normal reverse-bias conditions. As shown in Fig. 13.8, each power supply is isolated by a diode whose cathodes form a common voltage supply node for a circuit. Under normal operating conditions, each diode is forward biased, because the respective power supplies are functioning. When one supply fails, its associated diode becomes reverse biased, thereby preventing the failing supply from pulling power from the functioning supply and causing the system to fail. These diodes are often called *OR-ing diodes,* because they perform a logical OR function on the power supplies.

Diode OR-ing circuits are also seen in battery-backup applications in which it is desired to keep a low-power static RAM chip powered by a battery when the main power supply is turned off. A typical scenario is a higher-voltage operating supply (e.g., +5 V) and a lower-voltage data-retention bat-

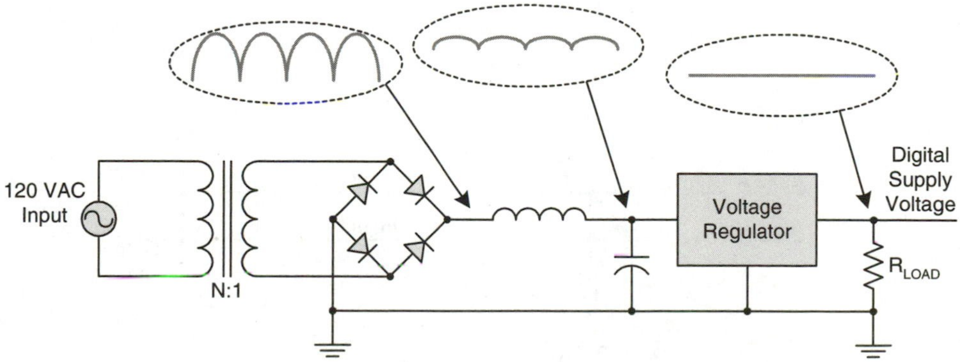

FIGURE 13.7 AC-to-DC power supply.

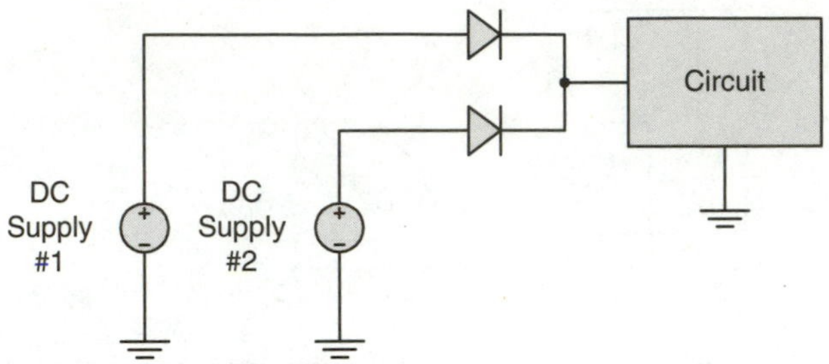

FIGURE 13.8 Power supply OR-ing diodes.

tery (e.g., +3 V) are each connected to an SRAM via independent OR-ing diodes. Under normal operation, the operating supply forces the battery's diode into reverse bias, preventing the battery from supplying power to the SRAM, thereby extending the battery's life. When power is turned off, the battery's diode becomes forward biased and maintains power to the SRAM so that its data are not lost. Schemes like this are commonly employed in certain PCs and other platforms that benefit from storing configuration information in nonvolatile SRAM.

13.3 DIODES IN DIGITAL APPLICATIONS

Not only can diode logic functions be implemented for power supply sharing or backup, they are equally applicable to implementing certain simple logic tasks on a circuit board. Diodes can implement both simple OR and AND functions and are useful when either a standard logic gate is unavailable or when the amplitude of the incoming signals violates the minimum or maximum input voltages of other components. Figure 13.9 shows diodes implementing two-input OR and AND functions. Pull-down and pull-up resistors are necessary for the OR and AND functions, respectively, because the diodes conduct only when forward biased. When both diodes are reverse biased, the circuit must be pulled to a valid logic state. The value of the resistors depends on the input current of the circuit being driven but ranges from 1 to 10 kΩ are common.

The pull-down resistor in the OR circuit maintains a default logic level of 0 when both inputs are also at logic 0. Both inputs must remain below V_F = 0.7 V for the circuit to generate a valid logic 0-V level. When the input signals transition to logic 1, they must stabilize at a higher voltage

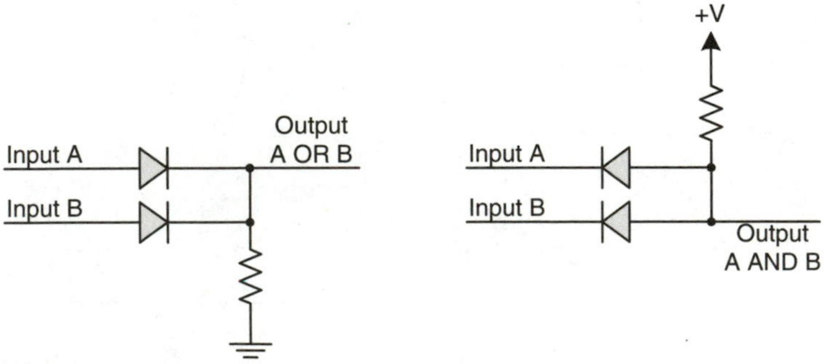

FIGURE 13.9 Diode OR and AND functions.

that is sufficient to meet the minimum logic-1 input voltage of the driven circuit. The value of the pull-down resistor should be high enough to limit the power consumption of the circuit but low enough to create a voltage that is comfortably below the driven circuit's logic-0 threshold. A CMOS input has a much lower input current specification than a TTL input. A typical TTL input has a low-level input current of under 0.5 mA, and it should be kept well below 0.8 V for adequate margin. A 1-kΩ pull-down resistor would create less than a 0.5-V drop under these conditions. This may be adequate for some designs, or a more conservative approach could be taken by using a smaller resistance, perhaps 470 Ω. When either input rises to its logic-1 voltage, this will be reflected in the circuit's output minus a diode drop. This places a restriction on the input voltages: they cannot exceed the maximum input voltage of the driven circuit by more than a diode drop. However, the input voltages can violate the minimum input voltage specification, because the diodes will be reverse biased under these conditions and thereby prevent the circuit's output voltage from dropping below 0 V, or ground.

Similarly, the AND circuit emits a logic-1 voltage when both diodes are reverse biased, because of the pull-up resistor. The diodes are reverse biased whenever the input voltages are near or above the logic supply voltage, +V. This enables the circuit to perform the AND function for input signals that would otherwise violate the maximum input voltage of the driven circuit. When either input transitions to a logic 0, that input's diode becomes forward biased and drags the output voltage down to the input level plus a diode drop, driving a logic-0 out. For the AND circuit to function reliably, a guarantee must be made of meeting the maximum logic-0 input voltage for the driven circuit. Using a normal silicon diode with $V_F = 0.7$ V may make this impossible if the input voltage is not guaranteed to go below 0 V. Therefore, a low-V_F Schottky diode such as a BAT54 may be required that exhibits $V_F < 0.4$ V at low currents. Just as the OR-circuit provides input voltage protection at logic 0 but not logic 1, the AND-circuit provides input voltage protection at logic 1 but not logic 0. The input voltage must not fall below the driven circuit's minimum input voltage by more than a diode drop.

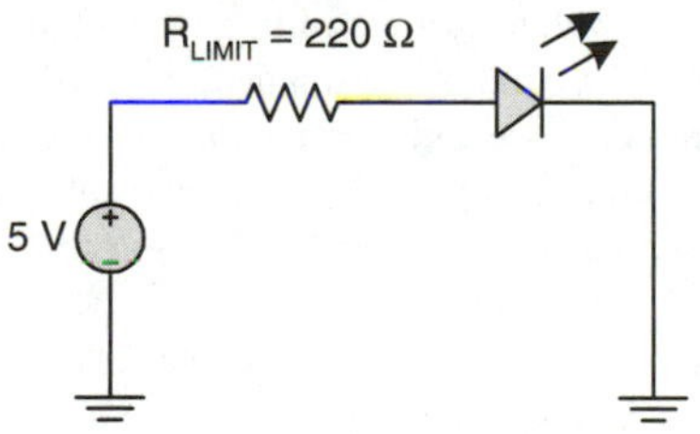

FIGURE 13.10 LED circuit.

Finally, perhaps the most visible types of diodes are *light emitting diodes* (LEDs). LEDs are constructed from various semiconductors and metals that emit visible or invisible light when forward biased. An LED is graphically distinguished from a normal diode by drawing representations of light or photons next to the diode symbol as shown in Fig. 13.10. LEDs exhibit forward voltages that are substantially higher than normal diodes, typically in the range of 2 to 3 V. Whether an LED is wired to the power supply to provide a "power on" indicator, or it is connected to the output pin of an IC, it should be *current limited* using a series resistor. Each LED has its own specifications for V_F and I_F. The current-limiting resistor should be chosen to provide the required current given the supply voltage and V_F.

In Fig. 13.10, the diode is assumed to have $V_F = 2$ V, and the supply voltage is 5 V. The resistor therefore drops 3 V and allows 13.6 mA of current to flow through the circuit. Allowing more current to flow through the LED will make it glow brighter but will also cause it to dissipate more heat. Most small LEDs emit sufficient light at currents ranging from 10 to 30 mA. In situations where power savings are critical, less current may be possible, depending on the desired light intensity.

LEDs are available in a wide range of colors. At first, only shades of green, red, and yellow were commonly found. Blue LEDs became widely available in the late 1990s, allowing full-color red-green-blue (RGB) displays. Common household remote control units rely on infrared LEDs. Ultraviolet LEDs are available as well for special applications.

13.4 BIPOLAR JUNCTION TRANSISTORS

Transistors are silicon switches that enable a weak signal to control a much larger current flow, which is the process of amplification: magnifying the amplitude of a signal. *Bipolar junction transistors* (BJTs) are a basic type of transistor and are formed by two back-to-back pn junctions. Figure 13.11 shows the general BJT structures and their associated symbolic representations. The BJT consists of three layers, or regions, of silicon in either of two configurations: NPN and PNP. The middle region is called the *base*, and the two outer regions are separately referred to as the *collector* and *emitter*. As will soon be shown, the base-emitter junction is what enables control of a potentially large current flow between the collector and emitter with a very small base-emitter current. A BJT's construction is more than simply placing two diodes back to back. The base region is extremely thin to enable conduction between the collector and emitter, and the collector and emitter are sized differently according to the fabrication process.

Currents in an NPN transistor flow from the base to the emitter and from the collector to the emitter. The relationship between these currents is defined by a proportionality constant called beta (β, also known as h_{FE}): $I_C = \beta I_B$. Beta is specific to each type of transistor and is characterized by the manufacturer in data sheets. Typical values for beta are from 100 to less than 1000. The beta current relationship provides a quick view of how a small base current can control a much larger collector current. A higher beta indicates greater potential for signal amplification. Because the base-emitter junction is essentially a diode, it must be sufficiently forward biased for the transistor to conduct current (V_{BE} = 0.7 V under typical conditions). A PNP transistor functions similarly, although the polarities of currents and voltages are reversed.

When a transistor circuit is designed, care must be taken not to overdrive the base-emitter junction. Like any other diode, it presents very low impedance beyond its forward voltage. Without some type of current limiting, the transistor will overheat and become damaged. Transistors are biased using resistors placed at two or three of its terminals to establish suitable operating voltages. Figure 13.12 shows a common NPN configuration at DC with a current limiting resistor, R_B, at the base and a voltage-dropping resistor, R_C, at the collector. The emitter is grounded, establishing the base voltage, V_B, at 0.7 V. R_B sets the current flowing into the base and thereby controls the collector voltage, V_C. As R_B increases, V_C increases, because less current is pulled through the collector, reducing the voltage drop across R_C. In this example, the base current, I_B, is (5 V – V_B) + R_B = 0.43 mA. Assuming a beta of 100, the collector current, I_C, is 43 mA, and $V_C = 5\ V - I_C R_C = 2.85$ V.

The transistor is limited in how much current it can drive by both its physical characteristics and the manner in which it is biased. Physically speaking, a transistor will have a specified maximum power dissipation beyond which it will overheat and eventually become damaged. In this circuit, the transistor's power dissipation is $V_{CE} I_C + V_{BE} I_B$, although the dominant term is between the collector and emitter, where the great majority of the current flow exists. Using this small simplification, the transistor is dissipating approximately 2.85 V × 43 mA = 123 mW.

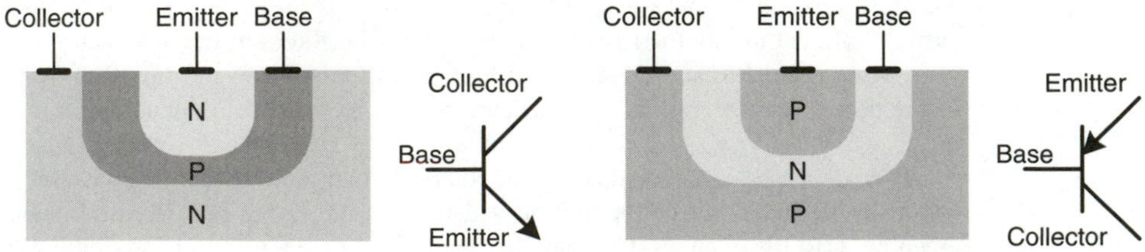

FIGURE 13.11 NPN and PNP BJT structures and graphical representations.

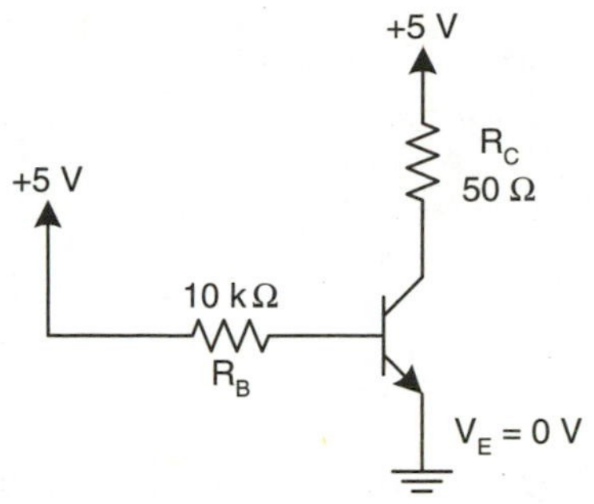

FIGURE 13.12 NPN DC topology.

Assuming that a transistor is not operated beyond its physical limitations, the bias configuration places an upper limit on how much current flows into the collector. A BJT has three modes of operation: *cutoff*, *active*, and *saturation*. In cutoff, the transistor is not conducting, because the base-emitter junction is either reverse biased or insufficiently forward biased. The collector is at its maximum voltage, and the base-collector junction is reverse biased, because no current is flowing to create a voltage drop through R_C or its equivalent. When the base-emitter junction is forward biased, the transistor conducts current and V_C begins to drop. The transistor is in active mode. As long as the base-collector junction remains reverse biased, increasing base current will cause a corresponding increase in collector current, and the transistor remains in active mode. If I_B is increased to the point at which the base-collector junction is forward biased (increased I_C causes V_C to approach V_E), the transistor enters saturation and no longer can draw more current through the collector. Saturation does not damage the transistor, but it results in a nonlinear relationship between I_B and I_C, nullifying the effect of beta. If R_C is increased or decreased, saturation occurs at lower or higher I_C, respectively. Amplifier circuits must avoid saturation to function properly because of the resulting nonlinearity. When used in a purely digital context, however, transistors can be driven from cutoff to saturation as long as the power dissipation specifications are obeyed.

13.5 DIGITAL AMPLIFICATION WITH THE BJT

With a basic knowledge of BJT operation, an NPN transistor can be already be applied in a useful digital application: driving a high-current LED array with a relatively weak output pin from a logic IC. Typical digital output pins have relatively low current drive capabilities, because they are designed primarily to interface with other logic gates that have low input current requirements. CMOS logic ICs tend to exhibit symmetrical drive currents in both logic 1 and logic 0. A CMOS output may be rated for anywhere from several milliamps to tens of milliamps. Bipolar logic ICs tend to exhibit relatively low drive current at logic 1, often less than one milliamp, and higher currents of several milliamps at logic 0. The use of bipolar logic is widespread, and it is often advantageous to take advantage of the greater drive capability of the logic-0 state. Figure 13.13 shows a logic output connected directly to an LED with each of the two possible polarities. The active-low configuration turns the LED on when the output is logic 0, and the active-high turns the LED on when the output is logic 1.

If neither the logic-1 nor logic-0 current capabilities are sufficient for the load, a simple transistor circuit can solve the problem. A typical 74LS logic family output pin is specified to source 0.4 mA at 2.7 V when driving a logic 1. Assuming a minimum beta of 100, an NPN transistor can be used to

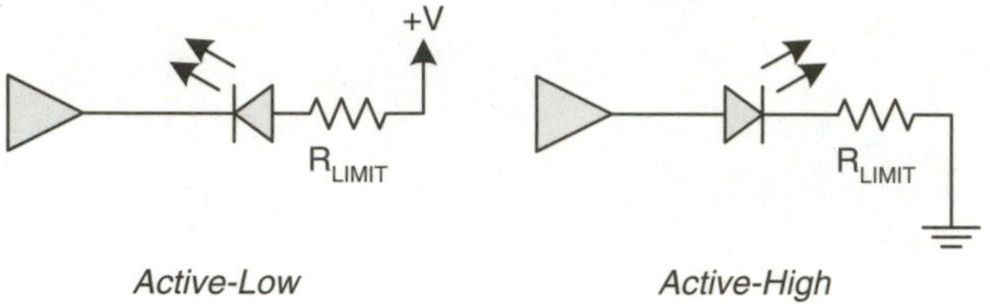

FIGURE 13.13 Direct connection of logic output to LED.

drive two 20-mA LEDs when a 74LS logic pin goes high. This circuit is shown in Fig. 13.14. Given the 74LS output specification, the base resistor is chosen to pull maximum current from the logic IC: 0.4 mA. $V_B = 0.7$ V, because it is known that $V_B = V_E + 0.7$ V and $V_E = 0$ V. The logic output is at least 2.7 V, resulting in a voltage drop across R_B of 2 V. Selecting $R_B = 4.7$ kΩ will pull slightly more than 0.4 mA, but this will not damage the logic IC; it may just cause the output voltage to sag a little below 2.7 V.

With the transistor's base circuit determined, attention turns to the collector. When the transistor is conducting, voltage is dropped by R_C, the LEDs, and the transistor. A range of values for R_C may be chosen that will allow the transistor to conduct the full 40 mA desired. Smaller values of R_C will result in higher collector-emitter voltage, V_{CE}. Larger values of R_C will reduce V_{CE} until the transistor enters saturation, after which increasing R_C will decrease I_C. Higher V_{CE} will result in higher transistor power dissipation for a constant current of 40 mA. It is usually preferable to minimize the power dissipated by active components such as transistors, because prolonged heating reduces their life span. Heat in general reduces a circuit's reliability. Typical manufacturers' specifications for an NPN transistor's saturation V_{CE}, $V_{CE\,(SAT)}$, are 0.3 V at moderate currents. $V_{CE\,(SAT)}$ can be lowered to less than 0.2 V by injecting more current into the base. Assuming an LED $V_F = 2$ V, R_C will have to drop 5 V – 2 V – 0.3 V = 2.7 V at 40 mA, yielding an approximate value of $R_C = 68\ \Omega$. The 2N2222 and 2N3904 are two widely available NPN transistors suitable for general applications such as this LED driver. These transistors have been around for a long time and are produced by multiple manufacturers. They are leaded devices whose surface mount equivalents are the MMBT2222 and MMBT3904, respectively.

If the load requirement is substantially increased, this NPN transistor circuit with $\beta = 100$ will be insufficient, because more base current will be required, violating the bipolar logic-1 output specification. Aside from trying to find a single transistor with a higher beta, there are alternative solutions that use two transistors instead of one. The idea is to keep the existing NPN transistor that is directly driven by the logic output and have it drive the base of a second transistor instead of directly driving

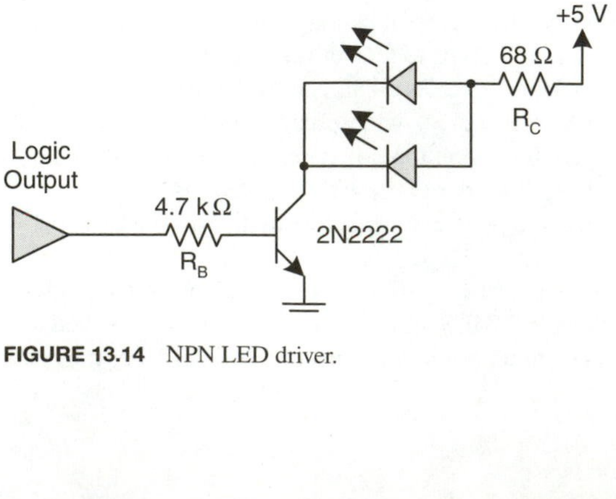

FIGURE 13.14 NPN LED driver.

the load. The collector current of the first transistor, $\beta_1 I_{B1}$, becomes the base current of the second transistor and is multiplied to yield a final load current of $\beta_2 I_{B2}$, which equals $\beta_1\beta_2 I_{B1}$.

A two-stage NPN/PNP transistor circuit is shown in Fig. 13.15. A PNP transistor's pn junctions have the opposite polarity as compared to an NPN. The base-emitter junction is forward biased by applying a higher voltage to the emitter than to the base. Thus, the PNP circuit topology is flipped in comparison to the NPN. This example assumes that a current of 500 mA is required to drive an array of LEDs. Standard convention is to label transistors with the letter Q and then append a number to uniquely identify each device. At I_{C2} = 500 mA, Q2 should be saturated to minimize V_{CE2}; therefore, the voltage drop across R_{C2} is 5 V – $V_{CE2\ (SAT)}$ – V_F = 2.7 V, the same as in the previous example. With the current higher than before, R_{C2} is selected to be 5.6 Ω, which is fairly close to the calculated value of 5.4 Ω. A common general-purpose PNP transistor that would be suited to an application such as this is the 2N3906 or its surface mount equivalent, the MMBT3906.

Assuming that β_2 = 100, $I_{B2} = I_{C1}$ = 5 mA, and so I_{B1} = 0.05 mA. Practically speaking, I_{B1} does not have to be set to such accuracy, because the circuit is ultimately current limited by R_{C2}. R_{B1} can be conservatively selected to guarantee that at a minimum of 0.05 mA is injected into the NPN transistor's base. If I_{B1} turns out to be greater than 0.05 mA, slightly more power will be dissipated by both transistors, but heating will not be a problem at these submilliamp current levels. Selecting R_{B2} = 22 kΩ (logic-high output = 2.7 V) lightly loads the bipolar logic output with approximately 0.1 mA. This results in Q1's collector current being greater than 10 mA, assuming β_1 = 100. When the circuit is conducting, V_{CE1} is held at 4.3 V by the fixed V_{EB} of Q2 equal to 0.7 V. Therefore, Q1 dissipates slightly more than 43 mW.

The power dissipated by the PNP transistor is 500 mA × 0.3 V = 150 mW. This may not sound like much power, but a small transistor will experience a substantial temperature rise at this sustained power level. For the most part, this level of heating will not cause a problem in most circuits, but care should be taken to analyze the thermal characteristics of semiconductor packages when designing circuits. These issues are discussed later in this book.

Considering this mixed NPN/PNP circuit, the question may come to mind as to why the logic output cannot directly drive the base of the PNP transistor, simplifying the circuit by removing the NPN transistor. A bipolar logic output certainly has the drive strength at logic 0 to sink sufficient current. The problem is that a bipolar output is not specified with a high enough logic-1 voltage level. For the Q2 to be turned off, V_{B2} must be driven higher than 4.3 V, a requirement that is outside the guaranteed specifications of a typical bipolar logic device. If a CMOS output were used, the situation would be different, because most CMOS drivers are guaranteed to emit a logic-1 voltage that is much closer to the system's positive supply voltage.

An alternative dual-transistor circuit is two NPN transistors arranged in what is commonly termed a *Darlington pair.* As shown in Fig. 13.16, a Darlington pair connects the emitter of the first stage

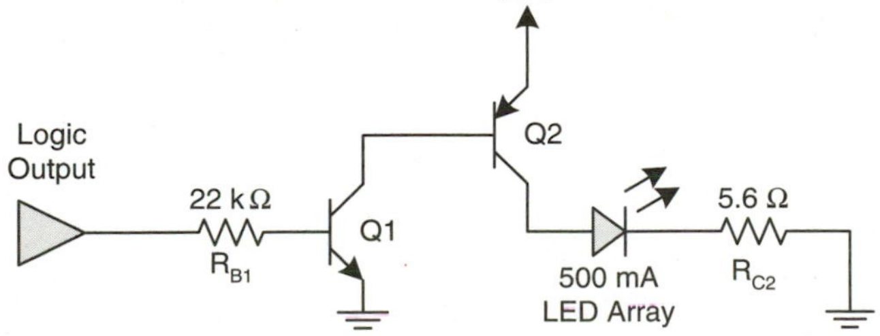

FIGURE 13.15 NPN/PNP LED driver.

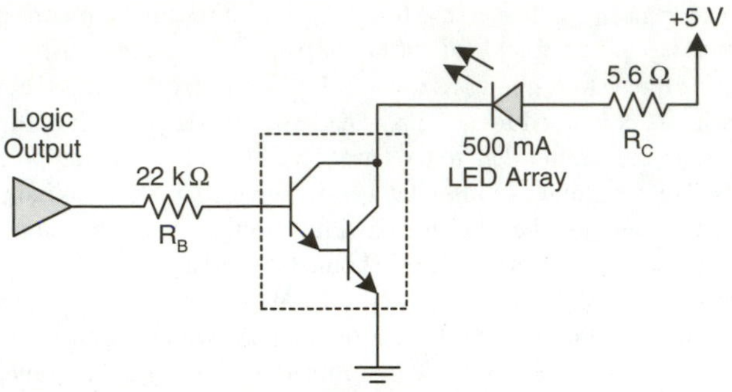

FIGURE 13.16 NPN Darlington pair LED driver.

transistor to the base of the second. As before, the betas multiply, but now V_{BE} is doubled to 1.4 V, because the two base-emitter junctions are in series. The higher V_{BE} reduces the base current from the previous example to roughly 0.06 mA, because the same value of R_B, 22 kΩ, is used. However, with an overall beta of approximately 10,000, the LED array is adequately driven.

13.6 LOGIC FUNCTIONS WITH THE BJT

The preceding circuits are binary amplifiers. In response to a small binary current input, a larger binary output current is generated. Looking at these circuits another way reveals that they are very basic logic gates: inverters. Consider the now familiar circuit in Fig. 13.17a. Rather than driving a high-current load, the output voltage is taken at the collector. When a TTL logic 0 is driven in, R_C pulls the output up to logic 1, 5 V. When a TTL logic 1 is driven in, the transistor saturates and drives the output down to logic 0, $V_{CE\,(SAT)}$. Logic gates in bipolar ICs are more complex than this, but the basic idea is that a simple inverter can be constructed with discrete transistors if the need arises. This discrete inverter can be transformed into a NOR gate by adding a second transistor in parallel with

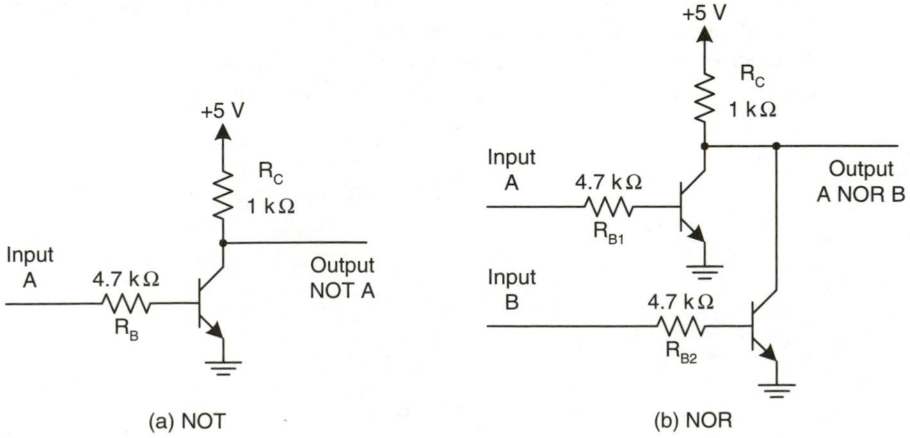

FIGURE 13.17 NPN NOT and NOR gates.

the one already present. In doing so, the output is logic 1 whenever both inputs are logic 0. As soon as one input is driven to logic 1, the accompanying transistor pulls the common V_C output node toward ground and logic 0. An advantage of creating logic gates from discrete transistors is that incompatible voltage domains can be safely bridged. In this example, the logic output is 5 V compatible, while the input can be almost any range of voltages as long as the transistor's specifications are not violated. This voltage conversion function would not be possible over such a wide range with normal bipolar or CMOS logic ICs.

It has been previously mentioned that TTL, or bipolar, logic outputs are asymmetrical in their 0 and 1 logic level drive strengths. A TTL output, shown in Fig. 13.18, consists of a *totem pole* output stage and a driver, or buffer, stage that passes the 0/1 logic function result to the output stage. It is called a totem pole output stage, because the vertical stack of two transistors and a diode somewhat resembles the layering of carvings on a totem pole. Classic TTL logic is composed entirely of NPN transistors, because PNP transistors are more difficult to fabricate on an integrated circuit.

When Q1 is turned on, its emitter voltage, V_{E1}, is fixed at 0.7 V by Q2. Q1 is driven into saturation by the logic circuit (not shown), which brings its collector voltage, V_{C1}, down to $V_{CE\,(SAT)} + V_{E1}$, which is approximately equal to 0.9 V. The saturation voltage of Q1 is approximately 0.2 V, because a sufficient current is injected into the base of Q1. This causes Q2 to saturate as well, driving an output level of $V_{CE\,(SAT)}$, 0.2 V or less. A logic 0 is driven strongly, because there is a direct, low-impedance path to ground through the saturated Q2. Turning our attention up to Q3, the transistor's base-emitter junction would normally be forward biased when $V_{C1} = V_{B3} = 0.9$ V and $V_{E3} = 0.2$ V. The presence of the diode prevents this from happening by increasing the forward-bias threshold by 0.7 V. Therefore, Q3 is in cutoff when Q2 is in saturation.

The resistors surrounding Q1 pull its collector and emitter to the respective voltage rails when Q1 is turned off by the logic circuit. This causes Q2 to turn off and Q3 to turn on by raising V_{B3} to 5 V. When $V_{B3} = 5$ V, $V_{E3} = 4.3$ V, and the output voltage is dropped an additional 0.7 V by the diode to 3.6 V when the load current is small. For small currents, Q3 remains in the active mode, and the output voltage is a function of the drop across the 1.6-kΩ resistor, V_{BE3}, and V_{DIODE}. As the load current increases, Q3 saturates, and the output voltage becomes a function of the drop across the 130 Ω resistor, $V_{CE\,(SAT)}$, and V_D. The impedance from the 5-V supply to the output is greater than for the logic-0 case, which is the reason that TTL logic drives a weaker logic 1.

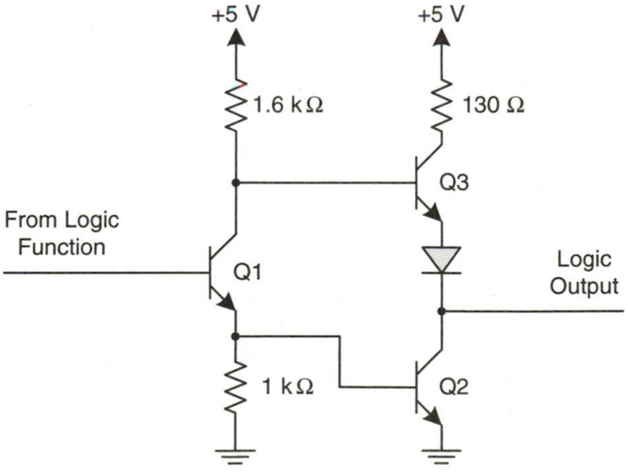

FIGURE 13.18 TTL driver.

From studying NPN transistors and the bipolar logic circuits that they form, it becomes apparent why the common terms V_{CC} and V_{EE} have come to represent the positive and negative supply voltages in a digital circuit. Bipolar logic has the collectors of its NPN transistors connected to the positive supply voltage and its emitters connected to the negative supply voltage—usually ground, but not always.

13.7 FIELD-EFFECT TRANSISTORS

Metal oxide semiconductor (MOS) technology represents the vast majority of transistors used to implement digital logic on integrated circuits ranging from logic devices to microprocessors to memory. Most *field-effect transistors* (FETs) are fabricated using MOS technology and are called MOSFETs. (The JFET is an exception to this and is briefly mentioned at the end of this chapter. This discussion uses the term *FET* in reference to a MOSFET.) Figure 13.19 shows the general structure of an *enhancement-type* FET. Just as BJTs are available in NPN and PNP according to the doping of their regions, there are n-FETs and p-FETs, referred to as NMOS and PMOS technologies, respectively. A FET consists of two main conduction regions, one called the *source* and the other the *drain*. In an n-FET, the source and drain are both N-type silicon. A channel of oppositely doped silicon separates the source and drain. Without any external influence, there is no conduction across the channel, because one pn junction is always reverse biased. A third terminal, the *gate*, is the control element that enables conduction across the channel. The gate is insulated from the rest of the FET by a thin layer of silicon dioxide (SiO_2). As the gate voltage is increased relative to the source voltage in an n-FET, the electric field developed at the gate causes a portion of the channel to change its electrical properties. The channel begins to behave as if it were doped the same way as the source and drain, enabling current to flow between the source and drain.

Whereas a BJT's conduction between emitter and collector is a function of its base current, a FET's conduction is a function of the gate-source voltage, V_{GS}, and the drain-source voltage, V_{DS}. An n-FET begins to conduct when V_{GS} exceeds the *threshold voltage*, V_T. In a typical circuit configuration, the drain is at a higher voltage than the source and current flows from drain to source. Current flowing into the drain, I_D, equals current flowing out of the source, I_S, because current cannot flow into or out of the insulated gate ($I_G = 0$). For a fixed $V_{GS} > V_T$, the relationship between I_D and V_{DS} is a curve that starts out nearly linear and then begins to taper off as V_{DS} increases, as shown in Fig. 13.20. The region in which I_D increases with V_{DS} is called the *triode region*. For small V_{DS}, the V_{GS}-induced channel presents very little resistance, and I_D increases almost in a linear manner with respect to V_{DS}. As V_{DS} and I_D increase, the resistance of the induced channel begins to increase, causing the curve's slope to decrease. At a certain point, the FET saturates and can conduct no more current even as V_{DS} continues to increase. The saturation voltage, $V_{DS(SAT)}$, equals $V_{GS} - V_T$. Increasing V_{GS} increases the saturation point, enabling more current to flow through the transistor.

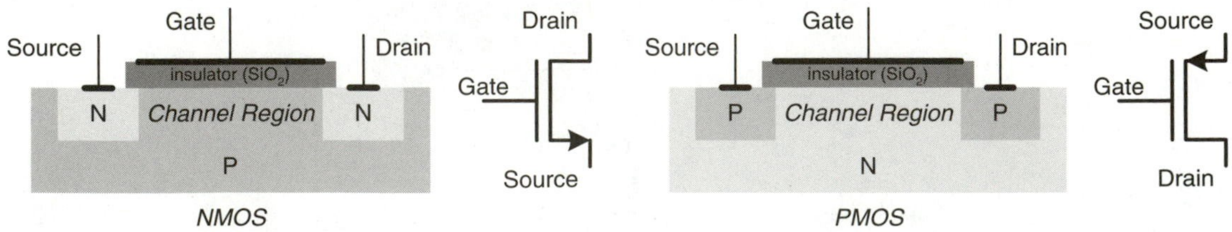

FIGURE 13.19 NMOS and PMOS enhancement-type FET structures and graphical representations.

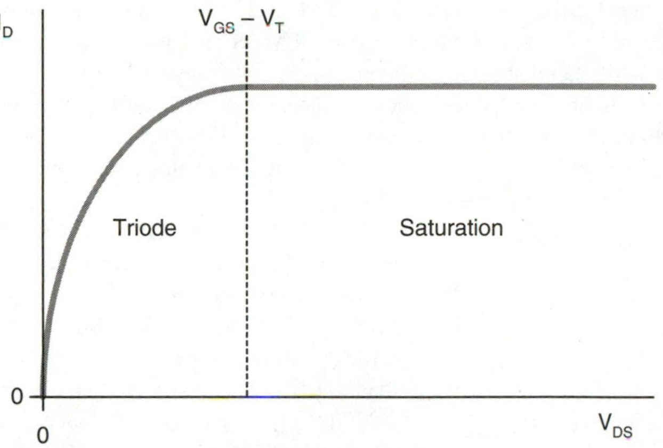

FIGURE 13.20 Enhancement-type n-FET I_D vs. V_{DS} for fixed $V_{GS} > V_T$.

The I_D/V_{DS} curve can be mathematically calculated, but the formulas require knowledge of specific physical parameters of a transistor's fabrication process. When integrated circuits are designed, such information is critical to device operation, and manufacturing process parameters are at an engineer's disposal. Data sheets for discrete FETs, however, do not typically provide the detailed process parameters required for these calculations. Instead, manufacturers provide device characterization curves in their data sheets that show I_D/V_{DS} curves for varying V_{GS}. An example of this is the graph contained in Fairchild Semiconductor's 2N7002 NMOS transistor data sheet and shown in Fig. 13.21. PMOS transistors function in the same manner as NMOS, although the polarities are reversed. The source is typically at a higher voltage than the drain, and V_{GS} is expressed as a negative value.

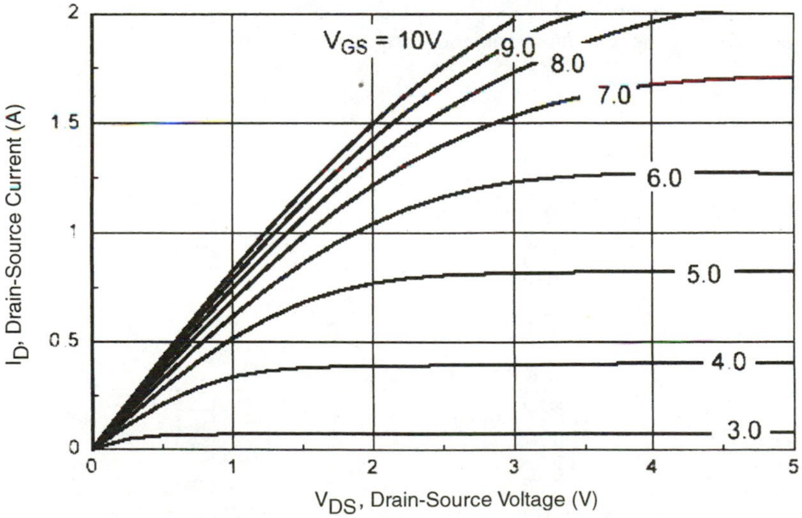

FIGURE 13.21 2N7002 I_D/V_{DS} graph. *(Reprinted with permission from Fairchild Semiconductor and National Semiconductor.)*

In the context of integrated circuits, NMOS and PMOS transistors are combined in close proximity to form *complementary-MOS*, or CMOS, circuits. CMOS circuits are dense, consume relatively little power, and are easily fabricated on a silicon chip. As with any real-world device, FETs contain parasitic properties, including capacitance between the gate and the source and drain, C_{GS} and C_{GD}. This capacitance, although small, imposes a load on the circuit that drives the gate during switching. However, when a FET is held static, such as in a DC or low-frequency circuit, the load on the driving circuit is essentially zero.

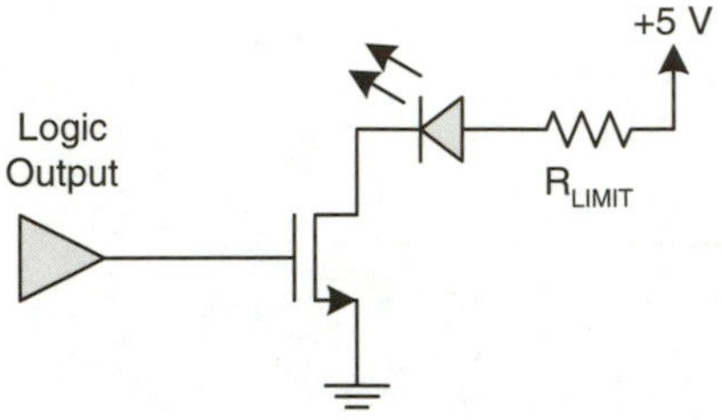

FIGURE 13.22 NMOS transistor LED driver.

The 2N7002, a widely available FET, can be connected to drive a load as shown in Fig. 13.22 using the previous LED example. A gate resistor is not needed, because there is no DC path to limit the current through. An appropriate current limiting resistor in the load's path may or may not be required. If the 2N7002's $I_{DS\ (MAX)}$ at the driven V_{GS} is within the load's specifications, the FET will serve as the current limiting element. A logic output, especially one operating at a low voltage (e.g., 3.3 V or less) may not provide sufficient V_{GS} to drive a heavy load. This must be verified ahead of time using manufacturers' data sheets.

Aside from the terminal capacitances already mentioned, FETs contain other parasitic characteristics that must be taken into account when designing certain types of circuits. One such characteristic develops when the FET's *body*, the silicon that forms the channel and that surrounds the source and drain regions, is kept at a voltage that differs from that of the source. A typical discrete FET does not suffer this problem, because the body and source are electrically connected within the package. ICs, however, consist of thousands or millions of FETs wherein the bodies of each FET are not all at the same voltages as their associated sources. This voltage, V_{BS}, degrades the ability of the channel to conduct current and causes a FET's V_T to increase, requiring a higher V_{GS} to conduct. More complete graphical representations of NMOS and PMOS FETs are shown in Fig. 13.23. Unlike the representations shown thus far, these explicitly show the FET's body. A discrete FET usually has its body and source connected as shown. For the sake of simplicity, many circuit diagrams showing discrete FETs use the basic representation, because it is known that $V_{BS} = 0$.

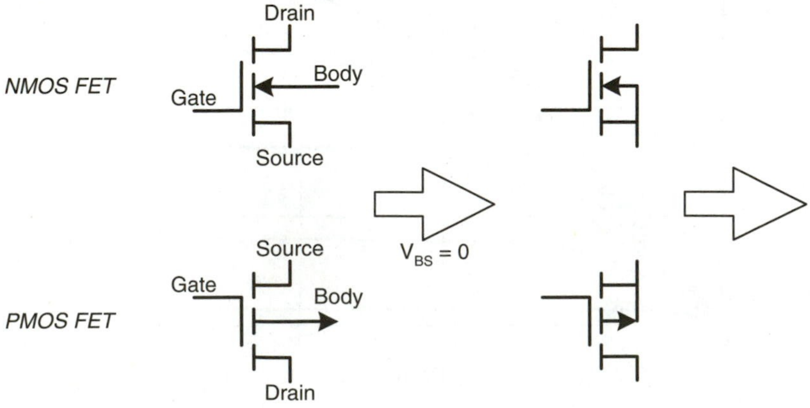

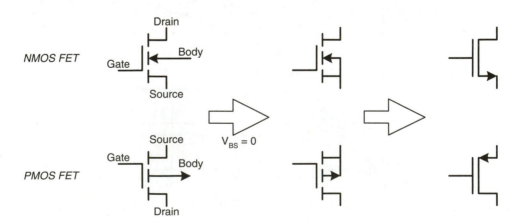

FIGURE 13.23 Graphical representation of FET body.

13.8 POWER FETS AND JFETS

Discrete FETs are used in a variety of applications. In digital systems, FETs are often found in power supply and regulation circuitry because of the availability of low-resistance devices. A key parameter of a FET used in power applications is its channel resistance between the source and drain, R_{DS}. Per the basic power relationship, P = I^{2}R, a FET with low R_{DS} will waste less power and will therefore operate at a cooler temperature. Power FET circuits can either handle more current without overheating or can be made to run cooler to extend their operational life span. It is not difficult to find power FETs with R_{DS} well below 10 mΩ. In contrast, a BJT exhibits a $V_{CE\ (SAT)}$ that dissipates significant power at high currents (P = IV). The saturation voltage also increases with I_C, causing more power to be dissipated in high-power applications.

When FETs are constructed as part of an IC, they are built in a lateral configuration atop the silicon substrate with a structure similar to what was shown earlier. Discrete FETs, however, are often constructed in a vertical manner known as *double-diffused MOS*, or DMOS, where the source and drain are on opposite sides of the silicon chip. The DMOS structure is shown in Fig. 13.24. DMOS surrounds the source with a thin region of oppositely doped silicon that serves as the body through which the conducting channel is induced by a voltage applied to the gate. Around the body is the substrate, which is doped similarly to the drain. Discrete FETs are constructed in this manner, because the thin channel between the source and the substrate provides low R_{DS} and, consequently, high current capacity with reduced power dissipation.

DMOS FETs are constructed in a manner that electrically connects the source and body regions, as shown by the metal source contacts in Fig. 13.24, so that a parasitic NPN transistor does not arise and cause problems. A consequence of this technique is that a parasitic diode is formed between the body and drain. Because the body is connected to the source, this is actually a source-drain diode with the anode at the source and the cathode at the drain for an n-FET. The diode is reverse biased under most conditions, because an n-FET's source is usually at a lower voltage than the drain. Therefore, V_{DS} would have to approach –0.6 V for conduction to occur. Figure 13.24 also shows the graphical representation of an n-type DMOS transistor with a source-drain diode. If an n-FET is designed such that the source is always connected to ground and the drain can never drop below 0 V, the diode has no effect. In less obvious configurations, the biasing of this inherent diode should be taken into account to ensure that current does not flow through an unintended path.

A potentially dangerous characteristic of FETs is a consequence of the insulated gate's high input impedance. If a FET circuit does not create a path to the gate through which stray electrical charge can drain away, a high voltage can build up and cause permanent damage to the device. The gate is a small capacitor, and it is known that V_{CAP} = Q ÷ C. Therefore, a small charge accumulating on a very

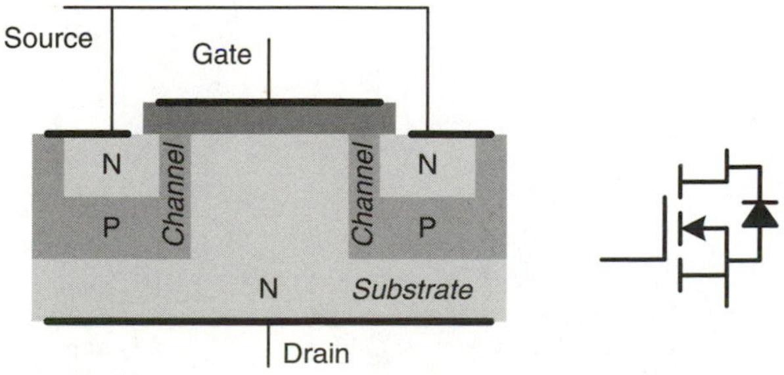

FIGURE 13.24 DMOS n-FET simplified structure and graphical representation.

small capacitor results in a high voltage. If the voltage exceeds the breakdown voltage of the gate's insulating silicon dioxide, it will break through and cause permanent damage to the FET. This is why MOS devices are particularly sensitive to static electricity. MOS components should be stored in conductive material that prevents the accumulation of charge on FET gate terminals.

Thus far, we have covered only enhancement-type MOSFETs. *Depletion-type* MOSFETs are built with a similar structure, but the channel region is doped with n-type silicon (in the case of an n-FET) so that the transistor conducts when $V_{GS} = 0$. Instead of defaulting to an open circuit, it conducts instead through the physically implanted channel. If positive V_{GS} is applied to a depletion-type n-FET, the channel is enhanced, and it can conduct more current as V_{DS} increases. However, if V_{GS} is made negative, the channel is depleted, and less current is conducted for a given V_{DS}. Schematic symbols for depletion-type MOSFETs are shown in Fig. 13.25. These devices are used in integrated circuits and are less common in discrete form.

Another type of FET, the *junction FET* (JFET) is not a MOS device and bears some similarity in structure to a BJT. As shown in Fig. 13.26, a JFET does not contain an insulated gate and does contain a physically implanted channel. Like a depletion-type MOSFET, a JFET conducts when $V_{GS} = 0$, and decreasing V_{GS} depletes the channel. As a result of the lack of gate insulation, the gate-drain and gate-source junctions will conduct when forward biased, thereby negating the transistor's operation. In the context of integrated circuits, JFETs are used mainly in bipolar analog processes, because they provide a higher input resistance as compared to BJTs.

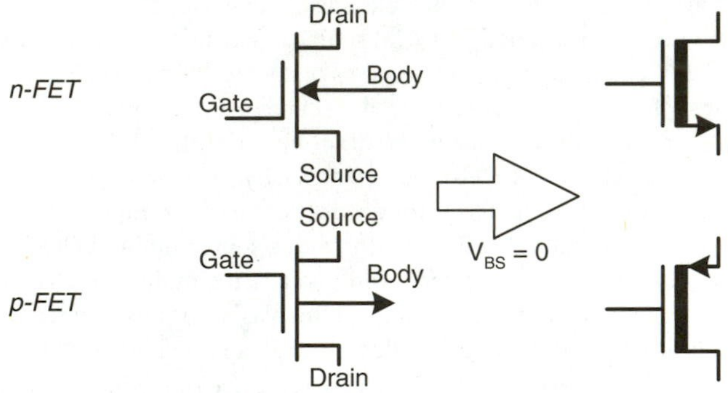

FIGURE 13.25 Depletion-type MOSFET graphical representations.

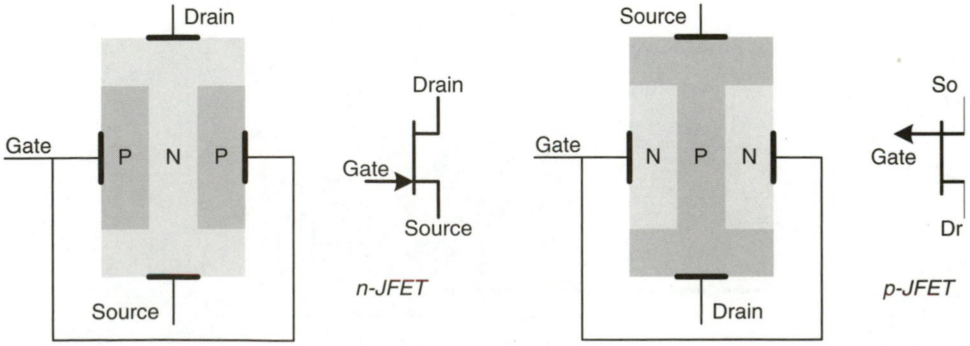

FIGURE 13.26 JFET structure and graphical representation.

CHAPTER 14
Operational Amplifiers

Transistors are the basic building blocks of solid-state amplifiers. Designing an amplifier using discrete transistors can be a substantial undertaking that requires theory outside the scope of this book. Operational amplifiers, or *op-amps,* exist to make the design of basic amplification circuits relatively easy. A digital engineer can use an op-amp to construct a general-purpose amplifier or active filter to preprocess an analog signal that may serve as input to the system. Comparators, which are based on op-amps, are useful in triggering events based on a signal reaching a certain threshold.

A key benefit of the op-amp is that it implements complex discrete transistor circuitry within a single integrated circuit and presents the engineer with a straightforward three-terminal amplifier that has well defined specifications and that can be externally configured to exhibit a wide range of characteristics.

Op-amps are presented here from three basic perspectives. First, the device is introduced using an idealized model so that its basic operation can be explained clearly without involving too many simultaneous details. The ideal op-amp is a very useful construct, because many real op-amp circuits can be treated as being ideal, as will be demonstrated later. Fundamental op-amp circuit analysis is stepped through in detail as part of the ideal presentation. The second section brings in nonideal device behavior and discusses how the idealized assumptions already introduced are affected in real circuits. The remainder of the chapter walks through a broad array of common op-amp circuit topologies with step-by-step analyses of their operation. The last of these presentations deals with the op-amp's cousin, the comparator, and explains the important topic of hysteresis.

14.1 THE IDEAL OP-AMP

The design of amplifiers is normally most relevant in analog circuits such as those found in audio and RF communications. An amplifier is an analog circuit that outputs a signal with greater amplitude than what is presented to it at the input. Amplification is sometimes necessary in digital systems. Amplifiers are often found at interfaces where the weak signal from a transducer (e.g., microphone or antenna) must be strengthened for sampling by an analog-to-digital converter. Even if a signal has sufficient amplitude, it may be desirable to scale it for better sampling resolution. For example, if an analog-to-digital converter accepts an input of 0 to 5 V and the incoming signal swings only between 0 and 3 V, 40 percent of the converter's resolution will be wasted. An amplifier can be used to scale the signal up to the full 5-V input range of the converter.

Solid-state amplifiers are constructed using transistors integrated onto a silicon chip or discrete transistors wired together on a circuit board. Amplifiers range greatly in complexity; complete AC analysis theory and its application to discrete amplifier design are outside the scope of this book. However, the design of many general-purpose amplifiers is made easier by the availability of prebuilt components called *operational amplifiers* (op-amps). Op-amps are so common that they are considered to be basic building blocks in analog circuit design. An op-amp may contain dozens of transistors, but its external interface consists of two differential inputs and an amplified output as shown in Fig. 14.1. The positive, or *noninverting*, and the negative, or *inverting*, inputs form a differential voltage, $v_D = v_+ - v_-$, that is amplified by a certain gain, A, at the output so that $v_O = Av_D = A(v_+ - v_-)$. When discussing the AC signals on which amplifiers operate, lower-case letters are used to indicate voltages and currents by convention to distinguish them from DC voltages and currents that have already been shown to use capital letters.

Many of the implementation details of how an AC signal is amplified within the op-amp are hidden from the circuit designer, requiring only an understanding of how the op-amp behaves from an external perspective. It is best to first explore an ideal op-amp's operation and then take into account the real-world deviations from the ideal model as necessary when designing a real circuit. An ideal op-amp has the following characteristics:

- *Infinite input impedance.* No current flows into or out of the inputs
- *Infinite open-loop gain.* This may sound confusing, but most op-amp circuits employ feedback that reduces the infinite gain to the desired level. $A = \infty$ simplifies op-amp circuit analysis, as will soon be shown.
- *Infinite bandwidth.* The op-amp's gain is constant across frequency from zero to infinity.
- *Zero output impedance.* The op-amp's output will always be equal to Av_D regardless of the load being driven.

These fundamental assumptions provide the engineer with a very flexible amplifier component that can be customized by surrounding circuitry to suit a wide range of applications. Perhaps the first question that comes to mind is how an amplifier with infinite gain can be made useful. The trick is in creating a *closed-loop* circuit that provides feedback from output to input to control the gain of the overall circuit. Without a feedback path, an *open-loop* op-amp circuit would, in fact, exhibit very high gain to the point of grossly distorting most types of signals. Consider the basic *noninverting* closed-loop op-amp circuit in Fig. 14.2. While the signal is injected into the positive input, the op-amp's output feeds back to the negative input through the resistor network formed by R1 and R2.

Based on the assumption of infinite input impedance, a basic resistor divider expression for v_- can be written based solely on the output voltage, v_O, and the two resistors.

$$v_- = v_O \frac{R1}{R1 + R2}$$

Knowing that $v_O = Av_D = A(v_+ - v_-)$, this expression can be used to reveal a relationship between the input voltage, v_I, and v_O.

$$v_O = A(v_+ - v_-) = Av_I - A\left[v_O \frac{R1}{R1 + R2}\right]$$

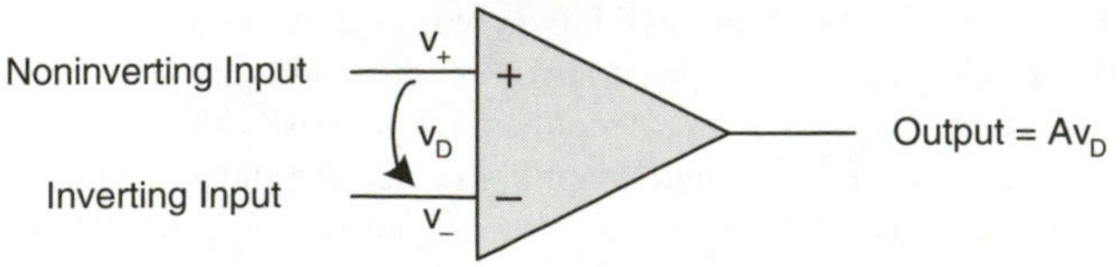

FIGURE 14.1 Op-amp graphical representation.

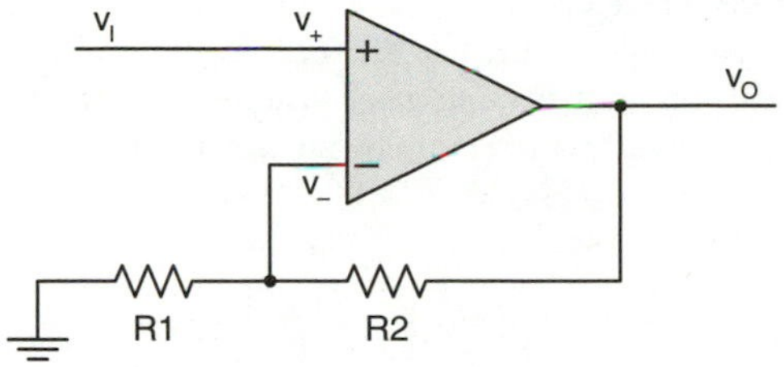

FIGURE 14.2 Noninverting op-amp circuit.

This relationship can be simplified based on the assumption of infinite gain. Dividing both sides of the equation by $A = \infty$ causes the lone v_O term on the left-hand side to disappear because $v_O \div \infty = 0$.

$$0 = v_I - \frac{v_O R1}{R1 + R2}$$

Finally, the input and output terms of the equation can be separated onto separate sides of the equality to yield a final simplified relationship between v_I and v_O as follows:

$$v_I = v_O \frac{R1}{R1 + R2}$$

$$v_O = v_I\left(\frac{R1 + R2}{R1}\right) = v_I\left[1 + \frac{R2}{R1}\right]$$

This shows that, despite the ideal op-amp's infinite gain, the circuit's overall gain is easily quantifiable and controllable based on the two resistor values. A noninverting op-amp circuit can be used to scale up an incoming signal for a purpose such as that already mentioned: using all of the available resolution of an analog-to-digital converter. In this example, a transducer of some kind (e.g., temperature sensor or audio input device) creates a signal that ranges from 0 to 3 V, and the analog-to-digital converter that is sensing it has a fixed sampling range from 0 to 5 V. To take full advantage of the sampling range, it is desirable to apply a gain of 1.667 to the input signal so that it swings from 0 to 5 V. This is accomplished using the noninverting circuit shown in Fig. 14.3. R1 and R2 are chosen arbitrarily as long as they satisfy the ratio R2:R1 = 2:3. Values of 2.2 and 3.3 kΩ provide feedback in the desired ratio with relatively low maximum current draw on the order of 1 mA.

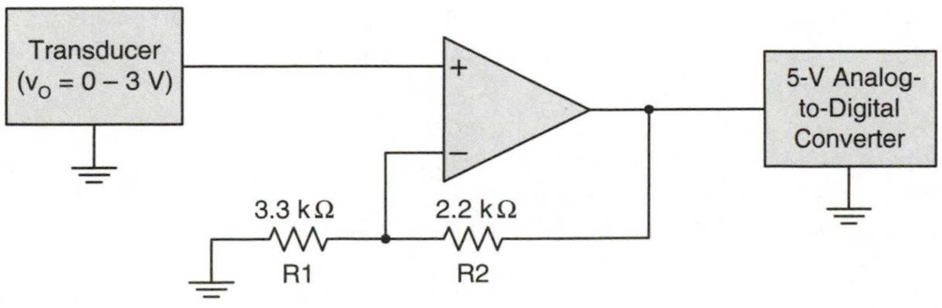

FIGURE 14.3 Scaling up an analog-to-digital converter input signal.

If a circuit such as this is implemented in reality, it would be helpful to understand the effect of finite op-amp gain on the circuit's overall gain to ensure that the circuit operates as intended. To do so, we can turn back to the previous derivation of op-amp gain. Instead of dividing the equation by $A = \infty$ early in the derivation, the input and output terms are immediately separated and re-expressed as follows:

$$v_O + A\left[\frac{v_O R1}{R1 + R2}\right] = Av_I = v_O\left[1 + A\frac{R1}{R1 + R2}\right]$$

$$v_O = v_I\frac{A}{1 + A\dfrac{R1}{R1 + R2}} = v_I\frac{A\dfrac{R1 + R2}{R1}}{\dfrac{R1 + R2}{R1} + A} = v_I\frac{A\left[1 + \dfrac{R2}{R1}\right]}{1 + \dfrac{R2}{R1} + A}$$

This relationship is more complex than the ideal case, but it can be seen that, as A approaches infinity, the expression simplifies to that which has already been presented. A wide variety of op-amps are manufactured with differing gains and electrical characteristics. One of the most common op-amps is the LM741, a device that has been around for decades. The LM741 has a minimum voltage gain of 20 V/mV, or A = 20,000 V/V.* For small circuit gains where R2 ÷ R1 is much less than A, the LM741 will provide an overall circuit gain that is extremely close to the ideal. Using the previous example, the gain expression becomes

$$v_O = v_I\frac{20000\left[1 + \dfrac{2.2}{3.3}\right]}{1 + \dfrac{2.2}{3.3} + 20000} = v_I\frac{33333}{20001.67} = v_I 1.6665$$

It can be observed that, for a real-world op-amp gain of much less than infinity, the ideal gain expression for an op-amp provides a very accurate calculation. As the gain desired from the circuit is increased, the denominator of the nonideal gain expression will increase as well, causing greater divergence between ideal and nonideal calculations for a given op-amp gain specification. Of course, the LM741 is not the only op-amp available. Newer and more advanced designs are readily available with gains an order of magnitude higher than that of the LM741.

The minimum gain achievable by the noninverting op-amp circuit is 1, or *unity gain*, when R2 = 0 Ω. There are instances in which a unity-gain buffer stage is desired. An example is the need to isolate a weak driver from a heavy load. While an ideal op-amp has infinite input impedance, a real op-amp has very high input impedance. Consequently, even a nonideal op-amp will present a light load to a driver. And while a real op-amp has nonzero output impedance, it will be much lower than the weak driver being isolated. As shown in Fig. 14.4, a unity-gain buffer is constructed by directly feeding the output back to the negative input. It can be observed from the previous circuit that when R2 = 0 Ω, R1 becomes superfluous.

A limitation of the noninverting op-amp circuit is that the minimum gain achievable is 1. When a gain of less than 1 is desired, a slightly different circuit topology is used: the inverting configuration. As shown in Fig. 14.5, the noninverting input is grounded, and the signal is injected into the negative input through R1. As before, R2 forms the feedback loop that stabilizes the circuit's overall gain.

---

* LM741 Single Operational Amplifier, Fairchild Semiconductor Corporation, 2001, p. 3.

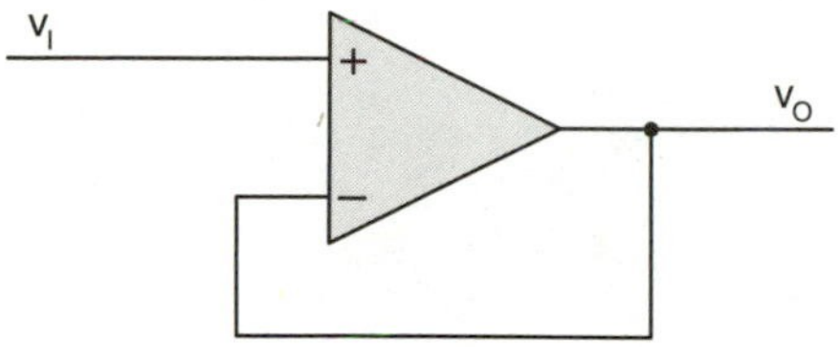

FIGURE 14.4 Unity-gain buffer.

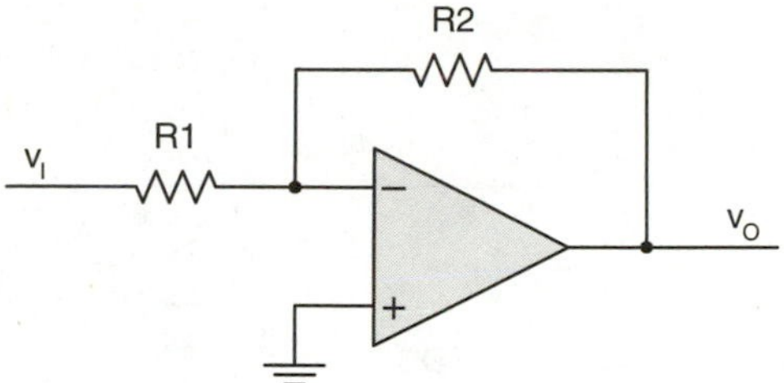

FIGURE 14.5 Op-amp inverting closed-loop circuit.

The inverting configuration's relationship between v_I and v_O can be derived using the resistor divider method shown previously for the noninverting circuit. Op-amp circuits can also be analyzed with an alternative method that provides a slightly different view of their operation. In some cases, mathematical analysis is made somewhat easier using one of the two methods. The alternative method uses the assumption of infinite gain to declare that the differential input voltage, v_D, equals zero. If $A = \infty$ and $v_O = Av_D$, it follows that $v_D = 0$ for a finite v_O. This assumption leads to an implied *virtual short circuit* between the op-amp's two input terminals. If $v_D = 0$, $v_+ = v_-$. The virtual short circuit tells us that if a voltage is applied at the positive terminal, it will appear at the negative terminal as well, and vice versa. Therefore, rather than expressing v_- as a resistor divider between v_O and v_I, each portion of the circuit can be analyzed separately. In such a simple circuit, this concept may not seem to have much advantage. However, the analysis of more complex op-amp circuits can benefit from this approach.

To demonstrate circuit analysis using the virtual short circuit approach, we begin by knowing that $v_- = 0$ V, because the positive terminal is grounded. Therefore, the voltage drop across R1 is known by inspection, and its current, i_1, is simply $v_I \div \mathrm{R1}$. We know from basic circuit theory that current cannot just disappear. Assuming that the op-amp has infinite input impedance, all of i_1 must flow toward v_O. Hence, $i_2 = -i_1$. The output voltage can now be determined using Ohm's law to show that the overall gain is controlled by the resistors when an ideal op-amp is used.

$$v_O = v_- + i_2 R2 = 0 - i_1 R2 = -v_I \frac{R2}{R1}$$

The inverting circuit can be designed with arbitrary gains of less than 1. However, both a positive and negative voltage supply are required to enable the op-amp to drive both positive and negative voltages. If a signal with a voltage range from 0 to 3 V is applied to an inverting circuit with a gain of 0.8, the op-amp will generate an output signal from –2.4 to 0 V. In some situations, this may be undesirable because of the requirement imposed by processing negative voltages. Fortunately, op-amps are very flexible, and the inverting configuration can be biased to center the output signal about a nonzero DC level. Consider the circuit in Fig. 14.6. Rather than grounding the positive input, a bias voltage is applied.

Using the virtual short circuit approach, an expression relating v_O, v_I, and V_{BIAS} can be derived in the same basic manner as done just before for the basic inverting configuration.

$$v_O = v_- + i_2 R2 = V_{BIAS} - i_1 R2 = V_{BIAS} - \frac{(v_I - V_{BIAS})R2}{R1} = -v_I \frac{R2}{R1} + V_{BIAS}\left[1 + \frac{R2}{R1}\right]$$

For a given gain controlled by R1 and R2, a bias voltage can be selected such that v_O sits at a nonzero DC voltage when $v_I = 0$ V. Assuming a desired gain of 0.8 and an output level of –2.4 V to com-

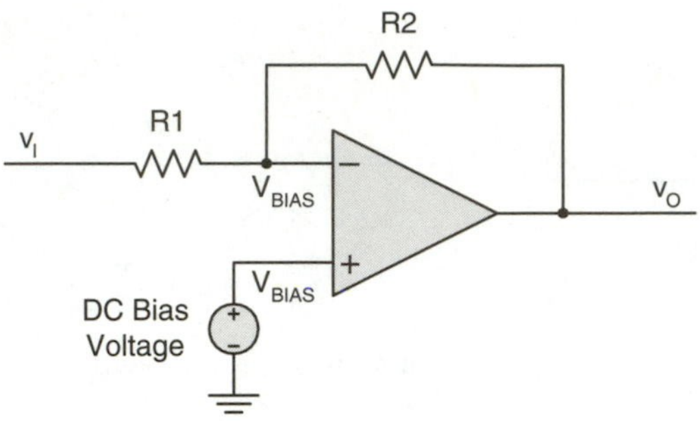

FIGURE 14.6 Biased inverting op-amp circuit.

pensate for, V_{BIAS} = 2.4 V ÷ 1.8 = 1.33 V. The specific bias voltage required may not be readily available. In these circumstances, a voltage divider may be employed at the positive input to produce an arbitrary bias voltage from a common supply such as +5 or +3.3 V. This scheme enables an op-amp to be used in the inverting configuration with a single supply voltage as long as the bias voltage is sufficient to produce a minimum output voltage of zero.

14.2 CHARACTERISTICS OF REAL OP-AMPS

Real op-amps are subject to the same laws of physics that cause nonideal behavior in all types of devices. The degree to which these deviations negatively affect a system varies according to the specific circuit. Op-amp manufacturers provide detailed data sheets to accompany their products, because many parameters must be characterized to enable proper circuit design. Fairchild Semiconductor's venerable LM741 serves as a useful example with which to explore real op-amp specifications. A portion of the LM741 data sheet is shown in Fig. 14.7.

Nonideal input characteristics are specified in the top portion of the LM741's data sheet. *Input offset voltage*, V_{IO}, is a DC error introduced by the op-amp's internal circuitry that manifests itself as an applied differential voltage between the two inputs. Assuming an ideal op-amp, V_{IO} appears as a voltage source applied by the external circuitry. If 0 V is applied to the op-amp, the input voltage is actually equivalent to V_{IO}. If a real op-amp is powered on in an open-loop configuration, its very large, though not infinite, open-loop gain can cause the output voltage to saturate because of the nonzero input voltage due to V_{IO}. The LM741's open-loop gain, shown as G_V in the data sheet, is between 20 and 200 V/mV. With V_{IO} = 2 mV, the output voltage is forced to its limit. Of course, op-amps are not generally used in an open-loop configuration. Figure 14.8 shows a common op-amp configuration in which the circuit's two inputs are grounded. The inputs are grounded to simplify analysis of V_{IO} effects in the absence of an input signal. When an input signal is present, the V_{IO} effects are added to the input/output transfer function. V_{IO} is represented as a separate voltage source applied to an ideal op-amp. Because V_{IO} appears as a differential voltage, the op-amp's output is approximated by the idealized closed-loop noninverting gain relationship: $V_O = V_{IO}$ (1 + R2 ÷ R1). This is essentially the same situation as the biased inverting op-amp circuit shown in Fig. 14.6, although the bias voltage is undesired. Therefore, the input offset voltage is amplified along with signals that are applied to the circuit.

It can be difficult to compensate for input offset voltage, because each individual op-amp has a different polarity and magnitude of V_{IO}, making it impossible to design a circuit with a fixed compensation factor suitable for all devices. Additionally, V_{IO} changes with temperature. Depending on

Electrical Characteristics

(V_{CC} = 15V, V_{EE} = - 15V. T_A = 25 °C, unless otherwise specified)

Parameter	Symbol	Conditions	LM741C/LM741I			Unit
			Min.	Typ.	Max.	
Input Offset Voltage	V_{IO}	$R_S \le 10K\Omega$	-	2.0	6.0	mV
		$R_S \le 50\Omega$	-	-	-	
Input Offset Voltage Adjustment Range	$V_{IO(R)}$	V_{CC} = ±20V	-	±15	-	mV
Input Offset Current	I_{IO}	-	-	20	200	nA
Input Bias Current	I_{BIAS}	-	-	80	500	nA
Input Resistance (Note1)	R_I	V_{CC} =±20V	0.3	2.0	-	MΩ
Input Voltage Range	$V_{I(R)}$	-	±12	±13	-	V
Large Signal Voltage Gain	G_V	$R_L \ge 2K\Omega$, V_{CC} =±20V, $V_{O(P-P)}$ =±15V	-	-	-	V/mV
		$R_L \ge 2K\Omega$, V_{CC} =±15V, $V_{O(P-P)}$ =±10V	20	200	-	
Output Short Circuit Current	I_{SC}	-	-	25	-	mA
Output Voltage Swing	$V_{O(P-P)}$	V_{CC} = ±20V, $R_L \ge 10K\Omega$	-	-	-	V
		V_{CC} = ±20V, $R_L \ge 2K\Omega$	-	-	-	
		V_{CC} = ±15V, $R_L \ge 10K\Omega$	±12	±14	-	
		V_{CC} = ±15V, $R_L \ge 2K\Omega$	±10	±13	-	
Common Mode Rejection Ratio	CMRR	$R_S \le 10K\Omega$, V_{CM} = ±12V	70	90	-	dB
		$R_S \le 50\Omega$, V_{CM} = ±12V	-	-	-	
Power Supply Rejection Ratio	PSRR	V_{CC} = ±15V to V_{CC} = ±15V $R_S \le 50\Omega$	-	-	-	dB
		V_{CC} = ±15V to V_{CC} = ±15V $R_S \le 10K\Omega$	77	96	-	
Transient Response: Rise Time	T_R	Unity Gain	-	0.3	-	μs
Transient Response: Overshoot	OS	Unity Gain	-	10	-	%
Bandwidth	BW	-	-	-	-	MHz
Slew Rate	SR	Unity Gain	-	0.5	-	V/μs
Supply Current	I_{CC}	R_L= ∞Ω	-	1.5	2.8	mA
Power Consumption	P_C	V_{CC} = ±20V	-	-	-	mW
		V_{CC} = ±15V	-	50	85	

Note:
1. Guaranteed by design.

FIGURE 14.7 LM741 data sheet. *(Reprinted with permission from Fairchild Semiconductor and National Semiconductor.)*

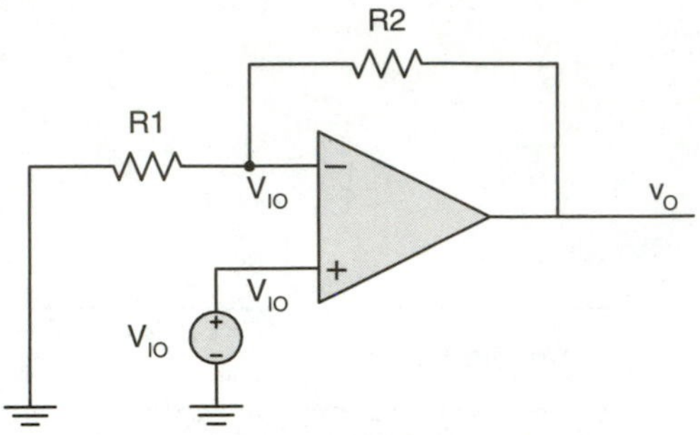

FIGURE 14.8 Analysis of input offset voltage.

the accuracy of the op-amp, the manufacturer will specify V_{IO} at room temperature (25°C) and at other temperature thresholds and provide coefficients or charts that relate the change in V_{IO} to temperature. Some op-amps, including the LM741, include compensation inputs that can be used to null-out the input offset error with external resistors. There is a limit to the practicality of this approach, however, resulting from temperature effects and the variation between components.

V_{IO} may not cause much trouble in an op-amp circuit with small gain, because a few millivolts of offset may be relatively insignificant as compared to several volts of actual signal. However, high-gain op-amp circuits that handle inputs with amplitudes in the millivolt range may be substantially degraded by V_{IO} effects. It may be desirable to AC-couple the op-amp's input when the frequencies of interest are high enough to make AC coupling feasible. The inverting circuit in Fig. 14.9 uses a capacitor to block a DC path to the op-amp's negative input. Because V_{IO} is a DC offset, the gain of the circuit is analyzed at DC to determine the amplification of V_{IO}. The impedance of the capacitor is ideally infinite at DC, resulting in unity gain for V_{IO}. For this circuit to function properly, the input frequency must be sufficiently high to not be attenuated by the highpass filter created by the series resistor and capacitor. The circuit exhibits a cutoff frequency where $f_C = 1 \div 2\pi C\text{R1}$. Therefore, frequencies much larger than f_C will be amplified by the idealized gain factor, –(R2 ÷ R1), when the capacitor essentially becomes a short circuit between the input and R1. The result is high gain at the frequencies of interest, yielding volts of output magnitude with only millivolts of V_{IO} error resulting from unity gain at DC.

Another parameter that relates to input offset voltage is the *power supply rejection ratio* (PSRR). The PSRR relates changes in the supply voltage to changes in V_{IO}. As the power rails fluctuate during normal operation, they influence the internal characteristics of the op-amp. PSRR is expressed in decibels as

$$PSRR = 20\log\frac{\Delta V_{CC}}{\Delta V_{IO}}$$

The LM741's minimum PSRR is specified at 77 dB. To ease arithmetic for the sake of discussion, we can round up to 80 dB and calculate PSRR = 10,000. This means that for every 1-V change in the supply voltage, V_{IO} changes by 100 μV. We know that V_{IO} is amplified by the gain of a DC-coupled circuit from the example in Fig. 14.8, indicating that a high-gain circuit is more susceptible to PSRR effects. PSRR declines with increasing frequency, making an op-amp more susceptible to power supply fluctuations.

Nonzero input currents are another source of inaccuracy in op-amp circuit, because real op-amps do not have infinite input resistance. Small nonzero currents flow into and out of the op-amp's in-

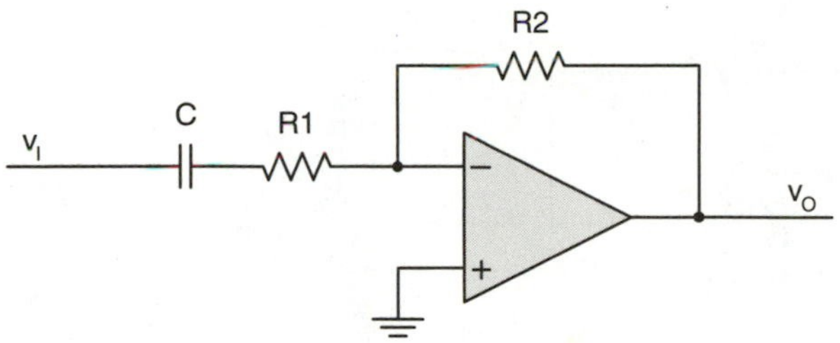

FIGURE 14.9 Mitigating V_{IO} with AC coupling.

puts. This current is the *input bias current*, I_{BIAS}. Ideal analyses of the preceding circuits assume that all currents flowing through R1 flow through R2 as well. This is clearly not true when the input impedance is finite. The result is that undesired voltage drops are created as input bias currents are drawn through resistors in an op-amp circuit. Consider the basic circuit in Fig. 14.10 with the inputs grounded to isolate I_{BIAS} effects independently from any input signal. The positive terminal remains at 0 V despite nonzero I_{BIAS}, because there is no resistance between it and ground (wire resistance is negligible). Therefore, the negative terminal is also at 0 V according to the virtual short assumption. While there is no current flowing through R1, because there is no voltage drop across it, I_{BIAS} flows from the output through R2. This results in a nonzero output voltage of $V_O = I_{BIAS}$R2 despite the fact that the circuit's inputs are grounded.

As seen in the LM741 data sheet, I_{BIAS} is measured in nanoamps for bipolar devices. CMOS op-amps specify I_{BIAS} in picoamps because of higher impedance MOSFET inputs. Practically speaking, many op-amp circuits with resistors measuring several kilohms or less do not have to worry about I_{BIAS} effects, because the product of nanoamps and resistance on the order of 10^3 Ω is on the order of microvolts. Of course, as circuit gains increase and allowable margins of error decrease, I_{BIAS} effects start to cause trouble. The problem is compounded by the fact that I_{BIAS} is not identical for each op-amp input as a result of slight physical variations in the circuitry associated with each input. A specification called *input offset current*, I_{IO}, is the difference between the two input bias currents. As seen in the LM741 data sheet, I_{IO} is smaller than I_{BIAS}, making I_{BIAS} the more troublesome characteristic.

Finite input bias current effects can be minimized by matching the induced offset voltages developed at each input. In Fig. 14.10, I_{BIAS} flowing into the positive terminal does not cause a voltage drop, which forces a corresponding I_{BIAS} to flow through R2 and create a nonzero output voltage. This circuit is augmented as shown in Fig. 14.11 by adding a resistor, R3, to the positive terminal.

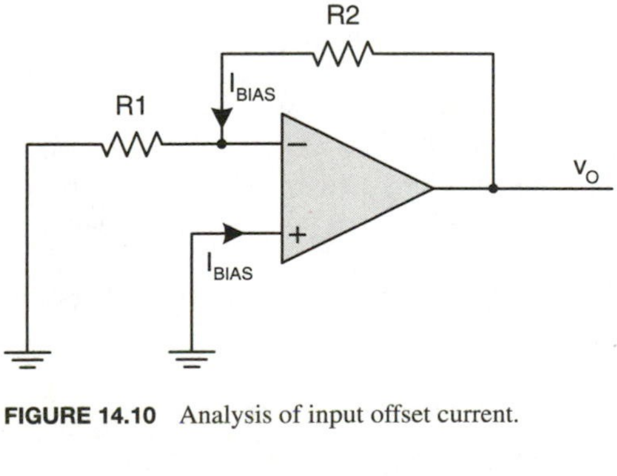

FIGURE 14.10 Analysis of input offset current.

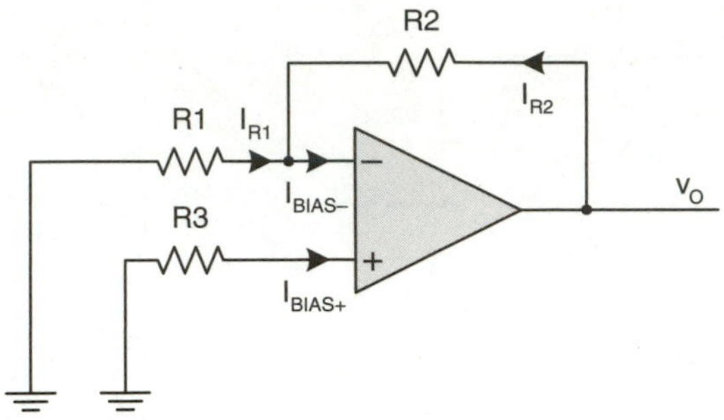

FIGURE 14.11 Mitigating I_{BIAS} with matched input resistor.

I_{BIAS+} now causes a voltage drop across $R3$, which is reflected at the negative terminal by reason of the virtual short. $V_- = V_+ = 0 - I_{BIAS+}R3$. Therefore, the current through $R1$ is $I_{R1} = I_{BIAS+}R3 \div R1$. The currents flowing through $R1$ and $R2$ into the negative terminal must equal I_{BIAS-}, because current does not simply disappear. Per node analysis, the sum of the currents entering a node must equal the sum of the currents leaving that node. This relationship can be restated to solve for the current through $R2$.

$$I_{R2} = I_{BIAS-} - I_{R1} = I_{BIAS-} - I_{BIAS+}\frac{R3}{R1}$$

Knowing I_{R2} enables the final expression of the output voltage,

$$V_O = V_- + I_{R2}R2 = (-I_{BIAS+})R3 + \left[I_{BIAS-} - I_{BIAS+}\frac{R3}{R1}\right]R2$$

By temporarily assuming that $I_{IO} = 0$, $I_{BIAS+} = I_{BIAS-}$. This reduces the output voltage expression to

$$V_O = I_{BIAS}\left[-R3 + R2 - \frac{R2R3}{R1}\right] = I_{BIAS}\left[R2 - R3\left(1 + \frac{R2}{R1}\right)\right]$$

This expression shows that the effects of I_{BIAS} can be compensated for by choosing $R3$ such that $V_O = 0$. Solving for $R3$ when $V_O = 0$ yields the parallel combination of $R1$ and $R2$,

$$R3 = \frac{R2}{1 + \frac{R2}{R1}} = \frac{R1R2}{R1 + R2}$$

A similar compensation can be designed for the case of an AC-coupled circuit, but it differs by taking into account the fact that no current flows through the coupling capacitor at DC. The circuit in Fig. 14.9 can be augmented by adding a resistor to the positive input terminal as shown in Fig. 14.12.

Unlike in the previous DC circuit, all of I_{BIAS-} flows through the feedback resistor, and the output voltage is

$$V_O = V_- + I_{BIAS-}R2 = V_+ + I_{BIAS-}R2 = -I_{BIAS+}R3 + I_{BIAS-}R2$$

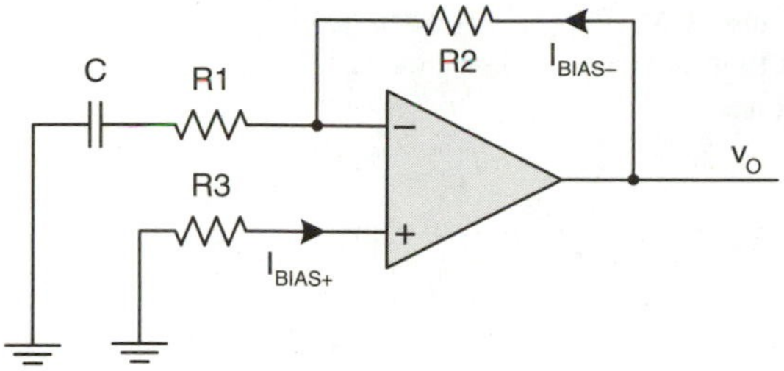

FIGURE 14.12 Mitigating I_{BIAS} in an AC coupled circuit.

If the input bias currents are assumed to be equal, selecting $R3 = R2$ nulls out the effects of I_{BIAS} in the AC coupled circuit.

Unfortunately, I_{IO} remains a problem in both the DC and AC circuits, although it is a problem of less magnitude than that of uncompensated I_{BIAS}. In Fig. 14.10, it was observed that uncompensated I_{BIAS} results in $V_O = I_{BIAS}R2$. Once I_{BIAS} has been compensated for by the inclusion of a resistor in the positive input terminal, I_{IO} effectively becomes the uncompensated I_{BIAS} term, resulting in $V_O = I_{IO}R2$. Again, I_{IO} as specified in nanoamps and picoamps may not present much problem for many circuits in which the feedback resistor, $R2$, is relatively small.

Continuing down the LM741 data sheet, additional physical realities are observed. Input voltage levels are limited by the chip's supply voltages. As the input levels approach the supply voltages, the op-amp's ability to deliver ideal performance diminishes. Supply voltages also relate to the output voltage swing of the op-amp. Many op-amps are unable to drive *rail-to-rail*. In other words, if an op-amp's supply voltages are ±15 V, the range of output voltages may be only ±14 V under lightly loaded conditions (10 kΩ in the LM741's case). As the output current demand increases, the op-amp's guaranteed drive level is diminished. Modern op-amps are available with a wide range of output drive strengths and characteristics. Rail-to-rail op-amps guarantee the ability to drive to within millivolts of either voltage supply at their rated drive current.

An op-amp is a differential amplifier with an ideal transfer function of $v_O = Av_D$. According to this simple relationship, only the voltage difference between the positive and negative terminals has any bearing on the output voltage. The same v_D results when $v_+ = 0.02$ V and $v_- = 0.01$ V as when $v_+ = 10.02$ V and $v_- = 10.01$ V. Ideally, the *common-mode* voltage, that voltage applied to both terminals, has no effect upon the output voltage. The first and second pairs of input levels have common-mode voltages, v_{CM}, of 0.01 and 10.01 V, respectively. Both pairs of inputs may be within the specified input range of a particular op-amp, but each pair will cause somewhat different behavior. The reason for this is that a real op-amp does not contain a perfect differential amplifier. The internal amplifier has some sensitivity to the common-mode signal that is applied to it.

If the inputs of an op-amp are tied together and a common-mode voltage is applied, there will be some nonzero output response even if the circuit has already been compensated for input offset errors. An op-amp has a finite common-mode gain associated with it such that $v_O = A_{CM}v_{CM}$ independent of the differential gain, A. Manufacturers define a *common-mode rejection ratio* (CMRR) that specifies the ratio of differential to common mode gain,

$$CMRR = 20\log\frac{A}{A_{CM}}$$

CMRR is expressed in decibels to more easily represent large ratios. A high CMRR indicates that $A >> A_{CM}$, which means that undesired effects due to common-mode input voltages are reduced.

Fairchild's LM741 has a minimum CMRR of 70 dB when $|v_{CM}| < 12$ V. The total output voltage of an op-amp is the sum of the differential and common-mode gains: $v_O = Av_D + A_{CM}v_{CM}$. If CMRR is $A \div A_{CM}$, then $A_{CM} = A \div CMRR$ and

$$v_O = Av_D + \frac{A}{CMRR}v_{CM} = A\left[v_D + \frac{v_{CM}}{CMRR}\right]$$

This relationship shows how higher CMRR yields behavior that is closer to the ideal differential amplifier. CMRR is specified by manufacturers at DC and is sufficiently large that most applications operating at low frequencies are not adversely affected by common-mode input voltages within the stated specifications. As with many specifications, CMRR gets worse as the frequency increases. Graphs are typically included in data sheets that relate CMRR to frequency.

Figure 14.13 shows how finite CMRR affects inverting and noninverting circuit topologies differently. An unbiased inverting circuit has the op-amp's positive input grounded, which results in a DC potential of 0 V at the negative input as well because of the virtual short assumption. This creates a common-mode input voltage of approximately 0 V, thus minimizing the effects of CMRR. If the inverting circuit is biased, a common-mode voltage approximately equal to the bias voltage results, with CMRR impact. Similarly, a noninverting circuit injects a signal into the positive input, resulting in a common-mode voltage approximately equal to the input signal.

Finite CMRR influence on a noninverting circuit can be quantified as a function of v_D without v_{CM}, because $v_D \approx v_{CM}$. To isolate the CMRR effects, the real noninverting op-amp circuit can be

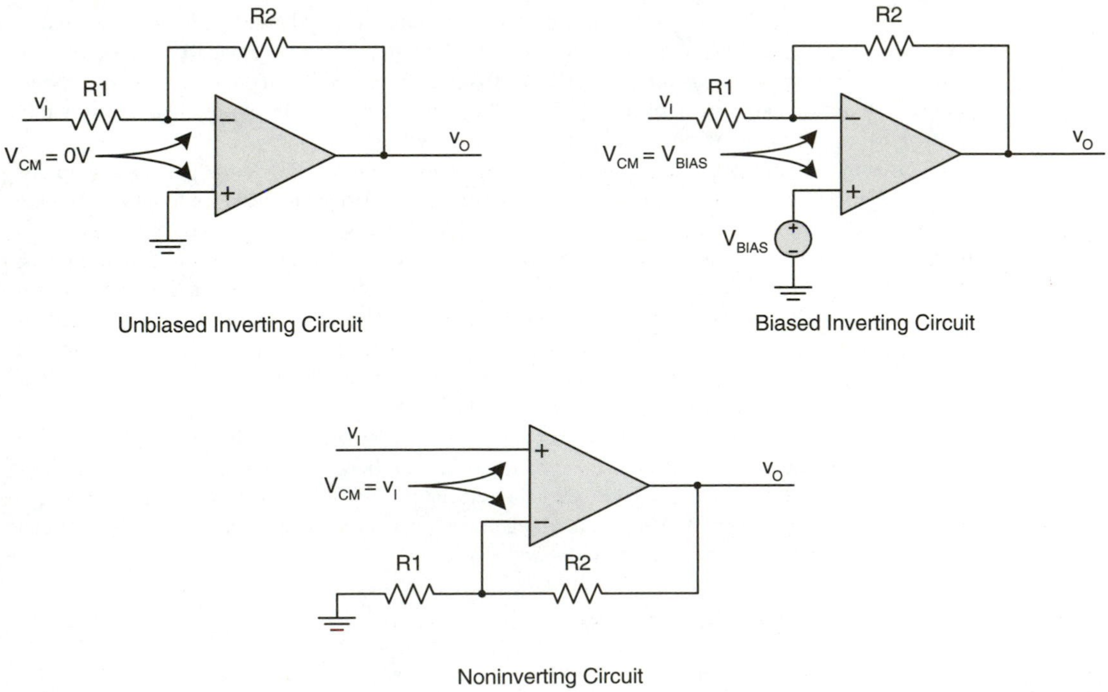

FIGURE 14.13 Common-mode input vs. circuit topology.

modeled as the combination of an ideal op-amp with an offset voltage caused by finite CMRR, v_{CMRR}, as shown in Fig. 14.14. An ideal op-amp multiplies the input voltage, v_{CMRR}, by its differential gain, A, to yield an output voltage. Likewise, the nonideal op-amp being modeled multiplies the common-mode input voltage, $v_{CM} \approx v_D \approx v_I$, by its common mode gain to yield an output voltage. Therefore, $Av_{CMRR} = A_{CM}v_I$. This equivalency can be restated as $v_{CMRR} = A_{CM}v_I \div A$, which is actually a function of CMRR: $v_{CMRR} = v_I \div CMRR$.

With CMRR now modeled as an input that is a function of CMRR and the actual input voltage, the ideal op-amp model can be treated as a block to which the actual input signal is presented as shown in Fig. 14.15.

The total output voltage is the sum of the input voltage and v_{CMRR} passed through the ideal op-amp as separate terms,

$$v_O = [v_I + v_{CMRR}]\left[1 + \frac{R2}{R1}\right] = \left[v_I + \frac{v_I}{CMRR}\right]\left[1 + \frac{R2}{R1}\right] = v_I\left[1 + \frac{1}{CMRR}\right]\left[1 + \frac{R2}{R1}\right]$$

As with input offset voltage, CMRR effects are more pronounced at higher circuit gains.

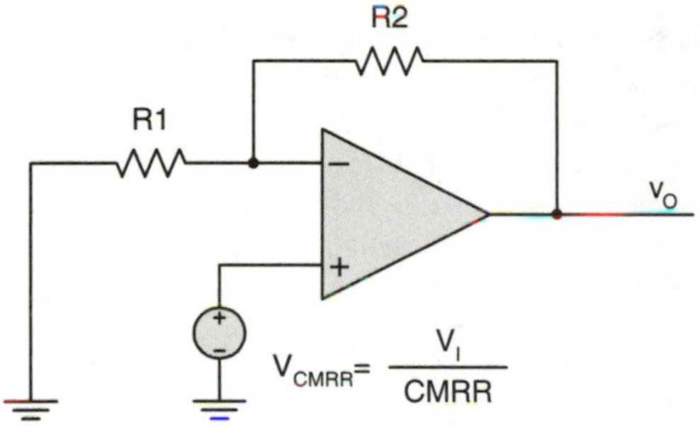

FIGURE 14.14 Modeling CMRR effects in a noninverting circuit.

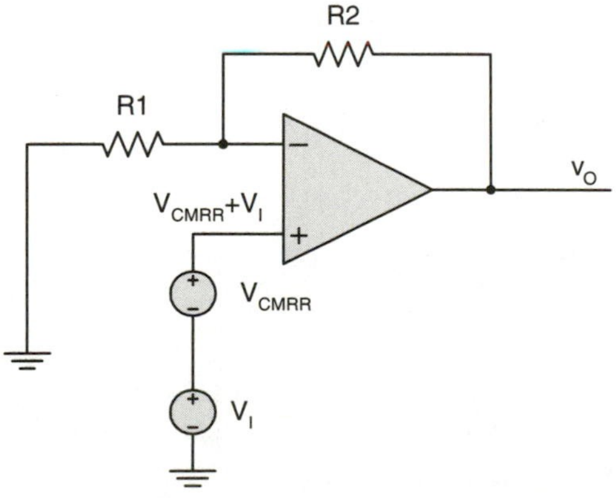

FIGURE 14.15 CMRR model with signal input.

14.3 BANDWIDTH LIMITATIONS

An op-amp's ability to amplify a signal is a direct function of the frequencies involved. The maximum frequency at which an op-amp provides useful performance can be characterized in multiple ways. Manufacturers provide a variety of related specifications to assist in this task. First, there is the *gain-bandwidth product* (GBW), also referred to as *bandwidth* (BW). As the name implies, GBW specifies a gain at a certain frequency. As the frequency of operation is decreased, the gain increases, and vice versa. Most op-amp data sheets provide either a bandwidth number, sometimes specified as the unity-gain bandwidth, or a chart that relates gain to frequency. The large-signal voltage gain of the LM741 remains relatively constant for frequencies only under 10 Hz (not kHz!). It is called "large-signal," because the response is essentially at DC; "small-signal" would denote practical operating frequencies that are typically orders of magnitude higher. Near 10 Hz, the gain rolls off at the rate of –20 dB per decade and reaches unity gain, 0 dB, near 1 MHz. Above 1 MHz, the LM741's gain rapidly drops off. Keep in mind, however, that this is the LM741's open-loop response. When a closed-loop circuit is formed, the gain is substantially lowered, thereby increasing the circuit's usable bandwidth over which the gain remains constant.

Figure 14.16 shows an approximate gain versus frequency curve for an older bipolar op-amp such as the LM741. The gain of the open-loop op-amp is approximately 100 dB below 10 Hz. Observe that, if the closed-loop gain is set at 10 (20 dB), the constant gain portion of the curve drops by approximately 80 dB, which correlates to four decades of frequency. Therefore, the closed-loop bandwidth increases to 10 Hz multiplied by four decades, or 100 kHz. Put another way, 100 dB equals 100,000. The gain-bandwidth product is approximately 100,000 × 10 Hz = 1 MHz. If a closed-loop gain of 10 is used, the bandwidth is 1 MHz ÷ 10 = 100 kHz.

Two other interrelated frequency metrics are *slew rate* and *full-power bandwidth*. These specifications enable an engineer to determine whether a signal will be distorted by the op-amp based on the signal's frequency and amplitude. Slew rate defines the rate at which the op-amp can change its output voltage. The LM741's specified slew rate is 0.5 V/μs. If the desired output signal has a component that changes voltage faster than the slew rate, the op-amp will not be able to fully reproduce the signal and is said to be *slew-rate limited*. An example of slew-rate limiting is shown in Fig. 14.17, where the desired sine wave output, a pure signal with a single frequency, is converted to a triangle waveform that transitions at the op-amp's maximum slew rate.

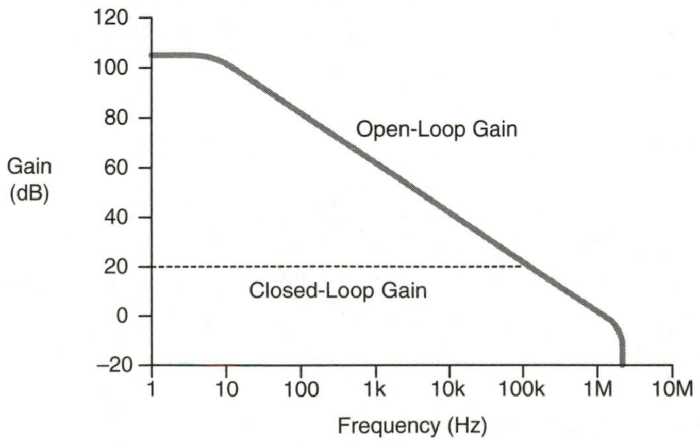

FIGURE 14.16 Typical gain vs. frequency curve.

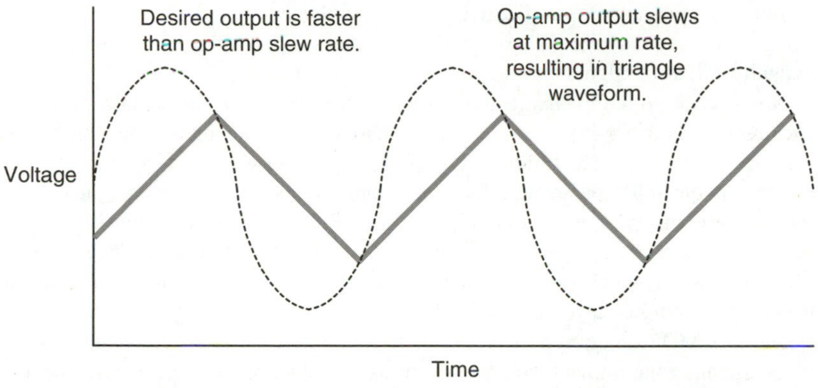

FIGURE 14.17 Slew-rate-limited output.

Higher output frequencies are possible when output amplitudes are reduced, because the slew rate is a function of both time (frequency) and amplitude. For a sine wave of amplitude V_{MAX}, its slew rate is determined by computing the maximum value of its first-derivative, which is equal to $2\pi fV_{MAX}$. Therefore, the op-amp's slew rate, SR, can be combined with this expression to calculate the highest output frequency that will not be slew-rate limited.

$$f_{MAX} = \frac{SR}{2\pi V_{MAX}}$$

Manufacturers may specify an op-amp's full-power bandwidth, which is calculated with this equation by using the device's slew rate and maximum rated output voltage swing. In the case of the LM741, which is rated at ±14 V of typical output swing, its full-power bandwidth is approximately 5.7 kHz, indicating that it can reproduce a 5.7 kHz sine wave at its full output range. Op-amps with lower operating voltages and higher slew-rates resulting from modern semiconductor process improvements can achieve full-power bandwidths in the tens of megahertz.

Each type of op-amp has different frequency, noise, and power characteristics. Op-amps are well behaved at DC and low frequencies. General-purpose op-amps are well suited to applications in the kilohertz range such as processing audio signals or transducer signals with limited bandwidths. Newer op-amps have much improved high-frequency characteristics, enabling them to be used for video and other more demanding applications. The gain-bandwidth product and slew-rate (or full-power bandwidth) specifications can be used to determine if a particular op-amp will handle signals of the desired frequency and amplitude. If this important first hurdle is passed, an op-amp's data sheet should be more carefully inspected to quantify the degradations in such characteristics as input impedance and CMRR resulting from frequency.

14.4 INPUT RESISTANCE

Op-amps are often used to amplify, or buffer, weak electrical signals that are generated by transducers such as a microphone or photodiode. These transducers often have very high output impedances that translate into an inability to drive even light loads. Current drive capability may range from a few milliamps down to nanoamps. When trying to buffer a signal that is measured in nanoamps, the amplifier's input impedance is critical; it must be high enough to enable the weak input signal to develop a sufficient voltage. Otherwise, the amplifier's input will overload the transducer, severely attenuating whatever signal may be present.

Noninverting op-amp circuits present the highest input impedance to a signal, because the feedback resistor network is connected to the negative input, leaving the input signal connected only to the positive terminal. The LM741's typical input resistance is 2 MΩ with an I_{BIAS} of 500 nA. These are acceptable input parameters for many applications. However, it might otherwise make more sense to use an inverting configuration for a high-gain amplifier in which nonideal characteristics such as I_{BIAS} can be compensated for as discussed previously. Figure 14.18 shows an example that uses two op-amp stages to achieve high input impedance with a unity-gain noninverting circuit followed by a high-gain inverting circuit that is compensated for I_{BIAS}.

Newer CMOS op-amps have much higher input impedance specifications than older bipolar devices, enabling them to buffer extremely weak transducers. As one example, the Texas Instruments/Burr-Brown OPA336 is specified with a maximum I_{BIAS} of 60 pA and an input resistance of 10^{13} Ω.* A photodiode is a weak transducer that is used in many optical communications applications ranging from IR or UV remote control devices to laser communications systems and fiber optic transceivers. Such applications are usually digital and therefore require a saturated on/off output rather than linear amplification. Because of their weak output (nanoamps to microamps), these transducers are best used as current sources to develop a usable voltage across a high resistance rather than attempting to directly measure a voltage across their terminals. Figure 14.19 shows several circuits that use high R_{IN} and low I_{BIAS} op-amps to amplify transducer currents in the nanoamp range. Essentially all of the transducer current, i_{TRANS}, is passed through the resistors because of the op-amps' low I_{BIAS}. For every microamp passed through a 1-MΩ resistor, 1 V is developed across that resistor.

Circuit (a) utilizes the noninverting configuration to amplify an input voltage of several hundred millivolts (corresponding to i_{TRANS} of several hundred nanoamps) by approximately 1,000. Circuit (b) is simpler, because i_{TRANS} is used to directly establish the output voltage of the op-amp. Circuit (c) is an improvement that increases the gain by raising the voltage at each op-amp input in addition to causing a voltage drop across the feedback resistor. If 200 nA is conducted by the transducer, v_+, and hence v_-, rise to 0.2 V. At the same time, 0.2 V is developed across the feedback resistor, resulting in a total output voltage of 0.4 V. Even thought these last two circuits use the negative op-amp input, they are not inverting circuits, because analysis shows current being drawn from the output node by the transducer through the feedback resistor, causing a positive voltage drop with respect to ground.

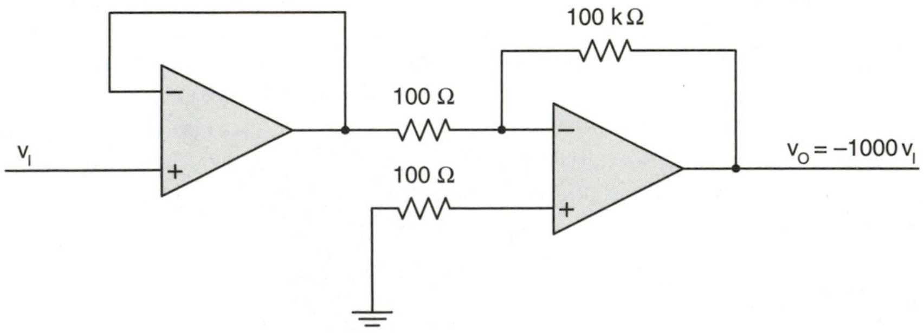

FIGURE 14.18 Two-stage low-input-impedance/high-gain amplifier.

* OPA336, Texas Instruments Incorporated, 2000, p.2.

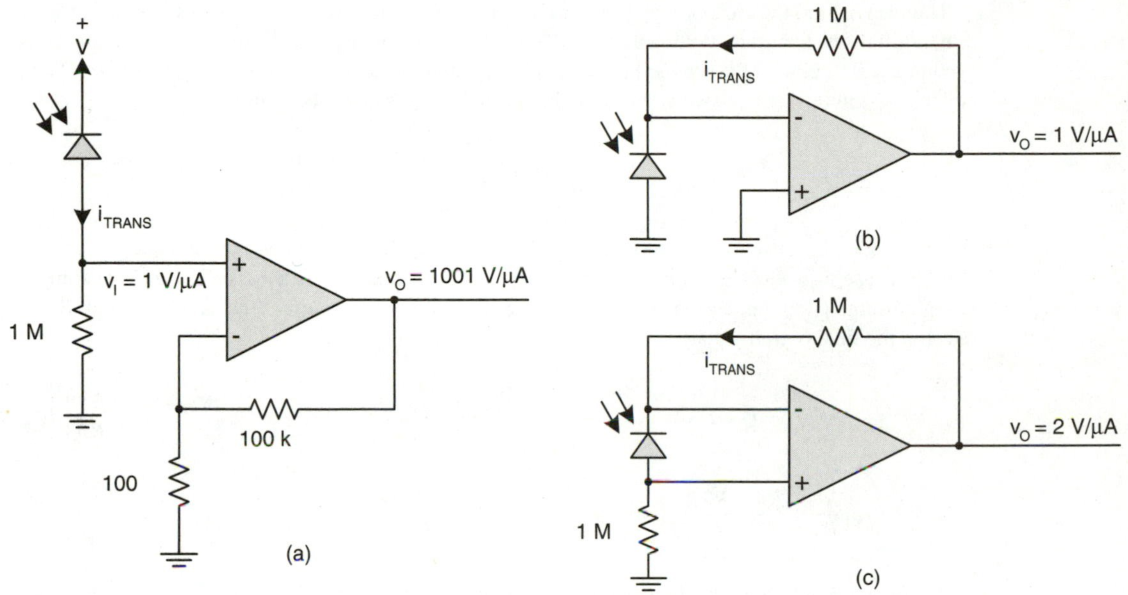

FIGURE 14.19 Amplification of very weak transducer signals.

A signal applied to a noninverting circuit directly drives the op-amp's positive input and, in doing so, establishes a voltage at both the positive and negative inputs due to the virtual short. Likewise, an inverting circuit's op-amp input voltages are established by the bias applied to the positive terminal. Referring back to the standard inverting circuit in Fig. 14.5, a voltage drop is developed across the input resistor, R1, that sets the input current: $i_{IN} = (v_{IN} - v_{\_}) \div R1 = v_{IN} \div R1$. Therefore, the input resistance seen by the signal is equal to this series input resistor. This basic circuit places a practical ceiling on the input resistance, because the circuit's gain is inversely proportional to that input resistor. If this ceiling is reached, and still higher input resistance is necessary, the basic inverting circuit can be augmented as shown in Fig. 14.20 with a "tee" topology in the feedback path.

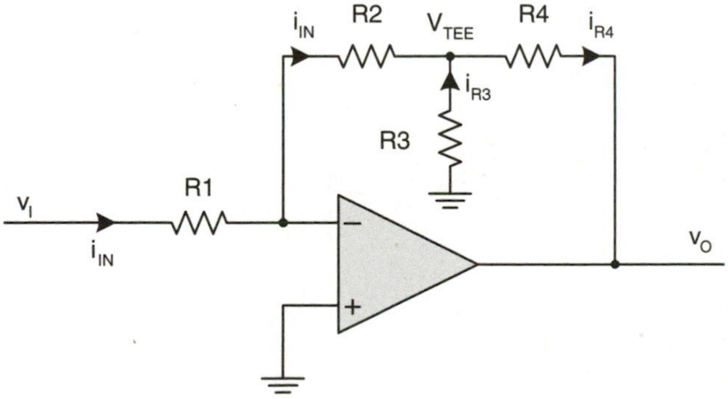

FIGURE 14.20 Inverting circuit with higher input resistance using tee feedback topology.

The idea behind the tee topology is to achieve a higher voltage drop across the resistor network by drawing current from an additional source other than the signal input. In this case, that source is ground via $R3$. As with the standard inverting circuit, input current, i_{IN}, flows through $R1$ and $R2$ and creates v_{TEE} in the same way that v_O would be created in the standard circuit.

$$v_{TEE} = v_{-} - i_{IN}R2 = 0 - \frac{v_{IN}}{R1}R2 = -v_{IN}\frac{R2}{R1}$$

This works, because i_{IN} has nowhere to go but through R2, assuming an approximately ideal op-amp. Because v_{TEE} has already been established, there is a voltage drop across $R3$ with an accompanying current. The input current and i_{R3} combine to form the total current that flows through R4 to develop the final output voltage,

$$v_O = v_{TEE} - (i_{IN} + i_{R3})R4 = v_{TEE} - \left[i_{IN} + \frac{0 - v_{TEE}}{R3}\right]R4 = -v_{IN}\frac{R2}{R1} - \left[\frac{v_{IN}}{R1} + v_{IN}\frac{R2}{R1R3}\right]R4$$

$$= -v_{IN}\left[\frac{R2}{R1} + \frac{R4}{R1} + \frac{R2R4}{R1R3}\right] = -\frac{v_{IN}}{R1}\left[R2 + R4\left(1 + \frac{R2}{R3}\right)\right]$$

A combination of high circuit gain and high input resistance is now achievable using resistors not exceeding a practical limit of 1 MΩ. The circuit in Fig. 14.21 produces a gain of approximately –1,000 with an input resistance of 1 MΩ.

14.5 SUMMATION AMPLIFIER CIRCUITS

It is becoming clear that op-amps are very flexible analog building blocks. Aside from basic amplification, op-amp circuits can be designed to perform mathematical operations. Decades ago, analog computers were built around op-amps constructed from vacuum tubes rather than integrated circuits. These computers were capable of advanced functions including multiplication, division, exponents, and logarithms. Analog computers operate on voltage levels rather than bits and bytes.

Although the typical digital system may not benefit much from complex analog computation circuits, basic op-amp summation circuit analysis may prove useful in designing for various combinations of input signals and bias voltages. Figure 14.22 shows a basic summation circuit built on the

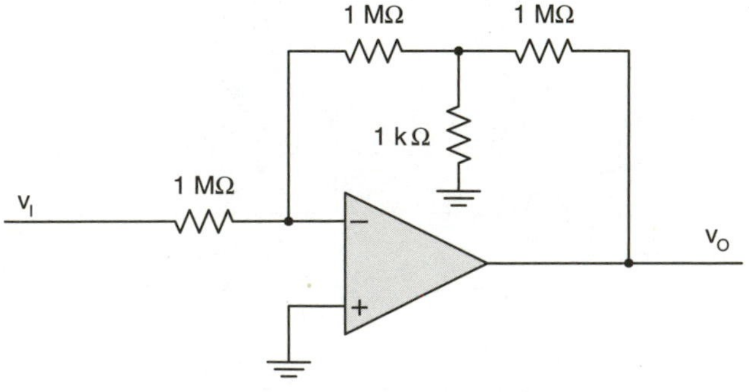

FIGURE 14.21 High gain/high input resistance inverting circuit.

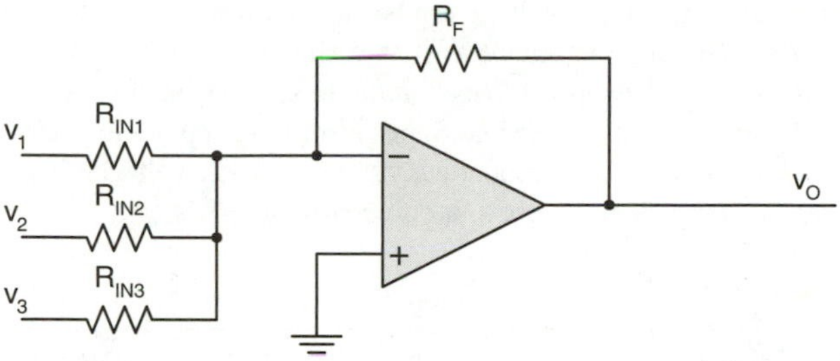

FIGURE 14.22 Summation circuit.

inverting topology. Each input is unaffected by the others, because the negative terminal is always at 0 V by reason of the virtual short. Therefore, each input resistor contributes an input current component determined by Ohm's law. These currents add and collectively pass through the feedback resistor to create an output voltage.

A weighted summation can be designed by individually selecting input resistors according to the application requirements. If the input resistors are all equal, the gain ratios for each input signal are equal, and the summation is balanced.

$$v_O = -\frac{R_F}{R_{IN}}[v_1 + v_2 + v_3]$$

Similarly, a noninverting circuit can be used to combine multiple input voltages without the –1 factor as shown in Fig. 14.23. Although usually referred to as a *noninverting summer,* this is more of an averaging circuit than a summer. The input voltages are averaged by the input resistor network, and this average level is then multiplied by the gain of the noninverting circuit.

Analyzing the input resistor network can be difficult without a common circuit-analysis trick that relies on the principle of *superposition*. Superposition works with a linear transfer function that relates an output to multiple inputs multiplied by some gain factor. For example, two signals, V_1 and V_2, add together and are multiplied to yield an output voltage: $V_O = A(V_1 + V_2)$. By the principle of superposition, the input terms can be broken up, computed separately, and then recombined at the output to yield a final expression: $V_{O1} = AV_1$ and $V_{O2} = AV_2$, thus $V_O = V_{O1} + V_{O2} = A(V_1 + V_2)$. Superposition works only when the transfer function is linear.

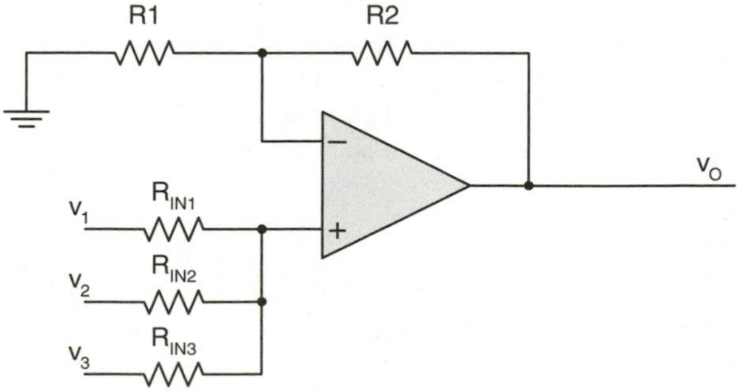

FIGURE 14.23 Noninverting summation circuit.

In circuit analysis terms, a single input voltage can be isolated from the others by tying each other input to ground and then evaluating the circuit as if that single input voltage were the only input present. This analysis method can be applied to each input in succession, and then the partial results can be summed to yield a final output expression. Superposition is applied to analyze the noninverting summer as follows. First, v_2 and v_3 are grounded, yielding a basic voltage divider expression for the op-amp input voltage that translates to a partial output voltage.

$$v_{O1} = Av_1\left[\frac{R_{IN2} \| R_{IN3}}{R_{IN1} + R_{IN2} \| R_{IN3}}\right] = \left[1 + \frac{R2}{R1}\right]v_1\left[\frac{R_{IN2}R_{IN3}}{R_{IN1}R_{IN2} + R_{IN1}R_{IN3} + R_{IN2}R_{IN3}}\right]$$

To ease calculation for the sake of discussion, assume that $R1 = R2 = R_{IN1} = R_{IN2} = R_{IN3} = 10\ \text{k}\Omega$. Under these conditions,

$$v_{O1} = 2v_1\frac{1}{3} = v_1\frac{2}{3}$$

This same treatment is performed separately for v_2 and v_3, and then each partial output voltage is summed.

$$v_O = \frac{2}{3}[v_1 + v_2 + v_3]$$

It can be observed that the noninverting summer is really just amplifying the average input voltage by setting $v_{IN} = v_1 = v_2 = v_3$. This results in an output voltage of $2v_{IN}$, which is exactly how the noninverting circuit behaves in the absence of input resistors with a single input signal.

Inverting and noninverting summer circuits can be combined by feeding input signals into both op-amp terminals with multiple resistors. This results in addition of the signals at the positive input and subtraction of the signals at the negative input. Analysis of these more complex circuits can be started using superposition to determine the voltage at the positive input. This reveals an expression for the voltage at the negative input, which enables computation of the currents flowing through the negative-side input resistors. With the current through the feedback resistor known along with the voltage at the negative terminal, a final expression for the op-amp output may be determined.

Analog addition and subtraction functions can be combined to form a difference amplifier. As its name implies, a difference amplifier emits a voltage that is proportional to the magnitude of the voltage difference between its two inputs. A simple difference amplifier can be constructed using a single op-amp as shown in Fig. 14.24.

This circuit can be analyzed in a couple of ways. The direct approach is to derive the voltage at the op-amp's two input terminals, find the current through the feedback resistor, and obtain a final expression for v_O. Alternatively, superposition can be used to isolate each input and then add the two partial results for a final answer. To provide a second example of analysis by superposition, this technique is used. Grounding v_{IN-} turns the circuit into a noninverting amplifier.

$$v_{O+} = v_+\left[1 + \frac{R2}{R1}\right] = v_{IN+}\frac{R2}{R1 + R2}\left[1 + \frac{R2}{R1}\right] = v_{IN+}\frac{R2}{R1 + R2}\left[\frac{R1 + R2}{R1}\right] = v_{IN+}\frac{R2}{R1}$$

Next, v_{IN+} is grounded, yielding an inverting amplifier.

$$v_{O-} = -v_{IN-}\frac{R2}{R1}$$

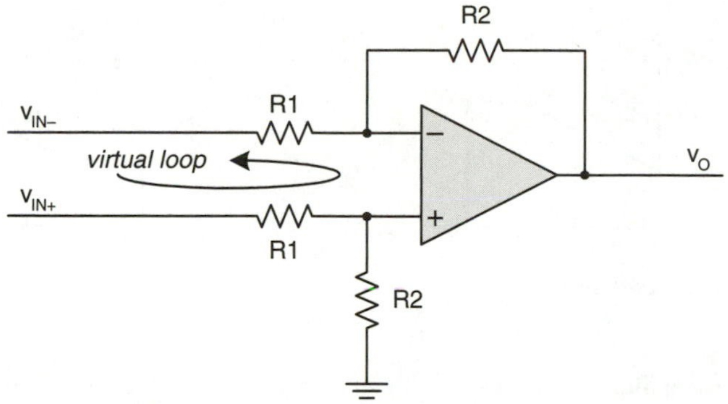

FIGURE 14.24 Single op-amp difference amplifier.

By superposition, the linear partial terms, v_{O+} and v_{O-}, are summed to yield a final output expression that clearly shows that this circuit is a difference amplifier.

$$v_O = v_{O+} + v_{O-} = v_{IN+}\frac{R2}{R1} - v_{IN-}\frac{R2}{R1} = \frac{R2}{R1}(v_{IN+} - v_{IN-})$$

A drawback of the single op-amp difference amplifier is that it has a relatively low input resistance. Using the virtual short concept, the two op-amp terminals are at the same voltage, thereby creating a virtual loop circuit consisting of the differential voltage input and the two input resistors, R1. Because this virtual circuit consists only of the input voltage source and the two resistors, the input resistance is observed to be equal to 2R1 by inspection. As with the basic inverting op-amp circuit, there is a practical ceiling imposed on input resistance caused by the circuit's gain and the range of resistances that are practical to use in a real circuit.

Many applications in which a difference amplifier is necessary involve weak signal sources such as an unbuffered transducer. To solve this problem, a more complex difference amplifier can be constructed with multiple op-amps to present a much higher input resistance. Usually called *instrumentation amplifiers*, these circuits commonly consist of three op-amps, two of which are configured in the noninverting topology for very high input resistance. The third op-amp is configured in the just-mentioned difference amplification topology. As with the example in Fig. 14.18, the noninverting op-amps buffer each half of the differential input signal, and the second op-amp stage performs the final difference function. If such a circuit is required in a digital system, it may be most practical to use an integrated instrumentation amplifier as manufactured by such companies as Linear Technology (e.g., LT1167) and Texas Instruments (e.g., INA332) rather than constructing one from discrete op-amps.

14.6 ACTIVE FILTERS

Active filters perform the same basic frequency passing and blocking function as passive filters, but they can simultaneously amplify the signal to form a filter that has unity or higher gain. This is in contrast to passive filters that achieve less than unity gain because of finite losses inherent in the components from which they are constructed. Op-amps can be used to implement active filters as long as their gain-bandwidth product is not exceeded. Figure 14.25 shows familiar first-order low-

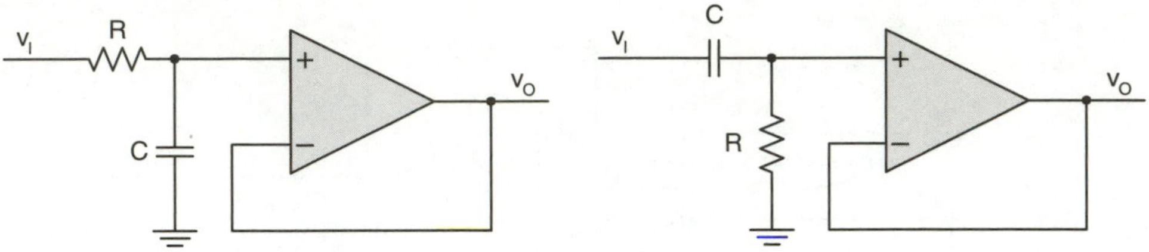

FIGURE 14.25 First-order active filters.

pass and highpass active filters implemented with op-amps. These simple filters buffer a passive filter with a noninverting op-amp stage. In this example, the configurations are for unity gain, although higher gains are possible. Because of the high input resistance of an op-amp, there is little signal loss through the series elements while operating in the passband. Unlike a passive filter whose characteristics are influenced by the load being driven, the op-amp isolates the load from the filter elements.

Filter design is a topic in electrical engineering that can get quite complex when very specific and demanding frequency response characteristics are necessary. Active filters add to this complexity as a result of nonideal op-amp characteristics. Although a complete discussion of filter design is outside the scope of this text, certain filtering tasks can be accomplished by drawing on a basic familiarity with common circuits. A common second-order topology used to implement active filters is the Sallen-Key filter. Sallen-Key lowpass and highpass filters are implemented with two resistors and two capacitors each for unity gain in the passband, as shown in Fig. 14.26. If higher gains are desired, two resistors can be added per the standard noninverting amplifier circuit topology. As with a passive second-order filter, the frequency response curve falls off at 40 dB per decade beyond the cut-off frequency.

The Sallen-Key lowpass filter operates by shunting the op-amp's input path to low-impedance sources at high frequencies. C1 shunts the signal to ground as in a passive filter, and C2 shunts the signal to the op-amp's output. The highpass filter operates in the reverse manner. At low frequencies, C1 blocks the incoming signal and allows R2 to feed the output back to the input. Simultaneously, C2 blocks the signal, which pulls the input to ground. The general expression for the cut-off frequency, f_C, is

$$f_C = \frac{1}{2\pi\sqrt{R1R2C1C2}}$$

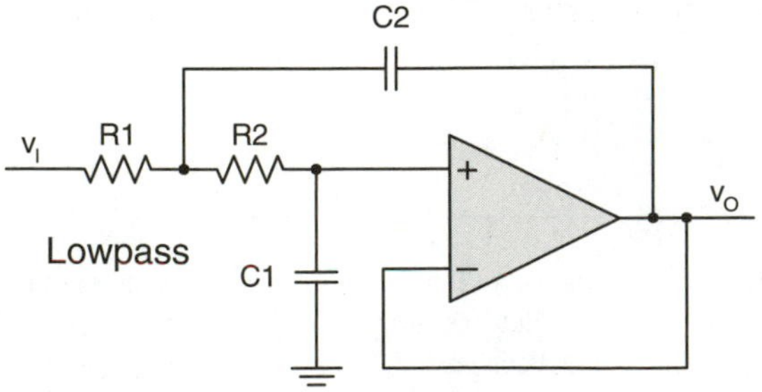

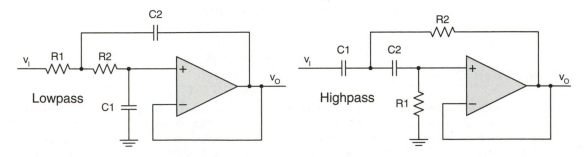

FIGURE 14.26 Sallen-Key second-order active filters.

In many cases, it is convenient to set $R1 = R2$ and $C1 = C2$, in which case f_C is given by

$$f_C = \frac{1}{2\pi RC}$$

Aside from providing amplification, a benefit of using op-amps to build filter stages is that multiple stages can be cascaded to implement more advanced filters. The presence of the op-amp in each stage isolates one stage's filter components from the other, thereby preserving the desired response of each filter circuit. Higher-order lowpass and highpass filters with sharper roll-off responses can be constructed in this manner. Alternatively, bandpass or band-reject filters can be designed by cascading lowpass and highpass stages as needed.

14.7 COMPARATORS AND HYSTERESIS

Each op-amp circuit discussed so far is closed loop because of the need to moderate the op-amp's large open-loop gain and, simultaneously, to increase the circuit's bandwidth to a practical frequency range. Analog electronics begins to turn digital when the concern is no longer about maintaining a linear relationship between a circuit's input and output signals. *Comparators* are a key crossover function between the analog and digital worlds. The job of a comparator is to assert its output when the input rises above a certain threshold and deassert its output when the input falls below a threshold. A comparator implements a binary step function as shown in Fig. 14.27. This transfer function is decidedly nonlinear and is produced using a differential amplifier with very large gain.

If an op-amp is operated in an open-loop topology, it can be adapted to serve as a comparator. Figure 14.28 shows an op-amp or comparator with its negative input connected to a fixed threshold voltage, V_{REF}, and its positive input being driven by the circuit's input signal. As long as the input signal remains below V_{REF}, the differential voltage is negative, and the output is driven to the lower voltage rail (ground, in this case). As the input signal rises above V_{REF}, a minute differential voltage is amplified by the tremendous open-loop gain, resulting in a nearly perfect step function.

Op-amps can be used as comparators, but dedicated comparator ICs have long been available. Comparators and op-amps share various internal similarities. Their common heritage is obvious

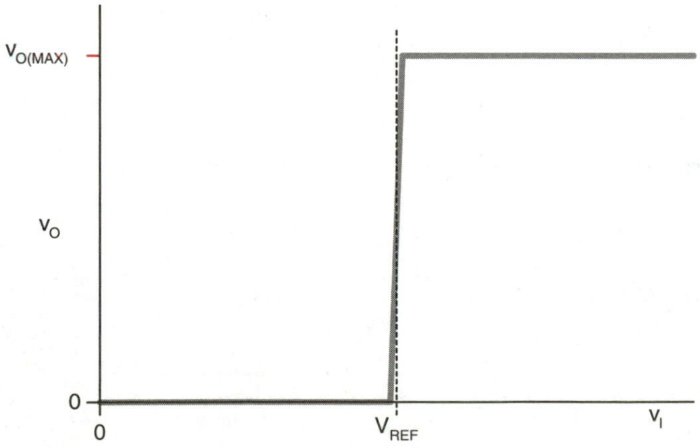

FIGURE 14.27 Generic comparator transfer function.

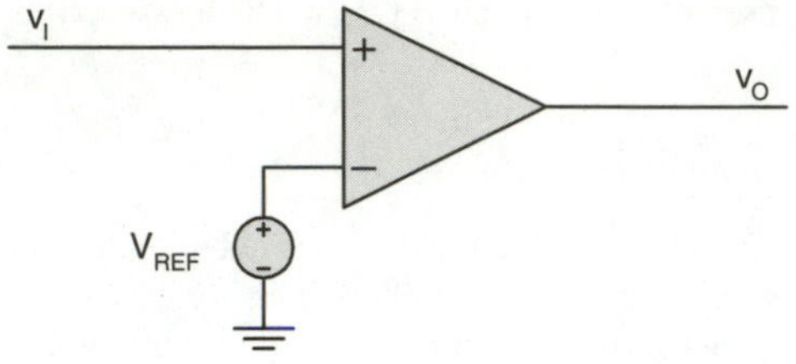

FIGURE 14.28 Open-loop voltage comparator.

when reading through a comparator's data sheet. Specifications such as V_{IO}, I_{BIAS}, I_{IO}, and gain are observed. However, op-amps and comparators are optimized for different characteristics. Op-amps are designed specifically for linear operation in closed-loop configurations. When operated open-loop, some op-amps may exhibit unstable output behavior. A comparator IC is specifically designed for open-loop operation with approximately rail-to-rail output behavior and fast switching times. Op-amp manufacturers discourage using op-amps as comparators for these reasons. Having said this, though, many engineers do adapt op-amps for use as comparators with success. If a quad op-amp such as the LM324 is already in a design that uses only three of its sections, it is tempting and economical to use the fourth section as a comparator if the need arises. In many cases, especially when working with mature bipolar devices, spare op-amps may be safely used as comparators.

The open-loop comparator in Fig. 14.28 is simple, but it has the drawback of being extremely sensitive to minute changes in the input voltage in the vicinity of V_{REF}. When v_I is either less than or greater than V_{REF} by some margin, the op-amp's output is clearly defined. However, when v_I is approximately equal to V_{REF}, the result is less clear. Because of the large open-loop gain, very small changes in v_I cause large changes in v_O. If v_I does not rise or fall monotonically around V_{REF} (consistently increasing or decreasing voltage without any temporary changes in the voltage curve's slope), the minute back-and-forth progression of the voltage curve will be greatly amplified and result in oscillation at the output. This is illustrated conceptually in Fig. 14.29. Note how very small voltage changes around V_{REF} cause v_O to swing from one extreme to the other.

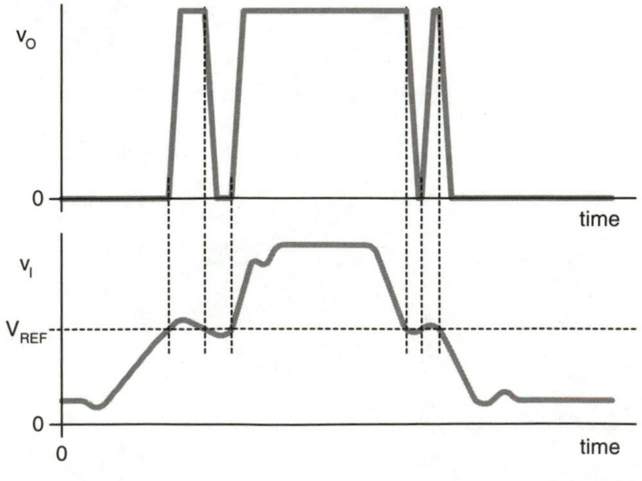

FIGURE 14.29 Unstable voltage comparison without monotonic input.

It can be difficult to achieve a perfectly monotonic signal because of ambient noise in a system. If the signal can be guaranteed to rise and fall quickly, the window of opportunity for noise to trigger an undesired response at V_{REF} is limited. This is how the majority of signals in a digital system operate. The rise and fall times are relatively fast, and the signals remain stable at logic-high and logic-low voltages. Problems do arise, however, when excessive noise or other signal integrity issues manifest themselves by causing nonmonotonic signal transitions. In a purely digital context, such problems can be solved with proper engineering solutions to reduce and shield noise and signal integrity problems. Most real-world analog signals do not behave in a clean binary fashion, which is why they often require analog circuits including op-amps and comparators to properly interface with digital systems.

Threshold comparison can be improved by adding *hysteresis* to an otherwise open-loop voltage comparator. Hysteresis is the application of two thresholds to stabilize a comparator so that it does not change its state with minute changes in the input voltage. Stabilization is desirable in situations wherein the applied voltage hovers near the threshold voltage for more than a brief span of time as seen in Fig. 14.29. Rather than a single threshold, separate low-to-high, V_{TLH}, and high-to-low, V_{THL}, thresholds are designed. V_{TLH} is higher than V_{THL}, as demonstrated in Fig. 14.30, where a much cleaner output is obtained as compared to the previous case. Note that the hysteresis created by the two thresholds prevents the comparator's output from returning to logic 0 when the input declines slightly after triggering a logic-1 output. Similarly, the input's nonmonotonic falling-edge does not cause the output to bounce, because the hysteresis is chosen to be greater than the local perturbation.

A *hysteresis loop* is a common means of representing the two distinct thresholds governing the input/output transfer function. Figure 14.31 shows that, when the input is starting from the low side, the high threshold is used to trigger an output state transition. As soon as the input crosses the high threshold, the output goes high, and the low threshold is now applied to the input comparison. Notice the advantage here. Once the input rises above the high threshold, a lower threshold is instantly substituted. This means that, if the input signal wanders and declines slightly, it is still above the lower threshold, and the output is unaffected. For the output to return low, the input must now fall below the low threshold. And as soon as this occurs, the high threshold is again activated so that the input must rise significantly before an output stage change will occur.

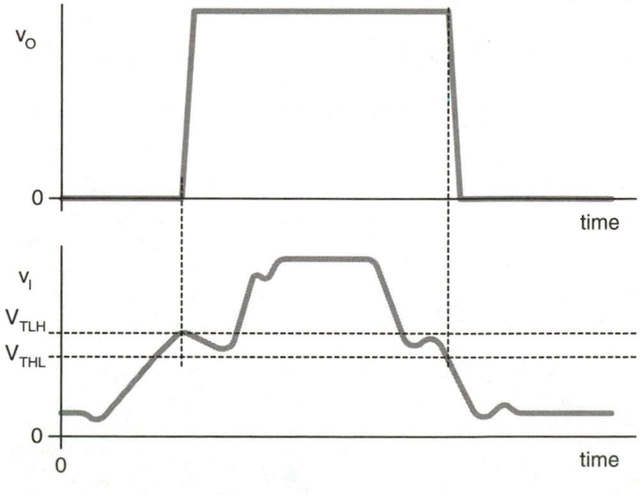

FIGURE 14.30 Effect of hysteresis on nonmonotonic input signal.

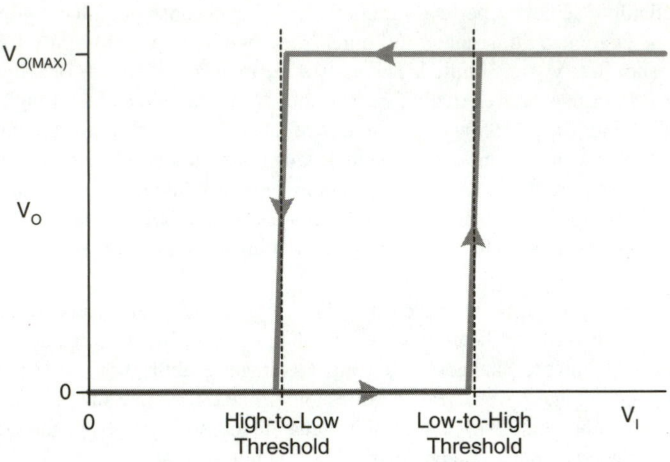

FIGURE 14.31 Hysteresis loop.

The degree of hysteresis designed into a comparator circuit is determined by the difference between the high and low thresholds. A small difference is less tolerant of noise. However, a larger difference has a more muted response, which must be considered in its effects on a system's behavior. Dual comparison thresholds that create hysteresis are implemented by applying positive feedback rather than negative feedback to an op-amp or comparator. Such a circuit is shown in Fig. 14.32. The circuit looks very similar to a conventional closed-loop amplifier, but the feedback is applied to the positive terminal rather than the negative terminal.

The positive feedback through R2 results in a voltage at the positive terminal that is determined by the basic voltage divider expression,

$$v_+ = v_O + (v_I - v_O)\frac{R2}{R1 + R2} = v_O\left(1 - \frac{R2}{R1 + R2}\right) + v_I\frac{R2}{R1 + R2} = v_O\frac{R1}{R1 + R2} + v_I\frac{R2}{R1 + R2}$$

Therefore, the output pulls v_+ down when it is low and pulls v_+ up when it is high. This means that, if v_I is increasing and trying to switch the comparator state from low to high, it must be raised to a higher voltage threshold to counteract the pull-down effect when the comparator's output is al-

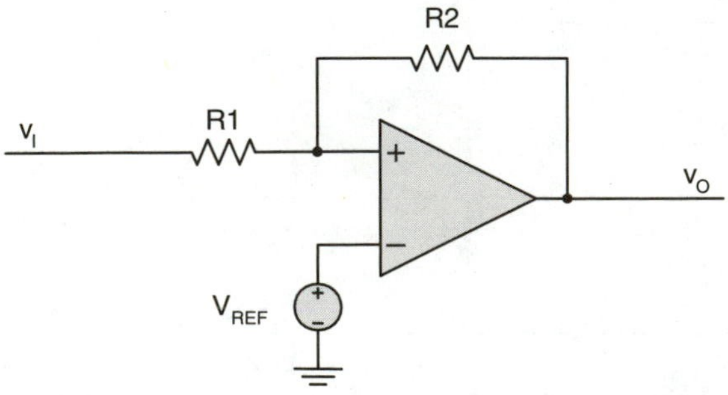

FIGURE 14.32 Hysteresis created by positive feedback.

ready low. Similarly, v_I must be reduced to a lower voltage threshold to counteract the pull-up effect when the output is already high.

Another benefit of hysteresis is that it "snaps" the comparator input voltage above or below V_{REF} once the output state transition is triggered. This forces what might otherwise be a gradually rising input signal into a sharp edge that improves the switching time of the comparator's output. Any time spent in a region of linear amplification as a result of $v_I \approx V_{REF}$ is minimized in favor of a rapid transition to saturation in either the high or low output state.

The voltage at the positive terminal must be brought to V_{REF} to cause a comparator state change. Therefore, the general relationships for the low-to-high and high-to-low thresholds, V_{TLH} and V_{THL}, can be written by substituting V_{REF} for v_+ and the appropriate threshold for v_I.

$$V_{TLH} = V_{REF}\frac{R1 + R2}{R2} - V_{OL}\frac{R1}{R2} \text{ and } V_{THL} = V_{REF}\frac{R1 + R2}{R2} - V_{OH}\frac{R1}{R2}$$

These relationships assume an ideal comparator that has zero input bias current. Real-world effects of I_{BIAS} can be reduced by using resistors that are not too large. If hundreds of microamps or a milliamp are designed to flow through the R1/R2 resistor network, the nanoamps of input bias current will likely introduce negligible error into the system.

To illustrate the application of hysteresis in a real circuit, consider a common bipolar comparator, the LM311, connected such that it drives an approximate output range of 0 V (V_{OL}) to 5 V (V_{OH}). It is desired to convert an incoming analog signal ranging from 0 to 3 V into a digital output. The halfway threshold of 1.5 V is chosen to differentiate between logical 1 and 0. However, it is known that the input signal can have noise of up to 500 mV peak-to-peak (p-p) superimposed on it. This means that a signal that is just over 1.5 V at one moment in time can abruptly drop to near 1 V a short while later. We do not want the LM311's output to swing every time a 500-mV noise spike happens to coincide with the rising and falling of the 3-V input signal. Therefore, at least 500 mV of hysteresis should be designed into the circuit.

It is decided to design in 600 mV of hysteresis for some added margin. Rather than establishing a single 1.5-V threshold for low-to-high and high-to-low triggering, two thresholds of 1.5 + 0.3 V and 1.5 – 0.3 V are desired, which combine to provide 600 mV of hysteresis. This configuration requires the input to rise above 1.8 V to be recognized as a logic 1 and requires the input to fall below 1.2 V to be recognized as a logic 0. Even if 500 mV of noise hits the input just after it reaches 1.8 V, the resulting 1.3 V level will be too high to trigger a return to logic 0.

The first design task is to calculate the ratio between the input and feedback resistors, R1 and R2. Recall that hysteresis is the difference between V_{TLH} and V_{THL}. The previously derived expressions for V_{THL} and V_{THL} can be simplified for this circuit, because V_{OH} and V_{OL} are already known.

$$V_{TLH} = V_{REF}\frac{R1 + R2}{R2} \text{ and } V_{THL} = V_{REF}\frac{R1 + R2}{R2} - 5\frac{R1}{R2}$$

Subtracting the two thresholds yields a relationship that can be used to solve for R1 and R2.

$$V_{TLH} - V_{THL} = V_{REF}\frac{R1 + R2}{R2} - V_{REF}\frac{R1 + R2}{R2} + 5\frac{R1}{R2} = 5\frac{R1}{R2} = 0.6 \text{ V}$$

It follows that R1 = 0.12R2. Because it is the ratio between R1 and R2 that is important rather than their actual values, these numbers can be multiplied by 10,000 to yield R1 = 1.2 kΩ and R2 = 10 kΩ, which are standard 5 percent tolerance resistance values. If more accurate 1 percent resistors are desired, R1 would be chosen as 1.21 kΩ, which is the closest 1 percent standard value.

With the resistors determined, the next step is to calculate the required reference voltage to achieve the desired thresholds. Using the slightly simpler expression for V_{TLH} provides the following result:

$$V_{TLH} = 1.8 = V_{REF}\frac{R1 + R2}{R2} = V_{REF}(1.12),\ V_{REF} = 1.61 \text{ V}$$

Because 1.61 V is not a particularly standard voltage, it can be obtained using a voltage divider resistor network. The math can be checked by verifying that the selected values for V_{REF}, $R1$, and $R2$ also produce the desired V_{THL}.

$$V_{THL} = V_{REF}\frac{R1 + R2}{R2} - 5\frac{R1}{R2} = 1.61(1.12) - 5(0.12) = 1.2 \text{ V}$$

Certain digital interface applications benefit from off-the-shelf 7400 logic gates with built-in hysteresis. A common industry term for a circuit employing hysteresis is the *Schmitt trigger.* Devices employing Schmitt triggers include the 74xx14 hex inverter and the 74xx132 quad NAND. The graphical symbol used to denote Schmitt trigger logic inputs is the hysteresis loop placed within a standard logic gate representation, two examples of which are shown in Fig. 14.33. The 74LS14 and 74LS132 Schmitt trigger devices are members of the common 74LS bipolar logic family. Each of these devices is specified with $1.4 = V_{TLH} = 1.9$ and $0.5 = V_{THL} = 1.0$.* These specifications guarantee a minimum hysteresis of 0.4 V, making them suitable for handling many noisy signals.

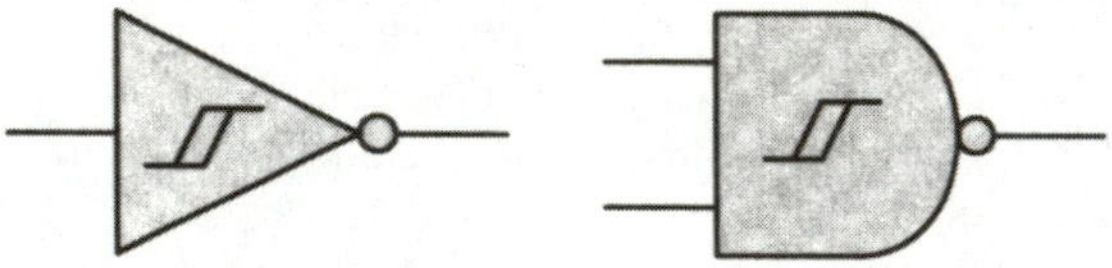

FIGURE 14.33 Graphical representation of Schmitt trigger inverter and NAND gate.

* 74LS14, Fairchild Semiconductor Corporation, 2000, p. 2.

CHAPTER 15
Analog Interfaces for Digital Systems

The intersection of analog and digital worlds has given rise to a tremendously broad range of applications for digital systems. Digital cellular telephones, enhanced radar systems, and computerized engine controls are just a few modern benefits enabled by data conversion circuits. Analog-to-digital and digital-to-analog converters enable computers to interact with the real world by representing continuous analog signals as sequences of discrete numbers.

The first portion of this chapter discusses topics including quantization, sampling rate, and the Nyquist frequency. These concepts form the foundation of data conversion and enable an engineer to evaluate the conversion requirements for individual applications. Specific analog-to-digital and digital-to-analog conversion techniques are presented in the next portion of the chapter. Selecting the correct data conversion IC is a combination of identifying the application requirements and then finding a device that matches these specifications.

Following the initial evaluation process, designing an effective data conversion circuit at the system level varies in complexity with how accurate the conversion must be. Some applications with moderate accuracy requirements can be implemented without much difficulty by following the manufacturer's recommended connection diagrams in their data sheets and application notes. More complex conversion circuits, such as those in digital radio transceivers or high-accuracy instruments, can present significant analog design tasks wherein noise reduction and stability over time and temperature are key challenges. While briefly discussed at the end of this chapter, these high-end applications require further reading into sampling theory and more advanced analog design skills.

15.1 CONVERSION BETWEEN ANALOG AND DIGITAL DOMAINS

Many digital systems interact with their environment by measuring incoming analog signals, such as sound from a microphone, and emitting other analog signals that have been processed in some manner, such as playing a CD on your computer's speakers. These functions are not natural to digital systems, because a binary signal can have only two discrete states, 1 and 0, whereas analog signals are continuous and exist at a wide range of voltages. Specific circuits and methods are necessary to convert between the two domains of discrete digital signals and continuous analog signals.

Analog-to-digital converters (ADCs) and *digital-to-analog converters* (DACs) bridge the gap between our decidedly analog world and the digital world of microprocessors and logic. Essentially, an ADC takes an instantaneous snapshot, or *sample*, of an analog input and converts the observed voltage into a string of binary digits—a number. A DAC performs the reverse operation of converting a discrete number into an analog output voltage. Let's first examine a conceptual ADC.

When an analog signal is applied to an ADC, the ADC evaluates the instantaneous sample of that signal against predefined high and low voltage extremes that define an allowable input range for that ADC. As shown in Fig. 15.1, the ADC overlays a range of numbers between these extremes, thereby dividing the overall voltage range into many small incremental ranges, or *quanta*. A switch is often used to convey the idea of an instantaneous sample being applied to the ADC; the switch is normally open and is closed for a brief instant so that the ADC can measure a voltage. Because of the binary nature of digital systems, practically every ADC divides the allowable input voltage range by some power of two. Each small voltage quantum is mapped to a unique number. The quantum index that most closely matches the observed sample is emitted from the ADC, enabling digital logic to comprehend the instantaneous amplitude of the applied signal as a discrete number. The conceptual example depicts an ADC with voltage extremes of +5 and 0 V and a resolution of 8 bits, indicating that the 5-V range is divided into 256 quanta. Each quantum represents a range of 5 V ÷ 255 = 19.61 mV. A digital sample with the value 0x00 would indicate 0 V, and a sample with value 0xFF would indicate 5 V. If a 0.75 V sample is converted from analog to digital, the sampled value will be either 0x26 (38_{10}) or 0x27 (39_{10}), depending on the rounding mechanism used, because 0.75 V is greater than 38 × 19.61 mV (0.745 V) but less than 39 × 19.61 mV (0.765 V). When an input voltage is presented that is outside the allowable range, the typical ADC returns a saturated sample value of either 0 or $2^N - 1$, depending on whether the input was too low or too high.

Like an ADC, a DAC divides a range of output voltages into many small quanta and operates on the concept of discrete samples in time. At the core of a conceptual DAC is a numerically controlled voltage source that emits a linear range of voltages corresponding to a linear range of discrete numerical samples applied to it. This is illustrated in Fig. 15.2, assuming a voltage source with a minimum output of 0 V and a maximum of V_{OMAX}. The range from 0 to V_{OMAX} is divided by the DAC's resolution, yielding the voltage range of each quantum as in an ADC. Whereas the sampling process in an ADC is modeled using a switch to capture an instantaneous voltage, the similar process in a DAC is modeled with a synchronous register that presents a new sample to the voltage source each sampling interval as regulated by the sample clock. Using the previous example of an 8-bit ADC where 0.75 V would be converted to 0x27, a similar 8-bit DAC could be setup with V_{OMAX} = 5 V. Under these circumstances, the sample value of 0x27 is converted to 0.765 V, because the quantum size remains 19.61 mV.

It is apparent from the preceding discussion that the sampling process is imperfect. The 0.75-V input level is rounded to a multiple of the quantum voltage magnitude, and the resulting digital sample

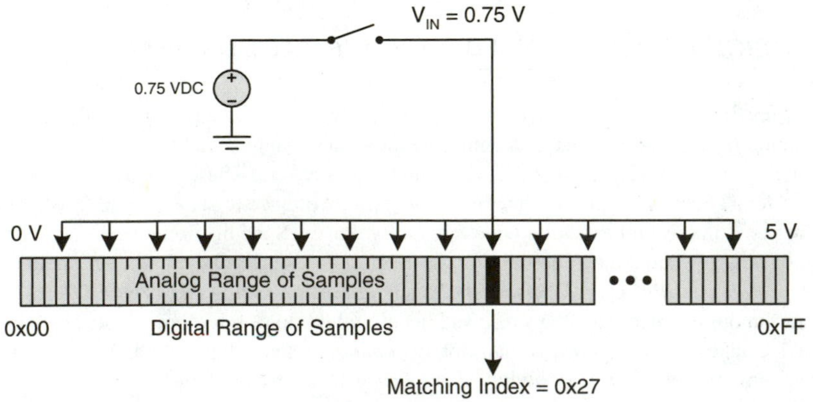

FIGURE 15.1 Conceptual analog-to-digital converter.

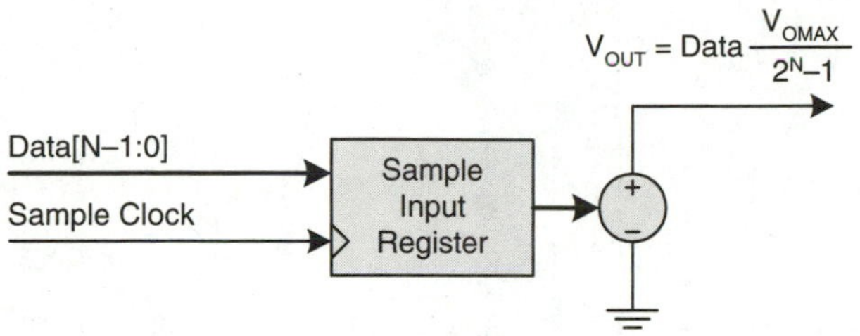

FIGURE 15.2 Conceptual digital-to-analog converter.

is converted back to analog with the rounding made permanent. The exact magnitude of the original analog signal is lost. The process of fitting a continuous analog signal into the closest matching voltage quantum is known as *quantization*. Quantization distorts the signal by skewing the actual voltage to a discrete level. In the preceding example, there would be no discernible difference between 0.75 and 0.76 V, because all allowable voltages are measured to a resolution of 19.61 mV. Quantization errors can be reduced by increasing the resolution of the digital samples. If 12-bit samples were converted instead of only 8 bits, the resolution would improve to 5 V ÷ $(2^{12} - 1)$ = 1.22 mV.

As reality must surely dictate, increased sampling resolution comes at a price. The question becomes how much resolution is required by a particular application. Sensing the temperature in a house for the purpose of controlling a furnace or air conditioner probably does not require more than eight bits of resolution. The useful range of household temperatures to measure may range from 10 to 40°C (50 to 104°F). Even widening this range to between 0 and 50°C (32 and 122 °F) would still provide a resolution of 0.2°C with an 8-bit ADC, plenty for the specified application. However, recording a musical performance with high fidelity may require 12, 16, or more bits of resolution.

Increased sampling resolution does not directly translate to an increase in sampling accuracy. Resolution and accuracy are related but not synonymous. Resolution indicates the granularity of samples, whereas accuracy specifies how reliably the conversion is performed. Accuracy in an ADC indicates how it selects the proper sample to represent the input voltage. For a DAC, accuracy indicates how stable a voltage is generated for each discrete sample. Accuracy is a fairly complex topic, because it involves many aspects of the ADC or DAC circuit, including ambient noise and filtering of that noise. If ±10 mV of noise is present in a 12-bit ADC circuit with 1.22 mV of resolution, the accuracy of the converted samples will be far worse than 1.22 mV, because the noise will randomly skew the voltage up and down as it is sampled. It follows that a well designed 8-bit ADC can provide more useful results than a poorly designed 12-bit ADC.

15.2 SAMPLING RATE AND ALIASING

The rate at which samples are converted is as important as the resolution with which they are converted. How fast should an analog signal be sampled: once per second, one thousand times per second, one million times? The necessary sampling rate is a function of the analog signal's frequency content. Higher frequencies change more rapidly and therefore must be sampled at a faster rate to be measured accurately. Per Fourier analysis, a signal's frequency content is evaluated by representing that signal as a sum of sine waves, each with a unique frequency and phase. The highest frequency sine wave sets the constraint that the sampling rate analysis must take into account.

Consider the 1-kHz sine wave shown in Fig. 15.3 that is being sampled by an ADC at 10 kHz. Each black dot represents a discrete voltage level that is converted into a digital sample. The equally

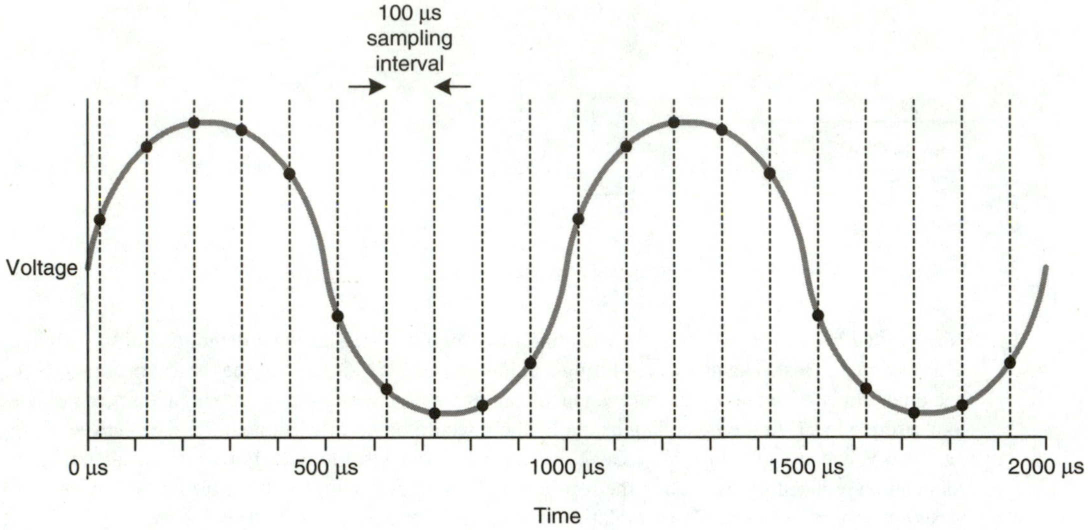

FIGURE 15.3 1-kHz sine wave sampled at 10 kHz.

spaced sampling intervals are shown at an arbitrary phase offset from the sine wave to illustrate the arbitrary alignment of samples to the incoming signal.

Discrete digital samples are reconstructed to generate an analog output that is a facsimile of the original analog input minus a finite degree of distortion. Figure 15.4 shows the 1-kHz sine wave samples as they are emitted from a DAC. On its own, the sampling process creates a stair-step output that bears similarity to the original signal but is substantially distorted. The output is not a pure sine wave and contains added high-frequency components at the sampling frequency—10 kHz in this case. This undesirable stair-step output must be passed through a lowpass filter to more closely reconstruct the original signal. A lowpass filter converts the sharp, high-frequency edges of the DAC

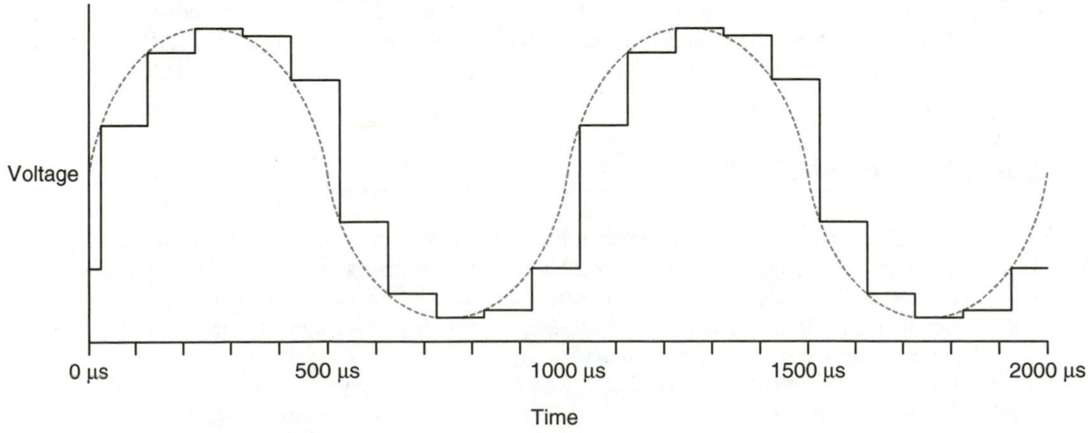

FIGURE 15.4 Reconstructed 1-kHz sine wave without filtering.

output into smoother transitions that approximate the original signal. The output will never be identical to the original signal, but a combination of sufficient sampling rate and proper filter design can come very close.

As the ratio of sampling rate to signal frequency decreases, the conversion accuracy becomes more coarse. Figure 15.5 shows the same 1-kHz sine wave being sampled at only three times the signal frequency, or 3 kHz. Consider how the stair-step output of a DAC looks with this sampling scheme. Here, a filter with sharper roll-off is required, because the undesirable frequency components of the DAC output are spaced only 2 kHz apart from the frequency of interest instead of nearly a decade as in the previous example. A suitable filter would be able to output a nearly pure 1-kHz sine wave, but the signal's amplitude would not match that of the input, because the maximum DAC output voltage would be less than the peak value of the input signal, as indicated by the position of the samples as shown.

Preservation of a signal's amplitude is not critical, because a signal can always be amplified. It is critical, however, to preserve the frequency components of a signal because, once lost, they can never be recovered. Basic sampling theory was pioneered by Harry Nyquist, a twentieth century mathematician who worked for Bell Labs. Nyquist's theorem states that a signal must be sampled at greater than twice its highest frequency to enable the preservation of its informational, or frequency, content. Nyquists's theorem assumes a uniform sampling interval, which is the manner in which most data conversion mechanisms operate. The *Nyquist frequency* is a common term that refers to one-half of the sampling frequency. A data conversion system is said to operate below the Nyquist frequency when the applied signal's highest frequency is less than half the sampling frequency.

Nyquist's theorem can be understood from a qualitative perspective by considering operation at exactly the Nyquist frequency. Figure 15.6 shows a 1-kHz sine wave being sampled at 2 kHz. There are a range of possible phase differences between the sampling interval and the signal itself. The worst-case alignment is shown wherein the samples coincide with the sine wave's zero crossing. Because each sample is identical, the observed result is simply a DC voltage.

If the sampling frequency is increased by a small amount so that the signal is less than the Nyquist frequency, it is impossible for consecutive samples to align themselves at the signal's zero crossings. The amplitude measurement may suffer substantially, but the basic information content of the signal—that it is a 1-kHz sine wave—will not be lost. As the prior examples show, operating with substantially higher sampling rates, or operating at substantially below the Nyquist frequency, increases the accuracy of the data conversion process.

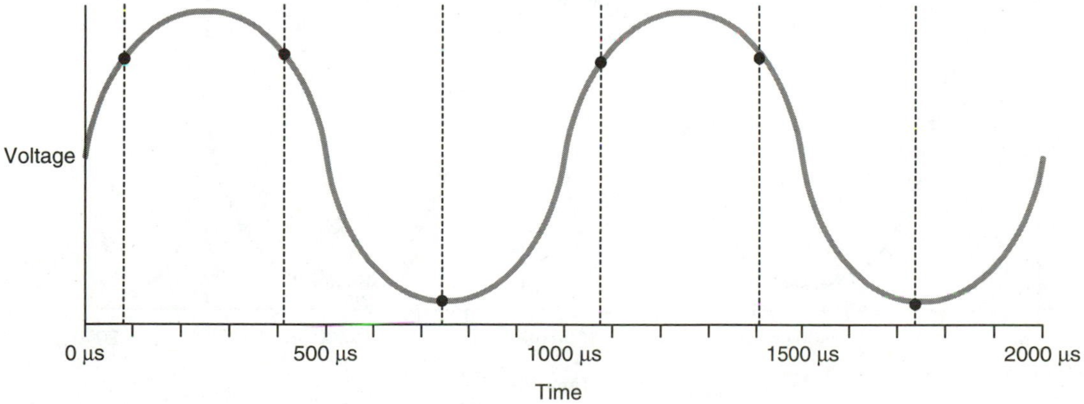

FIGURE 15.5 1-kHz sine wave sampled at 3 kHz.

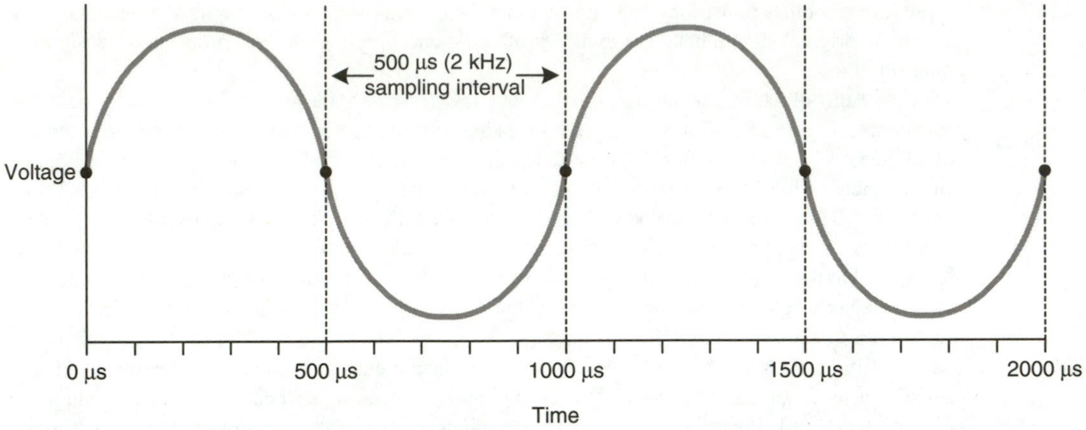

FIGURE 15.6 1-kHz sine wave sampled at the Nyquist frequency.

When a data conversion circuit is operated above the Nyquist frequency, *aliasing* develops, which distorts the information content of a signal. Aliasing occurs when a high-frequency signal is made to appear as a low-frequency signal as a result of inadequate sampling. Assume that a 1-kHz input is expected and a 3-kHz sampling rate is designed to satisfy the Nyquist theorem with some margin. For some reason, a 2-kHz signal enters the system. Figure 15.7 shows how this signal might be sampled. The 333 μs sampling interval is too slow to capture each 250 μs high and low phase of the 2-kHz signal. As a result, the samples convey information that is drastically different from the input. Instead of 2 kHz, the high-low-high interval of the samples indicates a 1-kHz signal! This 1-kHz aliased signal is the difference between the sampling rate and the actual signal frequency.

Aliasing applies to the analog-to-digital conversion process. When converting from digital to analog, the DAC is inherently self-limited by the sampling rate such that the highest frequency it can generate is half the sampling rate, or the Nyquist frequency. This behavior would result if consecu-

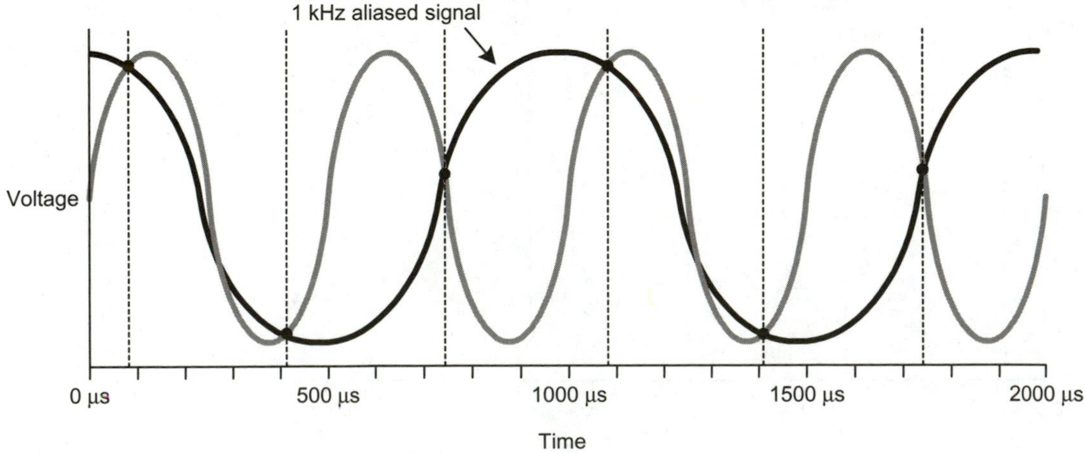

FIGURE 15.7 2-kHz sine wave sampled at 3 kHz with aliasing to 1 kHz.

tive samples alternated voltage levels about some DC level. A suitable lowpass filter on the DAC's output would remove the high-frequency edges and leave behind a sine wave of half the sampling frequency as shown in Fig. 15.8.

Lowpass filtering is a critical part of the data conversion process, because it removes unwanted sampling artifacts at DAC outputs and prevents aliasing at ADC inputs. The necessary slope of the filter's roll-off depends on how close unwanted high frequencies are to the Nyquist frequency.

15.3 ADC CIRCUITS

ADCs are available in a wide range of sampling rates and resolutions. The basic internal architecture of an ADC is straightforward, as shown in Fig. 15.9. A *sample and hold* (S/H) circuit captures a snapshot of the analog input signal so that the conversion circuit can work with a fixed sample over the conversion interval, which can be as long as the sampling interval. An ideal S/H circuit captures the input signal in zero time so that a true instantaneous sample is taken. In reality, the small capacitor that is used to hold the sample during conversion takes a finite time to charge through a finite switch resistance to the same voltage as the input. Once the conversion circuit maps the captured voltage to a digital sample, the digital interface conveys this information to the digital processor. ADC interfaces are available in both serial and parallel configurations.

A variety of basic analog-to-digital conversion circuits are used, based on the desired sampling rate and resolution. Three of the most common are *flash*, *successive-approximation*, and *sigma-delta*. A flash ADC, shown in Fig. 15.10, consists of a bank of parallel comparators, each fed by a unique incremental reference voltage. Each comparator's output represents one of the $2^N - 1$ possible outputs of the ADC. By process of elimination, the lowest quantum is not represented by a comparator, because it is implied if none of the comparators are at logic 1. Therefore, a 12-bit flash ADC

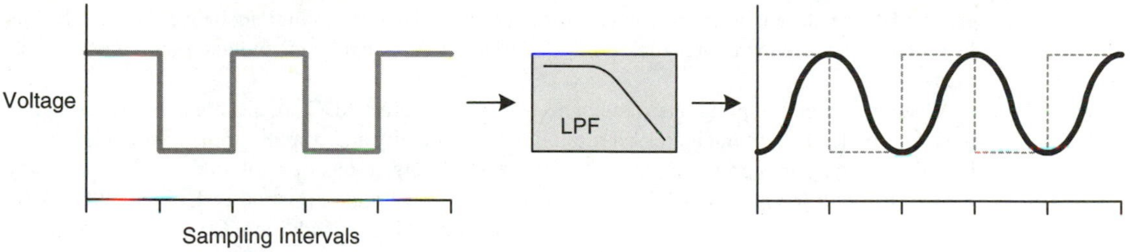

FIGURE 15.8 Maximum frequency output of the DAC.

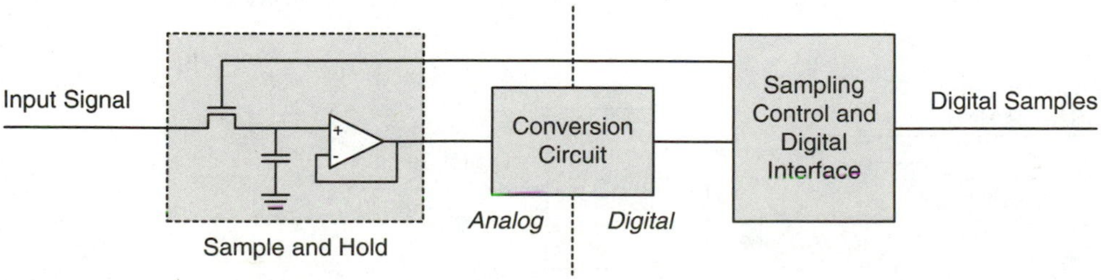

FIGURE 15.9 Basic ADC architecture.

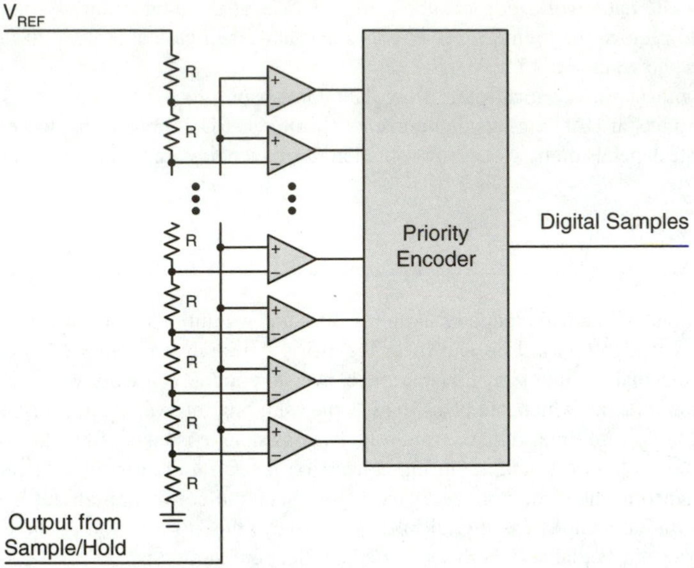

FIGURE 15.10 Flash ADC circuit.

requires 4,095 parallel comparators. When a voltage is applied to the flash circuit, one or more comparators may emit a logic 1. A priority encoder generates the final N-bit output based on the highest comparator that is emitting a logic 1. This circuit is called a flash ADC, because it is very fast; an analog input is converted to a digital sample in one step.

Flash ADCs are fast, but their complexity doubles with each added bit of resolution. Such ADCs are available with maximum sampling rates over 100 MHz and resolutions between 8 and 16 bits from manufacturers including Analog Devices, Intersil, National Semiconductor, and Texas Instruments.

When very high sampling rates are not necessary, alternative ADC circuits are able to accomplish the task with lower cost and increased resolution. High-quality audio applications commonly use 24-bit ADCs with sampling rates of either 48 or 96 kHz. Below 16 bits of resolution and 20 kHz, many inexpensive and low-power ADCs are available. successive-approximation and sigma-delta ADCs are manufactured by the same companies that make flash ADCs. In addition, Crystal Semiconductor offers a line of ADCs optimized for digital audio applications.

A successive-approximation ADC uses an internal DAC/comparator feedback loop to home in on the digital code that corresponds to the applied analog input. Figure 15.11 shows this feedback loop along with control logic that varies the code until the DAC output matches the input. Relatively

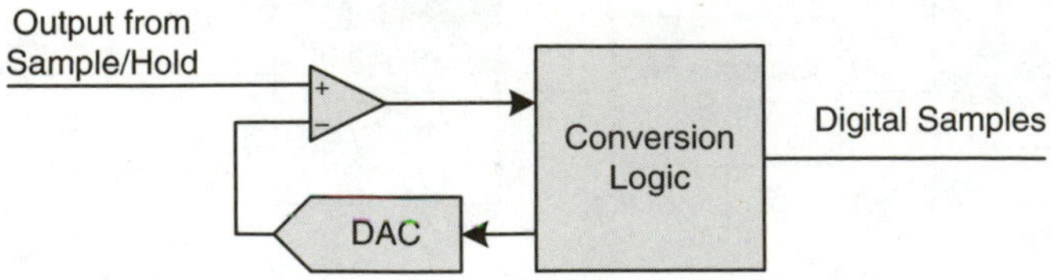

FIGURE 15.11 Successive-approximation ADC circuit.

dumb control logic could simply increment the code starting from 0 until the comparator's output changed from high to low. This would mean that an N-bit ADC would require up to 2^N cycles to perform a conversion. Instead, a successive-approximation ADC performs a binary search to accomplish the same task in only N cycles. A digital code of 0 is used as a starting point. Each bit in the code, starting from the most significant bit, is set, and the comparator's output is tested each time. If the output is low, the DAC voltage exceeds the input and, therefore, the bit that was set should be cleared. Otherwise, the bit is left set.

To illustrate how a successive-approximation ADC functions, consider an 8-bit ADC with a range from 0 to 5 V and an input level of 3 V. Each conversion quantum is 19.61 mV. Table 15.1 lists the eight sequential steps in performing the data conversion. In reality, the hypothetical ADC circuit may output 0x98 or 0x99 with a 3-V input, depending on the ambient electrical and thermal conditions. When an input is on the border between two quanta, slight changes in supply voltage, noise, and temperature can skew the result up or down by one digital code. The final result is chosen as 0x98, because the next code, 0x99, corresponds to a voltage that is slightly higher than the input voltage. This gets back to the concept of conversion accuracy. Manufacturers specify ADCs with worst-case accuracies. Additionally, the parameters of the circuit into which they are designed can further degrade the conversion accuracy.

TABLE 15.1 Eight-Bit Successive-Approximation Conversion Steps

Cycle	Test Code	DAC Voltage	Comparator Output	Resultant Code
1	**1**0000000	2.51	1	**1**0000000
2	1**1**000000	3.77	0	1**0**000000
3	10**1**00000	3.14	0	10**0**00000
4	100**1**0000	2.82	1	100**1**0000
5	1001**1**000	2.98	1	1001**1**000
6	10011**1**00	3.10	0	10011**0**00
7	100110**1**0	3.02	0	100110**0**0
8	1001100**1**	3.00	0	1001100**0**

A sigma-delta ADC over-samples the input at very coarse resolution: one bit per sample! To create a high-resolution sample, a typical sigma-delta ADC oversamples by 128 or 256 times the nominal sampling frequency and then passes the serial samples through a digital filter to create a usable set of N-bit samples at the nominal sampling frequency. The basic theory behind a sigma-delta ADC has been around for a long time, but its practical implementation is more recent because of its reliance on digital filter logic, which is now inexpensive to manufacture on an IC. Figure 15.12 shows a sigma-delta ADC incorporating a voltage summation stage, an integrator, a comparator, a 1-bit DAC, and a digital filter. The summation stage subtracts the DAC output from the input voltage. The integrator is a circuit that accumulates the resulting sum over time.

For a given input, the sigma-delta circuit will emit a serial set of samples with an average DC value over time that equals the input voltage. The integrator keeps track of the difference between the input and the DAC feedback voltage. When the comparator sees that this running difference exceeds 0, it causes a negative feedback through the DAC and summation stage. When the difference is

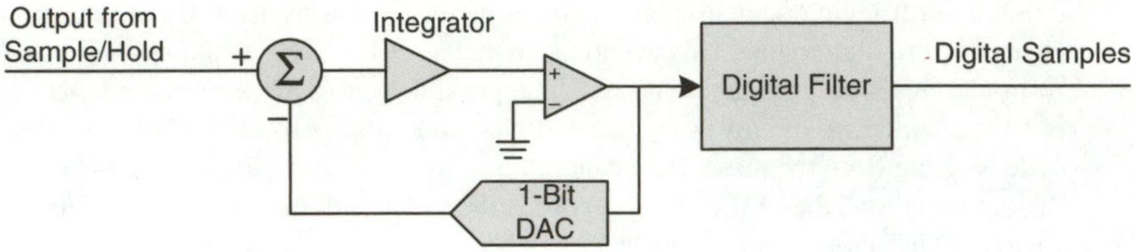

FIGURE 15.12 Sigma-delta ADC circuit.

less than 0, the comparator's output is 0, and there is no negative feedback, causing the running difference to increase once again.

The disadvantage of a sigma-delta ADC is that it requires a high oversampling frequency to function. However, this is not a problem at low frequencies, such as audio, where these converters are commonly employed. A sigma-delta ADC is able to deliver very high resolutions (e.g., 24 bits) with high accuracy, because most of its complexity is in the digital filter, and only a coarse single-bit conversion is performed. This means that the circuit is less susceptible to noise. Digital logic is much more tolerant of ambient noise as compared to delicate analog comparators and amplifiers.

A key advantage of sigma-delta ADCs for the system designer is that expensive lowpass filters with sharp roll-offs are not required. Since the actual sampling frequency is so much higher than the signal's frequency content, an inexpensive single-pole RC lowpass filter is generally sufficient. Consider a CD audio sampling application in which the maximum input frequency is 20 kHz and the nominal sampling rate is 44.1 kHz. A sigma-delta ADC might sample this signal at 128 × 44.1 kHz = 5.6448 MHz. Therefore, the Nyquist frequency is raised to approximately 2.8 MHz from 22 kHz! A first-order filter with f_C = 20 kHz would attenuate potentially aliasing frequencies by more than 40 dB. In contrast, a normal ADC would require a much more costly filter to provide the same attenuation where the passband and stopband are separated by only 2 kHz.

15.4 DAC CIRCUITS

Unlike an ADC, a DAC does not require a sample and hold circuit, because the instantaneous sample events are driven from the discrete digital domain where each clock cycle activates a new sample. A DAC consists of a digital interface and the conversion circuit. Two of the most common types of conversion circuits are the *R-2R ladder* and sigma-delta designs.

The R-2R ladder DAC uses the concept of current summation as found in an inverting op-amp summing circuit. Two resistance values, R and 2R, are connected in a multistage network as shown in Fig. 15.13 (using a four-bit example for the sake of brevity). This circuit is best analyzed using superposition: set one input bit to logic 1 (V_{REF}) and the others to logic 0 (ground). When this is done, the resistor ladder can be quickly simplified by combining parallel and series resistances, because all nodes other than the logic 1 input are at 0 V. Knowing that the voltage at the op-amp's negative terminal is also 0, the current through the resistor ladder can be determined and, therefore, the output voltage can be calculated.

After calculating the partial output voltage due to each individual input, the following overall expression for V_O is obtained:

$$V_O = -V_{REF}\left[\frac{D3}{2} + \frac{D2}{4} + \frac{D1}{8} + \frac{D0}{16}\right]$$

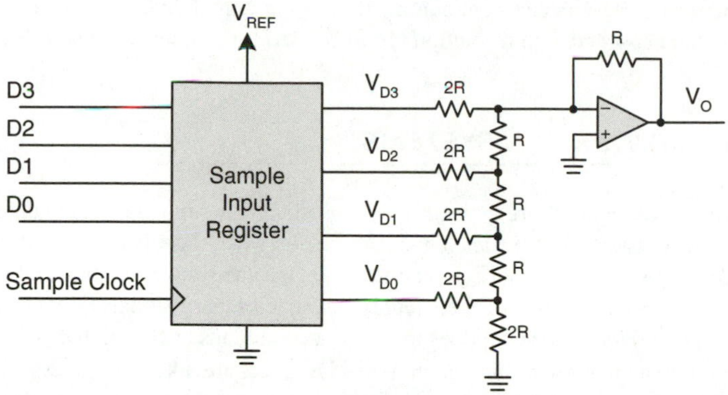

FIGURE 15.13 Four-bit R-2R ladder DAC.

This expression allows linear control of V_O ranging from 0 to within one least-significant bit position (1/16 in this case) of $-V_{REF}$. Similar results can be obtained with the basic op-amp summer circuit, but each input bit's resistor must be twice the value of the previous bit's resistor. This rapidly becomes impractical with 8, 12, 16, or more bits of resolution.

Real DACs modify this basic circuit to generate positive output voltage ranges. In actuality, the basic R-2R ladder concept is often used, but the exact circuit topology and current summation mechanism change to better suit circuit design constraints. Many R-2R ladder DACs are designed to emit the summed currents rather than a voltage and therefore require an external op-amp circuit to create the final desired signal. This requirement is not really burdensome, because a buffer of some type is usually needed to provided sufficient drive strength to the intended load. A DAC is designed to reconstruct analog signals with accuracy, but not to drive substantial loads. Therefore, even a voltage-output DAC may require a unity gain op-amp stage for the system to function properly. A current-output DAC can be connected in a manner similar to that shown in Fig. 15.14. Assuming that the output range is 0 to 10 mA, this circuit is capable of V_O from 0 to 4.99 V. The DAC is designed to sink rather than source current so that the inverting op-amp configuration produces a positive output voltage.

A sigma-delta DAC operating principle is similar to that of the sigma-delta ADC. An internal digital filter converts multibit samples (e.g., 16 bits) at the desired rate into single-bit samples at a high-frequency multiple of that rate. Over time, the average DC value of these high-frequency samples corresponds to the desired DAC output voltage. This high-frequency signal is filtered to yield a converted analog signal with minimal distortion. As with an ADC, a sigma-delta DAC has the advantage of requiring an inexpensive lowpass filter, because the sampling rate is so much higher than the fre-

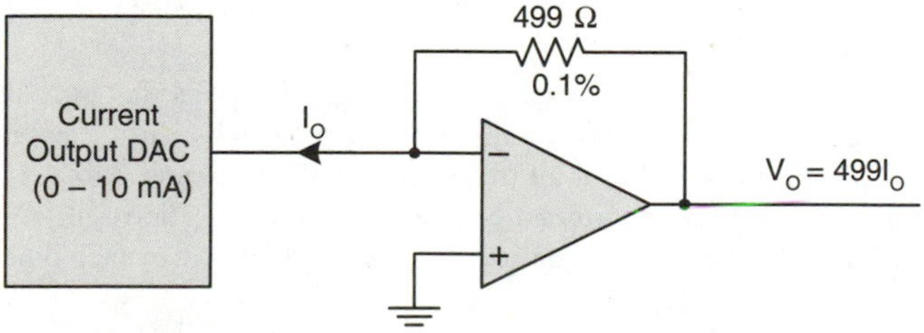

FIGURE 15.14 Buffered current-output DAC.

quencies generated. These devices are commonly used in digital audio equipment because of their low overall system cost and high resolution (16 to 24 bits) for frequencies below 100 kHz.

15.5 FILTERS IN DATA CONVERSION SYSTEMS

General noise-reduction filters are typically found on the power supplies feeding data conversion circuits, because less distortion of the analog signals results when there is less overall noise in the system. Purely digital systems are not immune to noise, but their tolerance threshold is much higher than for analog systems because of their highly quantized binary signals. It takes noise of greater magnitude to turn a 0 into a 1 than it does to distort a continuous analog signal.

In addition to general noise reduction, anti-aliasing filters are a key design aspect of data conversion systems. Filtering requirements dictated by the sampling rate and by the presence of undesired frequency content can be quite stringent. As the gap between the Nyquist frequency and the undesired frequencies decreases, more complex filters are necessary. The design of such complex filters requires a substantial set of analog design skills, the majority of which are outside the scope of this book. However, this chapter closes by identifying some of the issues that arise in anti-alias filtering so that you may be aware of them.

The first step in specifying an anti-aliasing filter is to identify how much attenuation is necessary in the stopband. An ADC or DAC can resolve an analog signal only to a finite resolution given by the number of bits that are supported. It is therefore unnecessary to attenuate high-frequency content beyond the point at which the circuit's inherent capabilities reach their limit. Attenuating unwanted signals to less than one-half of a voltage quantum renders them statistically insignificant. If a system represents an analog signal with N bits of resolution, this minimum attenuation, A_{MIN}, is given as

$$A_{MIN} = -20\log(2^{N+1}\sqrt{3})$$

One-half of a quantum is represented by the added power-of-two beyond that specified by the conversion resolution. The $\sqrt{3}$ term represents an allowance for the average quantization noise magnitude. As expected, a higher-resolution conversion requires greater attenuation, because the quantization noise is reduced. An 8-bit ADC, relatively low resolution by modern standards, requires a stopband gain of –59 dB to prevent distortion due to aliasing. A 12-bit ADC, quite moderate in resolution, requires attenuation of 83 dB.

Once the stopband gain is known, the filter's roll-off target can be determined by identifying the separation of the passband and stopband. Filter design is not a trivial process when minimal distortion is a design criterion. A real filter does not have perfectly uniform passband gain and therefore adds some distortion by attenuating some frequencies more than others even in the passband. When the passband and stopband are close together, a higher-order filter is called for to provide the necessary attenuation. Yet, filters with sharp roll-off can have the undesired side effect of exhibiting resonance around the cutoff frequency.

Additionally, filters do not simply change the amplitude of signals that pass through them but change their phase as well by introducing a finite time delay or phase shift. The problem is that filters do not have a uniform phase delay at all frequencies in the passband. Most real signals are not pure sine waves and therefore are composed of many frequencies. The result is distortion of the overall signal as the different frequency components are shifted relative to each other by slightly different delays.

The magnitude of these nonideal filter characteristics depends on each situation. A basic first-order RC filter is very well behaved, which is a key reason why sigma-delta conversion circuits are so popular. The limitation, of course, is that sigma-delta circuits can be used only at lower sampling rates.

If a sigma-delta circuit cannot be used for one reason or another, it may be practical to operate a conventional ADC or DAC at a significantly higher sampling rate than would be dictated by the frequency of interest. This allows a less complex filter to be used by increasing the separation between the signal and Nyquist frequencies. Digital signal processing techniques can then be used to filter the digital samples down to a more ideal sampling rate. The difference here is that the complexity of analog filter design is traded off against a faster data conversion system and some computational number crunching. As logic gates have become inexpensive and microprocessors increase in capability, DSP-based filter algorithms are often far superior to their analog implementations, because the nonideal characteristics of phase-delay, amplitude variance, and resonance are overcome by arithmetic manipulations. A major reason behind the proliferation of high-speed ADC and DAC ICs is the emergence of DSP technology in applications in which fully analog circuits formerly existed. DSP algorithms can implement complex and highly stable filters that are extremely costly to implement with analog components.

P • A • R • T • 4

DIGITAL SYSTEM DESIGN IN PRACTICE

CHAPTER 16
Clock Distribution

Clocks are inherently critical pieces of a digital system. Reliable operation requires the distribution of electrically clean, well timed clocks to all synchronous components in the system. Clocking problems are one of the last bugs that an engineer wants to have in a system, because everything else is built on the assumption of nearly ideal clocks. This chapter concentrates on the means of distributing low-skew and low-jitter clocks in a system. Most systems require the design of a *clock tree*—a circuit that uses an oscillator of some type to create a clock and then distributes that clock to multiple loads, akin to branches in a tree. Simple clock trees may have a single level of hierarchy in which the oscillator directly drives a few loads. More complex trees have several levels of buffers and other components when tens of loads are present.

Basic information on crystal oscillators is presented first to assist in the selection of a suitable time base from which to begin a clock tree. Once a master clock has been produced, low-skew buffers are the common means of replicating that clock to several loads. These buffers are explained with examples incorporating length matching for low-skew and termination resistors for signal integrity. Buffers are followed up with a discussion of phase-locked loops, commonly used to implement "zero-delay" buffers in clock trees. These devices become important when a system contains multiple boards and when there are special clocking sources other than a stand-alone oscillator. Low-skew and zero-delay buffers form the basis for most clock tree designs.

The second portion of the chapter discusses more advanced clocking concepts beginning with frequency synthesis. Originally conceived for analog and RF applications, frequency synthesis is an important part of many high-performance digital systems. It allows multiple clocks to be derived from a single time base and is how a leading-edge microprocessor operates many times faster internally than it does externally. Frequency synthesis can also be important when processing data between multiple interfaces that run at different frequencies.

Next, delay-locked loop technology is presented, because it can accomplish the same basic function as a phase-locked loop in many cases but is a purely digital circuit with resultant implementation advantages. The chapter concludes with a brief discussion of source-synchronous interfaces as a necessary alternative to conventional synchronous design when dealing with very high frequencies.

16.1 CRYSTAL OSCILLATORS AND CERAMIC RESONATORS

An electronic clock consists of an amplifier with a passive time-base element coupled into its feedback loop. It has previously been shown that a simple oscillator can be formed with just an RC time base and an inverter that serves as an amplifier. Simpler yet is connecting the output of the inverter directly to its input via a piece of wire sized to provide a certain time delay. While simple, neither of these approaches yields a sufficiently accurate clock source in most applications. Accurate timebases are created by exploiting the resonant properties of *piezoelectric* crystals cut to specific sizes. Piezoelectricity is the property found in certain crystals whereby slight changes in the crystalline structure result in a small electric field, and vice versa; exposing the crystal to an electric field causes slight deformation of its structure. Quartz is the most common piezoelectric crystal in use for oscillators, but other such crystals, both natural and synthetic, have this property. A solid object's natural resonant frequency is a function of its physical dimensions and its composition. A crystal such as quartz or certain ceramics has a predictable resonant frequency that can be finely adjusted by varying the size of the crystal slab. Furthermore, the resonant frequency is largely insensitive to variations in temperature and voltage.

A piezoelectric crystal of known resonant frequency can be incorporated with an inverting amplifier to yield an accurate clock. The amplifier drives one end of the crystal, and the other end feeds back to the amplifier's input. With proper circuit design, the crystal begins to resonate as it is driven and quickly settles into a continuous oscillation that both stimulates and is maintained by the amplifier. A generic crystal oscillator circuit found in many digital ICs is shown in Fig. 16.1. Note the crystal's graphical representation. The IC contains an internal crystal driver, which is a specialized inverter. Externally, a crystal and load capacitors are required. The load capacitors form an LC resonant circuit in concert with the crystal that is made to appear inductive. Manufacturers specify crystals with a particular load capacitance requirement for proper oscillation. The two capacitors are typically selected to be the same value, C. When this is done, the overall load capacitance, C_L, is 0.5 C plus any stray capacitance, C_S, in the circuit. Stray capacitance is often in the may range of several picofarads. A crystal specified with C_L = 18 pF might use 22-pF capacitors assuming $C_S \approx$ 7 pF.

The circuit shown is a digital oscillator, because it emits a square wave binary signal. Analog applications such as RF use a linear amplifier to drive a sine wave instead.

It is rare to find a circumstance these days in which an engineer must design a digital oscillator from scratch. Many embedded microprocessors and microcontrollers contain on-board driver circuits that require the connection of an external crystal, and usually the dual capacitors as well. When an IC does not contain an integrated oscillator, discrete crystal oscillators are the most common solution. A variety of companies, including CTS, ECS, and Ecliptek, manufacture off-the-shelf oscillators that include the crystal and driver circuit in a single package. These components typically have four terminals: power, ground, clock, and an optional clock enable.

Quartz crystals are ubiquitous, because they are inexpensive and provide a relatively high degree of frequency stability over time and temperature. Frequency tolerance between 50 and 100 parts per million (ppm), or better than 0.01 percent, is easily obtained. In contrast, most ceramic crystals, usually called *ceramic resonators,* are less expensive and have tolerances an order of magnitude worse

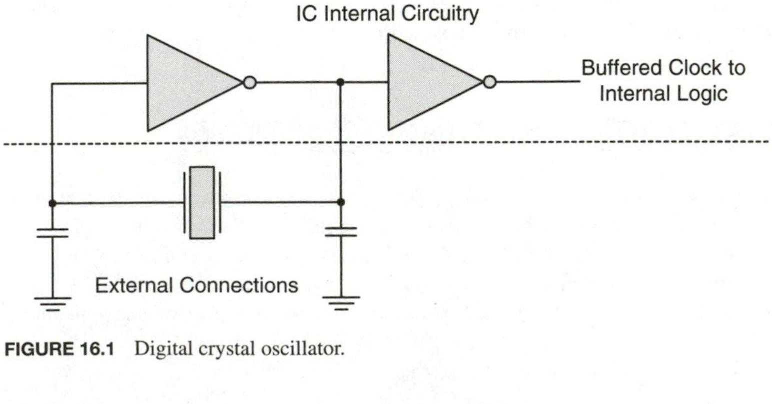

FIGURE 16.1 Digital crystal oscillator.

than common quartz crystals. Ceramic resonators are used in very low-cost applications wherein accuracy is forsaken for small cost savings. Most digital systems use quartz crystals that cost approximately $1.00, because they are reliable, they provide an accurate time base, and the crystal's cost is a small fraction of the overall system cost.

Some applications require tighter tolerances than normal crystal oscillators provide. More precise manufacturing techniques and control over materials can yield tolerances of approximately 1 ppm. Below this level, temperature control becomes a significant factor in frequency stability. So-called *oven-controlled oscillators* are specially designed to maintain the crystal at a stable temperature to greatly reduce temperature as a variable in the crystal's resonant frequency. Using this technique, oscillators are available with tolerances on the order of one part per billion! Conner-Winfield and Vectron International, among others, manufacture these high-accuracy oscillator products.

Most digital systems do not require clock accuracy better than 50 or 100 ppm. However, jitter is another clock stability characteristic of concern. Each oscillator circuit is subject to a certain amount of jitter based on the tolerance of its components. Aside from an oscillator's inherent jitter specification, ambient noise can couple into the oscillator and cause additional jitter. Therefore, it is desirable to attenuate ambient noise on the power supply that might otherwise couple into the oscillator circuit. It is common to find various types of LC filters on the power leads of crystal oscillators. A basic pi-type topology is shown in Fig. 16.2, consisting of a ferrite bead with capacitors on each side to attenuate high frequencies with small capacitors and provide lower frequency response with a larger capacitor. This circuit attenuates differential noise and provides the oscillator with a cleaner power supply relative to its ground reference.

16.2 LOW-SKEW CLOCK BUFFERS

Once a stable clock source has been established, the signal must be distributed to all components that operate on that clock. Clock distribution is critical to a digital system, because synchronous timing analysis assumes the presence of a reliable and consistent clock. Conventional synchronous buses running between ICs require a common clock so that they can work together with a known timing relationship. A single bit on a synchronous bus essentially consists of an output flop on one IC that drives an external wire, possibly passes through some combinatorial logic, and then is sampled at the input of a flop on another IC. Each IC on the bus should ideally see the same clock signal. In reality, there are slight skew variations between these individual clocks. Some clocks arrive a little sooner than others. Skew should be kept to a minimum because, like jitter, it reduces the synchronous timing budget.

Common clock signals are distributed in a low-skew manner by closely matching the delays from the clock source to all loads. Distribution delays are incurred as the clock passes through passive wires and active buffers. Consider the hypothetical clock distribution tree shown in Fig. 16.3. An oscillator drives a clock buffer, which drives five loads. All of the clock signals are point-to-point with

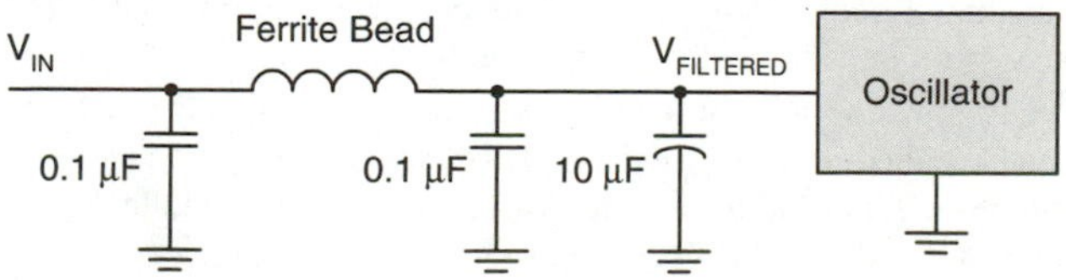

FIGURE 16.2 Oscillator LC pi power filter.

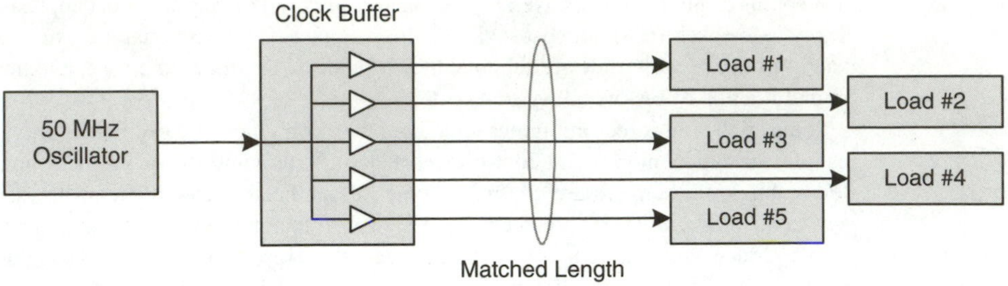

FIGURE 16.3 Single-buffer clock tree.

a single driver and a single load. It is assumed that each load is approximately the same as all other loads, which is usually the case when dealing with ICs. If one load is dramatically different from the others, it may cause its respective driver in the clock buffer to behave differently, and this could hinder the goal of creating a tree with balanced delays.

The absolute propagation delay from the oscillator to each load is immaterial in this case, because all that matters is that each load sees a clock with the same timing as the others. Therefore, the variables of interest are those that can potentially cause differential delays, or skew. The first such variable is the propagation delay variation from the buffer input to each output. Each output driver in an IC has slightly different physical properties that cause its propagation delay to vary within a narrow range. Low-skew clock buffers are specifically designed to minimize this variation. The acceptable amount of skew varies by application but generally declines with increasing frequency and with the use of faster devices, even at lower frequencies. Older low-skew buffers with pin-to-pin skews of better than 1 ns are still available for systems running at moderate frequencies. Newer buffers with skews of 200 to 50 ps are available for systems with very low skew budgets. Manufacturers of these buffers include Cypress Semiconductor, Integrated Circuit Systems, Integrated Device Technologies, and Texas Instruments. The second skew variable in this example is the variation in wire lengths connecting the buffer to each load. Wire delays are matched by matching wire lengths. Matching clock trace lengths when designing a PCB is one of the basics of proper clock distribution.

Signal integrity is tightly coupled with clock distribution concepts, because electrically clean clocks are a prerequisite for a digital system's reliable operation. Flops can be falsely triggered by nonmonotonic edges resulting from clocks that are improperly distributed in a way that causes their edges to become significantly distorted. Signal integrity is the topic of a later chapter, but a few basics are introduced here because of their relevance. The goal when distributing a clock is to prevent reflections on the transmission line that are formed by the wires that carry that clock signal. Reflections can result when a signal transitions rapidly relative to its propagation delay across the transmission line. It rises or falls so rapidly that one end of the wire is at a different voltage from the other end. Reflections distort a signal, because the instantaneous voltage on a wire is the sum of the driven signal and any reflected voltages that are present. A reflection occurs when a signal travels across an impedance discontinuity. Some energy passes through the discontinuity, and some is reflected back to its source. Terminating a transmission line with its characteristic impedance prevents reflections from occurring, because no impedance discontinuity exists.

One easy way of terminating a point-to-point wire that is always driven from one end is with *source termination*, also called *series termination*. This technique is illustrated in Fig. 16.4 as an enhancement to the single-buffer clock tree example. A resistor is placed at the driver and is sized such that its resistance plus that of the driver, Z_D, is approximately equal to the transmission line's characteristic impedance, Z_O. For this example, it is assumed that a transmission line with $Z_O = 50\ \Omega$ is implemented on a PCB and that the clock buffer's driver has $Z_D \approx 10\ \Omega$. Therefore, the series resistors

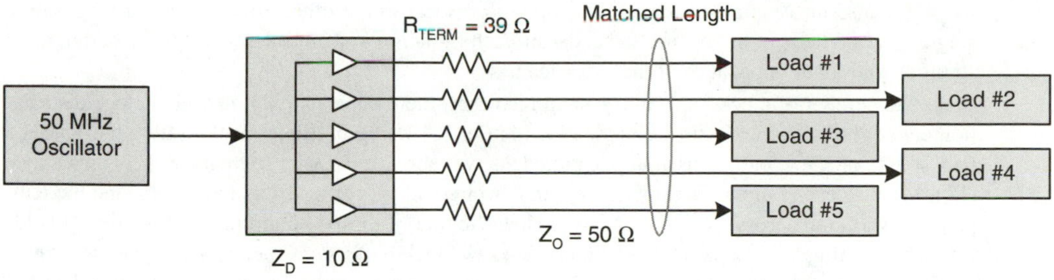

FIGURE 16.4 Series-terminated clock distribution.

are selected to be 39 Ω, the closest standard 5 percent value to $Z_O - Z_D$. In reality, it is difficult to precisely determine an output driver's impedance, because it is subject to variation with temperature and supply voltage. Trial and error may be necessary to select the best series termination value for a given clock buffer. Values between 39 and 47 Ω for 50 Ω for transmission lines are common.

Series termination does not prevent a reflection at the load, because the load is not terminated. The clock signal travels down a 50-Ω transmission line and then hits the high-impedance end load. The reflected energy travels back toward the source across the 50-Ω transmission line and then sees the 50-Ω load of the combined terminating resistor and the driver impedance. Therefore, the reflection is properly terminated and does not reflect back to the load and disrupt the clock signal. For series termination to work properly, the resistor should be placed as close as possible to the driver so that the resistances appear as a single lumped element rather than as two separate loads at high frequency.

Note that the wire between the oscillator and the buffer is not terminated. This is because the oscillator is usually placed very close to the buffer such that the unterminated transmission line is too short to cause reflections of significant amplitude. There are varying opinions on how long a transmission line can be before reflections become a problem. It is related to the signal's rise time. Many engineers use the rule that if a transmission line's delay is greater than one-fourth of the signal's rise time, it should be terminated. Electrical signals travel through a wire at approximately 6 in (0.15 m) per nanosecond. If a clock oscillator has a rise time of 1 ns, the wire between the oscillator and buffer should be no more than 1.5 in (3.8 cm) long.

Clock buffers have a number of outputs, and situations arise when one extra clock signal is needed. Rather than having to use a second clock buffer IC just for one more output, one pin can often drive two loads as shown in Fig. 16.5. A risk of this approach is that the more heavily loaded out-

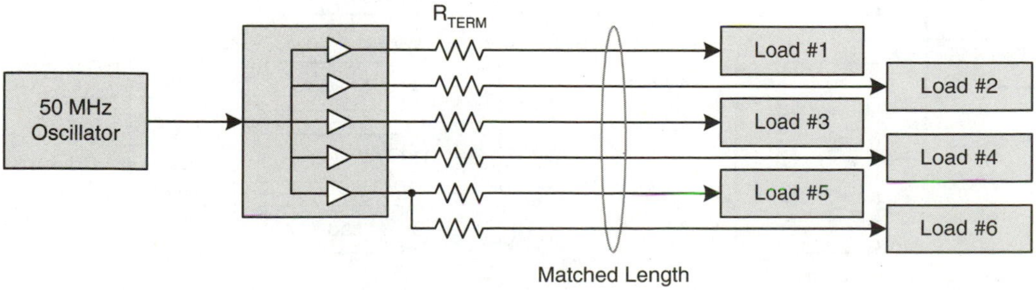

FIGURE 16.5 Single output driving two loads.

put will cause unacceptable skew to develop with respect to the other pins that are driving only a single load. Whether this occurs depends on each situation's skew tolerance, clock buffer drive strength, and input impedance of the clock load.

As the number of clock loads in a system increases, there comes a point at which clock distribution cannot be performed using a single clock buffer IC. Using multiple clock buffers introduces a new source of skew: part-to-part propagation delay variation. It is easier to manage skew on a single chip, because each output circuit has very similar physical properties to its neighbors, and the temperature variation across a small chip is very low. Neither of these assumptions automatically holds true when dealing with separate ICs. Some low-skew clock buffers are specially designed and fabricated for low part-to-part skew specifications. Newer devices are available with part-to-part skews of 0.5 ns, and more mature devices with 1 ns part-to-part skews are available as well. Figure 16.6 shows how two clock buffers might be connected to drive twice the number of loads as shown previously. Note that the oscillator directly drives both buffers in parallel without any termination, because it is assumed that the wire length is short enough to minimize reflections. If more than a couple of buffers are necessary to distribute clocks to the system, an intermediate buffer should probably be placed between the oscillator and final buffers so that the oscillator does not have to drive too many loads.

16.3 ZERO-DELAY BUFFERS: THE PLL

Low-skew buffers do an excellent job of distributing clocks with minimal relative delays when there is no constraint on the absolute propagation delay from the clock tree input to its outputs. However, there are circumstances where the absolute delay is as important as the relative delay. Many microprocessors with synchronous bus interfaces generate their own bus clock internally and emit this clock on an output pin. All bus signals must be timed relative to that output clock to meet the microprocessor's timing rules of t_{CO}, t_{SU}, and t_H. If the microprocessor communicates with just one pe-

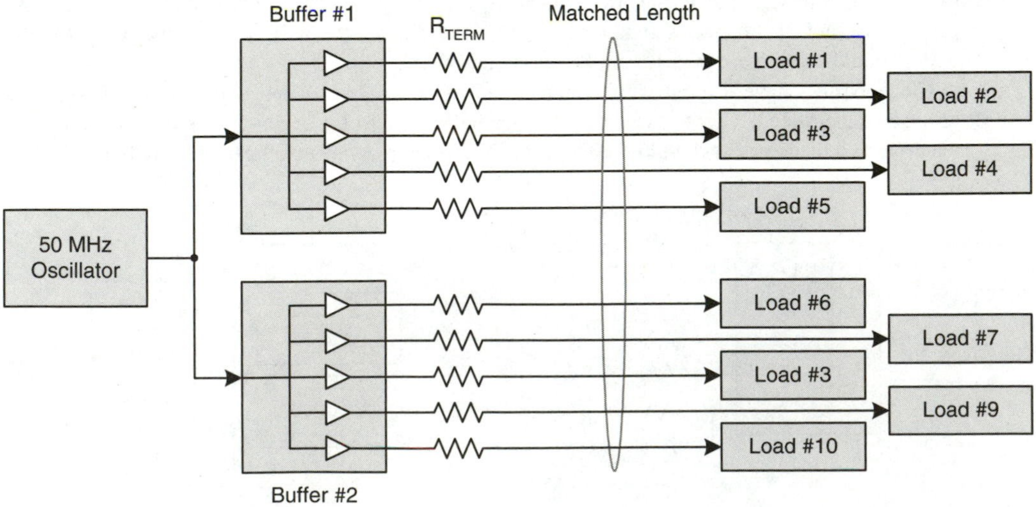

FIGURE 16.6 Dual-buffer clock tree.

ripheral IC that is located close to it, the peripheral's clock input can be driven directly by the microprocessor's clock output, and all should be okay. Chances are great, however, that the microprocessor must communicate with several peripheral and memory ICs that are located a significant distance away, as shown in Fig. 16.7. Placing a low-skew buffer between the microprocessor's clock output and the other ICs will not solve the problem, because the clocks at the loads will have a large skew relative to the microprocessor clock.

Another common clock distribution problem with an absolute delay constraint is driving common clocks to an expansion board. Consider the system shown in Fig. 16.8 in which a base board contains multiple ICs connected via a common synchronous bus. An expansion board can be plugged in to allow ICs on that card to communicate over the common bus on the base board. How can all the ICs on the base and expansion boards receive low-skew clocks? If the expansion board design is fixed ahead of time, a low-skew buffer tree on the base board can send the required number of clock signals to the expansion card and achieve length matching on both boards. This works as long as the expansion board's design never changes. A problem appears if a later design is intended to add more ICs with more clocks, or if the wiring of the expansion board can no longer maintain a length match with the wires on the base board. In all cases, as the number of individual clock loads increases on

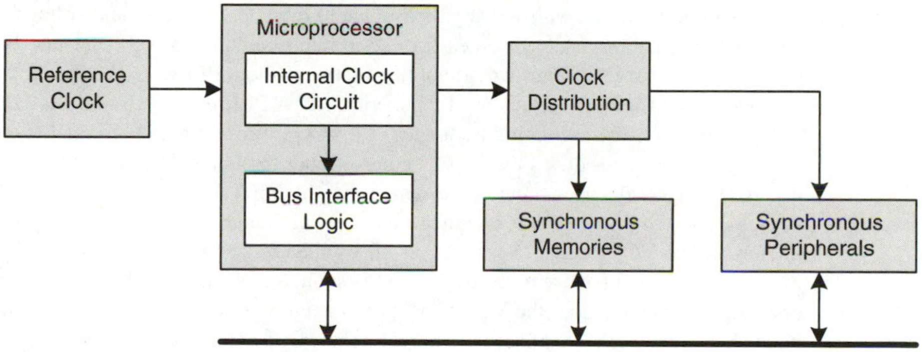

FIGURE 16.7 Microprocessor bus-clock master.

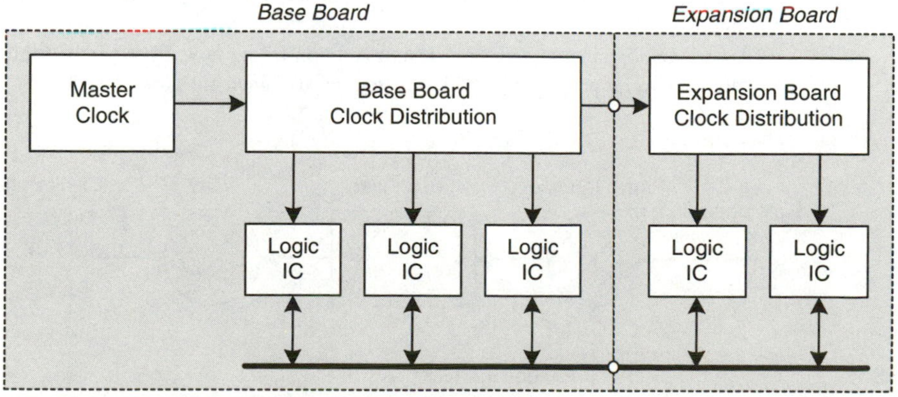

FIGURE 16.8 Clock distribution to an expansion board.

the expansion board, more pins or wires must be used to make the required connections. The increased connectivity increases cost yet does not improve the design's flexibility and tolerance of future changes.

A clock's purpose in a system is to provide a regular timing reference. Clock signals do not communicate information, because they are repetitive, and no single clock cycle has unique properties to set it apart from others. Therefore, a clock can be arbitrarily delayed without losing or gaining information. As with sine waves, periodic signals complete one full cycle over 360°. If a clock is delayed by an integer multiple of 360°, its delay cannot be detected, because each cycle is identical to all others, and its transition edges still occur at the same relative points in time. This important principle enables solutions to the clock distribution problems just posed and others like them. While it is physically impossible to construct a clock distribution circuit with zero delay, the end result of zero delay can be replicated by purposely adding more delay until an integer multiple of a 360° delay has been achieved.

Without a special trick, it would be extremely difficult to adjust a circuit for a perfect 360xN-degree delay. Propagation delays through semiconductors and wires have too much variation for this to be practical. The secret to achieving near perfect delays is with a closed-loop control system known as a *phase-locked loop*, or PLL. A PLL continually monitors a feedback clock signal and compares it against a reference clock to determine phase, or delay, errors. As soon as a phase shift is detected, the PLL can respond and compensate for that delay. This continuous compensation makes a PLL exceptionally able to deal with varying delays due to time, temperature, and voltage. PLLs are analog circuits and were originally invented for use in radio modulators and receivers. FM radios, televisions, and cellular telephones would not be possible without PLLs.

Figure 16.9 shows the structure of a generic PLL. A PLL is built around a *voltage controlled oscillator*, or VCO. The oscillating frequency of a VCO is proportional to its analog control voltage input. A phase detector drives the VCO's control input through a lowpass filter. When the reference and feedback clock edges occur at the same time, the input and output clocks are aligned by 360xN degrees, and the phase detector emits a neutral error signal, enabling the VCO to maintain its frequency. If the VCO begins to wander, as all oscillators do to a certain extent, the feedback clock edges begin to shift relative to the reference. As soon as this happens, the phase detector generates an error signal, which causes the VCO to slightly increase or decrease its frequency to move the feedback clock edge back into phase with the reference. One of the beauties of a PLL is that it can use an inexpensive VCO, because the feedback loop keeps the VCO at the correct frequency. A phase detector responds very quickly to phase errors, and the potential exists for an unstable control system resulting from rapid overcompensation. The lowpass filter stabilizes the control and feedback loop to average the phase detector output so that the VCO sees a more gradual control slew as compared to rapid swings.

PLLs are analog circuits and their design is a non-trivial task. Fortunately, numerous companies manufacture special-purpose digital PLLs that are designed for clock distribution applications. All

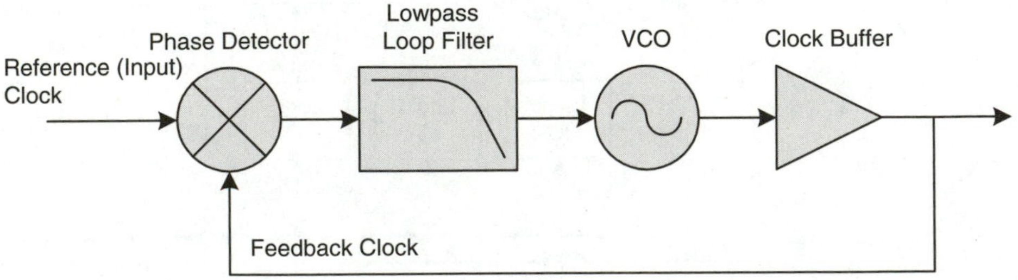

FIGURE 16.9 Generic phase-locked loop.

of the loop stability problems have been taken care of on a single IC, allowing engineers to use an integrated PLL as a tool and focus on the task of zero-delay clock distribution.

A zero-delay clock distribution circuit contains a PLL whose feedback path is chosen such that there is a nearly perfect phase alignment between the PLL reference clock and the clock loads. Figure 16.10 shows a general zero-delay clock distribution scenario. The feedback path has a low-skew relationship with the other clock loads. It is driven by the same low-skew buffer and is purposely routed on a wire that is length matched with the other clock loads. The result is that the feedback clock arrives back at the PLL at the same time that the other clocks reach their loads. Of course, a low-skew buffer has finite skew, and this skew limits the PLL's ability to align the output clocks to the reference clock. PLL feedback clocks are typically carried on wires that have been artificially lengthened by serpentine routing, because the PLL and buffer are usually close together while the clock loads are further away. It is actually the norm for a digital PLL to be integrated onto the same IC as a low-skew buffer, because the two functions are inextricably linked in digital clock distribution applications. Companies that manufacture low-skew buffers commonly offer integrated PLLs as well, often under the term *zero-delay buffer.*

Turning back to the microprocessor clock distribution example in Fig. 16.7, it is now apparent that a PLL with an integrated low-skew buffer would solve the problem. The microprocessor clock would drive the PLL reference, and matched-length wires emanating from the buffer would go to each clock load as well as the PLL feedback input. The resultant skew between the microprocessor's bus clock and the clock seen at each load is the sum of the propagation delay through the wire that connects the PLL to the microprocessor, the propagation delay through any small wire length mismatches, and the skew inherent in the PLL and buffer circuits. At an approximate signal conduction velocity of 6 in (0.15 m) per nanosecond, the skew due to wiring propagation delay is minimal if the PLL is located near the microprocessor and the output wires are matched reasonably well. As with any IC, the skew specifications for a zero-delay buffer should be compared against the system's requirements to select a suitable device.

The expansion card clock distribution problem in Fig. 16.8 can also be addressed using a zero-delay buffer as shown in Fig. 16.11. A normal low-skew buffer can be used on the base board, because the master clock is an oscillator with no skew requirement between that oscillator and the distributed clocks. All outputs from this buffer are length matched—including the wire going to the expansion board. As soon as the clock passes across the expansion connector, it drives a zero-delay buffer that has matched length outputs going to the expansion ICs as well as a feedback path back to the PLL. This ensures that the expansion ICs observe the same clock that was delivered across the connector. Attempts can be made to reduce skew introduced by the connector and routing to the zero-delay buffer by shortening the wire on the base board that drives the expansion connector. The wire should be shortened by the approximate distance represented by the connector and the short wire length on

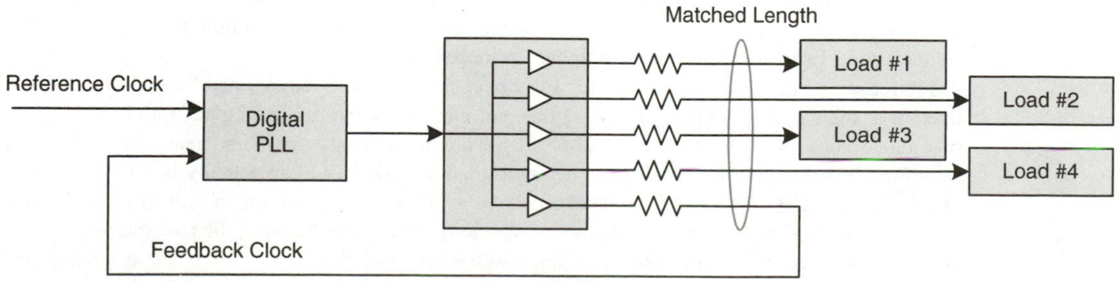

FIGURE 16.10 Zero-delay clock distribution.

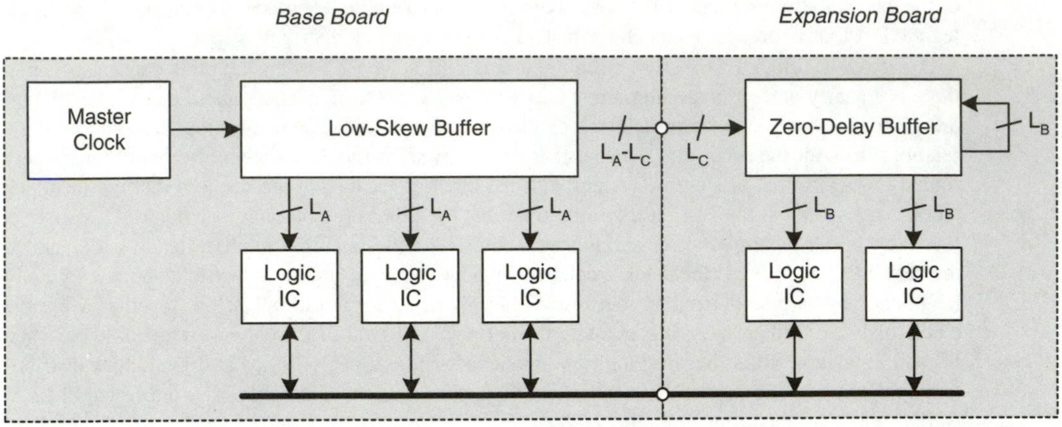

FIGURE 16.11 Zero-delay buffer on expansion card.

the expansion board between the connector and the zero-delay buffer. It is easier to constrain a short wire length leading to the zero-delay buffer, because these ICs are typically small and can be located near the connector, regardless of the board's overall layout.

The downside of PLLs is that they are sensitive to jitter and may actually add jitter to an otherwise clean clock. A typical PLL is only as good as the reference clock that is supplied to it, because the phase detector seeks to continually adjust the VCO to match the incoming reference. As such, a typical PLL will not reduce jitter. A loop filter with a long time constant may be used to reduce jitter, although other problems can result from doing so. Noisy power increases a PLL's jitter, because the analog VCO circuit translates noise into varying oscillation periods. Therefore, PLLs should have their analog power supply filtered with at least a series impedance and shunt capacitance as shown earlier for a crystal oscillator using an LC pi filter.

16.4 FREQUENCY SYNTHESIS

Digital systems can use PLLs for more than just zero-delay clock buffering. Arbitrary frequencies can be synthesized by a PLL based on a reference clock, and this arbitrary frequency can be changed in real time. Systems with analog front ends and digital processing cores use frequency synthesis for tuning radios and implementing complex modulation schemes. A common example of this is a digital cellular telephone. Purely digital systems make extensive use of frequency synthesis as well. Advanced microprocessors and other logic ICs often run their cores at a multiple of the external bus frequency. PLLs are used as clock multipliers in these ICs.

Frequency synthesis is possible with a PLL, because the phase detector and VCO do not have to operate at the same frequency. The phase detector cares only that its two inputs are phase aligned; they must be of the same frequency and phase for a neutral error signal to be generated. Subject to minimum and maximum operating frequency limitations, the VCO can actually be made to run at any frequency. Let's first consider the example of synthesizing an integer divisor of the reference clock. The PLL circuit in Fig. 16.12 places a divide-by-N counter between the true reference clock and the phase detector's reference input. The result is that the PLL is unaware of the original reference clock and instead sees the divided version, whose frequency it is able to match by adjusting the VCO until the phase error is zero.

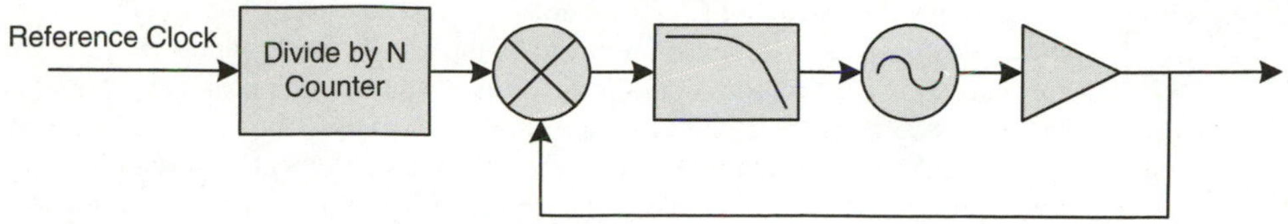

FIGURE 16.12 PLL clock divider.

Clock division by an integer divisor is not very interesting, because it can be done without the complexity of a PLL. Now let's look at integer multiplication, which does require a PLL. The circuit in Fig. 16.13 places a divide-by-M counter in the feedback path between the VCO output and the phase detector. Things start to get interesting here. Instead of observing the true VCO signal, the phase detector unknowingly gets a divided version of that clock. The phase detector has no knowledge of what signals it is seeing—just whether they are phase aligned. Therefore, the phase detector emits an error signal until the divided feedback clock matches the frequency and phase of the reference. The VCO must run at a multiple of M times the reference for the feedback clock to equal the reference, resulting in clock multiplication.

Integer multipliers and dividers are useful, but truly arbitrary ratios between input and output can be achieved when the reference and feedback dividers are joined into a single circuit. The circuit in Fig. 16.14 allows the output to run at a ratio of M ÷ N times the reference frequency by combining the principles of the aforementioned divider and multiplier schemes. Different frequencies can be synthesized in real time by periodically changing the M and N counters. A PLL has a finite lock time, the time that the loop takes to adjust to a new operating frequency. Any frequency synthesis scheme that requires real-time adjustment must take the lock time into account. If the counters are changed too rapidly, the PLL may spend all of its time hunting for the new frequency and never settling.

PLLs in general are capable of frequency synthesis when complemented with external counters, but most digital PLLs are not designed specifically for large multiplication ratios. Several factors in-

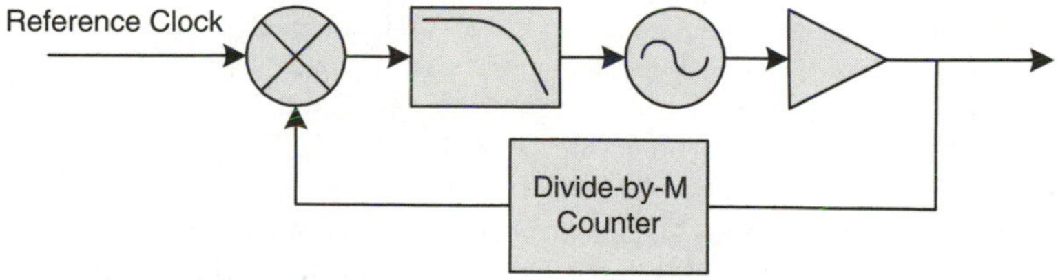

FIGURE 16.13 PLL clock multiplier.

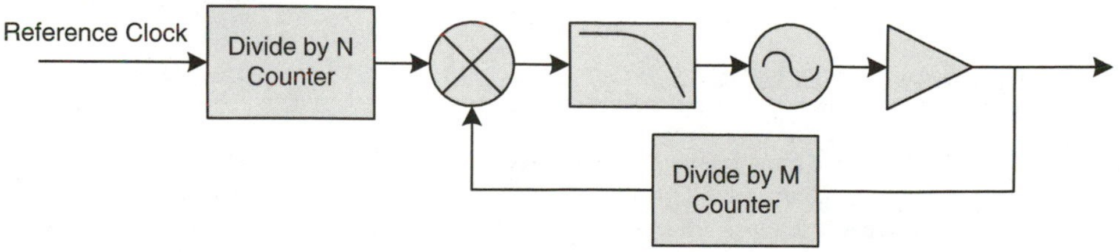

FIGURE 16.14 PLL clock synthesizer.

fluence the suitability of a PLL for arbitrary clock synthesis, including VCO stability and loop-filter characteristics. VCO stability is improved by filtering its power supply. Typical core clock synthesis applications in a microprocessor or logic IC may require multiplication by 2 or 4, which is relatively straightforward. As the multiplication ratio increases, the control path between the phase detector and the VCO is degraded, because fewer feedback adjustment opportunities exist for every VCO clock period. If a PLL is multiplying a reference by 100, the phase detector will only be able to judge phase errors on 1 percent of the VCO edges. There is the potential for VCO wander during the long intervals between phase detector corrections.

Most digital PLLs contain a lowpass loop filter integrated on chip that is matched to the typical operating scenario of a 1:1 input/output ratio or some small multiplication factor. Larger multiplication ratios may require a more complex off-chip filter that is specifically designed for the application's requirements. This is where PLL design in a digital system can start to get complicated. Fortunately, semiconductor manufacturers have documentation and applications engineering staff available to assist with such tasks. One set of devices that are well suited for digital frequency synthesis is Texas Instruments' TLC29xx family. The TLC29xx devices have separately connected phase detector and VCO sections such that an external loop-filter can be customized to the application.

It is not very common for a digital system to require complex multiplication ratios in a PLL, but the requirement does exist in applications that have multiple interfaces that run at different frequencies and must be phase locked to prevent data loss. One example is a digital video processor. The primary clock frequency of many digital video standards is 27 MHz; however, some newer high-definition video standards operate at 74.25 MHz. These frequencies are related by a factor of 2.75. If a digital video processor must perform some conversion between these two interfaces and do so in a manner that precisely matches their data rates, a PLL is necessary to lock one of the interfaces to the other. Otherwise, each interface would run on its own oscillator, and small amounts of frequency drift would soon cause one to get a little faster or slower than the other. No matter how accurate an oscillator is, it has a finite deviation from its nominal frequency. When two oscillators are paired with the expectation that they run at constant rates, a problem will eventually develop. As soon as a rate mismatch develops, the input and output data rates no longer match, and data is corrupted.

For the sake of discussion, assume that the 27-MHz interface is the master, and the 74.25-MHz interface is the slave. Because a counter cannot directly divide by 2.75, a ratio of integers must be calculated. The smallest pair of integers that yields $M \div N = 2.75$ is $M = 11$ and $N = 4$. This means that the PLL is essentially performing an 11× clock multiplication function, because the 27 MHz reference is divided by 4 to yield a 6.75-MHz PLL input.

The task of designing a PLL to implement large multiplication factors is decidedly non-trivial because of the problems of stability and jitter. Selecting an appropriate lowpass filter that keeps the loop from unstable oscillations and that adequately addresses VCO jitter can involve significant control systems theory and analog filter design skills.

16.5 DELAY-LOCKED LOOPS

PLLs have traditionally been considered the standard mechanism for implementing zero-delay buffering and clock multiplication. Their flexibility comes at a certain price for manufacturers of digital ICs, because PLLs are analog circuits that must be isolated from noisy digital switching power supplies for low-jitter operation. Although the problems of on-chip isolation have been addressed for a long time, the problem persists. Many large digital ICs are now manufactured with a purely digital *delay-locked loop*, or DLL, that produces similar results to a PLL. Instead of controlling a VCO to vary the phase of the output clock, a DLL contains a many-tap digital delay line through which the reference clock propagates. According to the detected phase difference between the reference and feedback clocks, a specific propagation delay can be picked in real time to align the edges. A DLL may reduce problems in certain designs as a result of the reduced noise sensitivity of purely digital circuits and the lack of a VCO. Figure 16.15 shows a basic digital DLL where the delay is programmed by selecting one of many delay line taps with a multiplexer. DLLs can also be designed as analog circuits by employing a voltage controlled delay line. However, the purely digital implementation is preferable due to its improved noise immunity.

A DLL must have sufficient delay granularity at each tap to be effective in minimizing skew between its input and output. If the incremental delay is 100 ps, the DLL can offer skew no better than ±50 ps. The phase detector within a DLL is able to dynamically adjust the delay line to compensate for changing propagation delays through active and passive elements over time, temperature, and voltage. In this regard, it is very similar to a PLL's dynamic compensation.

It is worth noting that different companies and engineers take differing positions on whether a PLL or DLL is superior for clock management. There are those who trust the time-proven PLL methodology and those who believe that a purely digital circuit is the cure-all for noise-related problems. In most situations, there is no choice, because a company that manufactures FPGAs or microprocessors has already made the decision and fabricated one solution. Custom IC design processes sometimes allow the customer to choose one over the other. In reality, DLLs and PLLs have both been demonstrated to work well in millions of individual units shipped, and one shouldn't be overly concerned unless working with very high-speed designs where picoseconds of jitter and skew can become significant problems.

16.6 SOURCE-SYNCHRONOUS CLOCKING

Clock distribution becomes a more challenging task as the number of low-skew loads increases and the operating frequency increases. Despite the low-skew technologies discussed so far in this chapter, there are practical limitations on how little skew can be achieved across a high-fan-out clock tree. When a system requires more low-skew clocks than a single buffer can drive, multiple buffers are required, and part-to-part skew becomes the limiting factor rather than output-to-output skew on a single buffer. Truly high-fan-out clock trees may require multiple levels of buffering, and each level adds to the overall skew. Zero-delay clock buffers can be used in place of low-skew buffers, but PLLs introduce jitter and have finite output-to-output skew as well.

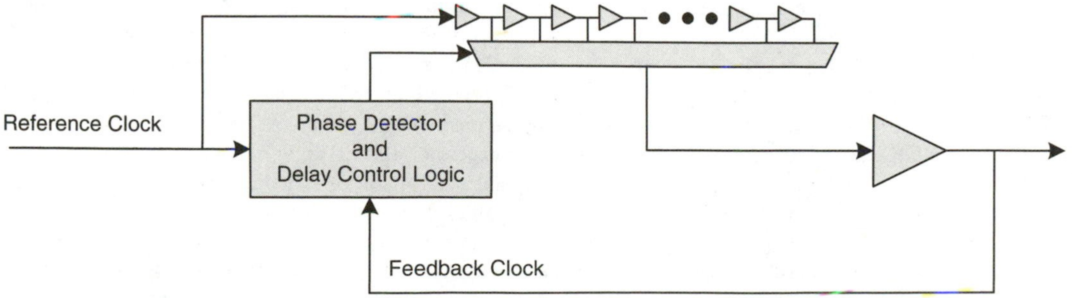

FIGURE 16.15 Generic delay-locked loop.

A typical high-fan-out clock tree has approximately 0.5 to 1.0 ns of skew resulting from multiple buffers and PLLs, each contributing several hundred picoseconds of skew and jitter. The question becomes how much skew a system can tolerate. Skew tolerance is a function of the clock frequency, the trace length separating devices, and IC timing specifications. If a clock tree's skew is relatively independent of frequency, it follows that skew becomes a greater percentage of the clock period, and therefore a greater concern, as the frequency increases.

There is no magic frequency at which clock tree skew and jitter become a dominant factor. Input hold time requirements are often the dominant restriction on clock skew. Evaluating hold time compliance requires knowledge of the source flop's output hold time, t_{OH}, which specifies how long the output remains unchanged after the active clock edge, usually the rising edge. Manufacturers do not always specify t_{OH}, requiring conservative estimates for proper analysis. A conservative estimate is generally 0 ns, or saying that the output can change immediately after the active clock edge.

Consider a hypothetical system using ICs with output specifications of output hold time (t_{OH}) = 2 ns and t_{CO} = 5 ns, and input specifications of t_{SU} = 2 ns, and t_H = 1 ns. The system bus wires have propagation delays ranging from 1 to 2 ns. Skew and jitter are handled as a combined entity in this example.

Below a certain frequency, the skew budget is limited by the hold time requirement,

$$t_{SKEW} < t_{OH} + t_{PROP(MIN)} - t_H$$

$$t_{SKEW} < 2 \text{ ns}$$

The hold time expression is derived knowing that the destination flop's input is delayed from the source flop's output by at least the minimum propagation delay of the interconnecting wires. Wiring delay plus the source flop's t_{OH} provides information on how long the destination flop's input remains static after the active clock edge. This static time minus t_H provides the margin with which t_H is met. Clock skew must be less than this margin for valid timing.

Clock skew as restricted by setup time is calculated by determining the worst-case delay from the active clock edge until the destination flop's input changes. This worst-case delay is the sum of the source flop's t_{CO} and the maximum propagation delay of the wiring. Next, t_{SU} must be added in to satisfy the destination flop. This entire delay restriction is subtracted from the actual clock period to yield the timing margin. Clock skew must be less than this margin. At an arbitrarily chosen frequency of 80 MHz (T = 12.5 ns), the skew budget as a function of setup time is greater than that of hold time,

$$t_{SKEW} < T - (t_{CO} + t_{PROP(MAX)} + t_{SU})$$

$$t_{SKEW} < 3.5 \text{ ns}$$

Readily available clock distribution ICs can be used in this system at 80 MHz and below without concern. At 100 MHz (T = 10 ns), the skew budget decreases to 1 ns, which is achievable, although with little added margin. This method of timing clearly runs out of gas at higher frequencies.

Traditional synchronous clocking does not work at high frequencies, because wire delays and clock tree skew do not scale with advancing semiconductor technology. *Source-synchronous* bus clocking architectures enable operation well in excess of 100 MHz by removing absolute wire delays and clock tree skew as variables in timing analysis. A source-synchronous interface distributes clock and data together as shown in Fig. 16.16 rather than separately as in a conventional synchronous clock tree design. The data source, or transmitter, emits a clock that is in phase with the data as if it were just another signal on the synchronous interface. In doing so, the clock and data have closely matched delays through the IC's output circuits and package. The signals are routed on the circuit board with length matching so that each wire that is part of the interface has approximately

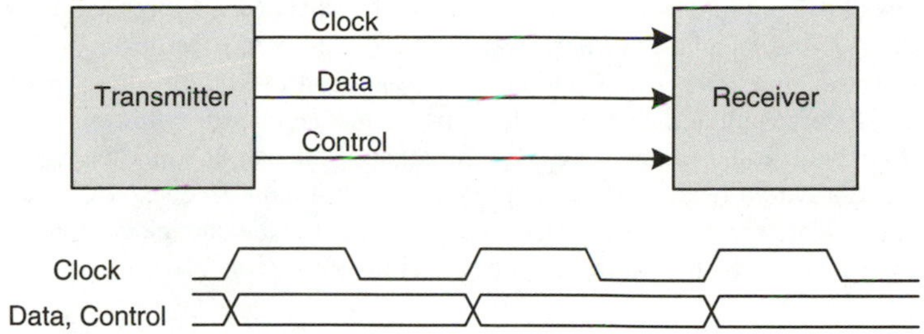

FIGURE 16.16 Source-synchronous bus.

uniform propagation delay. By the time the signals reach the receiving IC, clock and data are phase aligned within a certain skew error that is a function only of the transmitter's output-to-output skew and the delay mismatch of the PCB wiring. This skew is far less than the skew and wiring delay penalties of a conventional synchronous interface and may be on the order of 100 ps. Nearly the entire clock period, less allowances for finite skew and switching time, can be used to meet the input flops' setup and hold specifications, which is why high-performance memory (e.g., DDR SDRAM) and logic interfaces can run at hundreds of megahertz.

The clock and data timing relationship can be arbitrary as long as the skew is tightly controlled, because the receiver's input circuitry can implement appropriate delays based on the transition of data relative to clock to meet the flop's true setup and hold requirements. Two common source-synchronous timing relationships are shown in Fig. 16.17, where the clock is offset into the middle of the data valid window. In the case of single-data rate bus (SDR), this is akin to clocking off the falling edge. A double-data rate bus (DDR) would require that the clock be shifted by 90° relative to data such that the rising and falling edges appear in the middle of the valid windows so that data can be registered on both edges.

A disadvantage of source-synchronous bus architecture is the management of a receive clock domain that is out of phase with the core logic domain. A conventional synchronous design seeks to maintain a uniform clock phase across an entire design, whereas source-synchronous design explicitly gives up on this and proliferates disparate clock domains that have arbitrary phase relationships with one another. This trade-off in clock domain complexity is acceptable, because individual gates are cheap on multimillion-transistor logic ICs. A logic IC typically requires a FIFO and associated control logic to cross between the receive clock and internal clock domains.

Source-synchronous interfaces are often point-to-point, although this is not strictly necessary, because of the high speeds at which they usually run. It is electrically difficult to fabricate a multidrop

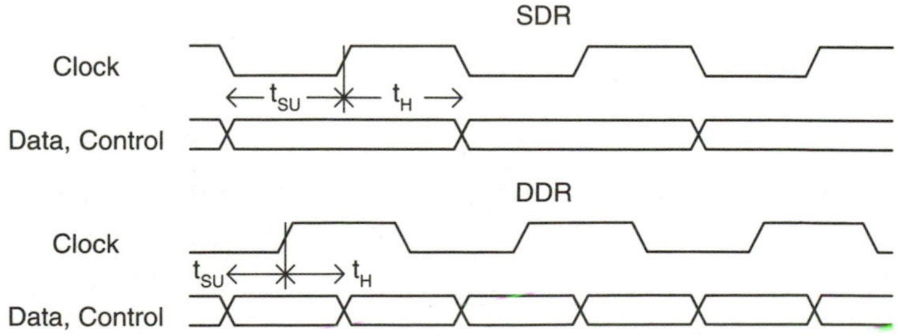

FIGURE 16.17 Typical clock/data timing.

bus that operates at hundreds of megahertz. Figure 16.18 shows a hypothetical system architecture that employs source-synchronous interfaces. There are three ICs, and each IC is connected to the other two via separate interfaces. Each interface consists of two unidirectional buses. Unlike the receive clock, the transmit and core clock domains are merged, because this example does not require the core to operate at a different frequency from that of the bus. It would be easy to insert a FIFO, too, if a benefit would result from decoupling the bus and core frequencies. Clock distribution for this system is relatively easy, because each IC requires the same core clock frequency, but the interfaces are self-timed, which removes clock tree skew as a concern.

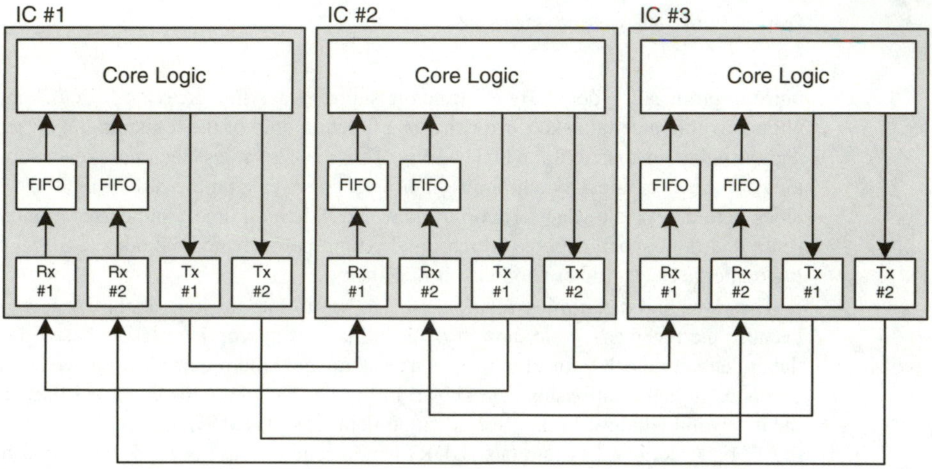

FIGURE 16.18 Source-synchronous system architecture.

CHAPTER 17

Voltage Regulation and Power Distribution

Power is an obviously critical component of a system, but some engineers have bad reputations for inadequately planning a system's voltage regulation and power distribution scheme. When power is an afterthought, luck sometimes works and sometimes does not. There is really no magic involved in designing an appropriate power subsystem, and this chapter seeks to explain the main issues that should be taken into consideration. The purpose of this chapter is to provide a broad understanding of power regulation concepts and how they can be applied to implement the power regulation and distribution circuitry in a digital system.

After many years of development and industry experience, we are fortunate to have at our disposal a broad array of off-the-shelf power regulation products. In concert with discrete semiconductors, a system's unique power requirements can be addressed with the appropriate tools. Many power requirements can be handled with off-the-shelf regulators. When situations arise that require a semicustom approach, the gaps can be filled in with a basic understanding of voltage regulation techniques.

Safety should be an overriding concern in all engineering disciplines, but safety is all the more critical when dealing with power circuits. Power circuits may carry potentially lethal voltages and currents and must be treated with a healthy respect. Resist the urge to quickly prototype a power circuit without taking the necessary time to study its thermal and electrical properties. Even if a power circuit appears to operate properly for a few minutes, an improper design can eventually overheat, with disastrous results.

This chapter is organized in two basic sections: voltage regulation and power distribution. First, basic voltage regulation principles are discussed to provide an orientation to the subject. Thermal analysis is then presented before any circuits are discussed, because thermal issues are fundamental to all power circuits. A safe and reliable circuit must be designed to handle its intended power load over its full range of operating conditions without overheating. The two subsequent sections deal with methods of regulating voltage and current, encompassing diode-shunt and transistor series-regulators. Off-the-shelf solutions are explored next to provide an understanding of the common integrated linear and switching regulators at an engineer's disposal.

Once basic regulation techniques have been presented, important issues in power distribution are discussed. Even the best regulator is useless without adequate means of conveying power to the load. Power distribution encompasses three basic themes: safety, reliability, and electrical integrity. Safety and reliability issues of handling AC power, fuses, and adequate wiring are discussed. Power regulation schemes employing precertified off-the-shelf AC-to-DC supplies are encouraged because of the improved safety of using known-good AC power modules. Finally, the use of low-inductance power planes and bypass capacitors is presented in the context of distributing power while preserving its electrical integrity. A brief power subsystem design example is used to illustrate these concepts.

17.1 VOLTAGE REGULATION BASICS

Power usually enters a system in a form dictated by its generation and distribution characteristics rather than in a form required by the components within that system. Plugging a computer into a 120 VAC wall outlet provides it with power in a form that was deemed efficient for generation and distribution in the early twentieth century. Unlike light bulbs, digital logic chips do not run very well if connected directly to 120 VAC, although they may emit a bright light for a brief period if connected in this manner! Rectifiers and the conversion between AC to DC has already been discussed, but this is only part of the solution to providing usable power to system components. Once the AC has been rectified to DC and filtered, the voltage probably does not meet the specifications of the circuit. A rectified power input may not only have the wrong DC level, it probably has a good deal of ripple artifacts from the original AC input, as shown in Fig. 17.1.

An ideal voltage regulator provides a constant DC output without ripple regardless of the input voltage's DC level and ripple. This applies not only to rectified AC power but also to power provided by batteries, solar cells, DC generators, and so on. Any time a system's power input does not provide the supply voltage required by its components, a voltage regulator is necessary to perform this conversion. Most digital systems require at least one voltage regulator, because it is rare to find a power source that provides the exact voltage required by digital and analog circuits. There are, however, some special-purpose ICs designed with wide supply voltage specifications so that they can be directly connected to batteries without an intermediate voltage regulator. On the flip side, there are systems that require multiple voltage regulators, because they contain circuitry with multiple supply voltage specifications. At one time, it was common to have a single +5-V digital supply. Now, it is not uncommon to have 3.3-, 2.5-, 1.8-, and 1.5-V supplies in myriad configurations.

Most voltage regulators are the *step-down* variety—they provide a constant output voltage that is lower than the input. There are many types of step-down regulators, as will be shown in the course of this chapter. Some applications require *step-up* regulators that provide a constant output that is higher than the input. These applications are often low-power battery-operated devices in which a 1.5- to 3-V battery is moderately stepped up to power a small circuit.

A voltage regulator must have access to some form of voltage reference to which it can compare its output to determine if the level is correct. Without a known reference, the regulator has no means of measuring its output. This is analogous to measuring an object with a ruler, which is a known length reference. Without the reference, the object's length remains unknown. The structure of a general voltage regulator is shown in Fig. 17.2.

The reference is powered directly by the arbitrary input voltage but maintains a constant output voltage subject to certain minimum operating conditions. A reference cannot maintain an output voltage higher than the input. Therefore, references usually provide low voltages that are compared against a scaled-down feedback voltage from the output. This enables the regulator to function over a wider range of input levels. The two resistors implement a basic voltage divider that drives a feedback comparator. These resistors are chosen to provide a feedback voltage equal to the reference at the desired output voltage. When the feedback voltage falls below the reference, the comparator sig-

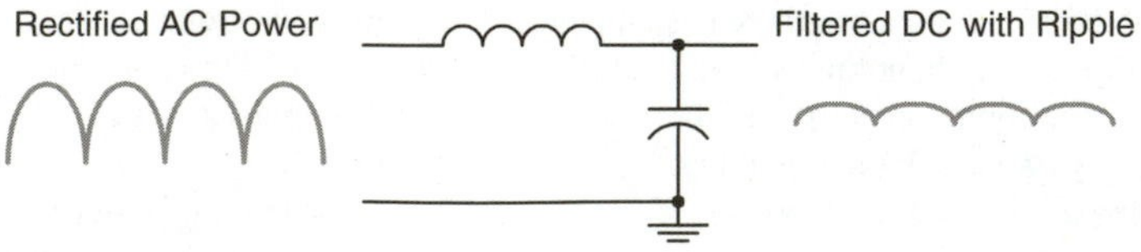

FIGURE 17.1 Ripple on a rectified AC power input.

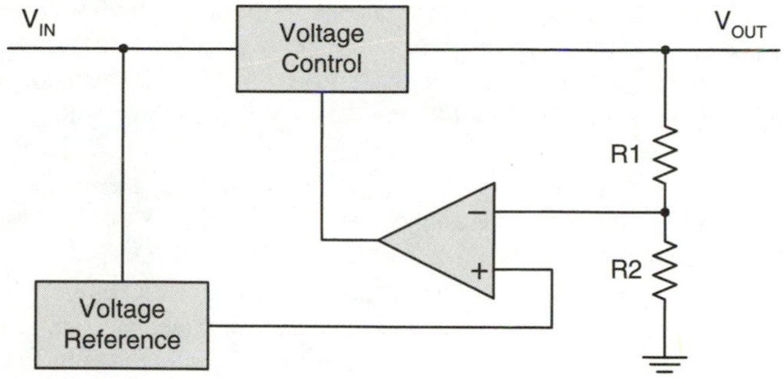

FIGURE 17.2 General voltage regulator.

nals the voltage control circuit to increase V_{OUT}. Similarly, once the feedback voltage rises above the reference, the voltage control circuit must reduce its output. This feedback loop is subject to stability problems unless special allowances are made in the design. An unstable voltage regulator can oscillate, which is very undesirable. Voltage regulator stability problems have long been dealt with in reliable and effective manners.

From the brief description of a regulator's operation, it is apparent that the accuracy of the reference directly affects the accuracy of the generated output voltage. Diodes of various types serve as voltage references, because semiconductor junctions exhibit nonlinear voltage drops that are not very sensitive to current changes—Ohm's law does not apply. However, current and temperature do affect this voltage drop, which causes variation. A standard small-signal silicon diode exhibits a drop of between 0.6 and 0.7 V at low currents. A crude regulator could be constructed using such a reference, but the degree of accuracy would be unacceptable for many applications. More specialized diodes with integrated compensation elements designed explicitly to serve as stable voltage references can achieve tolerances of 0.1 percent over temperature and part-to-part variations.

Voltage regulators are characterized using a wide range of parameters. Some of the metrics that might first come to mind are how much current can be supplied and what restrictions exist on the input and output voltages. Other specifications that are just as important involve the accuracy to which the output voltage is maintained and how the output is affected by a variety of ambient conditions.

The relationship of input and output voltages is usually the first order of business. A regulator has minimum and maximum input and output voltage limits as well as a *dropout voltage*, which is the minimum difference between input and output that can exist while guaranteeing a regulated output. When designing a particular regulator into a system, it is important to ensure that the regulator's specifications will not be violated and that any dropout voltage restriction will be met. Some types of switching regulators do not have a dropout voltage, but most regulators do require a certain minimum overhead between input and output. Output current and power specifications must also be observed. Each regulator has maximum current and power ratings beyond which it will overheat and damage itself and, possibly, surrounding circuitry. Some regulators have minimum current ratings, below which the device is incapable of guaranteeing a regulated output.

Adherence to these rules may sound obvious at first, but many an engineer has picked an inadequate voltage regulator as an afterthought, only to waste time and money later in solving problems of poorly regulated voltages and overheating regulator components.

Once the basic input/output conditions have been satisfied, regulation accuracy becomes the prime concern. Among others, there are three basic specifications: *line regulation*, *load regulation*, and *ripple rejection*.

Line regulation is the relationship between a change in the input (line) voltage and the output voltage and is expressed in either percentage or absolute magnitude terms. It is not uncommon to find regulators with line regulation on the order of 0.01 percent. Such a regulator would exhibit an output swing of only 0.01 percent over the full range of the specified input voltage.

Load regulation is the relationship between a change in the load current and the output voltage and is also expressed as either a percentage or a magnitude. Regulators are available with load regulation on the order of 0.1 percent. This means that, over the full range of specified output current, the output voltage will not change by more than 0.1 percent.

Line and load regulation are not the only parameters affecting the DC output level of the regulator. These parameters characterize the voltage control element and feedback amplifier in the regulator. The operation of these components is only as good as the quality of the voltage reference and the applied feedback voltage. High-quality DC power supplies can deliver 1 percent DC accuracy, meaning that the reference and feedback resistor network have a combined tolerance better than 1 percent. A 3.3-V, 1 percent supply guarantees an output range from 3.267 to 3.333 V. Most digital systems use 5 percent supplies, because such accuracy is typically easy to achieve, and most commercial logic ICs are specified with V_{CC}/V_{DD} tolerances of no more than 5 percent. This is where the common 4.75- to 5.25-V operating range comes from when dealing with 5-V logic. Some logic ICs are specified with V_{CC}/V_{DD} tolerances of 10 percent, although fewer ICs are found with such loose ratings these days because of the ease with which 5 percent supplies are implemented. It is easier for a semiconductor manufacturer or an analog circuit designer to build a system with a tighter supply voltage requirement, because the components do not have to operate over a wider range of conditions. Specifying all ICs with 1 percent requirements would make the vendors' jobs easier, but it would also result in more expense and complexity at the system level. Five percent is a good compromise at the current state of technology.

Finally, ripple rejection is the ability of a regulator to block incoming ripple from feeding through to the output. It is the ratio of the input ripple magnitude to the output ripple magnitude and is usually expressed in decibels, because it is a large number for any effective regulator. Ripple rejection specifications of 50 to 80 dB are common. These specifications are often measured at 120 Hz, because this is the ripple frequency resulting from a rectified 60 Hz AC input in North America and other parts of the world. Ripple rejection gets worse with increasing frequency, making a 120 Hz specification worst case when compared to the 100-Hz ripple that would be observed from a rectified 50-Hz AC input in Europe or Asia.

17.2 THERMAL ANALYSIS

Before heading straight into a discussion of voltage regulation techniques, it is appropriate to briefly introduce thermal analysis in the context of electronic components. Some energy will be dissipated as heat during the process of converting an input voltage to an output voltage, regardless of the type of regulator. All or most of the power in a typical system passes through voltage regulators, making them a concentrated point of heating. Without performing a basic thermal analysis, a power regulation component may be inadequately sized and cooled and could fail in a potentially dangerous manner. Melting circuits, fires, and even small explosions can result from poorly designed regulators. A given component, for example, can be designed to safely dissipate 10 W without heating excessively, but a similar component could fail while dissipating only 1 W if proper thermal design practices are not followed.

Power regulation and thermal analysis are topics so intertwined that one cannot separate them and be assured of a reliable and safe system. The best, most efficient regulator is worthless if it is operated in a manner that causes it to overheat and fail.

Four characteristics apply to basic thermal analysis: power dissipation, maximum device operating temperature, thermal resistance to ambient air, and the ambient air temperature. First, the amount of power dissipated by a device must be determined through circuit analysis. Second, each component is rated at a maximum safe operating temperature. This is the temperature of the device itself, not of the surrounding environment. Resistors have ratings up to 100°C, 200°C, and higher. Capacitors typically have specified operating temperatures below 100°C. Semiconductors have ratings according to their intended function. Diodes can be rated as high as 200°C, while many logic ICs have ratings under 100°C. When dealing with semiconductors, the internal operating temperature is referred to as *junction temperature*, the temperature of the silicon itself.

Thermal resistance to ambient air is an important parameter. It defines the ease with which heat is conducted away from the component and out to the ambient environment. A higher thermal resistance results in more heat buildup. Thermal resistance is designated by either "R" or the Greek letter theta, θ, and is related to power and heat using an analog of Ohm's law: $\Delta T = \theta P$. θ is expressed in units of °C/W. In other words, a rise in temperature results from a quantity of power multiplied by the thermal resistance through which the power is flowing. If a component is provided with a highly conductive thermal path to the ambient environment, its θ is low and, consequently, its temperature rise is low for a given power level.

In the semiconductor context, two thermal specifications are common: θ_{JC} and θ_{JA}, the thermal resistance of junction-to-case and junction-to-ambient, respectively. Semiconductors are often specified with at least one of these parameters, and sometimes both. If neither parameter is specified, the manufacturer will specify a maximum power dissipation at a particular ambient temperature along with derating information for each degree rise over the specified temperature. θ_{JA} is specified when the component is intended to be used free of any heat sink. Therefore, the component has a certain natural thermal resistance from the silicon to the surrounding air. θ_{JC} is specified when the component may be used with a heat sink, because a heat sink effectively becomes part of the package, or case, and enables an overall θ_{JA} calculation.

Finally, the ambient air temperature establishes a starting temperature for the thermal analysis. If the maximum ambient air temperature is 40°C, a component will definitely not run any cooler than 40°C. The higher the ambient temperature, the less headroom there is for heating before a component reaches its maximum operating temperature.

Thermal analysis for passive components and semiconductors is theoretically the same, but the former is usually less of a concern in practice. In a digital system, the passive components that tend to dissipate substantial power are resistors and chokes. Resistors and chokes are available in a wide range of packages and materials, depending on the intended application. They can be found with ratings of milliwatts or hundreds of watts. Resistors and chokes also have fairly wide operating temperatures. In most cases, therefore, the thermal analysis of a resistor consists of first determining the power that it will dissipate and then picking a device that is rated for some multiple of that calculated value. A common rule of thumb is two to three times the calculated power dissipation. One can be more conservative if desired.

Semiconductors are a different story, because their operating temperature ranges are lower, and they are small parts that can handle large quantities of power. As such, the power density of a semiconductor can be very high. Lots of power and high thermal resistance resulting from small physical dimensions can cause lots of trouble very quickly.

The thermal analysis of a semiconductor begins by establishing the maximum operating conditions: power dissipation, P_D, and ambient temperature, T_A. This allows an initial calculation of the junction temperature: $T_J = T_A + \theta_{JA} P_D$. If this initial result is less than the component's rated operating temperature, the analysis shows no problem. It is important to be realistic about the maximum T_A and P_D. A device that barely functions on a cool day may fail dramatically on a hot day when T_A and, consequently, T_J are higher.

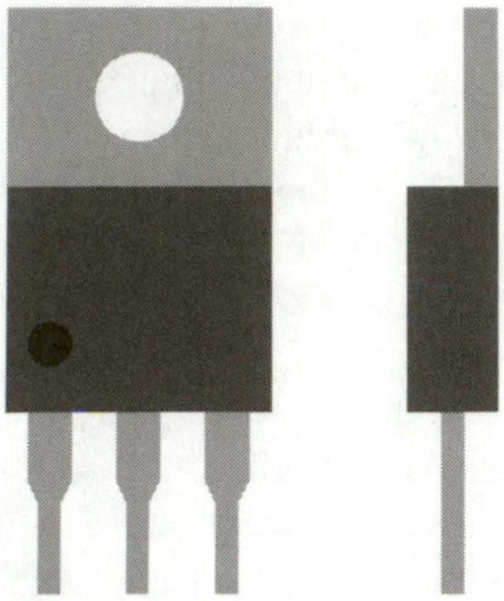

FIGURE 17.3 TO-220 package.

As an example, consider a device housed in a TO-220 package, shown in Fig. 17.3, with θ_{JA} = 65°C/W. The TO-220 is a common three-leaded package for power semiconductors including diodes, transistors, and integrated voltage regulators. If $T_{J(MAX)}$ = 150°C and $T_{A(MAX)}$ = 40°C, there is a 110°C temperature rise budget that corresponds to $P_D = \Delta T \div \theta_{JA}$ = 1.6 W (always round down for safety when dealing with thermal and power limits). If 1.6 W is insufficient for the application, there are several choices. The best option is to reduce power consumption so that heat is less of a problem in the first place. However, this is often not possible, given the design constraints. Failing that, either a larger package with lower θ_{JA} can be chosen, or the TO-220 can be attached to a heat sink to decrease the overall θ_{JA}. The TO-220 has a mounting hole to facilitate heat sink attachment.

Heat sinks are very common, because packages have practical limitations on how low their θ_{JA} can be. Thermal resistance is a function of materials and surface area. A small package has limited surface area. A good heat sink is constructed of a material that has low thermal resistance (most commonly, aluminum or copper) and that has a large surface area to radiate heat into the air. When a heat sink is attached, the device's θ_{JA} becomes irrelevant, and θ_{JC} becomes important, because it is the thermal resistance through the package to another material that matters. A TO-220's typical θ_{JC} is just 2.5°C/W—more than an order of magnitude better than θ_{JA}—because thermal conduction between attached metal surfaces is far better than between metal and air.

With a heat sink attached to the package, the heat sink's θ becomes relevant. Heat sinks are available in a wide range of sizes according to how much power needs to be dissipated. Off-the-shelf heat sinks for TO-220 packages can be found with thermal resistances under 3°C/W in still air. The junction between the heat sink and the IC package adds its own thermal resistance, which is often minimized via the application of thermally conductive grease. Assuming $\theta_{HEATSINK}$ = 3°C/W, the total θ_{JA} is now just 2.5 + 3 = 5.5°C/W. From a strictly thermal perspective, the device can now safely dissipate 110°C $\div\ \theta_{TOTAL}$ = 20 W! This solves the heating problem, but does not automatically mean that the semiconductor can handle this much power even when adequately cooled. Be sure to check the device's inherent current and voltage ratings to see if they become the limiting factor when heat is accommodated.

Semiconductors intended for lower-power applications do not always have associated thermal resistance specifications. Instead, they are rated for a certain power dissipation at a given ambient temperature, often 25°C—room temperature. If the ambient temperature exceeds the rating, a power derating curve or coefficient is provided that can be used to determine the safe power dissipation at a specific temperature. For example, a device may be rated for 400 mW at T_A = 25°C with a derating coefficient of 3.2 mW/°C. This means that, if the actual maximum T_A is 40°C, the device's power rating is reduced by $\Delta T_A \times$ 3.2 mW/°C = 15°C × 3.2 mW/°C = 48 mW, for an overall rating of just 352 mA.

It pays to be conservative when designing power regulation circuits, both in terms of current/voltage headroom as well as thermal headroom. The life span of most components is reduced as temperatures rise, so a cooler product is a more reliable product.

17.3 ZENER DIODES AND SHUNT REGULATORS

Perhaps the simplest type of voltage regulator is one formed by using a diode in a shunt circuit. A shunt regulator, shown in Fig. 17.4, consists of a device that conducts excess current to ground to maintain a fixed voltage between its two terminals. A series resistor, R_{DROP}, serves as the voltage-

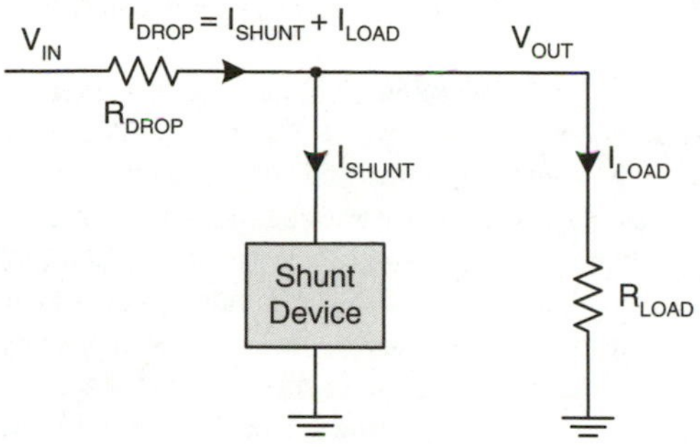

FIGURE 17.4 General shunt regulator.

dropping element. (The shunt device is typically a diode but can also be a more complex circuit that exhibits better characteristics.) When the load current increases, the shunt device draws less current to maintain the same V_{OUT}. The shunt device cannot feed current into the circuit, placing an upper limit on I_{LOAD} beyond which V_{OUT} will drop out of regulation. This concept was already touched on when diodes were discussed in an earlier chapter in the context of forming a voltage reference. A voltage reference may be a shunt regulator, and it is usually called upon to supply very minimal current.

Ordinary diodes do not make very practical shunt regulators, because their forward voltages are relatively small (0.3 to 0.7 V). Diodes are normally operated under forward bias conditions, but they can be made to conduct under reverse bias if the applied voltage exceeds their reverse breakdown voltage. Zener diodes are specifically designed to operate safely at their reverse breakdown voltage—the Zener voltage, V_Z. Furthermore, Zener diodes are manufactured with a wide range of voltages so that they can serve as shunt regulators or references. The common range of V_Z is from 2.4 to 33 V, with a typical tolerance of 5 percent, in increments of approximately 0.3 V at the low end, 1 V in the middle, and 3 V at the upper end of the range. Beyond their specified data sheet tolerance, the Zener voltage also varies with temperature and current. This voltage characteristic limits the accuracy to which a simple Zener-based regulator can function. As we continue, keep in mind the Zener's limitations and realize that, although it may not be well suited to directly regulating digital logic voltages, it can come in handy in less-restrictive power applications including battery charging, power supply clamping (e.g., preregulation), and coarse threshold comparison.

Figure 17.5 shows a Zener diode used in a shunt regulator circuit. Note that the cathode is at a higher voltage than the anode and that the Zener diode is identified with a Z-curve at the cathode. There are several common graphical representations for Zener diodes, but most involve some variation of a cathode Z-curve. This particular circuit uses the 5.1-V 1N4733A Zener diode with a 100-Ω

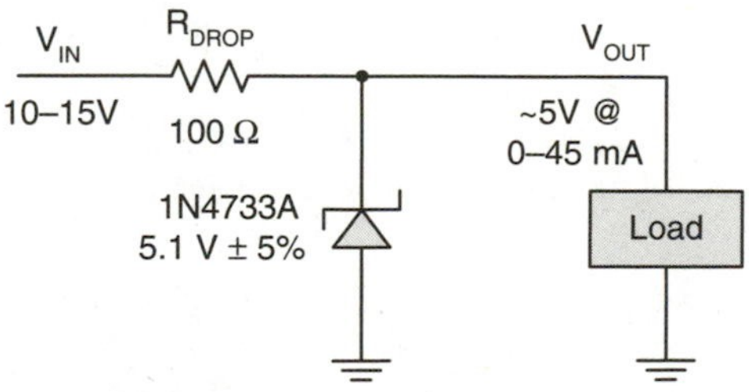

FIGURE 17.5 Zener diode shunt regulator.

series resistor to regulate an input of between 10 and 15 V down to approximately 5 V over a current range from 0 to 45 mA.

Aside from selecting a diode with the appropriate Zener voltage, a diode and resistor combination must be selected that can properly handle the circuit's worst-case power dissipation. Many types of regulators, including this one, operate by converting excess power into heat. The resistor is chosen to be as large as possible to minimize power dissipation while maintaining sufficient current through the diode to establish the desired voltage. In this situation, the resistor is bounded at a maximum value by its voltage drop at the maximum load current with a minimum input voltage, 10 V. The voltage drop in this situation is 4.9 V at 45 mA. 100 Ω is chosen, because it is smaller than the calculated value of 108 Ω, thereby providing a guaranteed minimum excess current for the diode to remain active: 4 mA. The circuit must allow the diode to conduct sufficient current at all times to remain active. Otherwise, the diode may take too long to begin conducting when the load current drops, thereby allowing V_{OUT} to exceed its specified range. The 1N4733A is characterized with Zener "knee" currents down to 1 mA, sufficiently below our 4-mA design point to remain active.*

Under this most favorable circumstance of minimum input voltage and maximum load current, the resistor and diode dissipate a minimum of 240 mW and 20 mW, respectively, for a total of just over a 0.25 W. If the input remains at 10 V, but the load current drops to its minimum value, 0 mA, the diode must shunt the full current to ground, resulting in 250 mW dissipated by the diode.

As the input voltage rises to its maximum of 15 V, the regulator must convert more power into heat. Now, the resistor must drop 9.9 V, burning 980 mW. This results in a current through the resistor of 100 mA, but the load is drawing only a maximum of 45 mA. Therefore, the diode must shunt the difference, 55 mA, to ground, thereby dissipating 280 mW. Unfortunately, this is not the worst-case scenario. The load is rated at a minimum current of zero, meaning that the diode must be able to shunt the full 100 mA that flows through the resistor when the input is at its maximum. This equates to a maximum diode power dissipation of 510 mW and an overall worst-case power dissipation for the regulator of nearly 1.5 W.

After calculating the maximum operating conditions of the regulator components, we must verify that they are within the manufacturers' specifications. The resistor dissipates nearly 1 W. Finding a resistor that can safely dissipate this quantity of power is fairly easy: a 2- or 3-W power resistor can be chosen. For the diode, a check of the data sheet confirms that the maximum current through the diode, 100 mA, is within the manufacturer's specifications of a continuous Zener (reverse bias) current of 178 mA.†

FIGURE 17.6 DO-41 package.

Next, we must ensure that the diode can handle 510 mW of continuous power dissipation at the circuit's maximum ambient air operating temperature, $T_{A(MAX)}$, which we will assume to be 40°C. The 1N4733A, manufactured in the axial-leaded DO-41 package shown in Fig. 17.6, has several relevant specifications:‡

- $P_{D(MAX)}$ = 1 W at 50°C, derate at 6.67 mW/°C above 50°C
- $T_{J(MAX)}$ = 200°C
- θ_{JA} = 100°C/W

According to the first specification, $P_{D(MAX)}$, we are operating safely. The power is approximately half the rated value, and $T_{A(MAX)}$ is under 50°C. Because the manufacturer also provides θ_{JA}, we can

* Zeners 1N4728A-1N4752A, Fairchild Semiconductor, 2001, p. 1.
† *Ibid.*
‡ *Ibid.*

perform a calculation to estimate how much margin we have. Recall that $T_J = T_A + \theta_{JA}P_D$. Therefore, $T_{J(MAX)} = T_{A(MAX)} + \theta_{JA}P_{D(MAX)} = 40°C + 100°C/W \times 0.51\ W = 91°C$. which is less than half the rated maximum.

Observe that if the junction temperature calculation used 1 W instead of 0.51 W and 50°C instead of 40°C, $T_{J(MAX)}$ would be calculated at just 150°C, well under the 200°C specification. This would ordinarily indicate that the device could handle more power at this temperature, yet the $P_{D(MAX)}$ specification disallows more power. Inherent limitations of a semiconductor may be more restrictive than the basic thermal resistance calculations would otherwise indicate. For this reason, it is important to look at all of the published specifications as a whole instead of relying on just a subset when determining the safe operating conditions for a semiconductor.

The shunt Zener regulator is fairly simple to construct but, aside from loose regulation accuracy, it also has the disadvantage that constant current is drawn from the input supply regardless of how much current the load is drawing. The diode establishes a fixed V_{OUT}, which results in a fixed current across the series resistor. In situations in which the V_{IN}/V_{OUT} differential is lower and the load current is specified over a smaller range, this type of regulator can be very useful—particularly at higher voltages for which other types of regulators get more complex. This narrower operating range enables a more efficient circuit design, because the series resistor dissipates less power by virtue of the lower voltage difference, and the diode dissipates less power because it has a narrower range of current slack for which it has to compensate when the load current falls to its minimum value.

17.4 TRANSISTORS AND DISCRETE SERIES REGULATORS

A significant inefficiency of the shunt regulator is that it must consume current that is not required by the load so as to maintain a fixed output voltage. Transistors can largely overcome this problem when used instead of a dropping resistor. Unlike a resistor that requires a current in proportion to the desired voltage drop, a BJT's collector-emitter voltage can be arbitrarily large without a proportional current flow. Consider the circuit in Fig. 17.7 with the same input/output ranges as the previous shunt regulator example, but using the TIP31 NPN power transistor as the series, or pass, element.

V_{OUT} is fixed by the transistor's base voltage less its V_{BE}, and the base voltage is fixed by a Zener diode reference of 5.6 V using a 1N4734A. Assuming $V_{BE} = 0.7$ V, $V_{OUT} \approx 4.9$ V. With the collector floating with changing V_{IN}, the input/output differential is taken up by V_{CE} without any proportional current requirement. The power dissipated by the pass transistor is dominated by $V_{CE} \times I_{OUT}$. Therefore, for constant V_{CE}, the transistor dissipates less power under light load conditions and more

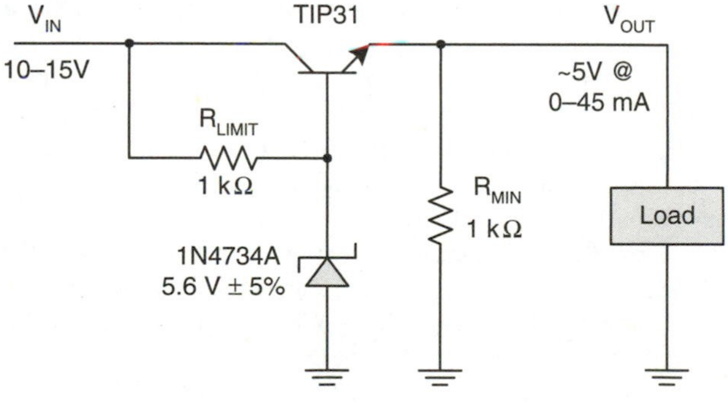

FIGURE 17.7 NPN series regulator.

power with heavier loads. V_{CE} cannot be reduced beyond the point of saturating the transistor because this would result in the inability to maintain regulation.

Note the shunt resistor R_{MIN} at the output. This resistor is present, because the minimum load current is zero. With zero load current, the transistor's V_{BE} circuit would not be complete, and the transistor would be in cut-off. Maintaining the transistor in the active mode enables it to respond quickly to changes in V_{CE}. A 1-kΩ resistor is added to the output in parallel with the load to guarantee a minimum output current of 4.9 mA so that the transistor remains active.

The regulator's power dissipation is a function of the input voltage and the load current. Table 17.1 summarizes the best- and worst-case power dissipation figures. Substantially lower power dissipation can be observed as compared to the shunt regulator discussed previously. The best-case power of 93 mW is less than 260 mW for the shunt circuit. The worst-case power of 669 mW is also much less than the shunt regulator's 1,490 mW. As expected, the majority of the power is dissipated by the TIP31 pass transistor when the load current and V_{CE} are at their maximum. The other components dissipate so little power that their thermal analysis is unnecessary in most situations.

TABLE 17.1 Series Regulator Power Dissipation

Component	V_{IN} = 10 V I_{LOAD} = 0 mA	V_{IN} = 15 V I_{LOAD} = 45 mA
R_{LIMIT}	19 mW	88 mW
Zener diode	25 mW	53 mW
R_{MIN}	24 mW	24 mW
Transistor V_{CE}	25 mW	504 mW
Total	93 mW	669 mW

The TIP31's maximum collector current is 3 A, well over our 49-mA operating point.* It is also rated for 2 W at an ambient temperature of 25°C and for 40 W at a case temperature of 25°C. These specifications imply that the device is intended to be used with a heat sink under more than lightly loaded conditions. Our present application has the transistor dissipating roughly one-quarter of the power rating without a heat sink at 25°C. However, if we want the circuit to be usable up to 40°C, some additional consideration is necessary. The TIP 31 data sheet does not provide detailed thermal resistance information, only a power derating curve versus case temperature and a maximum junction temperature of 150°C. Additional information is not needed if a heat sink is used, because the thermal resistance of the heat sink would be known, which would allow the determination of the case temperature for a given power dissipation. Knowing the case temperature would allow using the power derating curve to determine the safe operating limits of the transistor.

In the absence of information that applies specifically to our application, some estimates will have to be made. First, the transistor is being operated at one-quarter of its rating at 25°C. This probably provides sufficient margin at 40°C. Second, we can use thermal resistance information from other semiconductors packaged in a TO-220 package as a first-order approximation of the TIP31's characteristics. Other TO-220 devices have θ_{JA} = 65°C/W. Using this information, the transistor would experience a 33°C rise over ambient for $T_{J(MAX)}$ = 73°C. While this is only an approximation, it is less

* TIP31 Series, Fairchild Semiconductor, 2000, p. 1.

than half the rated $T_{J(MAX)}$ of 150°C. These estimates point in the right direction. However, the best approach is to confirm these findings with the manufacturer. Semiconductor manufacturers' application engineers are able to answer such questions and provide advice on matters that are not explicitly addressed in a data sheet.

As with the Zener shunt regulator, this circuit is relatively loose in its accuracy of V_{OUT} over varying temperature and load. The Zener voltage reference itself will drift with temperature, but its current is nearly static, so there will not be much drift with changing current. The transistor's V_{BE} changes with temperature and current as well. If these drawbacks have you wondering how accurate voltage regulators are ever constructed, the answer lies in various compensation schemes that involve more components. However, the biggest contributor to accurate voltage references and regulators is the integrated circuit, because an IC enables the pairing of transistors with closely matched physical and thermal characteristics. When transistors and diodes are fabricated on the same slice of silicon within microns of each other, they are nearly identical, and they operate at the same temperature. Close matching enables transistors and diodes to largely cancel out each other's undesired variations when arranged in specific configurations.

A series regulator can also be designed to provide a regulated constant current instead of constant voltage. Current sources are useful for battery chargers, among other applications. Figure 17.8 shows such a circuit using a PNP transistor. Once again, the variable input/output differential is taken up by the transistor's V_{CE}, although it is the output that is allowed to float with a constant current. A floating output voltage is necessary, because an ideal current source supplies constant current regardless of the impedance that it is driving. A higher load impedance results in a higher voltage output according to Ohm's law. This is the converse of a voltage source wherein constant voltage is desired at variable current. Real current regulators, of course, have limitations on the range of V_{OUT} for the circuit to remain in regulation, just as we have already observed that voltage regulators are subject to current limitations.

The TIP32 is chosen for this circuit, because it is a mature BJT with characteristics similar to the TIP31. Power circuits that require both NPN and PNP transistors sometimes use the complementary TIP31 and TIP32 pair. This current regulator functions by establishing a fixed voltage drop across R_{SET}, thereby establishing a fixed emitter current. Assuming negligible base current, the collector current drives the load with the same current. An emitter-base loop is established with a reference voltage provided by the 1N4728A Zener diode. Per loop analysis, $V_{ZENER} = I_{OUT}R_{SET} + V_{BE}$, assuming that the base current is negligible. When the Zener reference and V_{BE} are fixed, R_{SET} establishes the regulator's output current. R_{LIMIT} picks up the voltage difference between V_{IN} and V_{ZENER} and thereby serves as the Zener diode's current limiter.

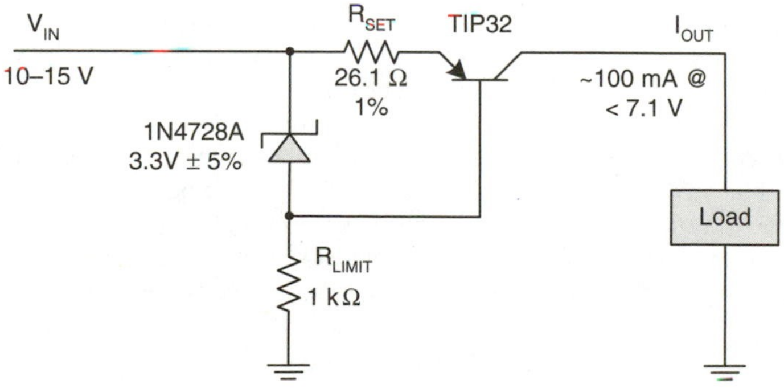

FIGURE 17.8 PNP current regulator.

The current regulator cannot supply a constant 100 mA for all values of V_{OUT}. As the load impedance is increased, there is a corresponding increase in V_{OUT}, which reduces the difference with respect to V_{IN}. This differential eventually forces the transistor into saturation when its collector-base junction becomes forward biased. Using the estimate of $V_{CE(SAT)}$ = 0.3 V provides us with a minimum voltage drop across the transistor for the regulator to maintain 100 mA. The voltage drop across R_{SET}, 2.6 V, must also be factored into the V_{IN}/V_{OUT} differential. The maximum V_{OUT} must be evaluated along with minimum V_{IN} to analyze the limiting scenario: $V_{OUT(MAX)} = V_{IN(MIN)} - 2.6\text{ V} - 0.3\text{ V} = 7.1\text{ V}$. This circuit can therefore be guaranteed to function properly up to 7.1 V. Of course, if V_{IN} is at the upper end of its range, V_{OUT} can be higher as well. But the worst-case scenario is what should be used when specifying the guaranteed parameters of a circuit.

Analysis of the regulator's dropout voltage shows that the drop across R_{SET} is the dominant term within the regulator circuit. If dropout voltage is a concern, the drop across the resistor should be minimized. This is why a low-voltage Zener has been selected rather than a 5- or 6-V diode. The drop across R_{SET} is the difference between V_{ZENER} and V_{BE}. Minimizing V_{ZENER} minimizes the drop across R_{SET}.

Calculating power dissipation to perform a thermal analysis requires bounding the output voltage at a practical minimum, because the transistor's power dissipation increases with increasing V_{CE} and, hence, decreasing V_{OUT}. R_{SET} has a constant voltage drop at constant current, so it has constant power dissipation of 0.26 W. For the sake of discussion, we can pick $V_{OUT(MIN)}$ = 0 V. Along with $V_{IN(MAX)}$ = 15 V, the power dissipation of the transistor is $I_{OUT}(V_{IN} - I_{OUT}R_{SET} - V_{OUT}) = 1.24\text{ W}$. The TIP32 and TIP 31 have equivalent power and thermal ratings. The transistor is rated at 2 W without a heat sink at an ambient temperature of 25°C, and a conservative design methodology might call for a heat sink on the TO-220 package in this case. A heat sink enables calculation of the package temperature for a given power dissipation and ambient temperature, which in turn enables us to take advantage of the manufacturer's power derating curve expressed in terms of package temperature.

A small TO-220 heat sink can provide a thermal resistance of 30°C/W with natural convection. Much lower values are achievable when a fan is blowing air across the heat sink and with a larger heat sink. A temperature rise of 37°C is attained with P_D = 1.24 W. Therefore, a 40°C ambient temperature would result in a transistor case temperature of 77°C. The TIP32's power derating curve shows that the transistor is capable of over 20 W at this case temperature.* The heat sink thereby enables a very conservative design with minimal cost or complexity.

17.5 LINEAR REGULATORS

Most voltage and current regulation requirements, especially in digital systems, can be solved with ease by using integrated off-the-shelf regulators that provide high-quality regulation characteristics. Constructing a regulator from discrete parts can be useful when its requirements are sufficiently outside the mainstream to dictate a custom approach. However, designing a custom regulator brings with it the challenges of meeting the load's regulation requirements over a potentially wide range of operating conditions. Power supplies for digital circuits have the benefit that the voltages are common across the industry. This has enabled semiconductor manufacturers to design broad families of integrated regulators that are preadjusted for common supply voltages: 5, 3.3, 2.5, 1.8, and 1.5 V. Manufacturers also offer adjustable regulators that can be readily customized to a specific output voltage. The result is the ability to treat regulators largely as "black boxes" once their overall charac-

* TIP32 Series, Fairchild Semiconductor, 2000, p. 2.

teristics have been quantified and accounted for. High-quality off-the-shelf regulators are inexpensive and provide a quality of regulation that is far more difficult to attain with discrete design.

One of the basic types of off-the-shelf regulators is the *linear regulator.* A linear regulator is an analog integrated circuit that essentially implements a more complex version of the transistor-based series regulator already discussed. Instead of one transistor, it includes more than a dozen. All of the functions of voltage reference, error feedback, and voltage control are included. For the sake of comparison, this is akin to buying a 74LS00 IC instead of constructing NAND gates from discrete transistors. Most linear regulators have three terminals: an input, an output, and either a ground or voltage sense input. Fixed regulators have a ground pin, and adjustable regulators use the third pin as a voltage feedback.

Perhaps the most classic and widely used fixed linear regulator family is the LM78xx, with common variants including the 7805 and 7812, which provide fixed outputs of +5 V and +12 V, respectively. These parts are available in a variety of packages with current ratings from 100 mA to 3 A. Small TO-92 packages, commonly used for low-power transistors, and small surface-mount packages house the 100 mA versions, whereas TO-220 packages and the surface mount equivalent DPAK house the higher current variants. The 78xx family provides line and load regulation performance in the range of 100 mV: over the possible range of input voltages and output loads, the output voltage will not change by more than 100 mV. Application of the 78xx family is illustrated in Fig. 17.9 using the example of a 7805. It is recommended that high-frequency bypass capacitors be placed at the input and output nodes to reduce noise and improve the overall circuit's high-frequency response. Although 0.33-µF and 0.1-µF values are the common manufacturer's recommendations, larger high-frequency capacitors can be used. Depending on the specific application, a reverse-bias protection diode is also recommended to prevent device damage when the system is powered off. The regulator is not designed to handle an output voltage that is higher than the input voltage by more than one diode drop. If the load has sufficient capacitance to briefly maintain a nominal voltage after power is turned off, current could flow through the regulator in reverse and damage it. The diode becomes forward biased in this situation, shunting current away from the regulator.

Complementary to the 78xx positive voltage regulator family is the 79xx negative voltage regulator family. These devices are very similar to the 78xx but require larger capacitors on the input and output because of their internal structure. At least 2 µF and 1 µF on the input and output, respectively, are recommended, along with a smaller 0.1-µF high-frequency capacitor on the output.

Aside from power dissipation, which is simply the product of the input/output differential and the load current, a main consideration when designing with linear regulators is their dropout voltage. Standard linear regulators such as the 78xx have dropout voltage specifications in the 2.5-V range. If a 5-V output is desired, at least 7.5 V on the input is required.

As digital supply voltages have steadily decreased, it has become common to find systems with multiple low-voltage supplies. For example, a system may use 3.3- and 2.5-V components. Higher dropout voltages can complicate an otherwise simple power subsystem design in such instances.

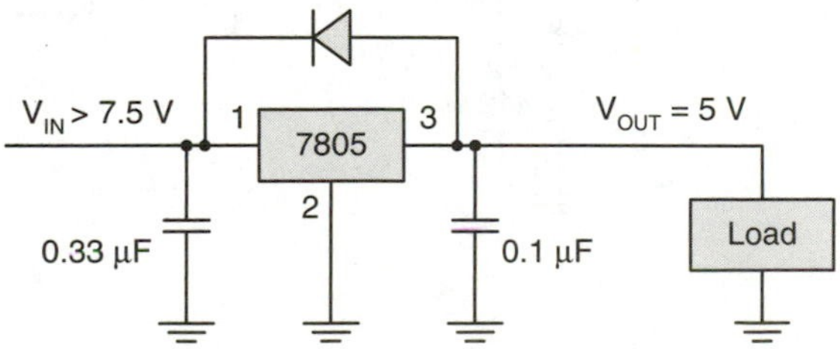

FIGURE 17.9 LM7805 circuit.

Clearly, a regulator with a 2.5-V dropout voltage is not capable of producing 2.5 V from 3.3 V. Furthermore, when the dropout voltage is on par with the output voltage, the regulator dissipates as much power as the load consumes! The semiconductor industry saw these problems emerging and developed families of *low dropout* (LDO) regulators with dropout voltages well under 0.5 V. There are a wide variety of LDOs on the market today, but they are no longer members of standard families such as the 78xx. Each manufacturer has its own product line. LDOs are advantageous not only because of more flexible input/output voltage relationships, but also because they enable lower power dissipation for a given load current. With a voltage differential of just 0.2 to 0.3 V, a 5-A regulator dissipates an order of magnitude less power than an older 2-V dropout device. These regulators are manufactured by companies including Fairchild Semiconductor, Linear Technology, Maxim, National Semiconductor, and Texas Instruments.

When an output voltage is desired that is not supported by a fixed regulator, adjustable linear regulators such as the ubiquitous LM317 and LM337 can be used for positive and negative voltages, respectively. These devices have an adjustment pin, instead of a ground pin, that is used to vary the output voltage. A basic LM317 circuit is shown in Fig. 17.10. Note the use of a reverse-bias protection diode as employed previously for the fixed regulator. Additionally, a larger capacitor is recommended at the output to improve stability caused by internal differences with the 78xx fixed regulator.

The real difference is in the LM317's voltage adjustment mechanism. An internal circuit maintains a fixed 1.25 V between the output and adjustment pins, resulting in a current through R1 determined by its resistance. This current is also referred to as the *programming current*. Assuming that negligible current flows into the adjustment pin, an accurate approximation is that the current through R1 passes through R2 as well, creating a voltage drop across R2. The drop across R2 plus the 1.25 V across R1 yields the regulated output voltage,

$$V_{OUT} = 1.25\left(1 + \frac{R2}{R1}\right)$$

Typical resistance values for R1 are relatively small, ranging between 120 and 330 Ω. Values for R2 are larger so that higher voltages can be developed as necessary. It is advantageous to select smaller values for R1 and R2 so that the assumption of zero current through the adjustment pin is more accurate. The actual adjustment pin current, I_{ADJ}, is a maximum of 100 μA.* In our example,

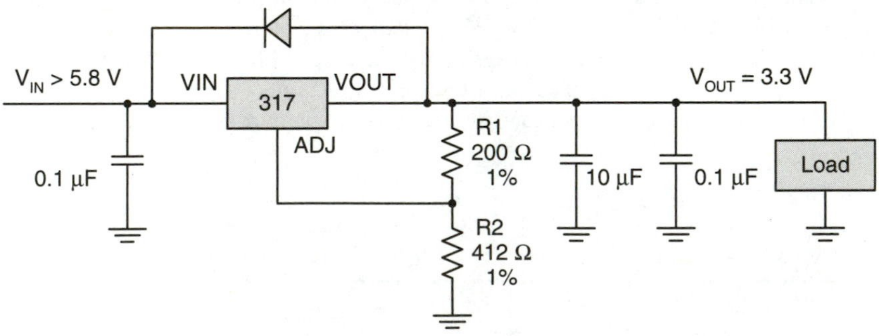

FIGURE 17.10 LM317 circuit.

* LM317, National Semiconductor, 2001, p. 2.

the current through R1 is more than 6 mA—more than two orders of magnitude larger than I_{ADJ}. The actual error imparted by I_{ADJ} = 100 μA is I_{ADJ}R2 = 41.2 mV, or roughly 1 percent of the 3.3-V output. This error can be reduced by increasing the programming current, which causes a corresponding decrease in R2.

The LM317 has two key requirements to maintain a regulated output. First, a minimum load current of 10 mA is necessary. This is partially addressed via the programming current—6 mA in our example. If the load cannot be guaranteed to sink the remaining 4 mA, a 120-Ω resistor can be substituted for R1 with an accompanying adjustment to R2. Second, the input must exceed the output by the LM317's dropout voltage, which is as high as 2.5 V, depending on load current and temperature. Dropout voltage decreases with decreasing load current and varies nonmonotonically with temperature.

A further improvement to the basic LM317 circuit is to add a 10 μF bypass capacitor between the adjustment pin and ground to filter ripple noise being fed back through R1. This improves the regulator's ripple rejection. When this capacitor is added, the same issue of safe power-off comes up, because the capacitor will hold charge and discharge through the adjustment pin. As before, the solution is a diode with its anode connected to the adjustment pin and its cathode connected to the output pin. This provides a low-impedance path from the bypass capacitor through two diodes to the input node.

Common three-terminal linear regulators can also be applied as current regulators. While both fixed and adjustable regulators are applicable, the adjustable regulators are more flexible in this role because of the low voltage differential maintained between the output and adjustment pins. In the case of the LM317, this 1.25-V difference allows a constant current to be established by a series output resistor as shown in Fig. 17.11. Note that the maximum output voltage for this circuit to maintain regulation is bounded by an overall dropout voltage that is the sum of the LM317's dropout voltage plus the 1.25-V drop across R_{SET}.

Because it is a series element, a linear regulator dissipates power equal to the product of the load current and the input/output voltage differential. At high currents and voltage differentials, the necessary cooling to prevent a regulator from overheating can be substantial. There is no way to reduce the overall power wasted by a linear regulator. More efficient switching regulators are often used in high-power applications. Yet there are times when the complexity of a switching regulator is undesirable, and absolute power regulation efficiency is not a prime concern. Under these situations, a small trick can be employed to reduce the power dissipated by the regulator by shifting a portion of that heat to a less thermally sensitive power resistor, thereby easing the regulator's cooling requirements.

Consider the example in Fig. 17.12, wherein a large voltage differential exists. The input is a 24-V supply with a 10 percent range. The output is a standard 5-V level for a small digital circuit drawing up to 500 mA. When the input voltage is at its maximum, a worst-case voltage differential of 21.4 V exists, which means nearly 11 W of power to safely dissipate. A large heat sink and/or moving air are normally required to maintain a safe junction temperature at this power level. Instead, a

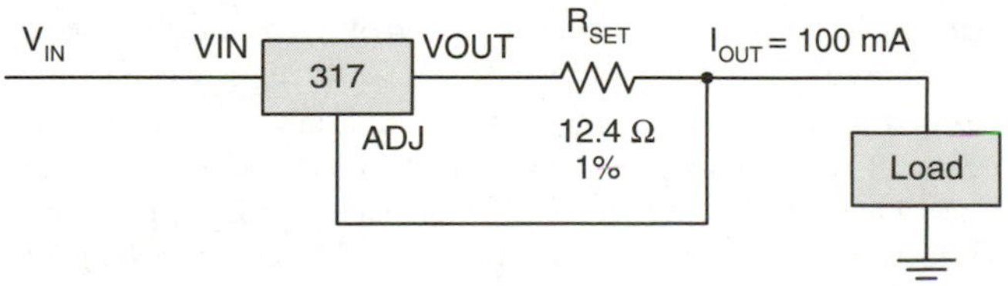

FIGURE 17.11 LM317 current regulator.

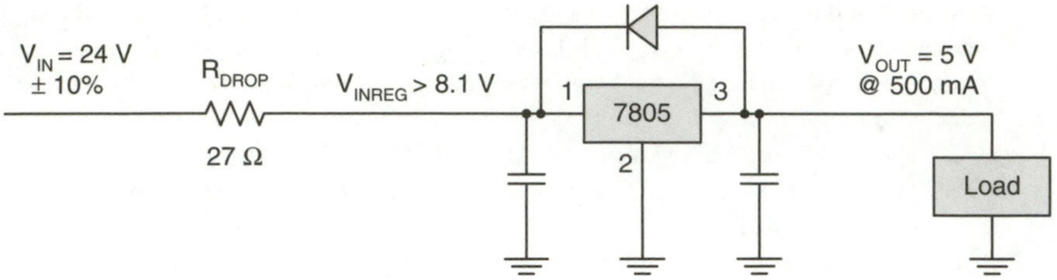

FIGURE 17.12 Shifting power dissipation to a resistor.

27-Ω dropping resistor is inserted before the regulator. The dropping resistor is sized based on the load current, the minimum input voltage, and the regulator's dropout voltage. Working with a conservative dropout voltage of 3 V and a minimum input level of 21.6 V, the resistor must drop a maximum of 13.6 V at 500 mA. The closest standard resistor value, 27 Ω, yields a drop of 13.5 V at maximum load with a power dissipation of 6.75 W. Finding a power resistor that can safely dissipate about 7 W is not a problem. This reduces the power dissipated by the LM7805 to just 3.1 V × 500 mA ≈ 1.6 W. However, the worst-case power dissipation occurs when the input is at its maximum of 26.4 V. The increased differential of 7.9 V results in a power dissipation of approximately 4 W, as expected. We did not save any power, but the regulator's share of the power has been decreased by more than 60 percent, and it is much easier to cool 4 W than 11 W.

Linear regulators provide the easiest and least troublesome manner by which to provide a digital circuit with clean power. They work without causing problems as long as proper thermal analysis is performed and they are kept from overheating. The main reason for not using a linear regulator is its inefficiency, particularly at higher input/output voltage differentials. It is quite convenient to use a linear regulator to derive 3.3 V from 5 V at currents under 1 A. Power dissipation becomes a bigger problem when dealing with several amps. When tens of amps are required, the thermal problem becomes critical, and a more efficient alternative must often be found in the form of a switching regulator.

17.6 SWITCHING REGULATORS

An ideal power regulator would achieve 100 percent efficiency by drawing only as much power from the source as required by the load. Assuming that the input and output voltages are unequal, the ideal regulator would convert a supply current into a different load current to maintain constant power under two different voltages. A linear regulator is nonideal, because it is inherently dissipative—it simply discards the excess power not required by the load, because it cannot convert an input current into an unequal output current. *Switching regulators* approach the ideal much more closely, because they are able to perform the current conversion process. Finite power losses in the switching regulator's components are what cause it to deviate from ideal efficiency. Depending on the input/output characteristics and the circuit used, efficiencies between 80 and 95 percent are realistically achievable.

Switching regulators operate by alternately applying and removing input power from a passive LC circuit. Figure 17.13 shows a simplified conceptual *step-down* regulator circuit, often called a *buck* regulator. The switching element alternately connects the inductor to the input and ground. When the switch selects the input, current flows through the inductor and charges the capacitor. During this time, the inductor develops a magnetic field resulting from the current passing through it. An inductor resists changes in current, because a current change disrupts its magnetic field. Whereas a

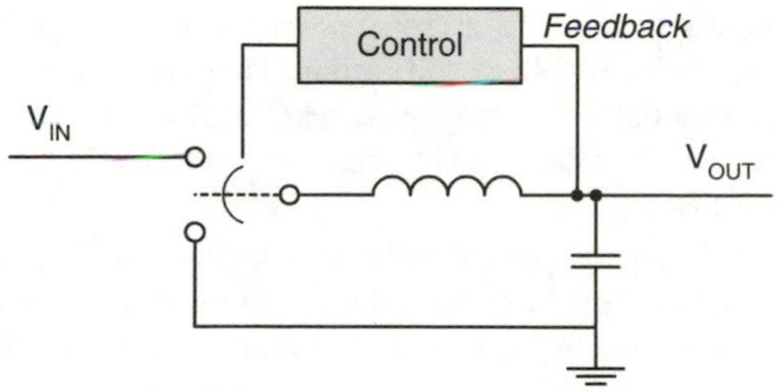

FIGURE 17.13 Conceptual step-down regulator.

capacitor opposes voltage changes by sinking or sourcing current, the inductor changes its voltage to maintain a constant current flow. When the switch selects ground, the inductor instantly flips its voltage so that it can continue supplying current to the capacitor. The inductor holds a finite amount of energy that must be quickly replenished by switching back to the input voltage. The role of the feedback and control circuit is to continuously modify the switching frequency and/or duty-cycle to maintain a fixed output voltage.

There are varying designs for switching circuits. Two common topologies often seen today are shown in Fig. 17.14. Most modern switching regulators employ power MOSFETs because of their low R_{DS}. Prior to the availability of power FETs, power BJTs were used as switches, and their finite $V_{CE(SAT)}$ resulted in higher losses than seen with modern FETs. When a single transistor is used as the switch, a diode serves as the ground shorting element, or rectifier. The diode is reverse biased when the transistor is conducting, and it becomes forward biased after the switching event causes the inductor's voltage to flip. The inductor changes its voltage to maintain a constant current flow, which causes its switch-side voltage to suddenly drop, and the diode clamps this dropping voltage to near ground. Substantial current flows through the diode and motivates the selection of a low forward-voltage Schottky diode (note the S-curve symbology for a Schottky diode). Power loss in this diode is a major source of switching regulator inefficiency. This has given rise to the dual-transistor switch circuit that replaces the diode with a FET as the main rectification element. When a transistor is used in this manner as a rectifier, the common industry term is *synchronous rectification*. The FET's low R_{DS} makes it a superior solution to the fixed voltage drop of the Schottky diode. However, a diode is still present to serve as a rectifier during the short but finite turn-on time of the bottom FET.

Minimal power loss in the switching transistors is a key attribute that enables high-efficiency switching regulators. A switch transistor is ideally either on or off. When off, the transistor dissipates no power. When on, there is minimum voltage drop across the transistor and, hence, minimal power

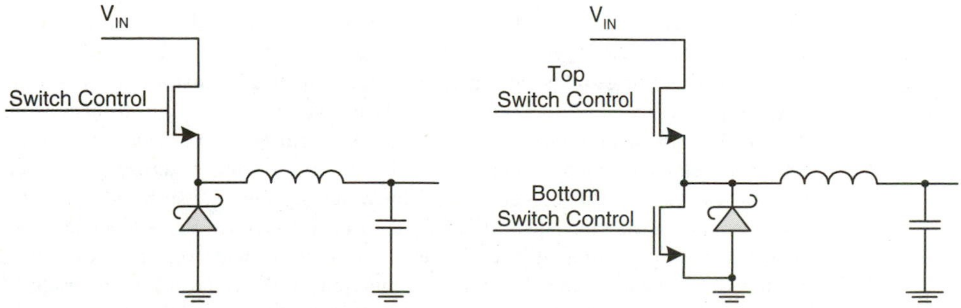

FIGURE 17.14 Single and dual FET switching circuits.

dissipation. Switching control circuits seek to turn transistors on and off as rapidly as possible to minimize the time spent in a region of higher voltage and current. The drive strength required to rapidly switch a large FET can be substantial, because a larger gate results in higher capacitance that must be charged to change the transistor's state. In reality, the finite on-off FET switching times are another significant source of power loss.

Aside from their high efficiency, switchers are extremely flexible, because they can do more than regulate down. *Step-up* regulators, often called *boost* regulators, can convert a lower input voltage into a higher output voltage. *Inverting* regulators can generate negative output voltages from a positive input. Switchers can also be designed for *buck-boost* operation, whereby an output can be generated from an input that is near the output level. This is a very useful feature for some applications, because it bypasses the problem of dropout voltage. Battery-powered portable devices commonly use step-up or buck-boost regulators because of the cost and size advantage of using fewer battery cells. Without a switcher, such a device would require more cells to meet the minimum dropout voltage of a linear regulator.

Each type of switching regulator has a different topology. Most contain FETs, diodes, inductors, and capacitors. Some low-current switchers are now available that do not need inductors. A complete presentation of switching power supply circuit design is unnecessary in the context of most digital systems, because semiconductor manufacturers have done an excellent job of producing integrated switching regulator ICs that are easy to use by following recommended circuit and connection diagrams. Perhaps the easiest way to use a switching regulator is to purchase an off-the-shelf module that already has the full circuit assembled and tested. Companies including Datel, Texas Instruments, and Vicor offer these ready-to-use modules.

More flexibility (and more work!) is possible when discrete ICs are used. Linear Technology, Maxim, and National Semiconductor offer wide varieties of switching regulator controller ICs that require the addition of FETs, inductors, capacitors, diodes, and resistors. (Some low-power regulator ICs incorporate the FET switch on-chip to reduce circuit complexity.) These circuits can be fully customized for individual applications, and manufacturers supply detailed example circuits and documentation to assist in the design effort. As with any semicustom design, more flexibility is possible in terms of the circuit layout to squeeze into small spaces. The trade-off is increased design effort to ensure that numerous discrete components operate properly.

Whatever implementation path is chosen, the incorporation of switching regulators into a system brings with it a set of issues that are either not present or not as potentially troublesome when using linear regulators. A switching regulator introduces noise and ripple because of its high-current on-off operation. Many switcher design issues revolve around minimizing this noise and ripple. The selection of the inductor and capacitors is very important. An IC vendor will usually recommend a particular inductor for a given switching frequency and output current. Because power inductors can get quite large, a current trend is to use smaller inductors at higher switching frequencies with the consequence of introducing higher-frequency noise into the system. If the high-frequency noise cannot be adequately filtered, a larger inductor may be used at a lower frequency.

Attenuating ripple and noise generated by the switching regulator is accomplished in large part by using high-quality capacitors with extremely low equivalent series resistance, or ESR. The capacitors' ESR must be minimized because of the high current spikes sent through the inductor by the switching element. Each current spike will develop a voltage across the capacitor's finite ESR. Higher resistance means higher voltage ripple. Ideally, ceramic surface-mount capacitors would be used because of their excellent high-frequency characteristics. Some switchers do use these capacitors, but their use is limited, because ceramics do not provide very high capacitances. High-power switchers require large capacitors rated at hundreds or thousands of microfarads. Tantalum and aluminum electrolytic capacitors are commonly used in switcher circuits. Normal electrolytics do not have adequate ESR for many switching applications, but manufacturers such as AVX, NIC, Nichicon, Panasonic, Sanyo, and Vishay have developed special-purpose lines of low-ESR electrolytic capacitors. Typical ESR for these capacitors is from 10 to 100 mΩ.

17.7 POWER DISTRIBUTION

Power regulation and distribution circuits inherently handle significant quantities of power in relatively compact volumes and are designed to provide power with minimal source impedance, making safety a prime concern. Minimal source impedance is highly desirable when evaluating a regulator's ability to supply high current at constant voltage. This characteristic also means that there is nothing preventing a short circuit or overload condition from drawing power beyond the circuit's specifications. When high power density and low source impedance are brought together, there is the potential for serious injury and damage if a component were to fail and cause a short circuit.

Reliability of power distribution is directly related to safety, because an unreliable circuit may fail in an unsafe manner. Yet even if the failure causes no damage, a failure is still quite undesirable. The reliability of a circuit is influenced by how heavily loaded its individual components are. Operating components at higher thermal loads (higher temperatures) degrades their longevity. Operating components at greater fractions of their rated current, voltage, and power also reduces their life span. Conservative designs use power components that are rated well in excess of their actual operating load.

Many electrical safety codes and standards have been adopted over time to minimize the adverse consequences of power circuit failures. In the commercial and residential AC wiring context, there are strict regulations requiring insulated connections, minimum wire gauge, maximum wire length, and fuses. The basic idea is to first reduce the chance of failure by employing conservative design practices. Adequate insulation reduces the likelihood of shorting. Limits on wiring ensure that the conductors can carry the desired load. Second, the inevitable failure's effects must be minimized. Engineering in general, but power engineering in particular, requires the anticipation of failure regardless of how unlikely that failure is. Murphy's law is a constant across all disciplines. Fuses are the main failure-handling mechanism in power distribution. A fuse is rated for a certain current and will blow when that current rating is exceeded. In the unlikely event of a short circuit, the fuse blows, and all power is cut off to that branch of the power distribution system. Fuses are the last line of defense, because they are simple, passive devices that deal with the problem in a quick and decisive manner.

Figure 17.15 shows a logical view of a power distribution system where the AC-to-DC rectification and preregulation functions are separated from the final DC-to-DC regulator that directly powers the load. Some power systems combine these elements into a single module, and others use a distributed approach as shown here. The AC-to-DC and DC-to-DC functions are logically separate

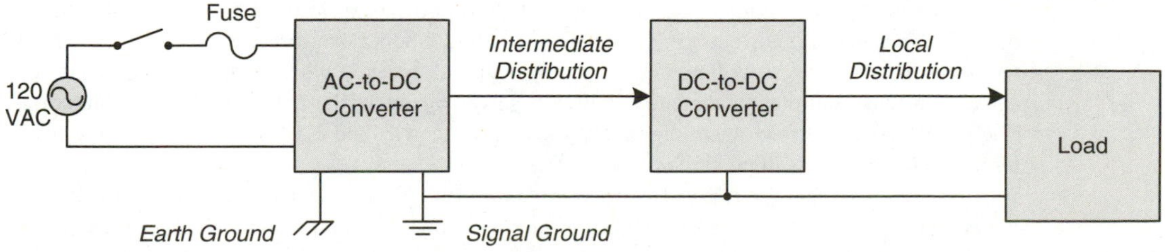

FIGURE 17.15 Overall power distribution.

entities for a couple of reasons. First, they are usually associated with different levels of safety standards. AC power supplies must conform to strict standards, because they have the potential to short circuit an AC wiring system with the resultant risk of serious injury and damage. Second, DC-to-DC converters are tailored specifically to the voltages required by a digital logic circuit, whereas AC power supplies can provide more standardized voltages given the widespread availability of DC regulators.

The example also shows two distinct ground nodes: Earth ground and signal ground. A grounded AC wall outlet or other connection provides an absolute 0-V Earth ground connection into which excess charge can drain. This path prevents the accumulation of charge to the point of damaging sensitive electronic components when a sudden electrostatic discharge may occur. It also ties separate pieces of equipment to the same ground potential so that they can be connected without adverse consequences. For example, if a printer and a computer are both connected via a cable, and each has a different ground potential, a ground loop may develop whereby unexpected current flows between the two dissimilar voltages. A ground loop can be disruptive to communication between the printer and computer by manifesting itself as noise on the cable. In contrast to Earth ground, signal ground is the return path to the voltage regulators. Signal and Earth ground are usually connected so that there is a uniform DC ground potential throughout a system. However, many styles exist for making this connection. Some engineers and situations favor connecting the two grounds at a single point, often in the AC power supply. Others favor connecting the grounds at many points throughout the system. Grounding is an important topic that is discussed later in the context of signal integrity.

Given their different treatment, different symbols are used for each type of ground. Purists have a valid argument in favor of consistent representation for each type of ground. When drawing circuit diagrams that must pass official certification standards, consistent representation may be enforced. In other circumstances, it is common for individuals to use whatever representation they prefer. When multiple ground symbols are present in a circuit diagram, the meaning of each symbol should be explicitly determined, because different engineers and companies use their own styles.

As with household AC wiring, strict safety standards should always be observed when using AC-to-DC power supplies. Underwriters Laboratories (UL) has become the de facto standard for safe product certification in the United States, and UL certification is widely considered a mandatory requirement despite not always being dictated by law. All reputable AC power supply vendors seek UL certification before offering their products for sale. Safe power supply design has long since become a well understood and economical practice, and there is no significant burden on the system design and no incentive—financial or otherwise—to not employ a certified design. When designing a digital system, it is almost always best to purchase a precertified AC power supply module from a third party. Well known power supply vendors include Artesyn, Astec, Cherokee, Condor, and Lambda. These power supplies incorporate standard safety features including overload protection devices that automatically shut down the supply when it gets too hot or is called upon to source too much current. Some supplies contain internal fuses and are ready to plug into a wall outlet. Others require that a separate AC wiring harness with fuse be connected.

Off-the-shelf power supplies are available in wide varieties of overall wattages and combinations of output voltages and currents. In many cases, a single module can directly provide a digital system's necessary voltage supplies. There are also many instances in which a digital system has voltage and current requirements that are not directly served by one of these off-the-shelf configurations. The proliferation of low-voltage digital supplies (e.g., 2.5, 1.8, and 1.5 V) has decreased the likelihood of a one-size-fits-all power supply.

Safety requirements for DC-to-DC converters are usually less stringent than for AC-to-DC converters, but the same concepts of fusing and overload protection can be applied. In applications in which a DC power connection is externally accessible by the end-user, fusing and current limiting precautions may be required for safety certification. Many switching regulator modules contain overload protection mechanisms that shut down in the event of excessive current draw. Safety standards for DC-to-DC converters are less uniform and strict, because it is often sufficient to rely on the safety cut-off mechanisms in the AC power module. If a DC regulator fails, the worst-case scenario is often that the AC supply's overload protection will be activated and prevent damage to other equipment and the AC wiring infrastructure.

Whenever power is distributed between two points, the conductors carrying that current must be adequately sized for the flow. Wires have current ratings based on their resistance, the ambient operating temperature, the maximum allowable insulation temperature, and the number of wires bundled together. It is important to conservatively specify wire capacity, because wire has a positive temperature coefficient, meaning that resistance increases with temperature. If a wire is operated beyond its safe capacity, a dangerous situation can develop in which heating increases resistance, which causes more heating in a self-destructive cycle. Wire manufacturers should always be consulted on the ratings for their products when selecting the necessary gauge wires for an application. The American Wire Gauge (AWG) standard provides a measure of wire size, with thicker wires indicated by smaller gauge numbers. Table 17.2 lists the resistance of solid copper wire per 1000 ft (304.8 m) at 25°C.* Current ratings are based on the allowable temperature rise over ambient, which is why a wire's environment directly affects its rating. AC wires enclosed in the walls of your home are rated more conservatively because of the risk of fire in confined spaces. The conservative use of lower-gauge wire results in less power loss and heating, with a resulting increase in safety and reliability.

TABLE 17.2 Resistance of Solid Copper Wire at 25°C

Wire Gauge (AWG)	Ω per 1000 ft (304.8 m)	Wire Gauge (AWG)	Ω per 1000 ft (304.8 m)
30	104	20	10.0
28	65	18	6.4
26	41	16	4.0
24	26	14	2.5
22	16	12	1.6

An example of a distributed power regulation scheme is shown in Fig. 17.16. This system uses an off-the-shelf AC-to-DC power supply to provide a 12-VDC intermediate power bus. Common intermediate voltages include 48, 24, 12, and 5 V. The advantage of using a higher voltage is less current flow through the intermediate distribution wiring for a given power level and hence lower resistive losses ($P_D = I^2R$) in that wiring. Lower voltages have the benefit of easier component selection because of the lower voltage ratings. When using a switching regulator, there is a compromise between very low and very high input voltages. Too high an input requires a small switching duty cycle, which results in higher losses as the transistors turn on and off more often relative to the time that they are in a static state. Too low an input causes higher current to be drawn from the source, which leads to higher I^2R losses in the regulator components. Often, 12 V is a good compromise between switching losses and easier regulator design. Additionally, some systems require 12 V for analog interface circuits or low-power motors such as a disk drive. This system can use whatever 12-V power supply is easily available, as long as its capacity is greater than 43 W.

* *The ARRL Handbook for Radio Amateurs,* American Radio Relay League, 1994, pp. 35–36.

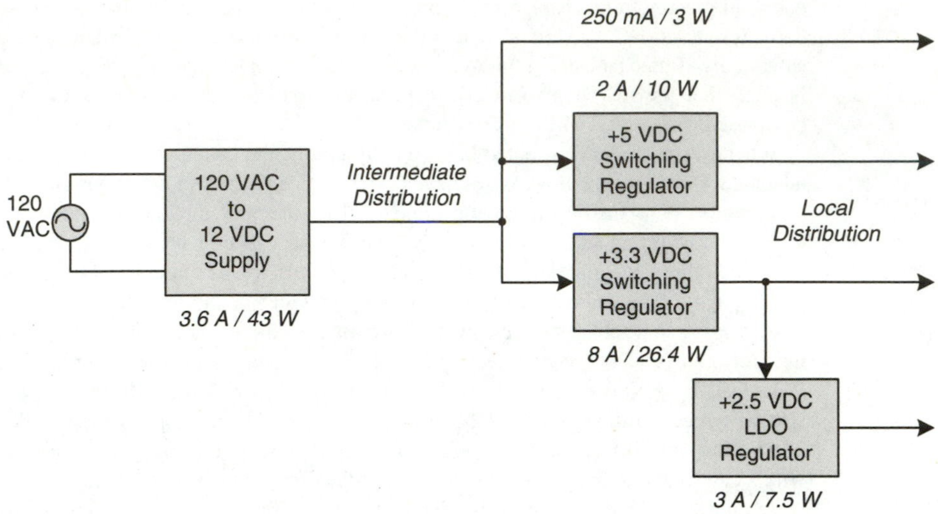

FIGURE 17.16 Distributed power regulation.

The fact that most systems can rely on an off-the-shelf AC power supply to provide an adequate level of safety enables a design style of selecting a certified AC supply with an adequate power rating and then using a custom combination of regulators to provide the desired logic-level voltages. The advantage to this approach is not requiring the perfect AC supply that provides every voltage at the correct current rating. Instead, an overall system power requirement is calculated and a supply is found that provides a suitable intermediate voltage at or above that level. In this example, the final voltages are needed as follows: 12 V at 250 mA, 5 V at 2 A, 3.3 V at 5 A, and 2.5 V at 3 A. It is decided to create the 5- and 3.3-V supplies with switching regulators and use a low dropout linear regulator to provide 2.5 V from 3.3 V, which is why the 3.3-V regulator must supply sufficient current for both the 3.3-V and 2.5-V needs. It is assumed that the switching regulators will operate with 90 percent efficiency and therefore require approximately 43 W from the AC power supply.

Taking a distributed approach means that, if the system specifications ever change to require different voltages, the proven AC supply is unaffected as long as the overall power level remains constant and the DC regulators can be adjusted as necessary. In many cases, adjusting the DC regulators is as simple as changing the value of a voltage-sensing feedback resistor.

A separate AC power supply module is often connected to the main circuit board with a wiring harness. In this case, wire must be chosen to safely carry 3.6 A at 12 V. Twenty-four gauge wire is adequate for the task and a more conservative choice is 22 gauge. The best results for distributing the regulated 5-, 3.3-, and 2.5-V supplies are achieved by using complete PCB planes or very wide copper paths to connect the regulator outputs to the various ICs and components that they serve. Distributing these final logic-level power supplies segues directly into electrical integrity issues.

17.8 ELECTRICAL INTEGRITY

An ideal power distribution system delivers a noise-free voltage supply at any current level required by the load. It performs this task regardless of how quickly the load's current demand changes and how much ambient noise is present. Electrical integrity is a measure of how close to the ideal a power distribution system really is. Factors that influence electrical integrity include the quality of the power conductors, the ability of the regulator to sense and respond to load variation, and the level of noise attenuation designed into the system. Electrical integrity should be considered only after the power distribution conductors have been adequately sized to carry the DC load with minimal voltage drop.

Assuming low enough resistance to adequately carry the current load, the quality of power conductors in the context of electrical integrity is a function of inductance and capacitance. Power is a signal like any other and is subject to degradation from noise created by components in the system, especially at high frequencies. Digital logic is infamous for creating high-amplitude and high-frequency noise because of the rapid binary switching characteristic of synchronous logic circuits. A sudden current spike is generated when a logic gate switches and charges its output node to a new state. A good power distribution system has low inductance so that the impedance between the power source and the switching load is minimal at the high logic switching frequencies. It also has sufficient capacitance near the load to attenuate the noise created by the switching event. The capacitance can be thought of as a tiny battery that briefly supplies the excess switching current until the power source can respond to the event.

Impedance at high frequency is minimized by increasing the surface area of a conductor, because high-frequency energy tends to travel along the surface of conductors according to a phenomenon known as *skin effect*. At DC, the entire cross section of a conductor carries current in a uniform distribution. As the frequency increases, the current distribution becomes nonuniform and moves out toward the conductor's surfaces. The inductance of a power supply wire is not going to be significantly reduced by moving to a lower gauge wire, because the surface areas of, for example, 22 and 18 gauge wires are not very different. A good way to distribute power in a low-inductance manner is with a solid sheet of metal. Modern PCBs are built up from individual copper layers, and it becomes very practical and cost effective to dedicate multiple layers as low-inductance power distribution planes.

Capacitance is distributed across a circuit board's power conductors using a variety of different capacitors. A typical system has a relatively small quantity of low frequency *bypass* capacitors, also called *bulk* bypass capacitors. Bypass capacitors are also called *decoupling* capacitors. Both terms refer to capacitors that help reduce power supply noise in a system. These are often aluminum or tantalum electrolytic capacitors, because electrolytics can pack a large capacitance into a small package. Standard aluminum electrolytic capacitors have a limited life span, because they use a wet electrolyte that gradually dries over time. Increased operating temperature shortens their life span. Tantalums use a solid electrolyte and are considered more reliable. New types of aluminum electrolytic capacitors have also been developed with extended life and temperature ratings.

The downside of electrolytics is their poor high-frequency response. Bulk bypass capacitors are placed at regular intervals on a circuit board to effectively lower the impedance of the voltage source (regulator). Values for bulk capacitors range from 10 to 1,000 μF. The specific quantity and type used and how they are placed varies greatly among system implementations. Some ICs recommend specific bulk bypass values. An engineer often errs on the side of conservatism and sprinkles 100-μF electrolytic capacitors around a circuit board so that there is one per supply voltage every few inches.

A larger quantity of high-frequency bypass capacitors are placed as close to the power pins of ICs as possible. These are ceramic capacitors, and they should be sized according to the expected noise frequencies. Historically, an individual 0.1-μF ceramic capacitor has been placed at each power and ground pin pair of each IC. It is important to minimize the inductance between each bypass capacitor and each power pin, because the resulting impedance at high frequency will limit the capacitor's effectiveness. This goal has given rise to the standard rules of placing capacitors as close as possible to their ICs and using surface mount components with much lower lead inductances as compared to leaded capacitors. Whereas a leaded component can have inductance of many nanohenries, Table 17.2 lists approximate inductances near 1 nH for various sizes of 0.1-μF surface mount ceramic capacitors.* Capacitors, resistors, and inductors are available in standard sized surface mount packages that are designated by their approximate dimensions in mils. Common sizes are 1210, 1206, 0805, and 0603. Larger packages such as the 1810 exist for handling higher power levels. Smaller packages such as the 0402 are used when space is at an absolute premium, but handling such packages whose dimensions are comparable to grains of sand requires special equipment.

TABLE 17.3 Surface Mount Capacitor Lead Inductance

Package	Inductance (pH)
1210	980
1206	1,200
0805	1,050
0603	870

The impedance of a 0.1-μF surface mount capacitor reaches a minimum value of under 100 mΩ at around 10 MHz and remains below 1 Ω from approximately 1 MHz to 100 MHz. This explains why 0.1-μF capacitors have been popular as bypass capacitors for so long: many digital systems have switching frequencies below 100 MHz. Above 100 MHz, 0.01-μF capacitors in 0603 packages become attractive because of their lower impedance at higher frequencies. It gets harder to reduce very high-frequency noise, because the inductance of surface mount bypass capacitors declines only to a certain point.

Having discussed the basic issues of power distribution, attention can be turned back to the example in Fig. 17.16. Distributing the regulated 5-, 3.3-, and 2.5-V logic-level supplies in a manner that ensures high-frequency electrical integrity requires minimal impedance between the regulator and the load. The consideration goes beyond DC resistance to include inductance, which has a more substantial impact on the conductor's impedance at high frequency. As the regulator and load are separated by higher impedance, the regulator's ability to respond to fluctuations in load current is degraded. Low-inductance distribution is achieved by using complete planes or very wide copper paths to connect the regulator outputs to the various ICs and components that they serve.

Power plane design is directly related to the electrical integrity of other signals in the system and will be covered in more detail later. The ideal situation is to devote entire PCB layers to serve as power planes for each separate voltage. This eliminates power plane cuts that can cause other signal integrity problems. Unfortunately, not all systems can afford the cost or, in some cases, the physical size of many power planes. In a situation like this, multiple voltages must share the same PCB layer. Figure 17.17 shows a hypothetical single power plane structure for the preceding example that requires distribution of 12, 5, 3.3, and 2.5 V. The shaded regions represent continuous copper areas. A ground return plane is required but not shown, because it occupies a second layer and is continuous across all power plane regions. Each system has its own unique power distribution flow governed by the grouping of components that require different supply voltages. This example assumes the com-

* Jeffrey Cain, *Parasitic Inductance of Multilayer Ceramic Capacitors*, AVX Corporation, p. 3.

mon situation in which most of the system runs at 3.3 V, but the inner core logic of certain ICs (e.g., a microprocessor) runs at 2.5 V. The remainder of the system voltages are used for peripheral and I/O functions. In many cases, the components cannot be cleanly separated into groups that use only one or two voltages. These situations may demand multiple power planes.

Figure 17.17 also shows the placement of 47-μF bulk bypass capacitors in each power section. Their exact placement is arbitrary, but they should be distributed across the board. Not shown are the high-frequency bypass capacitors, because their placement is a direct function of where the ICs are located. There are also additional capacitors that would be located within the bounds of the voltage regulators, because each regulator has its own recommendations for input and output capacitors.

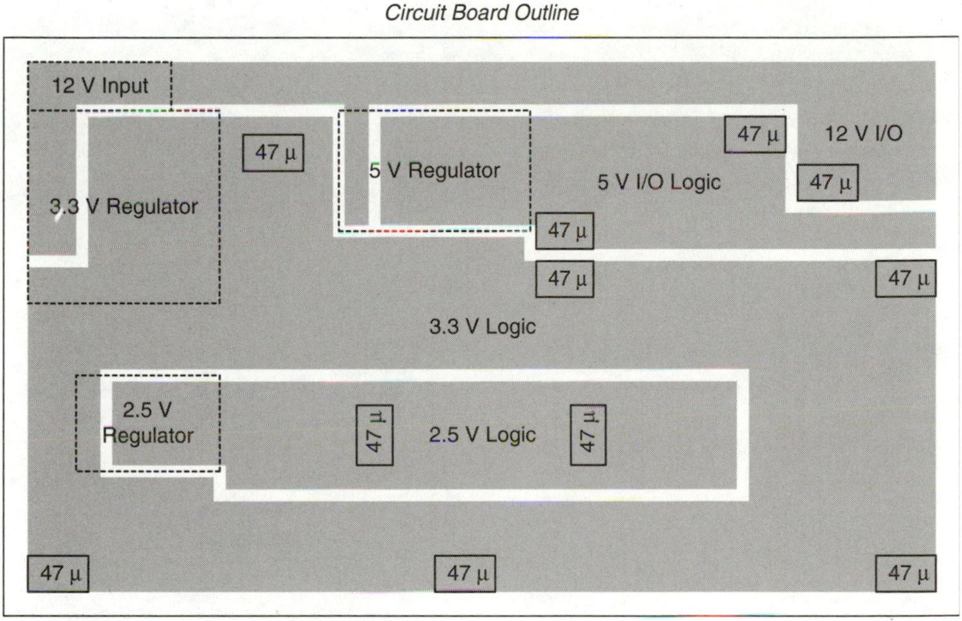

FIGURE 17.17 Multivoltage cut power plane.

CHAPTER 18
Signal Integrity

Getting high-speed digital signals to function properly is one of those areas that many people call *black magic.* It is so called, because fast digital signals behave like the analog signals that they really are, which is not an apparent mode of operation from a binary perspective. The typical story of woe is one in which a digital engineer continues to design faster circuits in the same way as slower circuits, and one day a system begins developing unexplained glitches and problems. From a digital perspective, nothing substantial has changed. On closer inspection with an oscilloscope, individual digital signals have mysterious transients and noise superimposed on them.

Signal integrity is the overall term for high-speed electrical design techniques that enable digital signals to function digitally in the face of physical phenomena that would otherwise cause problems. Many of the terms and techniques introduced in this chapter may sound familiar, because they have received increased scrutiny and coverage in the trade press and at conferences as a result of the steady increase in semiconductor operating frequencies. Signal integrity used to be a topic that many systems could ignore simply by virtue of their older technology and slower signals. That luxury has largely evaporated today, even for slow systems that unwittingly use ICs designed for high-speed operation.

A broad set of topics are discussed in this chapter with the goal of providing familiarity with signal integrity problems and general solutions to those problems. Transmission lines and termination are absolutely critical interrelated subjects, because they literally make the difference between working and nonworking systems. Transmission lines address head-on the reality that wires have finite propagation delay and are not ideal transparent conductors that ferry signals from point to point unchanged. From a purely functional perspective, proper transmission line analysis and design is the most important part of signal integrity, which is why these topics are presented first.

High-speed signals exist in a world of non-negligible electromagnetic fields that cause even small wires to act as antennas. These antennas are capable of both radiation and reception of noise. Crosstalk, electromagnetic interference, and electromagnetic compatibility are associated topics that hinge around the reality that electrical signals do not remain neatly confined to the wires on which they travel. The problems are twofold. First, excessive field coupling can cause a circuit to malfunction. Second, electronic products offered for sale in most countries of the world must comply with government regulations regarding their electromagnetic emissions. You don't want to bring home a new DVD player to find that it crashes your computer when you turn it on!

The chapter concludes with another related topic, electrostatic discharge. Static electricity is something that we are all familiar with, but its effects on a digital system are potentially disruptive and even destructive. Static electric discharges cannot be prevented in normal environments, but their effects can be reduced to the point of not causing problems.

18.1 TRANSMISSION LINES

Transmission lines, reflections, and impedance matching have been alluded to previously. The term *transmission line* can refer to any conductive path carrying a signal between two points, although its usual meaning is in the context of a conductive path whose length is significant relative to the signal's highest-frequency component. Circuits are normally drawn assuming ideal conductors whose lengths are negligible and assuming that the voltage at any instant in time is constant across the entire conductor. When a wire "becomes a transmission line," it means that it can no longer be considered ideal. An electrical signal propagates down a wire with finite velocity, which guarantees that a changing signal at one end will take a finite time to reach the other end. When a signal's rate of change is slow relative to the wire's delay, many nonideal characteristics can be ignored. Older digital circuits that ran at several megahertz with slow transition times were often not subject to transmission line effects, because the wire delay was short compared to the signal's rate of change.

A signal that changes rapidly forces one end of a transmission line to a significantly different voltage from other points along that conductor. At the instant this rapid change is produced by a driver, the signal has not yet reached the load at the end of the wire. Rather than observing current and voltage that are in proportion to the load impedance, they are in proportion to the characteristic impedance of the transmission line, commonly written as Z_O. Z_O is not a DC load; it represents the reactance developed by the conductors' inductive and capacitive characteristics. It is the impedance that would be observed between the two conductors of an infinitely long transmission line at nonzero frequency. When a high-frequency signal transitions before the driver sees the end load, it is as if the transmission line is infinitely long at that moment in time.

Transmission lines are composed of a signal path and a return path, each of which can be modeled using discrete lumped elements as shown in Fig. 18.1. The model shown is that of an unbalanced transmission line wherein all of the inductive and lossy properties are represented in one conductor. This is acceptable for many transmission lines in a digital system, because printed circuit boards commonly consist of etched wire conductors adjacent to ground planes that have negligible inductance and resistance. A balanced transmission line model, such as that representing a twisted pair cable, would show series inductance and resistance in both conductors. Analysis is simplified by assuming lossless conductors, which is often a suitable starting point in a digital system with moderate wire lengths. Using this simplification, the characteristic impedance is defined as $Z_O = \sqrt{L \div C}$.

Characteristic impedance is an important attribute, because it defines how a high-speed signal propagates down a transmission line. A signal's energy can fully transfer only between different transmission line segments that have equal Z_O. An impedance discontinuity results when two transmission lines are joined with differing Z_O. Impedance discontinuities result in some of a signal's energy being reflected back in the direction from which it arrived. This phenomenon is the crux of many signal integrity problems. An improperly terminated transmission line has the potential to cause reflections from each end of the wire so that the original signal is corrupted to the point of being rendered useless. A reflection coefficient, represented by the Greek letter gamma (Γ), that determines the fraction of the incident voltage that is reflected back from an impedance discontinuity is defined in the following equation:

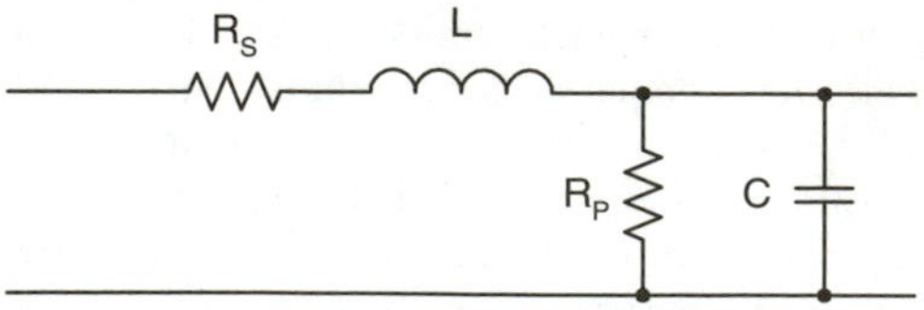

FIGURE 18.1 Lumped transmission line model.

$$\Gamma = \frac{Z_L - Z_O}{Z_L + Z_O}$$

Gamma is a dimensionless quantity that ranges from +1 to –1 and is a function of Z_O and the next segment's impedance, Z_L. It can be seen from this relationship that, when a transmission line is left open, $Z_L = \infty$ and $\Gamma = 1$: the entire voltage is reflected back to the source. This is the default situation for most digital signals, because a high-impedance logic input is effectively an open circuit from a transmission line perspective. Incidentally, a short-circuited transmission line, though not very desirable for digital signals, results in $\Gamma = -1$, because $Z_L = 0$, and causes the reflected signal to cancel the incident signal at the load. When a resistor equal to Z_O is placed at the end of a transmission line such that it connects the line's signal and return paths, $\Gamma = 0$, the line is said to be terminated, and no reflections are induced.

Transmission line reflections are perhaps best understood in a qualitative manner by looking at a time-domain view of a propagating signal. Figure 18.2 shows three scenarios that illustrate the benefits of termination and the effects of the reflection coefficient. The first scenario shows an edge injected into a 1 ns transmission line (about 6 in [0.15 m] long) that is improperly designed. Substantial bounce is present at the load. A high-speed clock is shown in the second scenario injected into the same transmission line. Note the bounce at both logic levels that can potentially create detection problems at the load. Poorly designed transmission lines can exhibit a wide variety of troublesome transient phenomena. Some may be better or worse than what is shown here. The third scenario shows the same clock injected into the same transmission line but with proper termination. The signal integrity is about as good as can be expected in a real circuit. There will always be small transients, even when a transmission line is properly matched, because of nonideal characteristics including stray inductance and capacitance in wires and termination components.

The need to minimize impedance discontinuities in modern printed circuit boards has given rise to the concept of controlled-impedance PCB design and manufacture. When a signal path and return path are arranged in various topologies, Z_O of the resultant transmission line is deterministic to an accuracy defined by the tolerance of the dimensions and electrical properties of the PCB materials. If all traces on a PCB are manufactured with similar Z_O, signals are subject to low reflection coefficients as they travel throughout the board. Three of the common PCB transmission line topologies are microstrip, symmetric stripline, and asymmetric stripline as illustrated in Fig. 18.3. These topologies assume one or two continuous ground planes adjacent to each signal layer. A continuous ground plane offers low sheet inductance with excellent high-frequency characteristics. Microstrip transmission lines are fabricated on the surface layers of PCBs where there is a single ground plane underneath the surface signal layer. A symmetrical stripline is evenly suspended between the two ground planes, while an asymmetric stripline has unequal spacings. Dual asymmetric striplines are implemented to achieve higher wiring density in a PCB versus symmetric striplines because of the higher ratio of signal to ground plane layers. Dual striplines have the potential for interference between adjacent signal layers, requiring more careful layout than with single striplines.

Z_O is a function of the trace geometry (width and thickness) and the relative permittivity and height of the dielectric, or insulator, that separates the ground planes from the traces. Relative permittivity, ε_r, also called the *dielectric constant,* quantifies the effect of an insulating material on the capacitance between two conductors relative to free space. The relative permittivity of free space is 1 by definition. Common insulators have ε_r ranging from 2 to 10. For microstrip and stripline topologies, Z_O is defined in the following equations:*

* *IPC-2141—Controlled Impedance Circuit Boards and High-Speed Logic Design,* Institute for Interconnecting and Packaging Electronic Circuits, 1996, pp. 11–14.

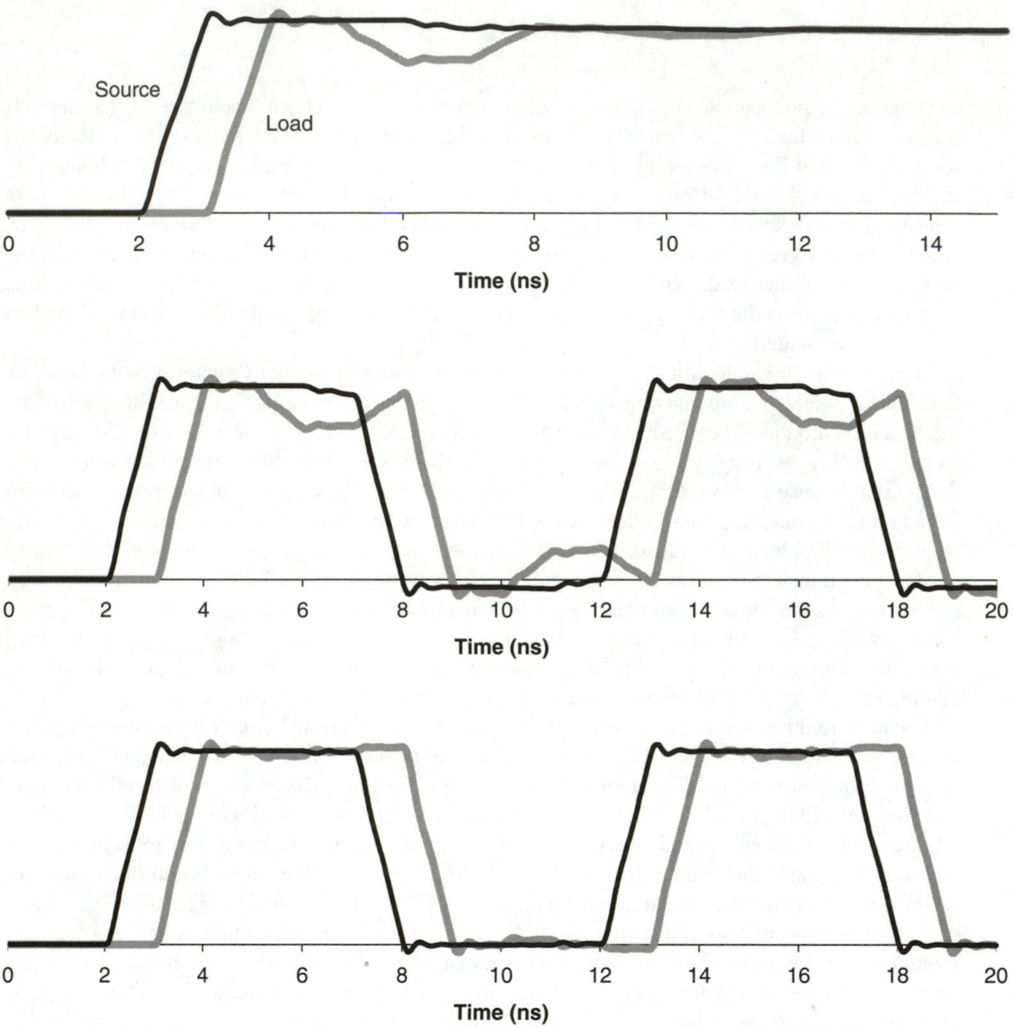

FIGURE 18.2 Transmission line reflections.

$$\text{Microstrip: } Z_O = \frac{87}{\sqrt{\varepsilon_r + 1.41}} \ln\left[\frac{5.98H}{0.8W + T}\right]$$

$$\text{Symmetric stripline: } Z_O = \frac{60}{\sqrt{\varepsilon_r}} \ln\left[\frac{1.9(2H + T)}{0.8W + T}\right]$$

$$\text{Asymmetric stripline: } Z_O = \frac{80}{\sqrt{\varepsilon_r}} \ln\left[\frac{1.9(2H_1 + T)}{0.8W + T}\right]\left[1 - \frac{H_1}{4H_2}\right]$$

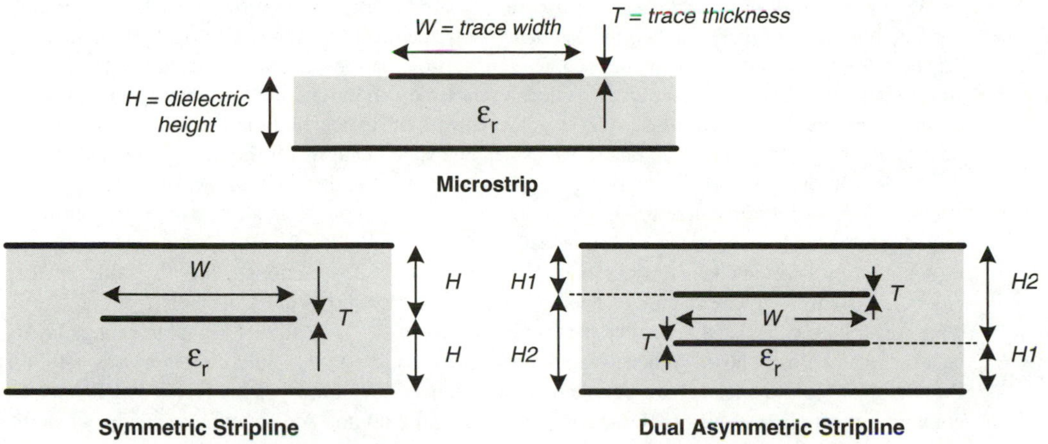

FIGURE 18.3 Microstrip and stripline transmission line topologies.

A type of fiberglass known as FR-4 is a common dielectric found in PCBs. Its relative permittivity is approximately 4.7. Half-ounce copper foil is typically used for signal traces and has an approximate thickness of 0.65 mils. Most PCBs are designed with Z_O between 50 and 75 Ω. These basic constraints lead to typical trace widths of between 4 and 10 mils and typical dielectric heights of 4 mils and higher. Dimensions outside of these ranges are perfectly acceptable and are justified by the requirements of each application. Choosing a trace width is usually based on the packaging technologies and component densities being used. Higher densities usually require finer traces to wire the circuits.

Actual PCB geometries are selected in concert with PCB vendors based on available materials and dimensional requirements. Many circuit boards must conform to a standard thickness so that they can plug into a system such as a PC. A common PCB thickness is 62 mils, roughly 1/16th of an inch. Figure 18.4 shows a sample stack-up using FR-4 and 0.5-oz copper that might be provided by a PCB vendor. This PCB has six layers configured as two power planes and four signal layers with $Z_O = 50\ \Omega$. Each signal layer uses 8-mil traces. The top and bottom signal layers are microstrips, and the internal signal layers are asymmetric striplines. The PCB vendor does not have much flexibility in assigning the dielectric thickness on the outer layers. Inner layers provide significant flexibility, because the asymmetric heights can be traded-off against each other to achieve the desired imped-

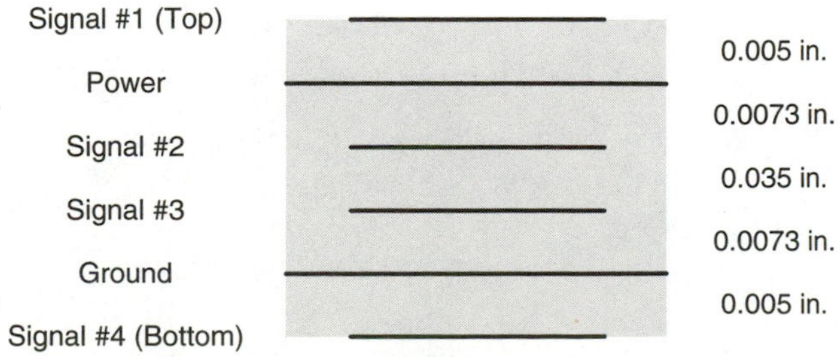

FIGURE 18.4 Sample six-layer PCB stack-up with 0.008-in traces and $Z_O = 50\ \Omega$.

ance with an overall PCB thickness. In this case, a desired thickness of 0.062 in is achieved by finding a combination of dielectric heights that match both the mechanical and electrical requirements.

The materials specifications mentioned here are approximations and vary between PCB vendors. Before beginning a PCB layout, always select a reputable vendor and confirm the layer stack-up to be sure that your desired impedance can be achieved with the expected dimensions. A vendor should be capable of quoting a stack-up that provides detailed specifications for the trace width and dielectric heights so that your layout can be executed knowing that the design goals are realistic. After accounting for various manufacturing variances, the vendor should be able to deliver controlled impedance traces to an accuracy of ±10 percent or better.

A signal must often change layers as it travels, because of obstructions caused by other devices and other groups of traces that are competing for their journey across the same area of circuit board. Vias are small metal-plated holes that are fabricated into the PCB to allow a signal to change layers as it is routed across a board. When a signal moves between layers, its return current moves with it if the two signal layers do not share a common return plane. For example, signal layers 3 and 4 share a common return plane in Fig. 18.4, but signal layers 1 and 4 do not. A discontinuity results when the return current is not presented with a low-impedance path to follow the signal's layer change. Discontinuities may result in nonideal transmission line behavior, such as reflections and increased noise radiation and susceptibility. Such discontinuities will be mentioned again later in the context of grounding and electromagnetic interference.

High-frequency signals in the gigahertz range can also be adversely affected as traces change direction. It is necessary for a trace to change direction so that it can route a signal to its load. However, the manner in which a corner is created has an effect on the transmission line's characteristic impedance. Recall that Z_O is a function of the trace's capacitance and inductance. A straight trace has approximately constant cross section throughout its length, resulting in constant Z_O. If a trace turns a 90° corner, its cross section changes as shown in Fig. 18.5. Changing cross section, or width, results in different capacitance and inductance. The ideal situation is for a trace to smoothly curve around corners to maintain constant cross section. Although this is usually done for the most demanding very high-speed circuits, it is uncommon for normal digital signals because of the burden that smooth curves place on conventional PCB design software. Imperfections due to corners are negligible for typical digital signals into the hundreds of megahertz. An alternative to a smooth curve is implementing a 45° compromise to provide less of a disruption as compared to a 90° turn, but not quite the cleanliness of a true curve.

Selecting the characteristic impedance for a PCB design is usually done according to industry conventions, because certain connectors and cables are specifically designed with 50- or 75-Ω impedances. Many coaxial cables, particularly in video applications, are 75 Ω, and it is convenient to minimize impedance discontinuities by matching the PCB to the cable. Most high-speed digital cir-

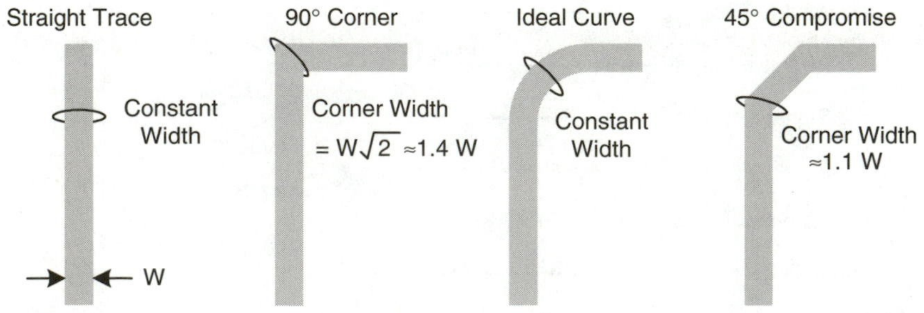

FIGURE 18.5 Effect of corners on trace impedance.

cuits use 50-Ω transmission lines, and some connectors are available with this impedance as well. There are advantages and disadvantages in using higher- or lower-impedance transmission lines. A lower-impedance transmission line requires a thinner dielectric between the signal and return paths. This smaller gap makes the transmission line less susceptible to radiating and coupling noise. It also allows multilayer circuit boards to be made thinner. The disadvantage of lower-impedance transmission lines is that they require higher drive current. Most ICs are capable of driving 50-Ω transmission lines, and high-speed design makes the trade-off of improved noise immunity worth the added power consumption. However, slower circuits may be more suited to 75-Ω transmission lines for power savings.

18.2 TERMINATION

In applying transmission line theory to a digital signal, one should first determine whether the combination of signal transition speed and wire length combine to merit transmission line analysis. Digital signals are characterized not only by their repetitive frequency but also by the frequency components in their edges. A signal with a rapid transition time will have high-frequency components regardless of how long the repetition period. Rules of thumb vary, but a common reference point is to treat a wire as a transmission line if the signal's rise time is less than four times the propagation delay of the wire. The common definition of rise time is the time that the signal takes to transition from 10 to 90 percent of its full amplitude. If the signal transitions slowly with respect to the wire delay, it can be assumed that all points on the wire transition together as the applied signal transitions. When a signal transitions quickly and the wire is long relative to this time, the wire should be treated as a transmission line. It is now common to see signals that transition in 1 ns, meaning that wires with propagation delays greater than 250 ps (about 1.5 in, or 38 mm) require transmission line analysis!

There are several basic termination schemes and numerous variant schemes that have been designed for special digital signaling standards. From our previous discussion of the reflection coefficient, the first obvious termination scheme is to place a resistor equal to Z_O at the end of a transmission line as shown in Fig. 18.6. This is known as *parallel termination*. Note that a transmission line is identified using a graphical representation of a coaxial cable. This does not mean that a transmission line must be implemented as such, but it conveys the idea that the wire length is non-negligible and cannot be treated as a single node of constant voltage as other wires are. A ground is explicitly connected to the transmission line return path to clearly indicate the circuit's return path. Assuming ideal conditions where Z_O is exactly known, $R = Z_O$, the output impedance of the driver is 0 Ω, and there is no stray capacitance or inductance at either end of the transmission line, parallel termination yields $\Gamma = 0$, and no reflections are present. Of course, these ideal conditions are never achieved, and real terminated transmission lines have some degree of imperfection. The goal is not

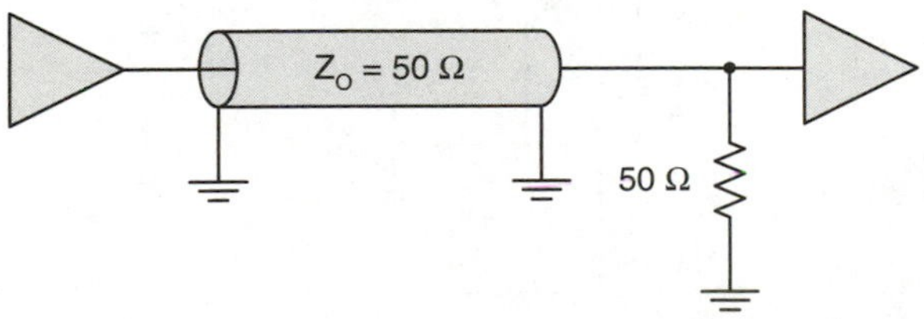

FIGURE 18.6 Parallel termination.

to accomplish the impossible but to achieve a transmission line topology in which the reflections are small enough to not degrade the load's valid detection of logic-1 and logic-0 states.

Parallel termination resistors should always be placed as close to the end of a transmission line as possible to minimize stub lengths between the terminator and the IC pin. Stubs appear as transmission lines of their own and can cause more reflections if not kept short. It can be difficult to squeeze termination components close to IC pins, but efforts should be made to achieve the best practical results.

Terminating a line at one end as shown in Fig. 18.6 is proper for a unidirectional signal, because the driver launches a signal into one end of the transmission line, and the termination is placed at the load end to prevent reflections. When both ends of the line are driven, as is the case with bidirectional buses, both ends require termination. The resistor at the driver end appears as a normal DC load, and the resistor at the far end serves as a terminator.

Situations commonly arise in which a bus has more than one load. A microprocessor bus must typically connect to several memory and peripheral ICs. The transmission line topology must be laid out carefully to minimize the potential for harmful reflections. The best scenario is to create a single, continuous transmission line terminated at each end that snakes through the circuit board and contacts each IC so that the stubs to each IC are of negligible length as shown in Fig. 18.7. When an IC at either end drives the bus, it drives a single transmission line that is terminated at the other end. When an IC in the middle drives the bus, it drives two equivalent transmission lines that are terminated at their ends. Graphically, the single transmission line can be drawn as shown using multiple segments connected by nodes that indicate tap points for individual ICs. Because nodes are drawn with the assumption of negligible length and constant voltage, they are conceptually transparent to the transmission line segments on each side. Keep in mind that not all buses require a perfect transmission line topology. Depending on their wire lengths and the switching times of the drivers, the wires may be regarded as idealized and not require special handling.

Many nonideal topologies exist in which no attempt has been made to shorten stubs and there is really no identifiable transmission line "backbone." Instead, the wiring is fairly random. A topology like this works either by virtue of the fact that the signal transition times are slow enough to not

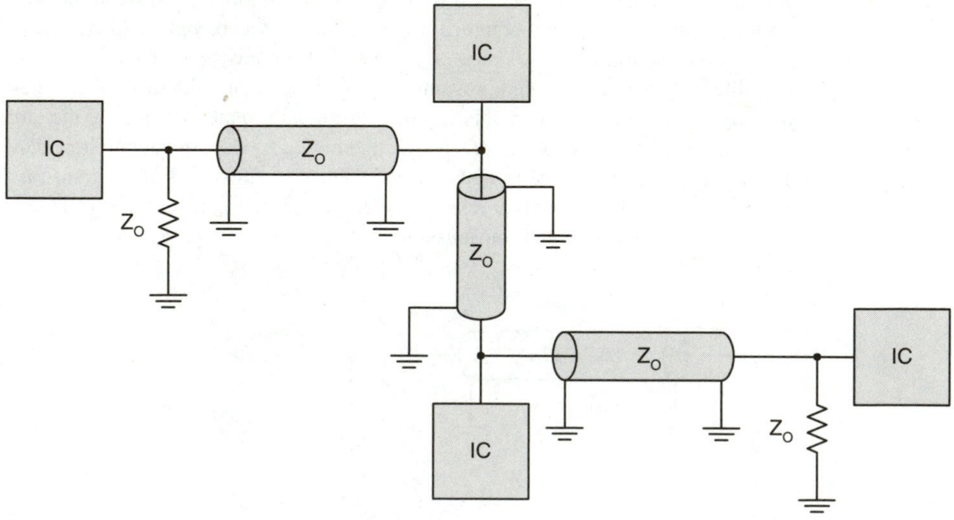

FIGURE 18.7 Multidrop transmission line topology.

make transmission line effects significant, or by plain luck. If the former, the circuit is perfectly valid, because there is no reason to cause wiring headaches if a wire does not have to be regarded as a transmission line. If the latter, luck can turn at any time with slight variations in components. This often explains why one or two units work in the lab, but a small manufacturing run has a high percentage of failures.

Parallel termination has the unfortunate consequence of heavily loading a driver because it is, after all, a small resistance connected directly across the signal and return paths. The driver sees a load of Z_O when it begins to switch its output but, once the line has stabilized, it still sees a load of $R = Z_O$. This is a substantial load with a 50- or 75-Ω transmission line. A 3.3-V driver would have to drive 66 mA into a 50-Ω load. If many bus signals need to be terminated, the power dissipation and stress on the driver ICs quickly mounts. This situation gets worse when terminating both ends of a bidirectional line.

There are several variants of parallel termination that seek to minimize the DC current drawn by the termination resistors. *Thevenin*, or *split termination,* operates using two resistors, each of which is chosen to be twice Z_O as shown in Fig. 18.8. One resistor terminates to ground and the other to the positive voltage rail. From an AC perspective, the positive voltage rail acts as a ground, and the parallel resistor combination appears as a single resistor equivalent of Z_O connected to ground. The benefit of split termination is that the maximum DC current drawn from the driver is halved, because a 100-Ω load is observed to the opposing voltage rail. When the driver drives to V_{DD}, the 100-Ω resistor to ground is the only load with a DC voltage drop across it. Split termination's disadvantage is that it always dissipates power, because the resistors establish a DC path between V_{DD} and ground.

The Thevenin concept can be carried further by employing a dedicated termination voltage rail, V_{TT}, that equals $V_{DD} \div 2$ and terminating to this rail using a resistor equal to Z_O as shown in Fig. 18.9. This scheme provides equivalent termination with less power dissipation, because the maximum voltage drop is $V_{DD} \div 2$ instead of V_{DD}. An additional benefit is that the terminator only dissipates power when the line is active. Bidirectional buses that are released to high-impedance states during idle will not load the termination resistor during that idle time.

Another variant of parallel termination uses a capacitor in series with the resistor to reduce power dissipation and is shown in Fig. 18.10. *AC termination* is suitable for clock signals or other DC balanced signals (e.g., 8B10B encoded signals), because such signals allow the capacitor to charge to $V_{DD} \div 2$ with the resultant benefits of reduced power dissipation as seen with $V_{DD} \div 2$ termination. Unbalanced signals remain at a static voltage for too long and charge the capacitor to either voltage rail. When a transition finally occurs, the full V_{DD} voltage drop appears across the terminator in a conceptually identical arrangement to basic parallel termination. The capacitor value must be chosen so that it maintains a nearly constant voltage during the balanced signal's period. A common 0.1-μF

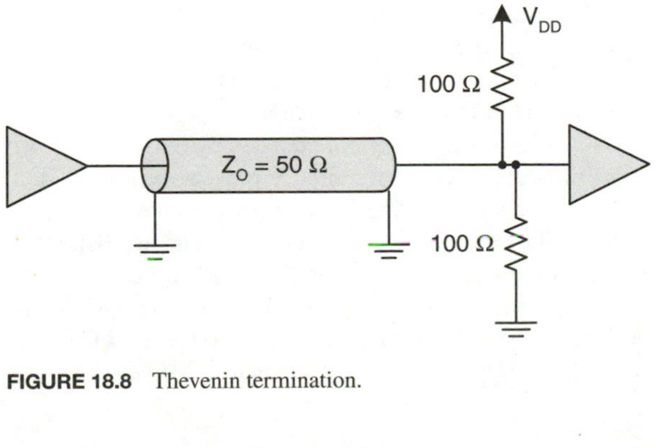

FIGURE 18.8 Thevenin termination.

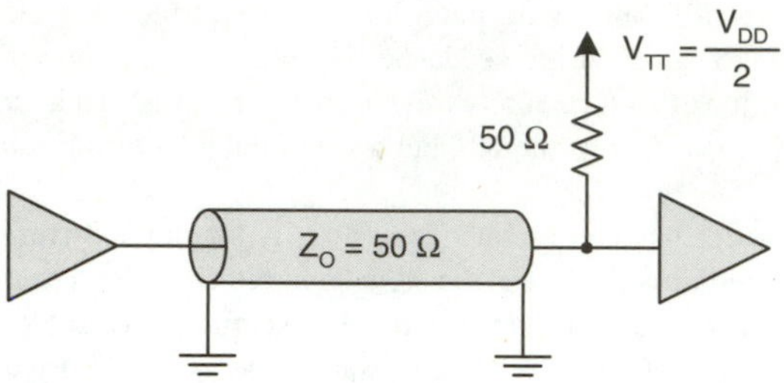

FIGURE 18.9 $V_{DD} \div 2$ termination.

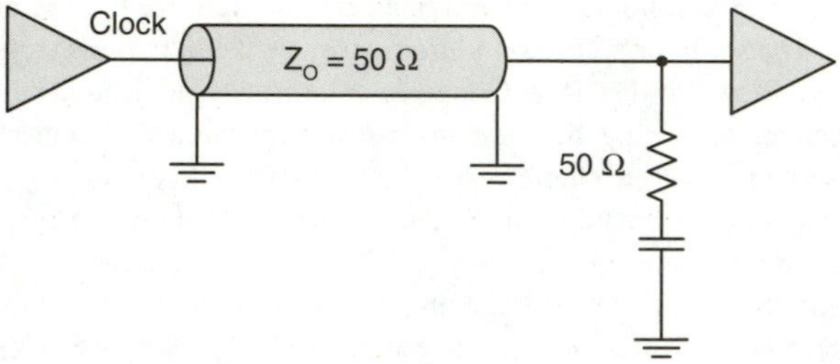

FIGURE 18.10 AC termination.

capacitor combines with a 50-Ω resistor to yield a time constant of 5 μs, which is suitable for most applications.

Some engineers do apply AC termination to normal unbalanced signals such as a data bus. The idea is to save power by presenting the termination resistor only during the high-frequency signal transition events and not during the longer static time. Smaller capacitors, perhaps 100 pF, are required in this case to provide low impedance at the high frequency edge but a low RC time constant so that it charges to the driven DC value quickly. Problems with this approach include added time for the wire to stabilize as a result of ringing and difficulty in matching the capacitor to the variation in driver's edge rate. AC termination may be the best solution for certain weak drivers with multiple loads, but some trial-and-error adjustment of the capacitor may be required in the lab.

Parallel termination can reduce power dissipation only to a point, because it is a shunt load to a DC voltage rail. Multidrop bus topologies require parallel termination, which prevents reflections from being formed at the transmission line ends. High-speed digital systems often employ point-to-point buses for reasons including source-synchronous clocking and transmission line stub reduction. Point-to-point buses also provide an advantage in not requiring parallel termination. *Series termination* can be used in these situations with its benefit of zero DC power dissipation. This technique was already briefly discussed in the context of clock distribution.

Series termination, also called *source termination*, operates by purposely creating a reflection at the load and then terminating that reflection at the source to prevent its re-reflection back to the load. A unidirectional point-to-point transmission line with series termination is shown in Fig. 18.11. When the driver first transitions, the transmission line presents its characteristic impedance as a load. In combination with the termination resistor, a voltage divider is formed, and only half the voltage propagates down the wire. When this half amplitude signal reaches the transmission line's unterminated end, a reflection coefficient of one causes the incident and reflected voltages to add with the result being the original full-amplitude signal. The receiving IC observes the full-amplitude signal.

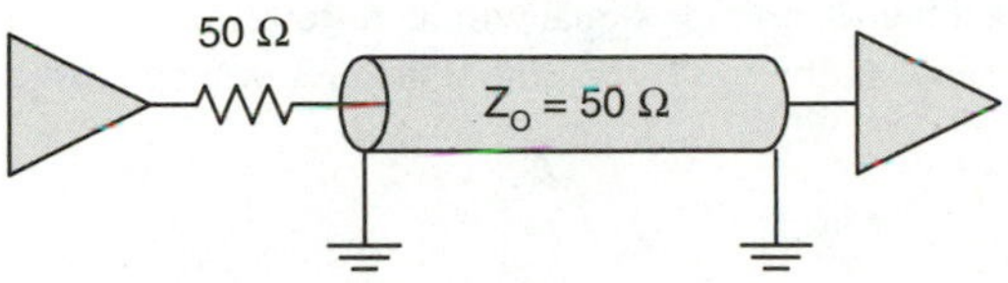

FIGURE 18.11 Series termination.

The reflected signal propagates back to the driver. This time, the transmission line end is terminated by the resistor connected to a power rail via the driver circuit itself, and the reflected energy is absorbed. There is no DC power dissipation with series termination, because the terminating resistor does not shunt the transmission line to a separate DC potential.

The reflection intentionally created in a series-terminated transmission line makes this scheme nonideal for high-speed multidrop buses, because it takes two round-trip times for the entire transmission line to stabilize. A 12"-in (0.3 m) bus would require approximately 4 ns for the transmission line to settle, which is a substantial fraction of the timing budget at speeds over 100 MHz.

Bidirectional point-to-point transmission lines can use series termination as well, with good results. Figure 18.12 shows a transmission line with series termination at each end. The mode of operation is the same as explained previously. When component A is driving, R1 serves as the series termination, and the signal propagates toward R2. R2 connects to the high-impedance input circuit at component B, effectively nullifying the presence of that resistor. A reflection is developed at the R2 end of the transmission line and is absorbed when it returns to the R1. Some delay and lowpass filtering of the signal may result because of the RC time constant formed by R2 and any stray capacitance at component B's input node. If the stray capacitance is up to 10 pF, the time constant is up to 500 ps—small, but non-negligible for very high-speed circuits.

Selecting the perfect series termination resistor is an elusive task, because it is difficult to characterize a driver circuit's actual output impedance. This finite impedance combines with the series resistor to yield the total termination impedance seen by the signal reflecting back from the load. A driver circuit's output impedance varies significantly with temperature, part-to-part variation, supply voltage, and the logic state that it is driving. It is therefore unrealistic to expect perfect series termination across time and multiple units manufactured. Some devices that are specifically designed for point-to-point transmission line topologies (e.g., certain low-skew clock buffers) contain internal series termination circuits that are designed to complement the driver's output impedance. In the remaining cases, standard resistance values are chosen with the understanding that an imperfect termination will result. A typical value of 39, 43, or 47 Ω can be chosen for an initial prototype build when using 50-Ω transmission lines, and the signal integrity can be evaluated in the laboratory. Switching to one of the other values may improve signal integrity by reducing the reflection coefficient.

Practically speaking, there is a common range for a driver's output impedance. If a 39 Ω termination resistor is chosen with the expectation that Z_{OUT} = 10 Ω, and the actual impedance under certain conditions is 3 Ω, the total termination will be just 42 Ω. For a 50 Ω transmission line, $|\Gamma| = 0.087$,

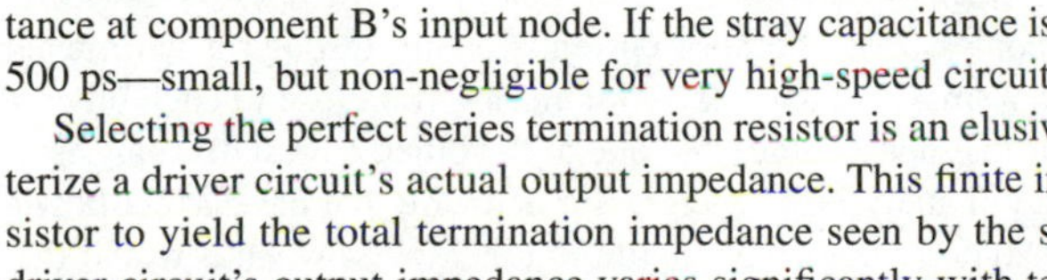

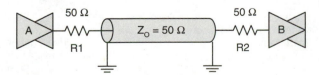

FIGURE 18.12 Bidirectional series termination.

and so 8.7 percent of the reflected half-amplitude signal will be re-reflected. This may not be a sufficient amplitude disturbance to cause problems. If it is, the transmission line must be given time for the reflections to diminish.

18.3 CROSSTALK

Initial transmission line analysis is typically performed with assumptions of ideal circumstances, including the assumption that the transmission line is independent of others. In reality, a wire acts as an antenna and is a radiator and receiver of electromagnetic fields. When two nearby wires couple energy between each other, the phenomenon is called *crosstalk* and is another source of signal integrity problems. Crosstalk is not always a problem, but the potential exists, and therefore circuit design and layout should be performed with its consideration in mind.

Energy can be coupled between nearby conductors either capacitively or inductively. High-frequency energy can pass through a capacitor, and a small capacitor is formed when two conductors are in proximity to one another. The capacitance between two wires is a function of their surface area and their spacing. When two wires are run parallel to one another on the same layer of a printed circuit board, their mutually facing surface area is relatively small. A dual stripline configuration, however, can present greater capacitive coupling problems, because wire traces may run parallel one on top of the other with significant surface area. A common PCB routing rule is to route adjacent dual stripline layers orthogonally whenever possible rather than parallel to each other as shown in Fig. 18.13. Minimizing the surface area of a wire that is in close proximity to the other wire reduces capacitive coupling.

Inductive coupling comes about because current flowing through a wire generates a magnetic field. Each wire is a very small inductor. If two wires are run close to each other, the two small inductors can couple their magnetic fields from one to the other. Crosstalk analysis uses the terms *aggressor* and *victim* to aid in analysis. The aggressor is a wire that has current flowing through it and is radiating an electromagnetic field. The victim is a nearby wire onto which the electromagnetic field couples unwanted energy. Because the intensity of the magnetic field is proportional to the current flowing through a wire, heavier loads will result in more coupling between an aggressor and nearby victims. Most crosstalk problems in a digital system are the result of magnetic fields, because of the high currents resulting from low-impedance drivers and fast edge rates.

Separation is an effective defense against crosstalk, because electromagnetic field coupling decreases with the square of distance. Doubling the separation between two wires reduces the coupling at the victim by 75 percent. Dielectric height in a PCB is another contributing factor, because the field intensity increases with the square of height between the aggressor trace and the ground plane. The dielectric ranges in thickness according to the desired characteristic impedance and width of the

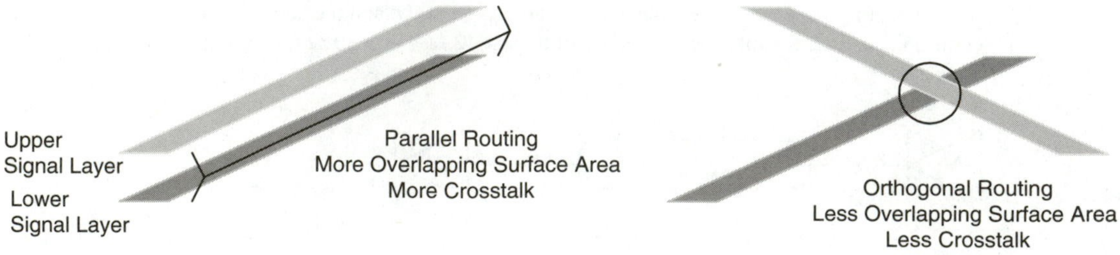

FIGURE 18.13 Dual stripline coupling reduction.

copper traces. As the dielectric gets thinner, the capacitance between the trace and ground plane increases, thereby decreasing Z_O. The trace's inductance must be increased to compensate for this decreased capacitance, indicating that a trace must be narrower to achieve 75 Ω versus 50 Ω with a certain dielectric height. Put another way, a 50-Ω transmission line has a lower dielectric height for a given trace width and, hence, weaker fields to cause crosstalk.

Crosstalk is a three-dimensional problem, and length is the third variable in coupling energy onto a victim trace. Traces that run parallel to each other are parallel antennas. Longer antennas can couple more energy. Therefore, many high-speed PCB designs enforce maximum parallelism rules that define the maximum distance that is allowable for two traces to run parallel to each other. Parallelism rules are related to separation rules. Traces that are spaced farther apart can tolerate longer parallel runs, because the separation reduces the aggressor's radiated field strength.

It is difficult to make generalizations about separation and parallelism rules, because every situation is unique. Furthermore, the formulas that define field strength and coupling coefficients are complex functions that incorporate dimensional and material parameters. A detailed crosstalk analysis requires the use of specialized software that simulates the coupling phenomena using detailed mathematical models of the driver and load circuits, the circuit board materials, and the three-dimensional arrangement of multiple traces. Such programs are known as field solvers and are available from a variety of vendors including Ansoft, Cadence, Innoveda, and Mentor Graphics.

When engineers do not have field-solving software at their disposal, very conservative design rules are often used to minimize the probability of excessive crosstalk. Sensitive signals are spaced apart from others. Parallelism is limited for signals whose spacing approaches the dielectric height where field coupling is greater. These rules are very coarse approximations, and application-specific rules need to be applied for very high-speed designs to avoid potentially thorny signal integrity problems in the prototype. If a great deal of time and money are invested in designing a leading-edge high-speed system, it is prudent to invest some time and money in appropriate signal integrity analysis tools.

One saving grace when dealing with crosstalk is that many signals do not have to be protected from each other if it can be shown that noise caused by crosstalk will not affect the received signal. Consider the case of a microprocessor bus's individual wires that are traveling close together between ICs. Inductive coupling, which is the dominant crosstalk contributor, occurs when current is flowing, because the magnetic field is proportional to current flow. Significant current flows during the switching time of the digital signal. Once the signal has stabilized, the load presents the driver with high impedance. This means that crosstalk between signals that transition together occurs mainly while they are switching and then quiets down after the lines have stabilized. A properly designed system does not sample signals until they have safely stabilized, at which point crosstalk within the group becomes negligible.

The wires that compose a single bus can usually be routed in close proximity to each other without regard for crosstalk within the bus. Clocks require special consideration, because one does not want a great deal of noise from switching bus wires to couple onto the clock wire and cause a false edge to appear. Likewise, the clock's transitions can potentially couple onto data wires and cause an incorrect observation at the receiver. Clocks are usually given preferential routing treatment aside from their low-skew distribution requirement. Clock traces are often routed before all others, and stricter minimum spacing rules may be used to minimize corruption of clock signals. Very conservative systems are designed by dedicating one or more entire PCB signal layers to clock routing so that the clocks can enjoy substantial trace-to-trace spacing minimums.

Although signals within a bus are usually not sensitive to mutual crosstalk, signals that transition asynchronously with respect to each other are prime candidates for trouble. Asynchronous signals, by definition, transition relative to each other without a defined timing relationship. If a system has two buses on different clock domains whose wires are mixed in close proximity with long parallel runs, it is probable that one bus will often transition during the assumed stable time of the other bus just as it is being sampled. Attention should be paid to the routing of asynchronous signals by separating them and minimizing parallelism.

18.4 ELECTROMAGNETIC INTERFERENCE

Electromagnetic interference (EMI) and crosstalk are closely related topics. It has already been seen that the quality and proximity of a ground plane has a significant impact on a transmission line and the fields that it radiates. Magnetic field strength is proportional to current and is also a function of the loop area defined by the conductors carrying that current. All circuits are loops, because current cannot flow unless a complete loop is formed from the source, through the load, and back to the source. Digital circuits are no exception to this rule, despite inputs typically having high impedance and low DC input currents. A high-speed system moves substantial currents around many individual loops formed by digital buses because of the rapid charging and discharging of transmission lines and input nodes. Even though each bit in a data bus may not be drawn with an explicit return path shown, that return path must exist for the circuit to function.

As the size of the loop formed by the signal's source and return paths increases, that wire radiates a magnetic field of greater strength and is more susceptible to picking up ambient magnetic fields. The general term for a circuit's sensitivity to coupling electromagnetic fields is *EMI susceptibility.* A circuit forms a loop antenna, and a loop antenna is more efficient with increasing area. Loop area and its effect on crosstalk have already been touched on in terms of dielectric height in a PCB. As the signal wire is brought closer to the ground plane, as it is in a lower-impedance design, the loop area and field strength decrease as illustrated in Fig. 18.14.

Minimizing loop area is a key reason why solid ground planes are necessary to high-speed signals. Consider the two scenarios, with and without ground planes, in Fig. 18.15. A ground plane allows a high-speed return current to flow through a path of least inductance that is directly under the signal trace. If the ground plane is replaced by a few discrete wires, those wires cannot lie directly under all signal paths. The return current is forced to take a path that is a greater physical distance from the signal path, with an associated increase in loop area and EMI susceptibility.

Vias along a set of PCB traces can inadvertently create larger loops. When a signal moves between PCB layers, its return current must follow. If a signal changes layers that are on opposite sides of the same ground plane, the return current is able to follow without interruption. Signals often move between layers that do not share the same return plane. In these cases, the return current finds the lowest-impedance path to the return plane that is adjacent to the new signal layer. This path may be direct, capacitive, or a combination of the two. If the new return plane is at the same DC level as the current plane (e.g., two different ground planes), the path of least impedance may be directly through a nearby via that connects the planes. If the new return plane is at a different DC level (e.g., a power plane serves as an AC ground), the return current capacitively couples between planes. Even if the planes share the same DC level, capacitive coupling may be the path of least impedance.

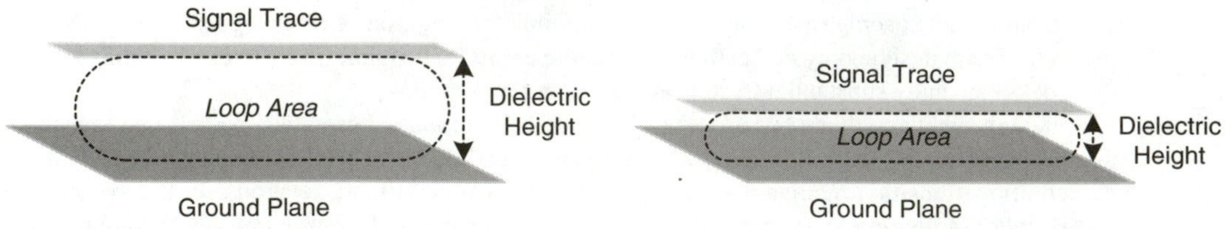

FIGURE 18.14 Loop area vs. PCB dielectric height.

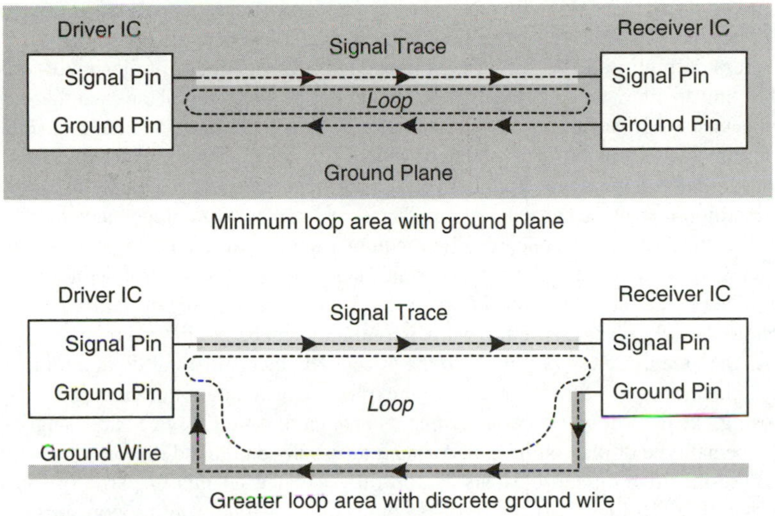

FIGURE 18.15 Greater loop circuit area without ground plane.

Capacitive coupling can occur through bypass capacitors that connect the two return planes and through interplane capacitance that is a function of the PCB dimensions and materials. Low-inductance capacitors are critical to provide adequate high-frequency coupling between planes. Impedance is a product of inductance and frequency. A bypass capacitor's total circuit inductance is the sum of the capacitor's inherent inductance and additional inductance caused by PCB traces and vias that connect it to the return planes. Typical surface mount capacitors in 0603 or similar packages can exhibit total inductance of approximately 2 to 3 nH. This inductance, combined with typical capacitance values of 0.1 or 0.01 μF, allow an impedance calculation at a given frequency. When multiple bypass capacitors are in close proximity to a via, they form a parallel combination with lower total inductance and higher total capacitance—both of which are desirable characteristics. The closer the bypass capacitors are to a via in question, the smaller the loop that is created for high-frequency return current between two planes. Nearby capacitors improve the high-frequency characteristics of the transmission line, whereas more distant capacitors increase the circuit's EMI susceptibility.

Discrete bypass capacitors are often the path of least impedance between return planes. As operating frequencies rise, however, the finite inductance of discrete components becomes more of a problem. A PCB may be constructed with planes separated by very thin dielectrics to provide significant interplane capacitance with negligible inductance. Capacitance increases with decreasing spacing between planes, and inductance decreases with greater surface area that a plane offers. High-frequency systems may require such construction techniques to function properly. These techniques can increase system cost by requiring more expensive thin dielectric materials and a greater number of PCB layers. Such costs are among many complexities involved in creating high-performance systems and must be considered when deciding on the practicality of a design.

The potential for via-induced EMI problems always exists. Conservative designs attempt to route sensitive and very high-speed signals with a minimum number of vias to reduce ground discontinuity problems. When vias are necessary, it is best to switch only between pairs of signal layers that are on opposite sides of the same ground plane.

Return path discontinuities may also be caused by breaks in power and ground planes. A PCB may contain multiple DC voltages (e.g., 5 and 3.3 V) on the same power plane to save money. Different voltage regions are created by splitting the plane. Return path discontinuities may result if the split plane is used as an AC ground for adjacent signal layers as shown in Fig. 18.16. As with a via, the return current finds the path of least impedance. An isolated split plane will force the return current to find an alternate path. Ideally, this should not be done, and planes should be continuous. Most engineers strive to never route a trace across a plane split so that potential signal integrity problems are minimized. If a plane does need to be split, it can be isolated from adjacent signal layers with additional solid ground planes. Alternatively, an adjacent plane spaced very close to the split plane can provide high interplane capacitance and therefore serve as a low-impedance path for the return current. Practical economic concerns often require engineers to employ nonideal approaches and still deliver a working system. An isolated split plane requires more careful trace layout to absolutely minimize the number of signals that cross the break. When signals must cross a break on a layer adjacent to the split plane, the return path discontinuity can be minimized by placing bypass capacitors across the break in close proximity to the traces. The mechanism at work here is the same where a signal changes layers through a via. If an explicit return path is not created by a capacitor of your choosing, the current will necessarily find its own path, which may be surprisingly long.

A special type of plane split, called a *moat,* is usually employed to isolate a small area of the PCB for noise-sensitive circuitry. Moats can be fully isolated islands or partially connected islands as shown in Fig. 18.17. When the moat is completely separate, routing concerns exist as for a split plane. A partially connected moat eases the routing burden somewhat by allowing a small "drawbridge" across which signals can travel with an unbroken return path. The idea behind a partially connected island is that stray currents do not flow across the island, because a less inductive path exists around the moat. Any noise that does make it into the island is attenuated somewhat as it passes

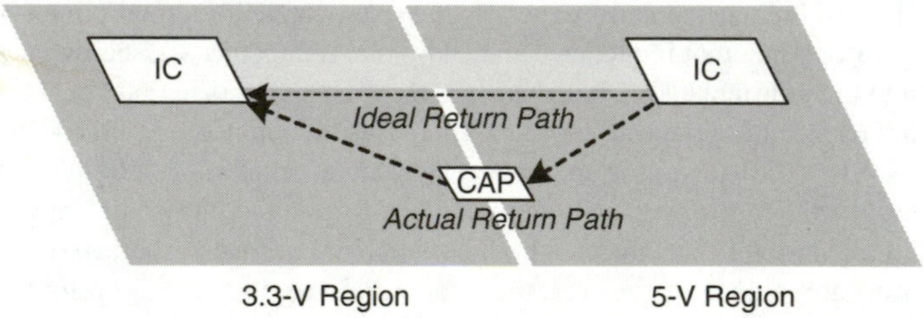

FIGURE 18.16 Split-plane return path discontinuity.

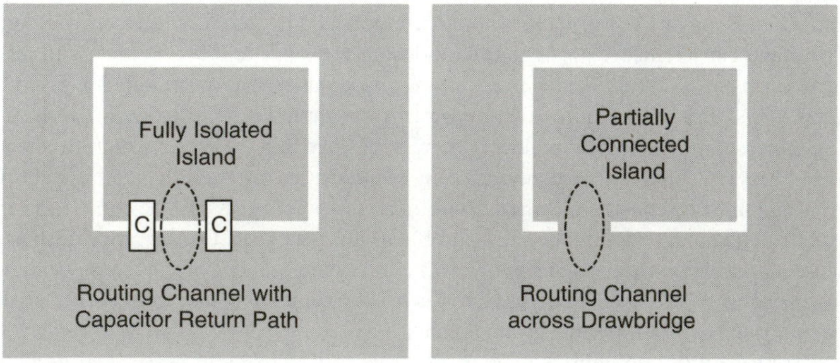

FIGURE 18.17 Moat routing.

through the thin drawbridge. A fully isolated island requires power and ground to be supplied through passive filters or some other connection.

Remember that a key goal in designing for signal integrity is to minimize loop area and return path discontinuities so that less energy is radiated from a wire when it is driven, and less energy is picked up as noise when other wires in the system are being driven.

18.5 GROUNDING AND ELECTROMAGNETIC COMPATIBILITY

By now it should be clear that grounding is a critical aspect of system design. Grounding becomes more important as speeds increase, because more intense electromagnetic fields are present, and higher frequencies radiate more efficiently from smaller antennas. *Electromagnetic compatibility* (EMC) is the ability of a system to peacefully coexist with other systems so that it neither malfunctions because of excessive EMI susceptibility nor causes other systems to malfunction as a result of excessive electromagnetic field radiation. In most situations, EMC means being a good neighbor and complying with governmental regulations on how much electromagnetic energy an electronic system can radiate. The Federal Communications Commission governs such regulations in the United States. Most digital systems applications are not particularly sensitive to ambient electromagnetic energy. The chances are pretty low that a computer will malfunction during normal use because of excessive ambient fields. Of course, this does not hold true in some demanding applications such as aerospace and military electronics.

Our discussion is concerned with basic techniques for reducing a system's radiated electromagnetic emissions in the context of complying with government regulations. EMI reduction through minimizing loop area and removing return path discontinuities is a fundamental starting point for EMC. Reasonable steps should be taken up front to minimize the energy that a circuit board radiates. If a sloppy design radiates significant energy, it may be difficult or impossible to effectively contain these fields to the point of regulatory compliance.

Electromagnetic energy can escape from a circuit board by radiating into space or conducting onto a cable. Radiated emissions can be blocked by enclosing the circuit board in a grounded metal enclosure. This is why many computers and other electronic equipment have metal chassis, even though the metal may be hidden under a plastic frame or bezel. Most metal enclosures are not perfect closed surfaces, because slots and holes are necessary for cables, switches, airflow, and so on. Enclosures also must be assembled and are often opened for service, so there are numerous seams, hinges, and joints that connect one sheet of metal to another. All openings in the metal are potential leakage points for radiation, depending on their size. A hole forms a slot antenna whose efficiency is a function of its size and the wavelength, λ, of energy being radiated. When engineers construct antennas, $\lambda \div 4$ and $\lambda \div 2$ are typical dimensions that radiate most efficiently. Clearly, slots and holes whose largest dimensions approach $\lambda \div 4$ are undesirable. Limiting chassis openings to be substantially smaller than $\lambda \div 4$, perhaps $\lambda \div 20$, is necessary.

Keep in mind that the frequencies that must be shielded are not just the highest system clock frequencies but also higher-order harmonics determined by the Fourier representation of a square wave. It is not uncommon to find energy violating emissions limits at the eleventh, thirteenth, or fifteenth harmonic of a digital clock. Therefore, it is best to make openings as small as possible. A rough starting point might be the assumption that, for a typical system, harmonics above 1 GHz will not be strong enough to cause problems. This may or may not be true, depending on the specific circumstance. The wavelength corresponding to 1 GHz in free space is 30 cm, and $\lambda \div 20 = 1.5$ cm, or about 0.6 in (1.5 cm). It is not as difficult as it may first appear to keep all openings smaller than that. Gaps for airflow are easily implemented using fine grilles formed in the sheet metal with holes far smaller than 0.6 in. Notice the grilles on computers, microwave ovens, and consumer electronics products.

The most troublesome gaps that cause EMC problems are improperly grounded connector shells and poorly fitted seams between metal surfaces. Many connectors are available with metal bodies or metal shields around plastic bodies. If these metal surfaces are securely mounted to the metal chassis at multiple points such that openings are substantially smaller than $\lambda \div 4$, there should be no gap at the connector to allow excessive radiation. Seams between individual sheets of metal, either movable or fixed panels, must be designed to meet cleanly without the slightest buckling. Imperceptible buckling at seams opens gaps that radiate unwanted energy. A well designed sheet metal chassis should have all of its fixed panels adequately riveted or welded to ensure uniform contact across the seam. Movable panels almost always require additional assistance in the form of conductive gaskets and springs. A gasket or spring serves as a flexible conductor that closes any electrical gaps between two metal surfaces that move over time and that may expand and contract with temperature changes. Gasketing is directly akin to the rubber washer in a sink faucet—minute gaps must be closed to prevent leakage.

Plastic enclosures can also be shielded by applying conductive coatings, although this is usually more expensive than a sheet metal chassis, which is a reason why many products use metal rather than all-plastic enclosures. Many small electronic products can get away with less expensive, more attractive, and lighter uncoated plastic packaging, because their circuits do not radiate excessive energy. This may be because of their relatively slow signals, careful circuit design, or a combination of both.

Unwanted high-frequency noise that may be conducted onto exterior cables should be filtered between the active circuitry and the cable connector. Passive differential and common-mode filters are discussed earlier in this book. High-frequency data interfaces often have standard means of dealing with noise, including application-specific off-the-shelf transformers and common-mode chokes. Lower-frequency interfaces such as RS-232 may be effectively filtered with a second-order LC filter using either a choke or ferrite bead for the inductive element. If noise is still able to couple onto internal wiring harnesses that lead outside the enclosure, the weapon of last resort may be a ferrite core. Ferrite cores are available in clamshell types whereby the ferrite fits around a cable, and in ring forms whereby the cable is wrapped several turns around the core. The ferrite increases the inductance of the cable, which increases its attenuation of high frequencies. When you see a computer monitor cable or some other type of cable that has a noticeable round bulge near one end, a clamshell ferrite has been added, because the equipment was unable to pass emissions regulations without it.

Filters can also be employed to attenuate higher-order harmonics of digital signals as they are distributed on the circuit board to reduce the strength of ambient electromagnetic fields on the board and within the enclosure. Clock distribution can account for a substantial fraction of unwanted emissions, especially at higher-order harmonics that radiate through small metal gaps. One technique is to insert lowpass filters at clock buffer outputs to attenuate energy beyond the fifth harmonic. A square wave substantially retains its characteristics with only the first, third, and fifth harmonics present. Unfortunately, component variation, mainly in capacitors, across the individual filters on a low-skew clock tree can introduce unwanted skew at the loads. Instead of an LC or RC filter at the source, inserting just a ferrite bead may provide sufficient high-frequency attenuation to substantially quiet a system. If it is unclear whether such filtering is necessary, the design can include ferrite beads as an option. Ferrite bead PCB footprints can be placed at each output of a clock driver in very close proximity to any series termination resistors that might already be in the design. If the ferrites are not needed, they can be substituted with 0-Ω resistors. Introduction of an extra 0-Ω resistor very close to the clock driver should not cause problems in most systems. For truly conservative situations, these scenarios can be modeled ahead of time with field-solver software.

The method by which a system's many ground nodes are connected has a major impact on EMC in terms of noise radiating from cables leaving the chassis. Conceptually, there is a single ground node that all circuits use as their reference. This is easy to achieve at DC, because resistance is the dominant characteristic that causes voltage drops, and solid sheets of metal have very low sheet resistances. Additionally, there are no EMC problems at DC, because there is no AC signal to radiate. Inductance becomes the problem in maintaining equipotential across an entire system's ground structure at high frequencies. Small voltage differences appear across a circuit board's ground plane despite its low sheet inductance. These differences can cause EMC problems despite having little to no effect on signal integrity. The ideal situation is to ground everything to the same point to achieve an equipotential ground node, but finite physical dimensions make this impossible.

Any opportunity for a cable to have a high-frequency potential difference with respect to the chassis is an opportunity for unwanted electromagnetic radiation. The basic idea in many systems is to take advantage of a chassis' sheet metal surfaces as a clean ground reference because of low inductance and negligible current circulation. If a circuit board is grounded to one face of the chassis, and all cables are grounded to that same face, the ground potentials in that region will be nearly equal, with less opportunity for radiated emissions.

A complete discussion of chassis grounding techniques for EMC design is beyond the scope of this presentation. If you anticipate having to pass governmental electromagnetic emissions requirements, further reading is recommended. Electronic products are tested and certified for regulatory compliance at licensed test ranges where it is also common to find EMC consultants to advise you on solutions to emissions problems. Like most design tasks, it is better to seek help before building a product than to wait until a problem arises, at which point it is usually more expensive and time consuming to resolve.

18.6 ELECTROSTATIC DISCHARGE

Electrostatic discharge (ESD) is another phenomenon related to EMC and grounding. Static electric discharges are common occurrences and have been experienced by everyone. An insulated object accumulates a static electric charge and holds this charge until it comes into close proximity with a conductor. The human body can easily accumulate a 15,000-V charge while walking on carpet. If a person with a 15-kV charge comes into close proximity with a conductor at a substantially different potential (e.g., Earth ground), the charge may be able to arc across the air gap and discharge into that conductor. Higher potentials can jump across greater distances between the charged body and nearby conductors. The problems with ESD are twofold. First, ESD can disrupt a circuit's normal operation by inducing noise that causes errors in digital signals. Second, ESD can permanently damage components if the event is strong enough and the circuit is not protected. CMOS logic is particularly sensitive to ESD because of a FET's high gate impedance and the possibility of punching through the thin gate dielectric if a high potential is introduced.

When an ESD event occurs, it can couple onto a system's internal wires by inductive or capacitive means. A discharge is a brief, high-frequency, high-amplitude event with current peaking on the order of 10 A at 300 MHz. When ESD occurs, a very strong magnetic field is generated by the fast current spike. This field can be picked up by wires some distance away, and the coupling characteristics are governed by the same EMI concepts discussed earlier. Larger loops and thicker dielectrics make a more efficient antenna for ESD. A discharge to a chassis' metal panel not only establishes a strong magnetic field, it also creates a capacitor wherein the panel accepts the high-frequency signal and then may capacitively couple this energy to nodes within the enclosure. ESD occurs so rapidly that normal ground wires have too much inductance to drain the charge before it can do damage. A typical chassis is grounded to Earth through the AC power cord. This connection prevents gradual charge accumulation to dangerous potentials, but it cannot be expected to drain ESD before a circuit is disrupted.

The basic mechanisms for dealing with ESD are to create an environment where the charge is spread out as quickly as possible, where circuits are designed to couple less energy, and where charge that enters the circuit is shunted away from sensitive components. Spreading the charge delivered by an ESD event minimizes concentrated currents at any one point and reduces the magnitude of resultant electromagnetic fields. The same grounding techniques used for EMC apply to ESD protection, because they establish continuous low-inductance grounded surfaces. Fewer gaps in a metal chassis are less restrictive to a high-frequency ESD pulse. Well grounded cables improve a system's ESD protection, because energy that makes its way onto a cable's shield can be rapidly conducted by the metal chassis instead of coupling onto the inner conductors and making its way into the circuit. A high-quality shield makes positive electrical contact with the metal chassis in many places—ideally continuous contact around its perimeter. In contrast, a thin wire, or pigtail, connecting a cable's shield to the chassis severely degrades the quality of the ground connection at high frequencies because of its high inductance. EMC and ESD engineers strictly avoid pigtails for this reason.

Objects such as switches that protrude from an enclosure should be well grounded to that enclosure so that charge can be dissipated by the enclosure rather than through internal circuits. Remember that the charge delivered by an ESD event must eventually find its way to ground, and it will find the path of least inductance to get there. You want that path to be through a system's ground structure instead of through its circuitry.

Despite its common use in portable consumer electronics, plastic is a less desirable chassis material as compared to metal, because it is unable to dissipate ESD on its own and it does not attenuate electromagnetic fields. Thin metallic coatings can help substantially with ESD and EMC issues, because conductive surface area is important at high frequencies because of skin effect. However, such coatings add to the cost of a product and may begin to flake off or wear over time.

The rules learned from EMC design—minimizing loop area with low-impedance transmission lines and ground planes—apply to ESD protection, because less energy will be coupled by a smaller loop. Older style single- and double-sided PCBs are more sensitive to ESD, because they typically have larger return path loops as a result of their lack of a continuous ground plane. Using wide traces to reduce the inductance of ground and power distribution can improve the situation somewhat, but there is a practical limit to how wide the power traces can become before routing other signals becomes impossible. One can minimize trace lengths by keeping passive components such as bypass capacitors and pull-up resistors close to ICs, and this may be one of the more significant circuit layout steps that can be taken to minimize ESD sensitivity in single- and double-sided PCBs.

A well designed system should restrict the bulk of potentially destructive ESD exposure to I/O circuitry. Inductive coupling can occur to any part of the circuit but, depending on distances and quality of PCB grounding, inductive coupling effects on internal digital buses may be mild enough to cause either no problems or only soft errors in which, for example, a memory read might return wrong data during the ESD event. Soft errors can certainly be problematic, because memory can get corrupted, or a microprocessor can crash after reading a corrupted instruction. If a system's operational environment subjects it to occasional ESD events, and inductive coupling is sufficient to cause soft errors, the solution is to somehow modify the circuit to reduce coupling or to become tolerant of soft errors via the use of redundancy or error-correcting codes. Inductive coupling might be reduced with multiple layers of shielding and better grounding. Redundant algorithms and error coding can take the pessimistic view that soft errors are expected every so often and therefore explicitly deal with them.

I/O components usually bear the brunt of ESD, because they must inherently connect to the outside world. If an ESD event occurs on an RS-232 cable, the RS-232 transceiver is going to be the first semiconductor to see that spike. This is where blocking and shunting high-frequency energy becomes important. A basic passive filter can be employed to attenuate high-frequency ESD events on an individual wire. Figure 18.18 shows several variations of ESD filtering. Many other solutions

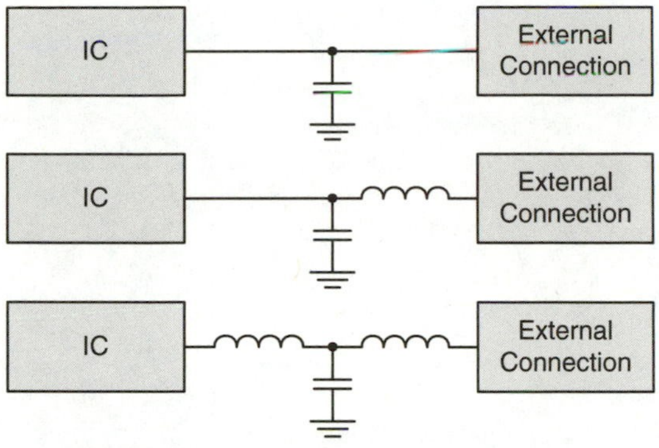

FIGURE 18.18 Various ESD filters.

are possible. Some scenarios rely on inductance of the wiring in combination with a single high-frequency capacitor to attenuate ESD. Others explicitly insert ferrite beads to add inductance to the circuit.

The example circuits shunt ESD to logic ground, but the ideal situation is to shunt the energy to chassis ground. Chassis ground presumably would have a less direct connection to the IC being protected. However, there is added complexity in separating ground regions on a circuit board, and the possibility then arises for unintended return paths to cause EMC problems. If logic ground is well attached to chassis ground at the connector, a low-inductance loop is formed from the cable, through a shunt capacitor, to logic ground, and finally to chassis ground. Inserting an inductor between the external cable connection and the shunt capacitor increases the inductance of this path, but the idea is to make this path less desirable for ESD so that it finds a lower-inductance path to ground and bypasses the wire leading to the IC altogether. Any ESD that makes it through the first inductor must be immediately presented with a low-inductance path to ground so that it does not conduct through the IC instead. A low-inductance path is created using a high-frequency surface mount capacitor connected to the inductor and ground nodes with the shortest traces possible. This filtering not only improves ESD immunity, it also reduces unwanted emissions that may cause regulatory compliance problems.

An additional ESD protection mechanism is a high-speed Zener diode that can respond quickly enough to clamp a node's voltage and shunt excess energy to ground. This semiconductor solution has the benefit of not being a lowpass filter, so it is therefore more applicable to high-speed interfaces such as Ethernet. Speed is a critical specification for an ESD clamping diode, because the initial ESD pulse typically ramps in less than 1 ns. A slow diode will add little or no ESD protection. Clamping diodes may be implemented with discrete components on a circuit board and are also implemented on some interface ICs for enhanced ESD tolerance. RS-232 transceivers are available from companies such as Linear Technology and Maxim with integrated shunt diode structures that provide up to 15 kV of protection, which is a standard threshold for ESD tolerance. RS-232 transceivers are prime candidates for ESD hardening, because the interface has become so inexpensive and common that it is used in many abusive environments. There are also ESD protection products specifically designed for high-speed interfaces, including Ethernet, that implement the same type of technology. They are made by companies such as California Micro Devices, Philips, and Semtech.

CHAPTER 19
Designing for Success

A host of details allow the major components of a system to function properly. Miscellaneous topics are often left out of many engineering discussions, because it is assumed that they will be covered elsewhere. This chapter attempts to gather into one place some of the remaining practical issues that make the difference between a smooth development process and one that is punctuated by a series of obstacles that waste time and detract from the operation of unique design elements that represent a system's true value.

Acquiring the necessary components and fabricating circuit boards is a mandatory step between design and testing. It is important to select technologies that are appropriate for both the application and your own resources. Practical considerations such as business relationships and support costs may constrain the choices of components and materials at your disposal. In extreme cases, it may not be possible to realize certain design goals with limited resources. In other situations, alternative implementation strategies may have to be employed to circumvent difficulties in obtaining the most ideal component for a given application. These topics are covered first, because they should be taken into consideration during initial conceptual and architectural definition phases of a project.

Next is a fairly simple topic near and dear to all digital systems: reset. Properly generating a reset sequence that allows a microprocessor to boot when power is applied is a task that is easy to accomplish once learned and is usually taken for granted once it becomes reliable. However, reset is a topic that requires introduction, because unpredictable reset behavior can become the Achilles' heel of an otherwise fine system.

The middle of this chapter discusses debugging strategies and how troubleshooting efforts can be reduced when proper design decisions are made ahead of time. Designing a system that can be more easily analyzed for problems does not initially occur to some engineers as a requirement. Some engineers learn this lesson the hard way after a system has already been assembled, a subtle problem arises, and there is no clear way to make a diagnosis. While head scratching and frustration over mysterious problems cannot be eliminated from the development process, proper consideration for debugging in the design phases of a project can greatly ease fault isolation.

The remainder of the chapter is devoted to support tools, both software and hardware, that assist with the development and analysis of circuits. Schematic capture software packages are almost universally known. An analog circuit simulation tool that many digital electronics engineers could employ to greater advantage is Spice. Spice is presented herein, along with real circuit examples to show how easily it can be used to answer questions that would otherwise be answered through trial and error in the lab. Once in the lab, test instruments become important partners in gaining visibility into the circuits' operation. A brief introductory discussion of common test equipment provides an orientation to what may be at your disposal once a prototype is ready for testing.

19.1 PRACTICAL TECHNOLOGIES

The combined semiconductor, electronics, and packaging industries develop many exciting and advanced technologies each year. An engineer may be tempted to use the latest and greatest components and assembly techniques on a new project, but careful consideration should go into making such decisions. Relevant constraints for any engineering organization are materials availability and cost, ease of manufacture, development resources, and general risk assessment. These constraints differ among organizations. A large company with extensive experience and support resources has a different view of the world from that of a one-man design shop building microcontrollers. This does not mean that a small organization cannot successfully utilize new technology. It does mean that all organizations must evaluate the practicality of various technologies using constraints that are appropriate for their size and resources.

Materials availability is often a problem, even for large companies, when dealing with cutting-edge technologies or newly introduced products. Cutting-edge technology is, by definition, one that pushes the limits of what is achievable at any given time. Pushing the limits in any discipline generally carries with it the understanding that problems may arise in the early stages of product release. New technologies may also carry higher initial costs while volumes and manufacturing yields are still low. Part of engineering is balancing the risks and benefits of new technologies. When you move into uncharted waters, an occasional setback is almost inevitable. Therefore, the new technology that one may read about in the trade press or see advertised in company literature is not necessarily ready for immediate use.

Aside from the general risk of new products, the economic strength that you represent as a customer has a significant impact on your ability to gain access to these products. If you are a semiconductor company that has just developed a new chip, and you have the staff to support only three initial customers, would you want three large customers or three small customers? Developing relationships with manufacturers and their representatives can assist you in determining when a new technology is practical to use and when it should be allowed to mature further. This applies equally to more mature products. Even components that have been shipping for some time may be subject to availability problems. The term *allocation* is well known to component buyers. In a tight market, vendors will preallocate their manufacturing capacity across a set of key customers to preserve successful business relationships. Even when a product is mature and being manufactured in high volume, a small customer may be unable to purchase it, because it is "on allocation." Allocation problems affect large companies as well in times of increased demand. The semiconductor industry tends to be quite cyclical, moving through phases of supply shortages and softness in demand.

Evaluating the risk of availability is an important step in the component selection process. More mature components are generally easier to obtain. The exceptions to this rule are ICs that have short production lives, such as some microprocessors and memory chips—especially DRAM. The microprocessor and memory markets are highly competitive, and products are sometimes phased out after just a few years. DRAM products are notorious for supply and obsolescence problems after their volumes peak within the first few years of introduction. There are certain bread-and-butter microprocessor and memory ICs that are supported for longer terms. These tend to be products for embedded markets in which semiconductor process technology changes at a slower pace than in the mainstream computer market.

Newer products are often available only through authorized distribution companies. Many mature products can be purchased from catalog distributors. Catalog distributors include Digikey, Jameco Electronics, JDR Microdevices, and Mouser Electronics. Larger engineering organizations with dedicated purchasing staff often prefer to deal with authorized distributors because of more direct access to manufacturers. A small organization may be able to satisfy all or most of its procurement needs with catalog distributors if mature technologies are acceptable.

Manufacturability should be taken into account when selecting components, because it may be impossible to assemble a system without expensive equipment. High-end ICs and electronic components utilize fine-pitch surface-mount technology almost exclusively because of the reduction in parasitics and the increase in signal density. Ball grid array (BGA) packages and 0402/0201 passive components are challenging to properly mount on circuit boards. They require accurate application of solder paste, precise positioning of components, and tightly controlled thermal profiles in solder reflow ovens. Most leaded surface mount components can actually be assembled by hand with a good quality soldering iron, although it may take some time and practice to do this. Ultra fine-pitch components are extremely difficult to assemble by hand in a reliable and repeatable manner, and BGAs are impossible to mount reliably without specialized equipment. These requirements should be taken into consideration according to the manufacturing facilities at your disposal. Organizations willing to spend the money necessary for high-end systems development routinely use BGAs and other advanced packages. Products that must be assembled either by hand (e.g., small prototype runs) or at lower-cost assembly shops that use older equipment must be designed with components that are compatible with these approaches.

More complex digital ICs may require significant development resources whose costs cannot be overlooked. When most digital systems were built from 7400 logic, the development tools required ranged from pencil and paper to a basic schematic capture program. The barrier to entry from a tools purchasing perspective was low and remains low for this class of design. Once a microprocessor is added, things get a little more involved. First, a microprocessor requires boot ROM that somehow must be programmed. EPROM—and now flash—programmers are common development tools in a digital systems lab. (Don't forget to socket the boot ROM rather than soldering it to the PCB, so that it can be removed for programming!) With microprocessors also come software development tools including assemblers and compilers. An assembler is the minimum software tool required to work practically with a microprocessor. Because of their relative simplicity, assemblers can be obtained for little or no cost for most microprocessors. Some manufacturers of embedded microprocessors give away assemblers to promote the use of their products. High-level language compilers (e.g., C/C++) for many microprocessors are also available at little or no cost, thanks to the GNU free software project.

Programmable logic devices require a whole other set of development tools that can get rather expensive, depending on the complexity of the design and which devices are being used. An HDL such as Verilog or VHDL is most commonly used to implement programmable logic, although some engineers still use schematic capture for smaller designs—a practice that was more common in the days of less-complex digital logic. Some PLD manufacturers also support proprietary design entry methods, although this support is more for legacy customers than for new mainstream business. The first step in HDL-based design after writing the logic is to simulate it to verify its operation prior to debugging in the lab. It is much easier to detect a problem in simulation, because all internal logic nodes can be probed, and the circuit can be run at an arbitrarily low speed to observe its operation. Lab debugging requires equipment such as oscilloscopes and logic analyzers to view logic nodes, and less visibility is generally available as compared to a simulation. The next step is synthesis, wherein the language constructs are converted into logic gate representations. Simulation and synthesis tools can cost thousands or tens of thousands of dollars, although some manufacturers offer very low-cost HDL design software for smaller designs. After synthesis, the logic must be mapped and fitted to the specific chip's internal resources. Mapping software is proprietary, based on the type of PLD being used, because each manufacturer's PLD has a different internal structure. This software is provided at low or no cost to customers.

Configuring PLDs once required expensive programming equipment, but most modern CPLDs and FPGAs are now in-circuit programmable via a serial or parallel interface. In fact, most CPLDs and FPGAs use EEPROM or SRAM technology, which makes them almost infinitely reprogrammable so that logic mistakes do not require replacing a potentially expensive IC. Smaller, more mature PLDs may still require dedicated programming hardware, which may be reason enough to avoid them if possible in favor of a small CPLD. The price difference between a PLD and a small CPLD is now slim to none.

For reasons previously discussed, signal integrity software packages may be necessary when designing high-speed digital circuits. These tools can be quite expensive, but the consequences of not using them can be even more costly in terms of wasted materials and time if a circuit malfunctions because of signal integrity problems. Before embarking on an ambitious high-speed design, make sure that signal integrity issues are either well understood or that the resources are available for proper analysis before fabricating a prototype. Certainly, not all designs require extensive signal integrity analysis. If it is known that the signal speeds and wire lengths are such that transmission line effects, crosstalk, and EMI can be addressed through conservative design practices, minimal analysis may be required. This determination generally requires someone with prior experience to review a design and make predictions based on previous work.

Risk assessment in choosing which components and technologies to employ is an important part of systems design. An otherwise elegant architecture can fall on its face if a key component or necessary development tools are unavailable. Therefore, be sure to make choices that are practical for both the application and the resources at your disposal.

19.2 PRINTED CIRCUIT BOARDS

The selection of appropriate technologies is a convenient segue into circuit construction, because the manner in which a circuit is assembled can have a great impact on the viability of the resulting prototype or product. Higher-speed circuits are more sensitive to construction techniques because of grounding and inductance issues. Most high-speed circuits can be fabricated only with multilayer PCBs, but more options are available for slower systems, especially in the prototyping phase of a project.

Circuit boards can be of either the printed circuit or manual point-to-point wiring variety. As already discussed, PCBs consist of stacked layers of copper foil that have been uniquely etched to connect arbitrary points in the circuit. The term "printed" refers to the standard technique of using photolithography to expose a chemically treated copper foil with a negative image of the desired etching. Similar to creating a photograph, the exposure process alters the photoresistive chemicals that have been applied to the foil so that the exposed or nonexposed areas are etched away when the foil is placed into a chemical acid bath. PCBs are an ideal technology, because they can be mass produced with fine control over the accuracy of each unit. Simple single- and double-sided PCBs can be manually fabricated using a variety of techniques, and the cost of having such boards professionally manufactured is low. Multilayer PCBs must be fabricated professionally because of the complexity of creating plated vias and accurately aligning multiple layers that are etched separately and then glued together. The major cost involved in designing a small PCB is often the specialized computer aided design (CAD) software necessary to create the many features that a PCB implements, including accurate traces, pads, and IC footprints. Low-end PCB CAD packages are available for several hundred dollars. High-end tools run into the tens of thousands of dollars.

Once a PCB is fabricated, it is assembled along with the various components to which it is designed to connect. Assembly may be performed manually or at a specialized assembly firm, almost all of which use automated assembly equipment. It is difficult to manually assemble all but relatively simple boards because of the fine-pitch components and the element of human error. Automated assembly equipment substantially increases reliability and improves assembly time for multiple boards, but at the expense of increase setup time to program the machines for a specific design.

Automated assembly is performed using one or two soldering techniques, depending on the types of components on a PCB. Before the widespread use of surface mount technology (SMT), the standard process was to program a pick-and-place machine to automatically insert leaded components into the correct holes in the PCB. The machine trims the leads as they are inserted so that a small length protrudes from the bottom, or solder side, of the PCB. When the PCB is fully stuffed, it is placed onto rails that carry it through a wave-soldering machine. A wave-soldering machine contains a bath of liquid solder over which the PCB is dragged. The metal component leads, in combination with the plated holes and pads of the PCB, wick molten solder up into the gaps, and the solder hardens as the PCB exits the molten solder wave. Wave soldering is a mature assembly technology that has worked well for decades.

Certain design rules are necessary for successfully wave soldering a PCB and preventing excessive solder from shorting adjacent leads. First, a PCB's top and bottom surfaces are finished with a thin layer of solder mask so that only metal intended for soldering is exposed. The molten solder will not adhere to the solder mask. Second, minimum lead spacing rules are followed that ensure that the lead gaps are too large for the solder to wick into them. Finally, rows of component leads are oriented in a single line perpendicular to the solder wave to minimize solder bridges, or shorts. If the leads of a DIP package are moved through the wave abreast, or in a parallel orientation, they are more prone to picking up an excessive quantity of solder in the many gaps formed between the leads. When the DIP is rotated 90° so that the lead rows are perpendicular to the wave, the solder can more freely travel down the narrow column and not get excessively stuck in one inter-lead gap. There are a variety of PCB layout tricks to reduce solder bridging in wave-soldered PCBs. If your PCB is expected to be professionally wave soldered, it is advisable to consult the assembly firm before designing the PCB. Manufacturing engineers at assembly firms are very knowledgeable about what works and what doesn't, because they work with this equipment on a daily basis. Each assembly firm has its own set of manufacturability guidelines that have been developed based on the tools at their disposal. Rules for one firm may not be fully in agreement with those of another. A half-hour conversation ahead of time can save days of headaches caused by myriad short circuits in an improperly designed PCB.

Wave soldering does not work for many SMT boards, because the small gaps in fine-pitch component leads act as efficient solder sponges, resulting in completely shorted boards. SMT PCBs use a solder paste reflow assembly process. The process begins by applying solder paste to the SMT pads through a special stencil. The stencil has openings in the exact locations where solder paste is desired. A pick-and-place machine then stuffs the PCB. Careful handling of the stuffed board is necessary, because adhesion of the solder paste is all that holds the parts in place. Many PCBs have components on both sides. In these situations, the bottom-side components are held in place with the solder paste's natural adhesion, which prevents them from falling off when they are upside down during the remainder of the assembly process. Once all SMT components have been loaded onto the PCB, the PCB is moved through a reflow oven with carefully controlled temperature zones to suit the solder, component, and PCB properties.

A PCB with both through-hole and SMT components usually requires a two-step process consisting of reflow followed by selective wave soldering. Selective wave soldering is accomplished with varying types of masks that are temporarily applied over the bottom-side SMT components so that they do not get shorted or stripped off the board in the solder wave. It is preferable to design a PCB without the need for a two-step process to save both time and money in assembly. An SMT PCB with only a few necessary through-hole components (e.g., connectors) may be more efficiently assembled by having a person manually solder the through-hole components after reflow.

As with wave soldering, the reflow process has its own set of manufacturing guidelines to increase the reliability of the manufacturing process. The intended assembly firm should be contacted before PCB design to fully understand its engineers' manufacturing rules. A standard rule for both wave soldering and reflow is to leave a minimum spacing, or *keep-out*, between the PCB edge and all components. These keep-outs make the PCB easier to handle and allow it to ride on rails through the pick-and-place, wave, and reflow machines. A typical component-to-edge keep-out is 0.2 in (5 mm). Very dense PCBs that cannot tolerate such keep-outs may require snap-off rails fabricated as part of the PCB. Such rails are not uncommon and are almost free, because they are formed by routing slots at the edges of a PCB as shown in Fig. 19.1. A related assembly rule is the inclusion of tooling holes at several locations around the PCB perimeter. These holes provide alignment and attachment points for the assembly machines.

Pick-and-place machines generally require assistance in perfectly aligning a high-pin-count SMT package to its designated location on the PCB. Most passive components and small multilead SMT packages can be automatically aligned once the PCB is locked into the machine. The alignment of packages such as QFPs and BGAs is aided with fiducial markers that are designed into the PCB in the vicinity of these IC locations. Fiducials are typically small circles or bull's eye symbols etched in the surface copper layer that can be optically detected by the pick-and-place machine as the dispenser head closes in on the desired location. Figure 19.2 illustrates fiducial placement around QFP

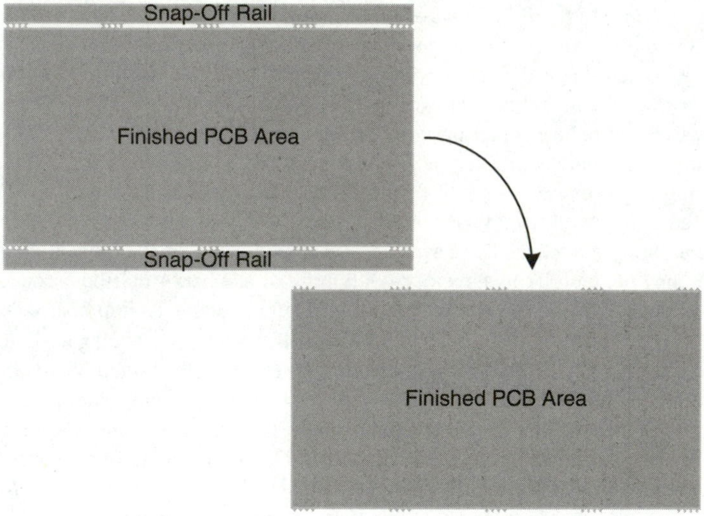

FIGURE 19.1 Removable PCB rails.

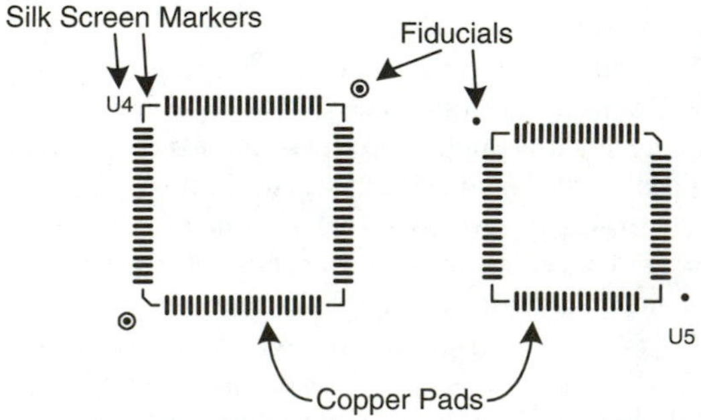

FIGURE 19.2 Fiducial markers for SMT alignment.

footprints. Also shown are visual inked silk-screen markers for people's benefit that are created to identify different components with unique reference designators and to draw shapes that assist in circuit assembly and troubleshooting. The fiducials do not have to be in a standard location. What is important is that their exact location is known and entered into the machine so that an absolute location can be determined once the fiducials are optically detected.

19.3 MANUALLY WIRED CIRCUITS

PCBs are an excellent solution for high-speed and production systems, but manual wiring may be suitable for building small quantities of lower-speed circuits. Manual wiring can be performed in a variety of ways and has the benefits of low material cost and less support infrastructure, and it is the quickest way to start building a circuit. The time and money spent designing a PCB can be amortized across multiple units built, but each manually wired circuit consumes the same construction resources. It is important to understand the limitations of manual circuit wiring, because it can take days of tedious work to build a single circuit board. For an engineer on a budget, it may be the cheapest and quickest way to prototype a lower-speed digital system.

Breadboard is an industry term for a blank circuit board that consists solely of equally spaced holes. A breadboard is generally made of fiberglass and has hundreds or thousands of holes aligned on 0.1-in (2.54-mm) centers to match the common pin spacing of DIPs and many other electronic components. The term breadboard comes from a loose similarity to a slice of bread that is permeated with holes. If the breadboard is designed for solder assembly, it is built with thin copper pads around each hole so that solder will adhere to the board. Some breadboards contain plated holes with pads on both sides of the board for increased solder adhesion. As shown in Fig. 19.3, some breadboards add thin mesh planes by taking advantage of the small spaces between copper pads. These planes are not as effective as a PCB's continuous plane, but they have less inductance than discrete wires. The presence of a mesh plane is not bad, but it should not lead to the false impression that high-quality transmission lines can be constructed.

Building a circuit from scratch on a breadboard is done by inserting components into the board and then running small wires, usually 30 gauge, point to point between their leads. DIPs and other components with preformed leads on 0.1-in centers can be directly inserted into the board. Some connectors, as well as TO-220 voltage regulators, transistors, and diodes, have a compatible 0.1-in lead pitch. Resistors, capacitors, and inductors with axial or radial lead configurations may require bending their leads so that they can lie flat on the board. An axial leaded component is most commonly in a cylindrical package with one lead protruding from the center of each face—the two leads

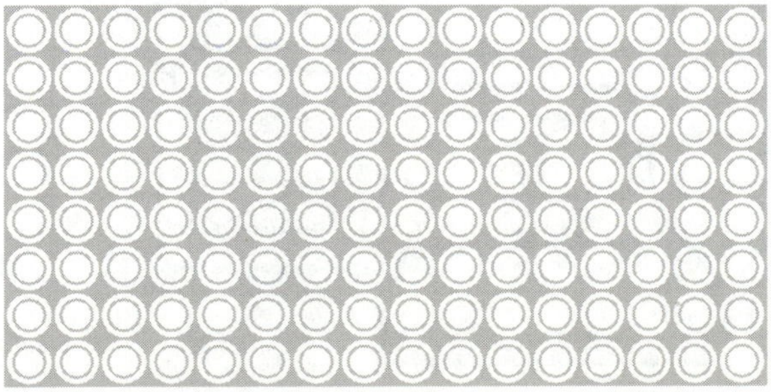

FIGURE 19.3 Breadboard with plated holes and mesh plane.

run through the component's axis. A radial leaded component is also cylindrical but has both leads protruding from the same face—the two leads are at equal radii from the component's center. Other components that do not have 0.1-in lead pitch also require lead bending: TO-92 packages (used for transistors and low-power voltage regulators) and common connectors including DB-9, DB-25, and RJ-45. These days, it may be impossible to find some desirable ICs in DIP packages. PLCCs are a popular low-cost surface mount package for lower-lead-count ICs. PLCC sockets are available with leads on 0.1-in centers, making it possible to prototype a circuit with PLCCs on a standard breadboard.

Sockets are desirable for manually wired circuits, not only for mating incompatible IC packages but also for component protection during soldering and for ease of component replacement. If a soldering iron is applied to an IC pin for too long, the IC may be damaged. Careful soldering is necessary to ensure that ICs are not damaged during construction. Additionally, static control measures must be taken when handling CMOS components because of the risk of damaging the MOSFET gate oxide. Both of these problems are eliminated when using IC sockets, because the sockets are not as sensitive to heat, and they are insensitive to ESD. A further benefit is gained if an IC is damaged during the initial prototype debugging process. Rather than having to unsolder all of the IC's pins, a socket allows a new IC to be easily installed.

Soldering a circuit on a breadboard can be done with minimal planning, but it is important to consider power distribution ahead of time. Adequate wires should be used so that components at one end of the board are not operating at a significantly different DC level from those at the other end. It may be prudent to run smaller gauge power bus wires around the board before beginning the more dense signal wiring; 24- or 26-gauge wire will ensure negligible DC voltage drop across a typical board. Implementing a mesh power distribution scheme will also marginally lower the inductance between IC power pins and bulk bypass capacitance. The absence of low-inductance power planes is a critical limitation of breadboards. Each IC should have adequate bypassing with ceramic disc capacitors arranged so that their lead length is minimized. A common scheme for bypassing a DIP IC is to directly connect the capacitor leads to the power pins. This often results in a diagonal configuration in which the capacitor runs from one corner of the IC to the other, because power pins in a DIP are often at opposing corners of the package. Pins 7 and 14 are ground and V_{CC}, respectively, for most 14-pin DIP logic ICs. Figure 19.4 shows a power distribution scheme for a breadboard in which the power buses are formed with meshed wires, high-frequency bypass capacitors are wired

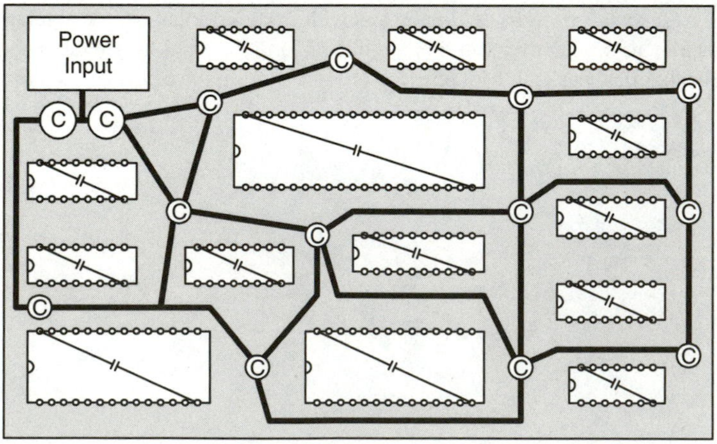

FIGURE 19.4 Breadboard power distribution.

directly to power pins, and low-frequency bulk electrolytic capacitors are arranged throughout the board. This hypothetical circuit contains 7400-type logic ICs in 14-, 16-, and 20-pin DIPs as well as VLSI memory and microprocessor ICs in 28- and 40-pin DIPs.

An alternative to conventional breadboards for temporary prototypes of small circuits is the *solderless breadboard*. A solderless breadboard isn't a fiberglass board, but a plastic frame in which many small spring-clips are embedded. Holes are on 0.1-in centers, and the spring clips are typically arranged in rows of five contacts separated by a gap, or channel, across which a standard 0.3-in (76.2 mm) wide DIP is inserted. Figure 19.5 shows a small solderless breadboard. Power distribution is often assisted by means of a continuous spring-clip bus running across the top and bottom of the breadboard. Connections are made by inserting solid wires between various spring clips. Since each spring clip has five contacts, a maximum of five connections can be made to a single node. If more are necessary, an unoccupied spring clip nearby must be used for the excess connections and then bridged to the other spring clip via a wire. Solderless breadboards are perfect for small experiments and are used in many academic lab settings, because solder irons and other assembly tools are not necessary. Clearly, solderless breadboards are not for every circuit. Aside from electrical integrity issues, they cannot accept common PLCC packages without a special breakout product that essentially converts a PLCC to a DIP. Nevertheless, substantial digital circuits can be prototyped on a solderless breadboard, including low-speed microprocessors with memory and basic peripherals.

Soldering is not the only way to prototype a digital system with permanent connections. *Wire-wrapping* is a technique that has been around for decades and was actually used for production assembly in many minicomputers and mainframes during the 1960s and 1970s. The wire-wrapping process establishes electrical connections by tightly wrapping small wire, typically 30 gauge, around square pins. Several turns of wire are made as shown in Fig. 19.6, resulting in a surprisingly durable electromechanical connection, even without the benefit of insulation that is often stripped off. Wire-wrapping requires the use of special IC sockets with long, square posts that protrude through the bot-

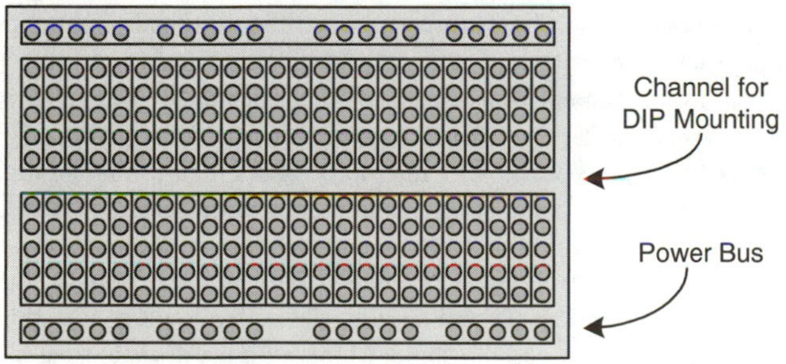

FIGURE 19.5 Solderless breadboard.

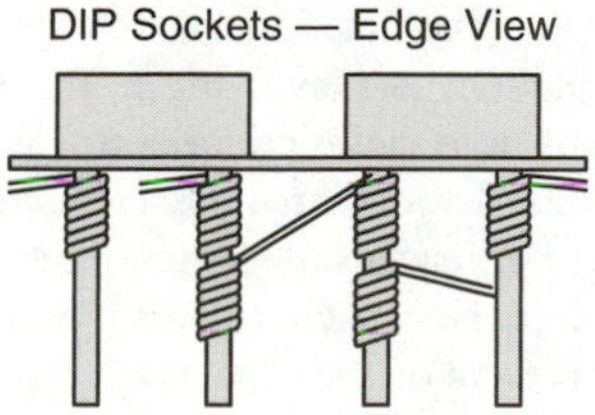

FIGURE 19.6 Wire-wrap connections.

tom side of a breadboard. Wire-wrap sockets are still available for DIPs in standard sizes for two or three wraps per post. The actual wrapping is accomplished with a special tool—either manual or automatic.

A benefit of wire-wrapping is that dense wiring can be achieved without the risk of melting through insulation with a hot soldering iron. Changes in connectivity are made by unwrapping a wire. However, if the wire to be unwrapped is at the bottom of a stack of other wires, those others may need to be removed as well. A fully manual wire-wrap process requires that a wire be cut to length and stripped at each end to expose approximately 1 in of bare wire, and the wrapping tool is turned around the square post at each end. Automatic wire-wrap guns are available that automatically strip and wrap wire, making the process move fairly quickly. The popularity of wire-wrapping has diminished significantly over time as clock frequencies have increased and PCBs have become much less expensive in small prototyping quantities. It may still be an appropriate construction technique for those who are less comfortable with a soldering iron and who require very dense wiring.

19.4 MICROPROCESSOR RESET

Almost all digital systems hinge around some type of microprocessor that controls the basic operation of specific peripherals. Smaller embedded systems may use microcontrollers that contain integrated microprocessor and memory elements. Others include discrete microprocessors. Regardless of the form, a microprocessor requires a clean initialization sequence for it to begin executing the program that has been designed for it. It is advisable to design a simple and reliable scheme for the microprocessor to boot so that initial system bring-up can proceed smoothly. Once a microprocessor successfully boots, it can be used to run software that helps with the remainder of system bring-up. Functions such as memory debug, accessing control bits in peripheral logic, and setting up a debugging console that can be accessed from a terminal program. All require that the microprocessor be alive. When a microprocessor doesn't boot correctly, it is difficult to make further progress, because the microprocessor is usually the gateway to the rest of the system.

Reset is the first hardware element, subsequent to stable power supplies, that a microprocessor needs to boot. Some microcontrollers have built-in power-on-reset circuits that guarantee a valid reset pulse to the internal microprocessor. Other microprocessors require that an external reset pulse be applied. While not complicated, generating a reliable power-on-reset has eluded more than one engineer. Dedicated power-on-reset ICs have become available in recent years that all but guarantee clean reset behavior once the power supplies become stable. At their simplest, these devices have three terminals (power, ground, and reset), and reset is held active-low for several hundred milliseconds after power passes a predetermined threshold. More complex devices have multiple power inputs for multivoltage systems, and the deassertion of reset occurs only when all power inputs have exceeded certain thresholds for a minimum time. Power-on-reset ICs are available from companies including Linear Technology, Maxim, and National Semiconductor.

When a dedicated power-on-reset chip is unavailable, the function can be implemented using discrete components in many configurations. Two simple schemes involving an RC circuit along with a discharge diode are shown in Fig. 19.7. Both circuits hold the microprocessor in reset for approximately 10 ms after power is applied. The first circuit uses just three passive components and starts out with RESET* at logic 0. As the capacitor charges, it reaches the logic-1 voltage threshold of the microprocessor. A diode is present to rapidly discharge the capacitor when power is removed. It becomes forward biased as V_{CC} drops and a charge is present on the capacitor. This ensures that the reset circuit will behave properly if the system is quickly turned on again and also prevents a capacitor discharge path through the microprocessor. Incomplete discharge is more likely with a larger RC

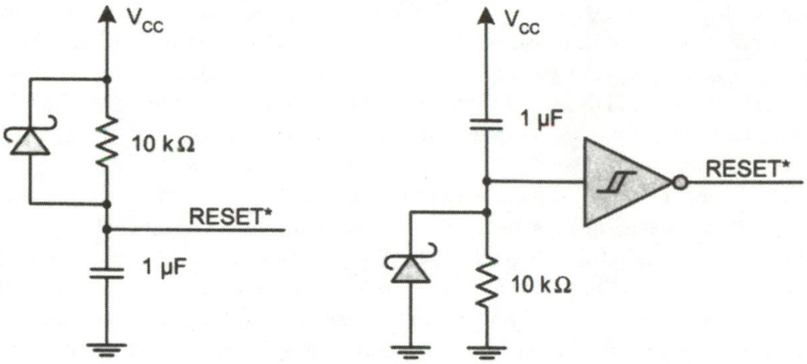

FIGURE 19.7 Discrete power-on-reset circuits.

time constant, as is required by some microprocessors. Better results are obtained using a Schottky diode, because its lower forward voltage discharges the capacitor to a lower voltage.

The second circuit is more robust, because it uses a Schmitt trigger to drive the microprocessor's input, guaranteeing a clean digital transition despite variations in the slope of the RC voltage curve. This is especially helpful when long RC time constants are required to generate long reset pulses as dictated by a microprocessor. A 74LS14 or similar Schmitt-trigger logic gate may be convenient to design into a system, and it can be used in places other than the power-on-reset circuit. Alternatively, a smaller voltage comparator can be used to implement the same function by designing in hysteresis. Before power-on, the inverter input node is at 0 V. At power-on, the voltage step of the power supply passes through the capacitor, because the voltage across the capacitor is initially 0 V and brings the input node to a logic-1 voltage, which in turn causes RESET* to be driven to logic 0. The resistor immediately begins pulling the voltage toward ground and eventually causes RESET* to be deasserted. The diode is present to clamp the inverter input node to ground during power-down. Clamping is desirable, because the resistor has already pulled the input node to ground and, without the diode, a negative V_{CC} step would force the input node to a negative voltage. When the diode is present, the input node remains near 0 V and is able to serve its intended purpose during an immediate power-on. The diode also prevents a large negative voltage from potentially damaging the inverter.

Power-on is not the only condition in which a microprocessor reset may be desired. Especially during the debugging process, it can be very useful to have a reset button that can quickly restart the system from a known initial state when software under development encounters fatal bugs. Many of the aforementioned power-on-reset ICs contain circuitry to *debounce* an external pushbutton. When a button is pushed, it may appear that a clean electrical connection is made and then broken when the button is released. In reality, the contact and release events of a button are noisy for brief periods of time as the internal metal contacts come into contact with each other. This noise or bounce may last only a few milliseconds, but it can cause a microprocessor to improperly exit its reset state. Debouncing is the process of converting the noisy edges of a pushbutton into a clean pulse. Filtering is a general solution for debouncing a noisy event and can be performed in an analog fashion or digitally by taking multiple samples of the event and forcing the bouncing samples to one state or the other.

19.5 DESIGN FOR DEBUG

"To err is human" is a truism that directly applies to engineering. The engineering process is a combination of design and debugging in which inevitable problems in the original implementation are detected and fixed. Hardware problems include logic mistakes, incorrect circuit board connections, and improperly assembled systems. Regardless of its nature, a problem must be isolated before it can be corrected. Bugs can be hard to find in a digital system, because time is measured in nanoseconds, and logic 1s and 0s are not visually distinct as they flow across wires. The debugging process is made easier by employing specific design elements and methodologies that improve visibility into a system's inner workings.

A basic debugging aid is the ability to probe a digital circuit with an oscilloscope or logic analyzer so that the state of individual wires can be observed. (A logic analyzer is a tool that rapidly captures a set of digital signals and then displays them for inspection. Test equipment is discussed in more detail later on.) Some circuits require little or no modification to gain this visibility, depending on their density, packaging technology, and operating speed. A circuit that uses DIPs or PLCCs exclusively can be directly probed with an oscilloscope probe, and clip-on logic analyzer adapters may be used to capture many digital signals simultaneously. As circuits get denser and use fine-pitch surface mount ICs, it becomes necessary to use connectors specifically designed for logic analyzer attachment. The correct type of connector should be verified with your logic analyzer manufacturer. Using logic analyzer connectors provides access to a set of predetermined signals fairly rapidly, because a whole connector can be inserted or removed at one time rather than having to use individual clips for each signal.

Logic analyzer probing becomes more of a challenge at high frequencies because of transmission line effects. Top-of-the-line logic analyzers are designed to minimally load a bus, and they include controlled impedance connectors to reduce unwanted side effects of probing. Depending on the trace lengths involved and the specific ICs, minimal impact is possible with speeds around 100 MHz. Careful PCB layout is essential for these situations, and it is desirable to minimize stubs created by routing signals to the connectors. At frequencies above 100 MHz, transmission line effects can rapidly cause problems, and special impedance matching and terminating circuitry may be necessary to achieve nonintrusive logic analyzer probing. Logic analyzer manufacturers have circuit recommendations that are specifically customized to their products.

The ideal scenario is to have logic analyzer visibility for every signal in a system. In reality, 100 percent visibility is not practical. More complex buses are more difficult to debug and, consequently, stand to benefit more from logic analyzer connectors. A simple asynchronous microprocessor bus, on the other hand, can be debugged with an inexpensive analog oscilloscope if a logic analyzer is not available. Whereas logic analyzers stop time and display a timing diagram of the selected signals over a short span of time, analog oscilloscopes usually do not have this persistence. The persistence problem can be addressed with a technique known as a *scope loop*. A scope loop is usually implemented in software but can be done in hardware as well, and it performs the same simple operation continuously so that an oscilloscope can be used to observe what has become a periodic event.

Debugging a basic microprocessor bus problem with a scope loop can be explained with a brief example. Figure 19.8 shows a portion of a digital system in which a RAM is connected to a microprocessor and enabled with an address decoder. In normal operation, the microprocessor asserts an address that is recognized by the decoder, and the RAM is enabled. One possible bug is an incorrectly wired address decoder. If the microprocessor is unable to access the RAM, the first thing to check is whether the RAM's chip select is being asserted. A single RAM access cannot be effectively observed on a normal analog oscilloscope, because the event may last well under a microsecond. Instead, a scope loop can be created by programming the microprocessor to continually perform RAM reads. The chip select can now be observed on an oscilloscope, because it is a periodic event. If the chip select is not present, the address decoder's inputs and output can be tested one by one to ensure proper connectivity. If the chip select is being asserted, other signals such as output enable or individual address bits can be probed. Oscilloscopes have at least two channels, and the second channel can be used to probe one other signal in conjunction with chip select to observe rela-

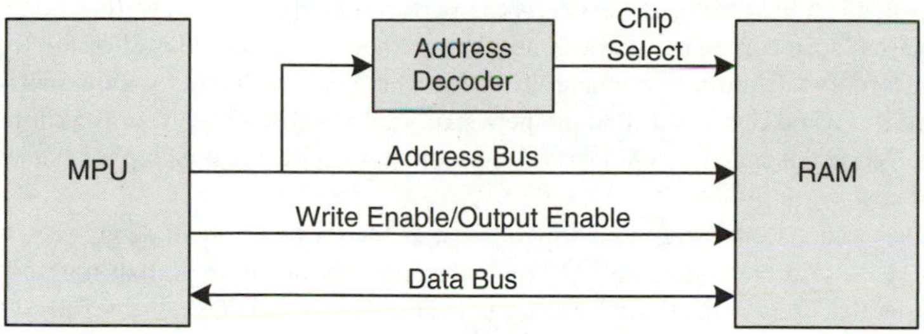

FIGURE 19.8 Microprocessor RAM interface.

tive timing. For example, if output enable is not asserted at the same time as chip select, the RAM will not respond to a read.

As a bus gets more complex, the two or four channels of a typical oscilloscope do not provide sufficient visibility into what is happening. This is where logic analyzers with dozens of channels are truly useful and why dedicated connectors are helpful in debugging digital systems.

Logic analyzer connectors alone can provide access to board-level signals only. Many digital systems implement logic within PLDs, with the result that many logic nodes are hidden from the board-level perspective. Proper simulation of a PLD can flush out many bugs, but others may escape detection until a real system is functioning in a lab environment. It is helpful to allocate unused pins on a PLD for test purposes so that internal nodes that are normally hidden can be driven out of the PLD and captured on a logic analyzer or oscilloscope.

19.6 BOUNDARY SCAN

When testing a newly fabricated system, there is generally the assumption of a properly wired and assembled circuit board. A high-quality manufacturing process should make this assumption realistic. The proliferation of BGA and very fine-pitch packaging has made PCB assembly a more delicate operation, because there is less room for error and less visibility to check for proper solder connections. BGAs are especially troublesome, because the solder ball connections are largely hidden from view. To make matters worse, a faulty BGA connection cannot simply be touched up in the lab with a soldering iron. The entire BGA must usually be removed from the board, cleaned, reprocessed, and then reattached. X-ray inspection machines are used to help verify proper BGA assembly, but these machines are imperfect and costly. If a prototype board initially arrives in the lab and fails to perform basic operations, the problem can be either assembly related or design related. This uncertainly lengthens the debugging process, because individual wires must be probed to verify connectivity because of the lack of visibility under a BGA package. Very fine-pitch leaded packages are also subject to inspection difficulty, because solder shorts can be hidden from view behind a screen of dense pins.

Members of the IEEE saw these assembly verification problems looming in the 1980s as packaging density continued to increase. The IEEE Joint Test Action Group (JTAG) was formed to address testability problems, and they developed the IEEE 1149.1 Standard Test Access Port and Boundary Scan Architecture. Because the full name of the standard is a mouthful, most people refer to IEEE 1149.1 simply as JTAG. JTAG is a simple yet powerful technology, because it places test resources directly into ICs and enables chaining multiple ICs together via a standard four- or five-wire serial interface. In essence, JTAG forms a long chain of shift registers whose contents can be set and read back through software. The placement of these shift registers, or test cells, determines the types of problems that can be debugged. Test cells can perform internal scan as well as boundary scan operations. Internal scan can help chip designers debug their logic, because all of the flip-flops in a chip can be incorporated into a scan chain. The scan chain can be loaded with a specific test state, the chip can be clocked once, and then the new state can be shifted out. This is akin to single-stepping software. While not speedy, JTAG provides excellent visibility into a chip's inner workings with little added cost.

Boundary scan is the more interesting side of JTAG for board-level debugging, because it allows the scan chain to drive and sense I/O pins independently of the logic that normally controls them. This means that the connections between two JTAG-equipped ICs can be fully verified electronically. One device can drive a wire, and the other can capture that driven state and report it back to the JTAG test program. JTAG is an excellent complement to BGA technology, because it can rapidly determine the correctness of otherwise hidden connections. Most connections can be verified only when the ICs at each end both contain JTAG logic. Connections from a JTAG-enabled IC to a memory IC often can be tested by writing data to the memory through the boundary scan chain and then reading it back in the same manner. JTAG logic is very common in BGA-packaged ICs, but not as common with other packages. If you are designing with BGA or very fine-pitch packages, JTAG support is a criterion that should be considered when selecting components. In some cases, there is a choice between vendors who do and do not support JTAG. In other cases, you may have to select a non-JTAG device and accept the reduction in testability.

The JTAG interface is designed to be easy to implement so that it does not place undue demands on already complex boards. Four primary signals compose the interface: test mode select (TMS), test clock (TCK), test data input (TDI), and test data output (TDO). A fifth signal, test reset (TRST*), is supported by some vendors but is not strictly necessary for JTAG operation. Devices are daisy chained by repeatedly connecting TDO to TDI until a chain has been formed with a set of TDI and TDO signals as shown in Fig. 19.9. TMS, TCK, and TRST* are bussed to all devices. One of the first questions that comes up in synchronous design is a circuit's tolerance for clock skew. JTAG can tolerate almost arbitrary clock skew, because inputs are sampled on the rising edges of TCK, and outputs are driven on the falling edges. Almost half a clock period of skew is permissible. If JTAG is run at several megahertz, the skew is unconstrained for all practical purposes, because hundreds of nanoseconds corresponds to wire lengths well in excess of what can fit onto a normal circuit board. JTAG's skew tolerance means that normally restrictive rules for clock distribution such as length

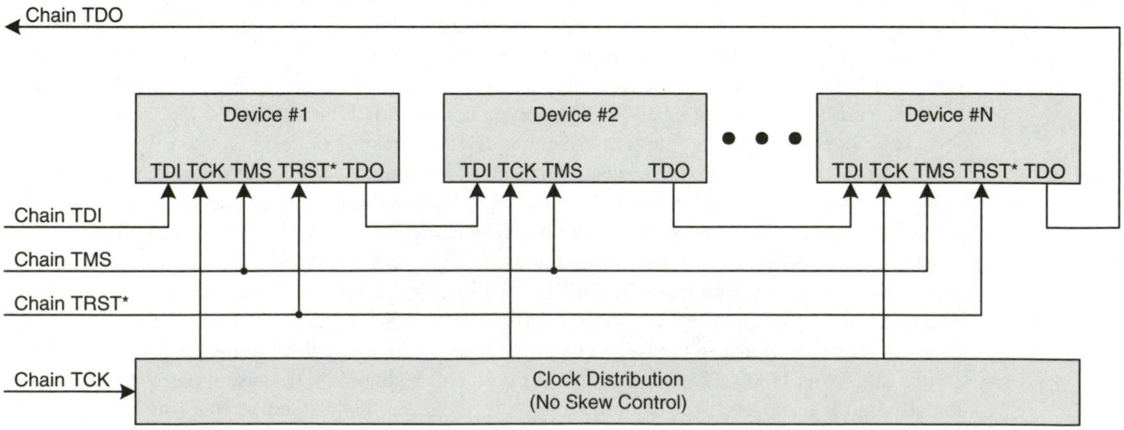

FIGURE 19.9 JTAG chain.

matching and low-skew buffers can be largely ignored. Of course, signal integrity considerations still apply so that TCK is delivered without excessive distortion to each IC. TRST* is active low and allows supporting devices to have their JTAG logic restarted. TMS is active high and places an IC into test mode by activating its JTAG controller.

JTAG scan chains are normally operated with special software designed to apply and check user-defined test patterns, or vectors. IC manufacturers who support JTAG provide boundary scan description language (BSDL) files that tell JTAG software how to interact with an IC. A standard procedure is to combine a PCB netlist (a file that lists each connection on the board in detail) with multiple BSDL files in a JTAG software package to come up with a set of test vectors. These vectors are then run through each board after it is assembled to verify connectivity between JTAG-equipped devices. Vendors of JTAG testing packages include Asset InterTech, Corelis, and JTAG Technologies. Verifying an assembled PCB with JTAG can save days of manual debugging when there may be problems in the assembly of BGA and fine-pitch components.

19.7 DIAGNOSTIC SOFTWARE

Software can help the debugging process by implementing scope loops, but the potential exists for much higher-level assistance. Special-purpose diagnostic programs are extremely helpful in detecting problems that could otherwise take much more time to isolate. The basic idea behind a diagnostic is for the software to thoroughly test elements of the hardware one at a time. A complete memory diagnostic, for example, would test every bit in every memory location. If the test fails, the nature of the failure provides valuable clues as to what is wrong. If there is a pattern to the failure, an improperly connected address or data bit may be the culprit. If the pattern is random, there may be a timing problem that causes marginal behavior over time or the device may be bad. Diagnostics are useful in several phases of development, including initial debug, extended reliability testing, troubleshooting damaged systems, and screening newly fabricated systems in either laboratory or manufacturing environments.

During initial debug, when little has already been shown to work, diagnostics can serve as both scope loops and sources of stimulation that exercise a wide range of functionality so that all features can be tested at least once. Memory tests are one of the most common classes of diagnostics, because a reliable memory interface is critical to any microprocessor-based system's operation. It is impractical to manually test millions of memory locations, but a program can easily perform this task in a quick, automated fashion. The specific patterns that a memory diagnostic uses should be chosen with knowledge of potential problems that may be uncovered. These patterns are written to memory in a complete write pass and then verified on a subsequent read pass. It is undesirable to test one memory location at a time, because certain problems can lie hidden from this approach. If the data bus is not connected, simple capacitance on the data wires might allow the microprocessor to "read" back a data value that was just written.

Aside from timing problems that can affect all transactions, circuit board connections are prime candidates for trouble because of potential errors during assembly or subsequent damage from excessive handling. A good diagnostic will pass only if there are no shorts or opens in the address, control, and data bus wiring. A shorted or open control signal will generally be easiest to find, because many, if not all, memory accesses are affected by these faults. Data bus wiring problems manifest themselves as a failure to read or write certain values to individual bit positions. An open data bit wire would result in reading a constant or random value, depending on the specific circuit. If a data bit is shorted to ground or power, the read value would be stuck at that logic level. If two data bits are shorted together, they will always assume the same value and will not be independently written.

Address bus wiring problems result in aliasing or an unexpected displacement within the memory range. In a 16-bit addressing range, a disconnected A[15] would float to a default logic level and result in the lower and upper 32,768 locations overlapping. If A[15] and A[14] were shorted together, the two bits would have only two logic states instead of four, causing the middle 32k of the address space to overlap with the upper and lower 16k regions.

It may be useful to have multiple memory diagnostics to help with different phases of debugging. Some engineers like basic walking-ones and walking-zeroes patterns to quickly determine if any data bits are stuck. As their names imply, walking-ones and walking-zeroes tests set all bits in a word to one state and then walk the opposite state across each bit position. An eight-bit walking ones test could look like this: 00000001, 00000010, 00000100, 00001000, 00010000, 00100000, 01000000, 10000000. These tests verify that each bit position can be independently set to a 0 or 1 without interference from neighboring bits.

Testing an entire memory after simple data bus wiring problems are resolved can be done efficiently by selecting a set of appropriate data patterns. Pattern selection depends on the diagnostic goals. Is it sufficient to verify only stuck address bits? Both stuck address and data bits? Or is it necessary to verify every unique bit in a memory array? Stuck data bits can be uncovered with a quick walking-ones and walking-zeroes test targeted to just a few memory locations. It is important to separate the write and read phases of a short diagnostic, and therefore use multiple memory locations, to guarantee that the microprocessor is not merely reading back bus capacitance. Isolating and verifying every bit in a memory array can be a more complex operation than first imagined, because specific memory design aspects at both the board and chip levels may require application specific patterns to achieve 100 percent coverage. A common memory diagnostic approach is to first test for stuck data bits in a quick test and then verify the address bus with a longer test and, in the process, achieve good, but incomplete, bit coverage. This level of coverage in concert with the testing that is performed by semiconductor manufacturers provides a high degree of confidence in the memory system's integrity.

Verifying address bus and decode logic integrity can be done with a set of ramp patterns. Each memory location is ideally written with a unique value, but this is not possible in systems where the data bus width is less than the address bus width. A microprocessor with 32-bit wide memory bus and 2^{20} locations can write a unique incrementing address into each word during a write phase and the verify on a subsequent read phase. A system with an 8-bit memory and 65,536 locations cannot write unique values in a single pass. An initial approach to verifying a 64-kB memory array is to write a repeating ramp pattern from 0x00 to 0xFF throughout memory and then read it back. This tests only the lower half of the address bus, because there are only 256 unique values written. Aliasing is a potential problem, because a 256-byte memory would appear identical to a 64-kB memory without further effort. The diagnostic requires a second pass to test the upper half of the address bus by writing the MSB of the address to each byte. Each pass is incomplete on its own, but together they are an effective diagnostic.

There are many memory diagnostic techniques, and engineers have their own favorites. The repeating, alternating pattern of 0x55 and 0xAA is popular, because it provides some verification that each bit position can hold a 1 or 0 independent of its nearest neighbors. This is in contrast to the pattern of 0x00 and 0xFF that could pass even if every data bit were shorted to every other data bit. However, 0x55 and 0xAA cannot isolate all data bus problems, because there is no restriction that shorts must occur between directly adjacent bits. Traces on a PCB are routed in many arbitrary patterns and shorts between nonadjacent bits are possible.

Each system requires a customized diagnostics suite, because each system has different memory and I/O resources that need to be tested. Serial communications diagnostics are common, because most systems have serial ports. A loop-back cable is connected to the serial port so that all data transmitted is received at the same time. The loop-back forms a complete data path that can be automatically tested with software. Pseudorandom patterns are sent out and checked on the receive side to verify proper serial port operation. The same basic test can be performed on many types of communications interfaces, including Ethernet.

Diagnostic software not only assists during initial debugging, it can also be used for extended stability and reliability testing where continual exercise of logic and data paths provides confidence in a design's integrity. Verifying proper memory accesses for a brief duration in a laboratory setting is insufficient for a production or high-reliability environment, because semiconductors are subject to variation over individual units, temperature, and voltage. Once a system appears to be functioning properly, it should be subjected to a *four-corners* test procedure. Four-corners testing subjects a system to the four worst-case permutations of temperature and voltage to ensure that there are no timing problems over the system's intended range of operation. It is called four-corners because temperature and voltage effects on timing can be represented graphically on Cartesian coordinates as shown in Fig. 19.10. As the temperature drops and voltage increases, the speed of a logic gate increases. Conversely, the gate slows down as temperature rises and voltage drops. Both extremes can cause problems if incorrect timing analysis was performed.

The first step in extended testing is often to run a diagnostic loop overnight and verify that there were no failures after millions of iterations. This is usually performed at room temperature (25°C) and nominal voltage conditions. With nominal conditions verified, a thermal chamber is used to maintain high and low temperatures for long durations as voltage is varied from high to low. Increasing confidence is gained when multiple systems are run through four-corners testing. Thermal chambers can be rented from a testing firm if you do not have the financial resources to acquire one. A common commercial operating temperature range is 0 to 50°C (32 to 122°F), and common digital voltage supplies are specified with a tolerance of ±5 percent. A diagnostic loop would be run for a long period of time at each of the four combinations of temperature and voltage. When multiple supplies exist in a system, a case can be made for more permutations of test cases. This can become quite burdensome with three or four different voltage rails. The decision must be made according to the requirements and context of each situation.

An investment of time in diagnostic development also pays dividends when manufacturing new circuit boards and systems by providing a ready-to-use testing regimen. Once a new system passes a

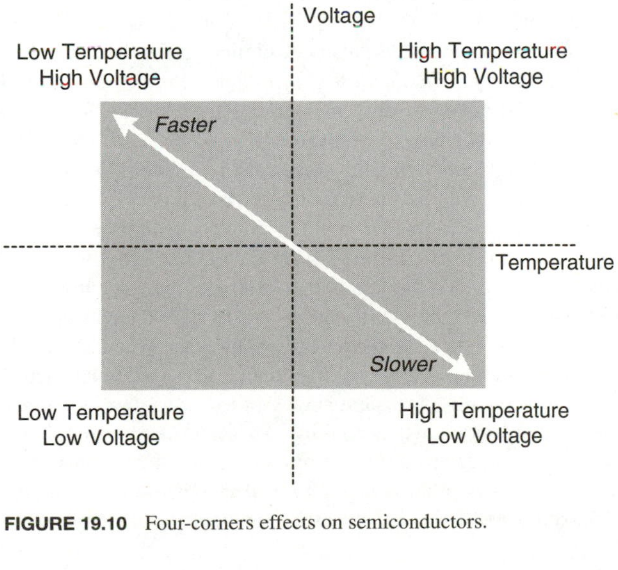

FIGURE 19.10 Four-corners effects on semiconductors.

comprehensive diagnostic suite, it can be packaged and shipped to a customer. Similarly, systems may come back from the field or from development work in the laboratory with damage from excessive handling, and diagnostics can be run to help isolate and fix the problems.

19.8 SCHEMATIC CAPTURE AND SPICE

Various types of computer aided design (CAD) software assist digital engineers in designing and implementing circuits and systems. Some types of CAD tools have already been discussed, including HDL simulators, PLD fitters, and electromagnetic field solvers. Perhaps the most universal CAD tool for electrical engineers of all disciplines is a schematic capture program. Schematic capture is to circuits what word processing is to text. At the most basic level, schematic capture programs allow the drawing and manipulation of graphical symbols. However, their utility extends to converting the information represented by the graphical connections into varying types of data formats used in subsequent stages of system development. Two of the most common and useful data conversion results are a *bill of materials* (BOM) and a *netlist*. A BOM is a complete summary of all components used in the schematic and generally has identical components grouped together so that one can quickly determine how many units of each component are used in a design.

A netlist is an exhaustive list of all the electrical connections in the schematic and is the means of transferring a schematic into a PCB layout program. With a schematic in hand, a circuit can be prototyped with either a manual wiring process or a tedious manual PCB design process. Both of these methods involve a person translating each drawn wire in the schematic into a wire on a circuit board. The assembly complexity and potential for errors increase as the design itself gets larger. The advantage of an automatically generated netlist is that errors are minimized as the entire design database is transferred from the schematic capture tool to the PCB layout tool. Each layout tool has its own proprietary data interchange format, so it is important for a schematic capture program to support the desired data format.

A wide variety of schematic capture programs are available, and they range in price from hundreds to many thousands of dollars, depending on their intended application. Major vendors include Cadence, OrCAD (owned by Cadence), Innoveda, and Mentor Graphics. At least one program, OrCAD, is available in a free student/demonstration edition that has limited features and can be used to draw circuits of moderate complexity.

Another CAD program that is applicable to many disciplines is the *Spice* family of analog simulators. Spice (or SPICE), an acronym for *simulation program with integrated circuit emphasis,* was originally developed at the University of California at Berkeley in the 1970s and has become a standard means of simulating circuit behavior. Many variants of Spice are available, and source code is available as well. Spice tools share a common basic syntax and feature set. Circuits have traditionally been represented manually in a netlist-like format for Spice processing. However, some vendors now enable schematic capture software to convert a drawn circuit into the Spice input format. PSpice is a well known variant that has been available for PC platforms for many years. It is now sold by Cadence, which has continued the practice of offering free student and demonstration versions. PSpice was used to evaluate some of the analog circuits and concepts discussed in this book.

Spice is a powerful tool. Two of its basic modes of operation are AC sweep and transient analysis. AC sweep performs frequency-domain analysis on a circuit and is used to characterize filters and the frequency response of general circuits. Transient analysis provides a time-domain view of an analog circuit and can be used to simulate a transmission line or view a filter's output in the time domain as it would be seen on an oscilloscope. Spice simulations are an excellent means of performing "what-if" evaluations of circuits while still in the design phase. Transmission line terminations can be evaluated to gauge signal integrity, and filters can be tested to determine the frequencies over which they are effective. The degree to which a Spice simulation matches reality depends on how closely the real conditions are modeled. Performing highly accurate simulations is a skill that requires a thorough understanding of circuit theory. However, useful first-order approximations of analog behavior can be readily achieved. A key source of divergence between simulation and reality in a digital design is the parasitic properties of wires and components that become significant at high frequencies. An idealized resistor or wire might require the explicit addition of parasitic inductance and capacitance to get a more accurate simulation.

A basic example of Spice simulation can be shown using the first-order RC filter in Fig. 19.11. This lowpass filter uses idealized components and has $f_C \approx 10$ MHz with a steady attenuation slope of 20 dB per decade.

Circuits are presented to Spice by uniquely naming or numbering each node and then instantiating circuit elements that reference those node names. Figure 19.12 shows the Spice circuit description for the idealized RC filter. Ground is represented as 0. Resistors and capacitors are designated with identifiers beginning with R and C, respectively. V denotes a voltage source, and this voltage source is specified with a 0-V DC component and a 1-V AC component. The `.AC` command instructs Spice to perform an AC sweep over a logarithmic (decade) range from 100 kHz to 1 GHz with 10 data points per decade. Note that the voltage source does not specify a frequency. This is because our simulation is an AC sweep that evaluates a range of frequencies. Finally, `.PROBE` instructs Spice to display the results graphically.

The expected filter transfer function in Fig. 19.13 is obtained following a brief simulation of the circuit description input.

Improving the simulation's realism can be achieved by introducing parasitic inductance in series with the ideal capacitor. The actual inductance varies with package type and the wiring scheme used.

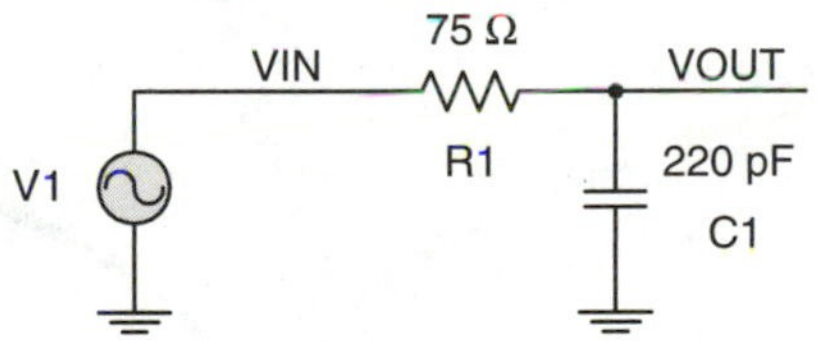

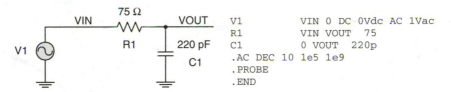

FIGURE 19.11 Idealized RC filter for Spice analysis.

```
V1      VIN 0 DC 0Vdc AC 1Vac
R1      VIN VOUT  75
C1      0 VOUT  220p
.AC DEC 10 1e5 1e9
.PROBE
.END
```

FIGURE 19.12 Spice circuit description for idealized RC filter.

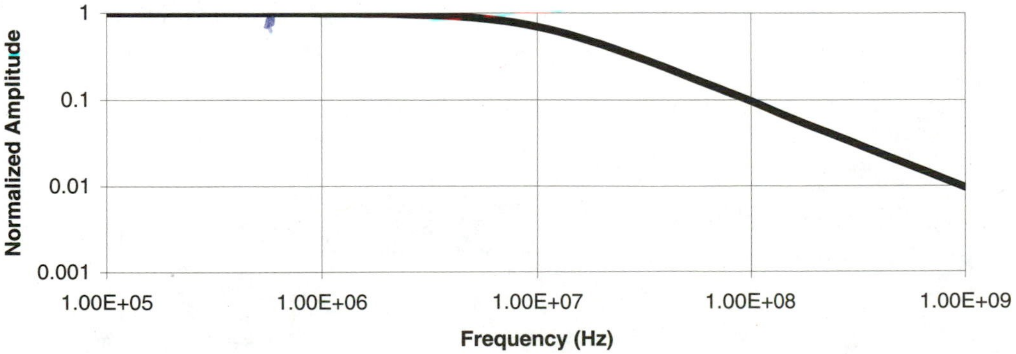

FIGURE 19.13 Bode magnitude plot for idealized RC filter.

For the sake of discussion, 2 nH is selected to observe the effects of parasitic inductance. The previous circuit description is modified as shown in Fig. 19.14. A new node is created, PARA, that is used to place the inductor between the output node and the capacitor.

Running the simulation with the added parasitic inductor yields a transfer function that has similar characteristics up to a decade past f_C, and then a notch is observed, after which the attenuation actually decreases because of the inductor's increasing impedance at high frequencies (Fig. 19.15). Increasing the parasitic inductance, as would be experienced with a leaded capacitor, would cause the notch to move to lower frequencies in simulation.

Transmission line analysis can be performed with Spice's transient analysis capability. Different termination circuits and transmission line topologies can be investigated in varying degrees of detail, depending on the accuracy of the models used. It may be difficult to generate an accurate model of an IC driver or receiver, but relatively good estimates can be obtained by combining multiple ideal elements. The circuit shown in Fig. 19.16 uses a pulse voltage source with a 5-Ω series resistor to simulate the driver's output resistance and an inductor to represent the IC package's lead inductance.

```
V1          VIN 0 DC 0Vdc AC 1Vac
R1          VIN VOUT 75
L1          PARA VOUT 2nh
C1          0 PARA 220p
.AC DEC 10 1e5 1e9
.PROBE
.END
```

FIGURE 19.14 Bode magnitude plot for RC filter with parasitic inductance.

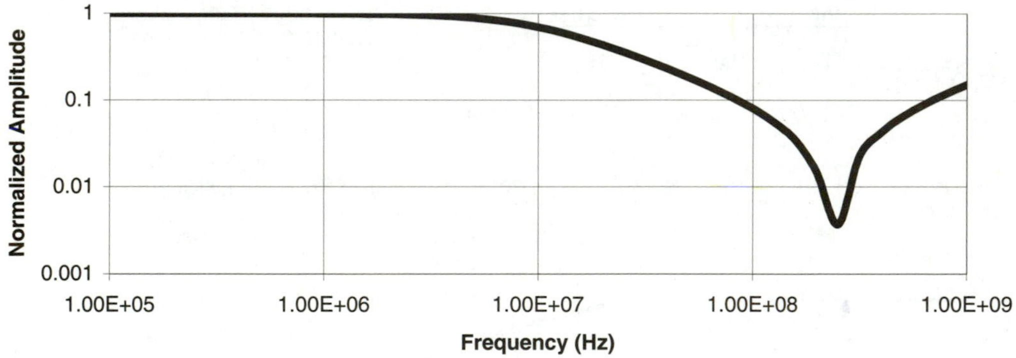

FIGURE 19.15 Bode magnitude plot for RC filter with parasitic inductance.

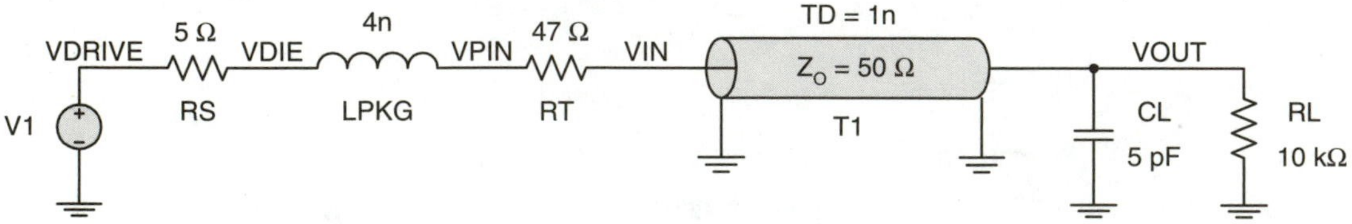

FIGURE 19.16 Transmission line circuit.

A 47-Ω series termination resistor connects the simulated driver to a 50-Ω transmission line model that has a delay of 1 ns, corresponding to a physical length of approximately 6 in (15 cm). The load is represented by a 10-kΩ resistance to ground and 5 pF of shunt capacitance.

Each node in the circuit is uniquely named so that the circuit can be represented in Spice's description format. Spice supports a variety of voltage and current sources. Voltage sources can emit constant DC levels; sine waves of varying phase, frequency, and amplitude; pulses; and other forms of circuit stimulation. The pulse voltage source is chosen for this transmission line analysis with a period of 10 ns, rise and fall times of 1 ns, and a 50 percent duty cycle to simulate a 100-MHz clock driver. Figure 19.17 shows the circuit description for the circuit in Fig. 19.16. Note the `+PULSE` specification following the voltage source declaration that specifies amplitude when off, amplitude when on (3.3-V driver is simulated), delay, rise time, fall time, high time, and period. The delay can be used to shift the pulse and is set here arbitrarily to 1 ns. Ideal lossless transmission lines are represented with a T prefix and are specified with ground nodes at the input and output as well as characteristic impedance and delay.

Following the circuit description is the `.TRAN` command that instructs Spice to perform a transient analysis for 120 ns and to display the results only after 100 ns. The first 100 ns are used to establish steady-state conditions so that the results are not affected by the circuit state at time zero when all elements are discharged. Waiting for a circuit to stabilize is not necessary in all cases and could have been omitted here with little effect. The first parameter to `.TRAN` is the step size, which is set to 0 to use the best simulation resolution. Selecting a larger step size shortens the simulation, which may be desirable for certain long analyses.

Figure 19.18 shows the plotted simulation results from the transmission line analysis showing minimal distortion. Little overshoot and undershoot are observed. How well this simulation matches a real laboratory observation depends on the accuracy of the circuit model and the accuracy of the observation. An ideal transmission line model is used in this example. If a good quality PCB with ground plane is used and stub lengths on termination components are minimized, the results should match relatively well. If the driver has significantly different output impedance, there will be less of a correlation.

One of the benefits of using Spice is the ability to try many different test cases to gain an understanding of how variance in certain circuit parameters affects the circuit's behavior. If a parameter is varied slightly, and the behavior changes dramatically, a warning should be recognized that the circuit is sensitive and small differences between the model and reality can cause trouble. Variation in driver output impedance, termination resistance, and transmission line impedance are useful parameters for experimentation. Fig. 19.19 shows the simulation results of the transmission line circuit with a 33-Ω termination resistor instead of a 47-Ω device. This simulation is done not to experiment with variance of purchased resistors but to examine the effects of a significantly different termina-

```
V1          VDRIVE 0
+PULSE 0 3.3 1n 1n 1n 4n 10n
RS          VDRIVE VDIE  5
RT          VPIN VIN  47
LPKG        VDIE VPIN  4n
T1          VIN 0 VOUT 0 Z0=50 TD=1ns
CL          0 VOUT  5p
RL          0 VOUT  10k
.TRAN  0 120ns 100n
.PROBE
.END
```

FIGURE 19.17 Spice circuit description for transmission line analysis.

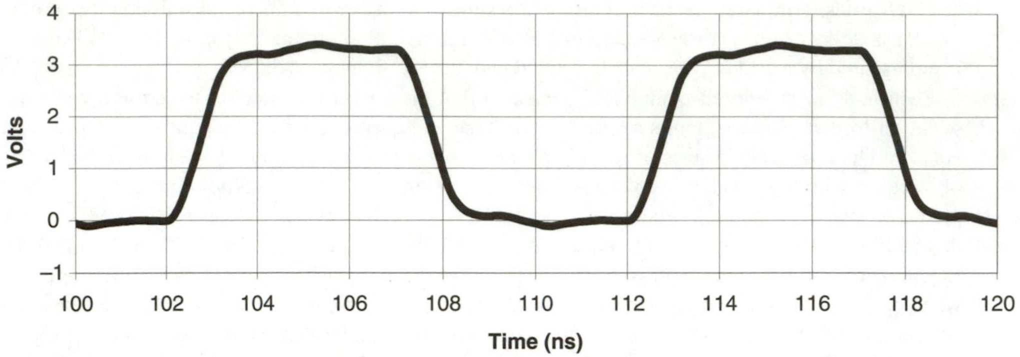

FIGURE 19.18 Transmission line with $R_T = 47\ \Omega$.

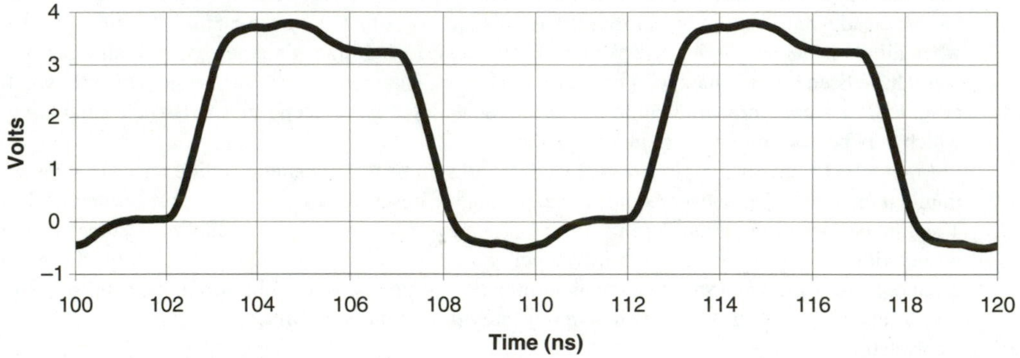

FIGURE 19.19 Transmission line with $R_T = 33\ \Omega$.

tion. As expected, there is greater distortion, because the transmission line is not as well terminated. Approximately 0.5 V of overshoot and undershoot are observed.

The preceding examples provide a brief glimpse of what is possible with Spice. In addition to passive circuits, semiconductors and active components can be modeled, enabling detailed simulations of analog amplifiers and digital circuits. Spice enables simulation to the desired level of resolution without forcing undue complexity. Small, quick simulations suffice for circuits with significant margins. More complex and detailed simulations that use highly accurate models are called for when operating near the limits of technology where margins for errors are very small.

19.9 TEST EQUIPMENT

Test equipment needs vary, depending on the complexity of a project. There are general types of test equipment used to debug and measure wide varieties of circuits, and there are very specialized tools designed for niche applications. This section provides a brief introduction to the more common pieces of test equipment found in typical engineering laboratories. As with any equipment, costs and capabilities vary widely. A 25-MHz analog oscilloscope may cost several hundred dollars, whereas a

1-GHz digital oscilloscope can cost tens of thousands of dollars. Both oscilloscopes perform the same basic function, but one performs that function at much higher frequencies and with more powerful measurement features. The following companies are among those who manufacture some or all of the test equipment mentioned in the following text: Agilent Technologies, B&K Precision, Fluke, Kenwood, LeCroy, and Tektronix.

Perhaps the most basic test tool is the *multimeter*, available in analog and digital varieties. A digital multimeter is called a *DMM*. Multimeters measure voltage, resistance, current, and sometimes capacitance, inductance, and diode forward voltages. Measuring voltage and current are passive functions, but the other capabilities require injecting current into a component under test to discern its voltage or other properties. DMMs are very common because of their low cost and ease of use.

Another relatively inexpensive test device is called a *logic probe* and is useful for engineers on a budget. A logic probe is placed onto a digital signal, and its indicators, visual and/or audible, tell whether a logic 0 or logic 1 is present and whether the line is clocking or pulsing high or low. Very basic debugging can be done with a logic probe in the absence of an oscilloscope. A logic probe, for example, can determine if a chip-select is being asserted to a device on a microprocessor's bus. Unfortunately, a logic probe cannot provide other useful information such as pulse duration and the relative timing of multiple signals.

Oscilloscopes have been mentioned in various contexts thus far. Their basic function is to display a time-domain view of voltage over very narrow time spans. Traditionally, oscilloscopes were fully analog instruments with cathode-ray tubes (CRTs) that required constant scanning to maintain a visible image. As such, a traditional analog oscilloscope is most useful with repetitive signals that could continually retrigger the trace moving across the CRT. Digital oscilloscopes, or digital storage oscilloscopes (DSOs), are hybrid analog and digital devices that sample an analog input and then load the digitized data stream into a buffer where it can be manipulated as a computer graphic. This memory feature allows digital oscilloscopes to capture single-shot events and display them indefinitely. Most oscilloscopes have at least two channels, allowing two signals to be displayed simultaneously and correlated with one another. More expensive oscilloscopes have four or more channels.

Once a waveform is displayed, analog and digital oscilloscopes enable various measurements to be taken, including frequency, amplitude, and relative timing between channels. Low-budget analog oscilloscopes may have only grid markings on the screen from which measurements can be manually estimated. More expensive oscilloscopes have built-in marker functions that automate the measurement process to varying extents.

The bandwidth of an oscilloscope is an important characteristic that indicates the frequency handling capabilities of its amplification and sampling circuits. An oscilloscope cannot be expected to provide meaningful observations when operated near the limits of its specified bandwidth. Keep in mind that digital signals contain high-frequency components created by sharp transition edges. A 100-MHz clock signal represents more than 100 MHz of bandwidth. Using an oscilloscope with greater bandwidth enables observations with less attenuation of higher frequencies. Rules of thumb vary on bandwidth versus actual signal frequency. Accepting an order-of-magnitude difference between a digital signal frequency and oscilloscope bandwidth yields good results. If this appears too expensive, a trade-off must be made between cost and the accuracy of measurements.

DSOs have another important defining characteristic: sampling rate. Sampling rate effects differ, depending on the capabilities of the oscilloscope and on whether the signal being measured is a single-shot event or repetitive, such as a clock. Some DSOs treat these two classes of signals differently. Single-shot events must be sampled in a single pass and are subject to a Nyquist limit imposed by the sampling rate. Measuring a signal with frequencies too near the Nyquist frequency (one-half the sampling rate) results in unacceptable distortion of the signal. Good results are attainable when measuring signal frequencies that are less than one-fifth to one-tenth of the sampling rate. Some DSOs are able to sample higher-frequency repetitive signals by combining the samples from many passes. These DSOs take relatively few samples on each pass, which is why a signal must be repetitive so that multiple passes are sampling essentially the same waveform. Sampling resolution is a closely related characteristic. It is common to find DSOs with eight-bit resolution, which is adequate for many digital probing needs because of these signals' binary nature. Certain analog applications may require finer resolution to take proper measurements, and such improvements come at a cost.

A relative of the oscilloscope is the logic analyzer, a device that is intended for purely digital test applications. Like a DSO, a logic analyzer captures signals at high sampling rates as they occur and then freezes them for human analysis for an arbitrarily long time. The principal differences are that a logic analyzer captures single-bit samples and, consequently, is able to work with dozens or hundreds of channels at the same time. High channel count enables a logic analyzer to capture complex buses in their entirety so that a complete picture of a bus's digital state can be displayed. An oscilloscope can show that a write-enable is coinciding with a chip-select. A logic analyzer can also show the specific data and address that are being transacted during that write operation. Logic analyzers are characterized by sampling rate, the number of channels supported, buffer depth, and triggering capabilities. At a basic level, a logic analyzer contains a large memory into which new samples are loaded every clock cycle. Faster, wider, and deeper memories cost more money than slower, narrower, shallower memories.

An important logic analyzer feature is its triggering capability. A simple type of triggering is one in which the logic analyzer waits for a specific pattern to present itself and then fills its entire buffer with the data following the pattern occurrence. If a serial port driver is being debugged, the logic analyzer trigger might be the address of a control register in the UART. Probing all signals on the microprocessor bus with the logic analyzer enables long instruction sequences to be correlated with hardware behavior to determine where a fault lies. When the microprocessor accesses a UART register, the logic analyzer triggers and stores all subsequent state information until the analyzer's memory is filled. The trigger point can usually be configured at an arbitrary offset in the analyzer memory. If the beginning of memory is chosen, the result is that the state following the trigger is captured. If the end of memory is chosen, the state leading up to the trigger is captured. Choosing a middle memory location captures the state both before and after the trigger event. Each debugging effort is best assisted by choosing a specific triggering option. If, for example, the wrong data is being transmitted from the UART, the state prior to the trigger event is likely to be useful, because it would show the microprocessor's instructions that presumably caused the data error. If, instead, the microprocessor appears to crash after transmitting data, the state after the trigger event may be more useful.

More complex digital buses and algorithms benefit from more powerful triggering capabilities. Many logic analyzers can trigger after a number of occurrences of the same pattern are observed or after several predetermined patterns are observed in specific sequences. In the previous UART debugging example, incorrect results may not occur until after many bytes have been transmitted. Alternatively, incorrect results may occur only after a certain byte sequence has been sent. A good analyzer can be configured to trigger on arbitrary combinations of patterns and occurrences.

APPENDIX A
Further Education

One of the exciting aspects of electrical engineering is that the state of the art changes quickly. Consequently, there is always the need to learn about new technologies, methods, and components. The modern engineer is fortunate to have a multitude of educational resources from which to draw. The Internet has made technical information more accessible than ever to anyone with a modem. Educational resources for engineers include

- Trade publications and subscription periodicals
- Technical books
- Manufacturers' web sites and publications
- Third-party web sites
- Colleagues and conferences

Trade publications and other periodicals are a good way to keep aware of current trends in the industry, because articles are often written about new and interesting technologies. Even advertisements provide an education, because manufacturers' claims can be compared against each other and against information acquired elsewhere. Many trade publications are funded by advertisements and can therefore offer free subscriptions to qualified subscribers with relevant professional responsibilities and technical needs. The following periodicals are recommended by the author:

- *Circuit Cellar* (paid subscription). The focus is on embedded systems with articles describing real implementations and discussions of how to make practical systems work.
- *EDN* and *Electronics Design* (free subscriptions). Articles are written by staff editors and professional engineers on topics from current trends to specific design implementation techniques and technical advice.
- *EE Times* (free subscription). This is a general electronics industry weekly news with articles on current trends and new technologies.

Technical books are available on practically every aspect of engineering. Chances are that you are already aware of this option, because you are reading this book right now! The following books are recommended as references for various topics that arise in digital electrical engineering:

- Computer architecture—*Computer Organization and Design: The Hardware/Software Interface*, David A. Patterson and John L. Hennessy, Morgan Kauffman.
- Signal integrity and PCB design—*High-Speed Digital Design: A Handbook of Black-Magic*, Howard W. Johnson and Martin Graham, Prentice Hall.

- Semiconductor theory (transistors, op-amps, etc.)—*Microelectronic Circuits,* Adel S. Sedra and Kenneth C. Smith, Saunders College Publishing.

The publications provided by manufacturers about their own parts and the technologies that they incorporate are, collectively, a treasure trove of information that is highly valued by this author. Semiconductor manufacturers have always published literature that describes the specifications of their devices in addition to application notes on recommended usage scenarios. The Internet has increased the accessibility of these publications to the point at which numerous manufacturers' web sites can be scoured in a short while to get information on a specific topic. Much can be learned by reading data sheets, even when all of the topics covered are not already understood. Application notes are highly beneficial, because they contain detailed descriptions of how a chip is actually used. The context and advice provided by application notes can fill in the questions that arise from reading a data sheet. While nearly all component manufacturers have web sites with useful information, the following companies stand out in the author's view because of the quality and comprehensive collection of technical information that is freely available to all visitors:

- Altera—FPGAs and CPLDs (www.altera.com)
- AVX Corporation—passive components (www.avxcorp.com)
- Fairchild Semiconductor—discrete semiconductors, logic, analog, and mixed-signal ICs (www.fairchildsemi.com)
- Linear Technology—analog and mixed-signal ICs (www.linear.com)
- Maxim Integrated Circuits—analog and mixed-signal ICs (www.maxim-ic.com)
- Micron Technologies—memory (www.micron.com)
- Microchip Technology—microcontrollers and nonvolatile memory (www.microchip.com)
- National Semiconductor—analog and mixed-signal ICs (www.national.com)
- Texas Instruments—discrete semiconductors, logic, analog, and mixed-signal ICs (www.ti.com)
- Xilinx—FPGAs and CPLDs (www.xilinx.com)

Many useful third-party web sites are maintained by generous and experienced members of the world's technical community. Almost every engineering topic imaginable can be found with a quick web search. While these sites can be very helpful, the information found should be correlated with other sources whenever possible. There is both good and bad information available on the web, and some sites do contain erroneous data that can cause much grief.

Numerous technical conferences exist in both broad and specialized areas of electrical engineering. Notices for these conferences can be found in trade publication advertisements.

INDEX

Q